ROTATING RELATIVISTIC STARS

The masses of neutron stars are limited by an instability to gravitational collapse, and an instability driven by gravitational waves limits their spin. Their oscillations are relevant to X-ray observations of accreting binaries and to gravitational wave observations of neutron stars formed during the coalescence of double neutron-star systems. This volume pulls together more than forty years of research to provide graduate students and researchers in astrophysics, gravitational physics, and astronomy with the first self-contained treatment of the structure, stability, and oscillations of rotating neutron stars. This monograph treats the equations of stellar equilibrium; key approximations, including slow rotation and perturbations of spherical and rotating stars; stability theory and its applications, from convective stability to the r-mode instability; and numerical methods for computing equilibrium configurations and the nonlinear evolution of their oscillations. The presentation of fundamental equations, results, and applications is accessible to readers who do not need the detailed derivations.

JOHN L. FRIEDMAN is a University Distinguished Professor at the University of Wisconsin–Milwaukee. A Fellow of the American Physical Society, he recently served as chair of its gravitational physics section. He has been on the editorial boards of *Classical and Quantum Gravity* and *Physical Review D*, and he was a divisional associate editor of *Physical Review Letters*. His awards include the Telegdi Prize and the Marc Perry Galler Award.

NIKOLAOS STERGIOULAS is an Associate Professor at the Aristotle University of Thessaloniki, Greece. He has a large number of publications in relativistic astrophysics and has released a widely used public-domain code for constructing numerical models of rotating relativistic stars. He has also served on the governing councils of the Hellenic Astronomical Society; the Hellenic Society on Relativity, Gravitation and Cosmology; and the Virgo Ego Scientific Forum.

CAMBRIDGE MONOGRAPHS ON MATHEMATICAL PHYSICS

S. J. Aarseth *Gravitational N-Body Simulations: Tools and Algorithms*†
J. Ambjørn, B. Durhuus, and T. Jonsson *Quantum Geometry: A Statistical Field Theory Approach*†
A. M. Anile *Relativistic Fluids and Magneto-Fluids: With Applications in Astrophysics and Plasma Physics*†
J. A. de Azcárraga and J. M. Izquierdo *Lie Groups, Lie Algebras, Cohomology and Some Applications in Physics*†
O. Babelon, D. Bernard, and M. Talon *Introduction to Classical Integrable Systems*†
F. Bastianelli and P. van Nieuwenhuizen *Path Integrals and Anomalies in Curved Space*†
V. Belinski and E. Verdaguer *Gravitational Solitons*†
J. Bernstein *Kinetic Theory in the Expanding Universe*†
G. F. Bertsch and R. A. Broglia *Oscillations in Finite Quantum Systems*†
N. D. Birrell and P. C. W. Davies *Quantum Fields in Curved Space*†
K. Bolejko, A. Krasiński, C. Hellaby, and M.-N. Célérier *Structures in the Universe by Exact Methods: Formation, Evolution, Interactions*
D. M. Brink *Semi-Classical Methods for Nucleus-Nucleus Scattering*†
M. Burgess *Classical Covariant Fields*†
E. A. Calzetta and B.-L. B. Hu *Nonequilibrium Quantum Field Theory*
S. Carlip *Quantum Gravity in 2+1 Dimensions*†
P. Cartier and C. DeWitt-Morette *Functional Integration: Action and Symmetries*†
J. C. Collins *Renormalization: An Introduction to Renormalization, the Renormalization Group and the Operator-Product Expansion*†
P. D. B. Collins *An Introduction to Regge Theory and High Energy Physics*†
M. Creutz *Quarks, Gluons and Lattices*†
P. D. D'Eath *Supersymmetric Quantum Cosmology*†
F. de Felice and D. Bini *Classical Measurements in Curved Space-Times*
F. de Felice and C. J. S Clarke *Relativity on Curved Manifolds*†
B. DeWitt *Supermanifolds, 2nd edition*†
P. G. O Freund *Introduction to Supersymmetry*†
F. G. Friedlander *The Wave Equation on a Curved Space-Time*†
J. L. Friedman and N. Stergioulas *Rotating Relativistic Stars*
Y. Frishman and J. Sonnenschein *Non-Perturbative Field Theory: From Two-Dimensional Conformal Field Theory to QCD in Four Dimensions*
J. Fuchs and C. Schweigert *Symmetries, Lie Algebras and Representations: A Graduate Course for Physicists*†
J. A. Fuchs *Affine Lie Algebras and Quantum Groups: An Introduction, with Applications in Conformal Field Theory*†
Y. Fujii and K. Maeda *The Scalar-Tensor Theory of Gravitation*†
J. A, H. Futterman, F. A. Handler, and R. A. Matzner *Scattering from Black Holes*†
A. S. Galperin, E. A. Ivanov, V. I. Orievetsky, and E. S. Sokatchev *Harmonic Superspace*†
R. Gambini and J. Pullin *Loops, Knots, Gauge Theories and Quantum Gravity*†
T. Gannon *Moonshine beyond the Monster: The Bridge Connecting Algebra, Modular Forms and Physics*†
M. Göckeler and T. Schücker *Differential Geometry, Gauge Theories and Gravity*†
C. Gómez, M. Ruiz-Altaba, and G. Sierra *Quantum Groups in Two-Dimensional Physics*†
M. B. Green, J. H. Schwarz, and E. Witten *Superstring Theory Volume 1: Introduction*†
M. B. Green, J. H. Schwarz, and E. Witten *Superstring Theory Volume 2: Loop Amplitudes, Anomalies and Phenomenology*†
V. N. Gribov *The Theory of Complex Angular Momenta: Gribov Lectures on Theoretical Physics*†
J. B. Griffiths and J. Podolský *Exact Space-Times in Einstein's General Relativity*†
S. W. Hawking and G. F. R. Ellis *The Large Scale Structure of Space-Time*†
F. Iachello and A. Arima *The Interacting Boson Model*†
F. Iachello and P. van Isacker *The Interacting Boson-Fermion Model*
C. Itzykson and J. M. Drouffe *Statistical Field Theory Volume 1: From Brownian Motion to Renormalization and Lattice Gauge Theory*†
C. Itzykson and J. M. Drouffe *Statistical Field Theory Volume 2: Strong Coupling, Monte Carlo Methods, Conformal Field Theory and Random Systems*†
C. V. Johnson *D-Branes*†
P. S. Joshi *Gravitational Collapse and Spacetime Singularities*†
J. I. Kapusta and C. Gale *Finite-Temperature Field Theory: Principles and Applications, 2nd edition*†
V. E. Korepin, N. M. Bogoliubov, and A. G. Izergin *Quantum Inverse Scattering Method and Correlation Functions*†
M. Le Bellac *Thermal Field Theory*†

† Issued as a paperback

Y. Makeenko *Methods of Contemporary Gauge Theory*†
N. Manton and P. Sutcliffe *Topological Solutions*†
N. H. March *Liquid Metals: Concepts and Theory*†
I. Montvay and G. Münster *Quantum Fields on a Lattice*†
L. O'Raifeartaigh *Group Structure of Gauge Theories*†
T. Ortin *Gravity and Strings*†
A. M. Ozorio de Almeida *Hamiltonian Systems: Chaos and Quantization*†
L. Parker and D. J. Toms *Quantum Field Theory in Curved Spacetime: Quantized Fields and Gravity*
R. Penrose and W. Rindler *Spinors and Space-Time Volume 1: Two-Spinor Calculus and Relativistic Fields*†
R. Penrose and W. Rindler *Spinors and Space-Time Volume 2: Spinor and Twistor Methods in Space-Time Geometry*†
S. Pokorski *Gauge Field Theories, 2nd edition*†
J. Polchinski *String Theory Volume 1: An Introduction to the Bosonic String*†
J. Polchinski *String Theory Volume 2: Superstring Theory and Beyond*†
J. C. Polkinghorne *Models of High Energy Processes*†
V. N. Popov *Functional Integrals and Collective Excitations*†
L. V. Prokhorov and S. V. Shabanov *Hamiltonian Mechanics of Gauge Systems*
R. J. Rivers *Path Integral Methods in Quantum Field Theory*†
R. G. Roberts *The Structure of the Proton: Deep Inelastic Scattering*†
C. Rovelli *Quantum Gravity*†
W. C. Saslaw *Gravitational Physics of Stellar and Galactic Systems*†
R. N. Sen *Causality, Measurement Theory and the Differentiable Structure of Space-Time*
M. Shifman and A. Yung *Supersymmetric Solitons*
H. Stephani, D. Kramer, M. MacCallum, C. Hoenselaers, and E. Herlt *Exact Solutions of Einstein's Field Equations, 2nd edition*†
J. Stewart *Advanced General Relativity*†
J. C. Taylor *Gauge Theories of Weak Interactions*†
T. Thiemann *Modern Canonical Quantum General Relativity*†
D. J. Toms *The Schwinger Action Principle and Effective Action*†
A. Vilenkin and E. P. S. Shellard *Cosmic Strings and Other Topological Defects*†
R. S. Ward and R. O. Wells, Jr. *Twistor Geometry and Field Theory*†
J. R. Wilson and G. J. Mathews *Relativistic Numerical Hydrodynamics*†

† Issued as a paperback

Rotating Relativistic Stars

JOHN L. FRIEDMAN

University of Wisconsin, Milwaukee

NIKOLAOS STERGIOULAS

Aristotle University of Thessaloniki

CAMBRIDGE
UNIVERSITY PRESS

CAMBRIDGE
UNIVERSITY PRESS

Shaftesbury Road, Cambridge CB2 8EA, United Kingdom

One Liberty Plaza, 20th Floor, New York, NY 10006, USA

477 Williamstown Road, Port Melbourne, VIC 3207, Australia

314–321, 3rd Floor, Plot 3, Splendor Forum, Jasola District Centre, New Delhi – 110025, India

103 Penang Road, #05–06/07, Visioncrest Commercial, Singapore 238467

Cambridge University Press is part of Cambridge University Press & Assessment,
a department of the University of Cambridge.

We share the University's mission to contribute to society through the pursuit of
education, learning and research at the highest international levels of excellence.

www.cambridge.org
Information on this title: www.cambridge.org/9780521872546

First published 2013

A catalogue record for this publication is available from the British Library

Library of Congress Cataloging-in-Publication data
Friedman, John L., 1945–
Rotating relativistic stars / John L. Friedman, University of Wisconsin, Milwaukee and
Nikolaos Stergioulas.
pages cm. – (Cambridge monographs on mathematical physics)
Includes bibliographical references and index.
ISBN 978-0-521-87254-6 (hardback)
1. Neutron stars. 2. Stellar oscillations. 3. Gravitational waves. 4. Stars – Rotation.
I. Stergioulas, Nikolaos, 1967– II. Title.
QB843.N4F75 2013
523.8′874–dc23 2012036771

ISBN 978-0-521-87254-6 Hardback

To Paula

To my parents, Christoforos and Evangelia

Contents

Preface

The masses of neutron stars are limited by an instability to collapse, and an instability driven by gravitational waves may limit their spin. Their oscillations are relevant to X-ray observations of accreting binaries and to gravitational wave observations of neutron stars formed during the coalescence of double neutron-star systems. This volume pulls together more than 40 years of research to provide graduate students and researchers in astrophysics, gravitational physics, and astronomy with a self-contained treatment of the structure, stability, and oscillations of rotating relativistic stars. Numerical and analytic work are both essential to the subject, and their interplay is emphasized in our treatment.

The book is intended for more than one audience: Readers who need to work through mathematical details of stellar perturbations and stability theory will find them here, in derivations and proofs of principal results. More commonly, a reader working in relativistic astrophysics will want the principal results of the theory but will need only a few of the derivations. The text is also designed to provide a coherent treatment for this second audience, with an exposition of the results preceding the more mathematical derivations. Although our primary concern is with rotating stars, we begin our discussion of oscillations and stability with spherical stars for completeness and to make the presentation accessible to readers with no previous knowledge of relativistic perturbation theory.

Those intending to work through the mathematical derivations should have a background comparable to a semester of gravitational physics at the level of MTW [480] or Wald [715]. The rest of the book should be accessible to students who have mastered Schutz's *First Course in General Relativity* [595], supplemented by the appendices here on Lie derivatives and integration.

We are indebted to all our long-term collaborators in a large number of joint publications, the main results of which are presented in this book. We are especially grateful for their contributions in several research areas that are presented here in an abridged form.

A number of colleagues and collaborators have read parts of the manuscript, made suggestions, answered our questions, and caught errors. For this we thank Marek Abramowicz, James Bardeen, Emanuele Berti, Demian Cho, Piotr Chrusciel, Matthew Glenz, Eric Gourgoulhon, Stephen Green, Brennan Hughey, Panagiotis Iosif, James Ipser, David Kaplan, Kostas Kokkotas, Lee Lindblom, Charalampos Markakis, Ben Owen, Andrea Passamonti, Eric Poisson, Bernard Schutz, and Kōji Uryū. We are indebted to Kate Valerius for typing parts of the manuscript. We thank the editors at Cambridge University Press for their support and encouragement and Aptara, Inc., for their careful typesetting. Finally, we would like to thank our families for their patience and support during the time of writing of this book.

List of symbols

This is a global glossary, restricted to symbols used in more than one place in the text. Local uses of symbols that appear within a page or two of their definition are in general not listed here. For example, h is used globally to mean specific relativistic enthalpy, and that definition is listed here; h is also used on a single page to mean the amplitude of a gravitational wave, and that local definition is not listed.

A	a generic constant
$A\ \ (A_\alpha)$	Schwarzschild discriminant (vector version)
A_α	electromagnetic vector potential
A_{ab}	trace-free part of extrinsic curvature
$\tilde{A}_{ab}$	related by conformal factor to A_{ab}
$\mathcal{A}, \mathcal{A}^a$	generic densities
a	constant in asymptotic metric of rotating star
	J/M in the Kerr geometry
B	metric potential of rotating star
B^α	magnetic field
$\mathcal{B}$	bag constant in quark interactions
b	constant in asymptotic form of metric
C	generic constant
C_μ	gravitational constraint
c	speed of light
c_s	speed of sound
$c(\tau), c(\lambda)$	path in spacetime
D_{lm}	mass multipole moment
D_a	covariant derivative of spatial metric γ_{ab}
$\tilde{D}_a$	covariant derivative of spatial metric $\tilde{\gamma}_{ab}$
d	exterior derivative
$d\sigma_\alpha, d\sigma_a$	$dS_\alpha/\sqrt{\lvert g\rvert}, dS_a/\sqrt{\gamma}$
dl	element of proper length
dS_α, dS_a	elements of (hyper)surface area
dV, d^4V	3- and 4-dimensional volume elements
E	energy
E_c	canonical energy
$E_{c,r}$	canonical energy in rotating frame
$E_{\alpha\beta}$	$G_{\alpha\beta} - 8\pi T_{\alpha\beta}$

E_{ab}	spatial projection of $E_{\alpha\beta}$
E^{α}	electric field
$\mathcal{E}$	injection energy
e	specific internal energy (per unit baryon mass)
$\mathbf{e}_{\hat{\mu}}$	contravariant basis vector of ZAMO
e_{ab}	metric on a sphere of radius r
F	the scalar $u^t u_\phi$
	a generic function
$\mathcal{F}$	function in Eulerian perturbation theory
$F^{\alpha\beta}$	electromagnetic field tensor (Faraday tensor)
f	a generic function
f_α	4-force per unit volume
G	Newton's constant
	occasionally a generic function
$G_{\alpha\beta}$	Einstein tensor
$G_{\alpha\beta\gamma\delta}$	metric expression appearing in $\delta(G^{\alpha\beta}\sqrt{\lvert g\rvert})$
g	specific Gibbs free energy
	determinant of spacetime metric
$g_{\alpha\beta}$	spacetime metric
H	thermodynamic quantity, $\ln h$ for homentropic fluid
	metric potential in slow-rotation approximation
$\mathcal{H}$	Hamiltonian density
H_0, H_1, H_2	potentials of polar metric perturbation of spherical star
h	specific enthalpy (per unit baryon mass)
$h_{\alpha\beta}$	metric perturbation: $\delta g_{\alpha\beta}$
h_0, h_1, h_2	potentials of axial metric perturbation of spherical star
I	moment of inertia
	action
$\mathcal{I}$	null infinity (scri)
J	angular momentum
	Jacobian of a diffeomorphism
J_c	canonical angular momentum
$J_{c,r}$	canonical angular momentum in rotating frame
J_{lm}	current multipole moment
J^α	generic current
j	specific angular momentum of fluid
	J/M^2, a dimensionless measure of angular momentum
j	$\rho\sqrt{\lvert g\rvert}u^t$
j_a	momentum density of fluid
	conduction current for heat flow
j^α	baryon current density
K	polytropic constant
	trace $K_a{}^a$ of extrinsic curvature
K_{ab}	extrinsic curvature

k	generic constant
	potential of a perturbed metric
k^α	helical Killing vector: $t^\alpha + \Omega\phi^\alpha$
L	angular momentum of free particle
	linear operator in perturbation theory
$\mathcal{L}$	Lie-derivative operator
$\boldsymbol{\mathcal{L}}$	the operator $\lvert g\rvert^{-1/2}\mathcal{L}\ \lvert g\rvert^{1/2}$
$\mathscr{L}$	Lagrangian density
l	label of rotation group representation, as in Y_{lm}
	proper length
M	gravitational mass
$M_{\rm ADM}$	ADM mass
$M_{\rm K}$	Komar mass
M_0	baryon mass of star
$\mathcal{M}$	manifold of fluid trajectories
m	angular eigenvalue, as in $e^{im\phi}$
$m(r)$	mass within Schwarzschild radial coordinate r
m_B	fiducial baryon mass: mass per nucleon of ^{12}C
m_e, m_s	electron mass, strange quark mass
m_n	complex multipole moments
m_0, m_2	potentials of a slowly rotating star
N	polytropic index
	baryon number
N	Brunt-Väisälä frequency
n	baryon number density
	neutron
	generic integer
n_e	electron number density
n^α	future-pointing unit normal to hypersurface
O	an order symbol
o	an order symbol
P	point of spacetime
	period of rotation
	point along a sequence of stars
$\mathcal{P}$	a parity transformation (diffeomorphism)
p	pressure
Q	scalar quadrupole moment
$\mathcal{Q}$	set of variables of perfect-fluid spacetime
Q_{ab}	tensor in Newtonian perturbation theory
q^α	heat-flow vector
$q_\alpha{}^\beta$	projection tensor orthogonal to u^α
R	circumferential equatorial radius of star
	3-dimensional Ricci scalar
^4R	4-dimensional Ricci scalar

$R_{\alpha\beta}, R_{ab}$	4- and 3-dimensional Ricci tensors
$\tilde{R}_{ab}$	Ricci tensor of $\tilde{\gamma}_{ab}$
r	radial coordinate
$\mathbf{r}$	position vector (Newtonian)
r_c	circumferential radial coordinate
S	2-dimensional surface
S^{ab}	stress tensor: spatial projection of $T^{\alpha\beta}$
s	specific entropy (per unit baryon mass)
s^α	entropy current
T	temperature
	rotational kinetic energy of rotating star
T_s	superfluid transition temperature
$T^{\alpha\beta}$	stress-energy tensor
$T^{a_1\cdots a_m}{}_{b_1\cdots b_n}$	a generic tensor
$\mathcal{T}$	a subspace of the space of trivial displacements
t	time coordinate
t^α	time-translation symmetry vector
U	effective potential in two-potential formalism
$U^{\alpha\beta\gamma\delta}$	tensor in Lagrangian perturbation theory
u	null coordinate
u^α	4-velocity of fluid or particle
$V^{\alpha\beta\gamma\delta}$	tensor in Lagrangian perturbation theory
V	3-dimensional region
v	magnitude of 3-velocity measured by a ZAMO
	null coordinate
$\mathbf{v}$, v^a	Newtonian fluid velocity
v^α	spatial part of fluid velocity u^α
	generic vector
W	gravitational potential energy of star
	symplectic structure (form) of perturbation
	Lorentz factor
$W^{\alpha\beta\gamma\delta}$	tensor in expression for $\Delta(T^{\alpha\beta}\sqrt{\lvert g\rvert})$
w	function occurring in Manko formalism
	function occurring in Eulerian perturbation theory
w^α	unnormalized tangent vector to fluid trajectory
	the vector hu^α
	generic vector
X	generic variable
x	coordinate
x^μ	spacetime coordinate
x_p	proton fraction
Y_k	fractional number density of kth species of particle
y	function in Eulerian perturbation theory
	coordinate

Z	Zerilli function
	redshift
z	coordinate
α	lapse
	imaginary part of mode frequency
β^a	shift vector
Γ	$d\log p/d\log\rho$; polytropic exponent
Γ_1	adiabatic index
${\Gamma^\lambda}_{\mu\nu}$	Christoffel symbol
γ	determinant of 3-metric
	Lorentz factor $1/\sqrt{1-v^2}$
γ_{ab}	3-metric
$\tilde{\gamma}_{ab}$	conformal 3-metric
${\gamma_a}^\alpha$	pullback of vectors on M to vectors on Σ
Δ	Lagrangian change
$\boldsymbol{\Delta}$	the operator $\lvert g\rvert^{-1/2}\Delta\lvert g\rvert^{1/2}$
δ	Eulerian change
$\boldsymbol{\delta}$	the operator $\lvert g\rvert^{-1/2}\delta\lvert g\rvert^{1/2}$
ϵ, ϵ_c	energy density, central energy density
$\epsilon_{\alpha\beta\gamma\delta}, \epsilon_{abc}$	normalized totally antisymmetric tensors
ϵ_{ab}	normalized antisymmetric tensor on sphere
ζ	metric potential of rotating star
$\boldsymbol{\zeta}$	Newtonian vorticity vector
ζ^α, ζ^a	generator of gauge transformation
ζ^a	generic vector field
η	coefficient of viscosity
$\eta_{\alpha\beta}$	flat Minkowski metric
η_{ab}	flat spatial metric
η^α	trivial Lagrangian displacement
Θ	step function
$\boldsymbol{\Theta}^\alpha$	vector density from variation of action
θ	spatial divergence of u^α
	angular coordinate
κ	coefficient of heat conductivity
	a generic constant
Λ	the Λ hyperon
λ	metric potential of spherical star
	parameter along sequence of stars
μ	metric potential of rotating star
ν	metric potential of rotating or spherical star
ξ^a	Lagrangian displacement
Π	the internal energy of a Newtonian star
Π_α	momentum conjugate to ξ^α in perturbed fluid

Π_a	momentum density of fluid
π	trace, $\pi_a{}^a$
$\pi^{\alpha\beta}$	momentum density conjugate to perturbed metric $h_{\alpha\beta}$
π^{ab}	momentum density conjugate to spatial metric γ_{ab}
$\pi^0, \pi^{\pm}$	pions
ϖ	cylindrical coordinate radius
ρ	baryon mass density
Σ	3-dimensional hypersurface
	the Σ hyperon
σ	the complex frequency of a mode
$\sigma_{\alpha\beta}$	shear tensor
τ	proper time along a trajectory
$\tau_{\rm b}$	bulk-viscosity damping time
τ_{GW}	gravitational wave damping time or growth time
$\tau_{\rm s}$	shear-viscosity damping time
Υ	scalar occurring in expression for fluid velocity u^α
Φ	metric potential of a spherical star
	Newtonian gravitational potential
Φ^I	one of a set $\{\phi^I\}$ of fields on spacetime
ϕ	angular coordinate
	potential in conformal factor $e^{4\phi}$ of 3-metric
ϕ^α	rotational symmetry vector
$\chi, \chi_{\rm s}$	diffeomorphisms describing fluid configuration
Ψ	velocity potential of irrotational fluid
ψ	diffeomorphism
	metric potential of rotating star
Ω	angular velocity of rotating star
Ω_K	Keplerian angular velocity
ω	frame-dragging potential of rotating star
$\bar{\omega}$	$\omega - \Omega$, used in slow-rotation formalism
$\omega_{\alpha\beta}$	vorticity tensor
ω_i	real mode frequency measured by inertial observer
ω_r	real mode frequency measured by rotating observer
ω^α	vorticity vector
$\boldsymbol{\omega}^{\hat{\mu}}$	covariant basis vector of ZAMO
∇_α	covariant derivative operator of spacetime metric $g_{\alpha\beta}$
$\nabla\!\!\!\!/_\alpha$	covariant derivative operator of flat metric $\eta_{\alpha\beta}$
$\nabla\!\!\!\!/_a, \boldsymbol{\nabla}$	covariant derivative operator of flat spatial metric η_{ab}
$\boldsymbol{\partial}_\mu$	contravariant basis vector associated with coordinate x^μ

Conventions, notation, and mathematical preliminaries

Units, metric and physical constants

Throughout the book, gravitational units, with $G = c = 1$, will be adopted in writing the equations governing stellar structure and dynamics, whereas numerical properties of stellar models will be listed in cgs units, unless otherwise noted. We use the conventions of Misner, Thorne, and Wheeler [480] for the signature of the spacetime metric $(-+++)$ and for signs of the curvature tensor and its contractions. Spacetime indices will be Greek, α, $\beta, \ldots$, whereas spatial indices will be Latin $a, b, \ldots$. (Readers familiar with abstract indices can regard indices early in the alphabet as abstract, whereas indices μ, ν, λ and i, j, k will be concrete, taking values $\mu = 0, 1, 2, 3$, $i = 1, 2, 3$.) Corresponding to a choice of coordinates, t, r, θ, ϕ, a vector u^α has components $u^t, \ldots, u^\phi$; its components along an orthonormal frame, $\{\mathbf{e}_{\hat{0}}, \ldots, \mathbf{e}_{\hat{3}}\}$, will be written $\{u^{\hat{0}}, \ldots, u^{\hat{3}}\}$. Parentheses enclosing a set of indices indicate symmetrization, and square brackets indicate antisymmetrization.

Numbers that rely on physical constants are based on the values $c = 2.9979 \times 10^{10}$ cm s^{-1}, $G = 6.670 \times 10^{-8}$ g^{-1}cm^{3}s^{-2}, $\hbar = 1.0545 \times 10^{-27}$ g cm^{2}s^{-1}, baryon mass $m_B = 1.659 \times 10^{-24}$ g, and $M_\odot = 1.989 \times 10^{33}$ g $= 1.477$ km.

Derivatives and integrals

The covariant derivative operator of the spacetime metric $g_{\alpha\beta}$ will be written ∇_α, and the partial derivative of a scalar f with respect to one of the coordinates – say r – will be written $\partial_r f$ or $f_{,r}$. Lie derivatives along a vector u^α will be denoted by $\mathcal{L}_{\mathbf{u}}$. The Lie derivative of an arbitrary tensor $T^{a\cdots b}{}_{c\cdots d}$ is

$$\mathcal{L}_{\mathbf{u}} T^{a\cdots b}{}_{c\cdots d} = u^e \nabla_e T^{a\cdots b}{}_{c\cdots d} - T^{e\cdots b}{}_{c\cdots d} \nabla_e u^a - \cdots - T^{a\cdots e}{}_{c\cdots d} \nabla_e u^b + T^{a\cdots b}{}_{e\cdots d} \nabla_c u^e + \cdots + T^{a\cdots b}{}_{c\cdots e} \nabla_d u^e. \tag{1}$$

Our notation for integrals is as follows. We denote by d^4V the spacetime volume element. In a chart $\{x^0, x^1, x^2, x^3\}$, the notation means

$$d^4V = \epsilon_{0123} dx^0 dx^1 dx^2 dx^3 = \sqrt{|g|}\, d^4x, \tag{2}$$

where g is the determinant of the matrix $\|g_{\mu\nu}\|$. Gauss's theorem (presented in Section A.3 of the Appendix) has the form

$$\int_\Omega \nabla_\alpha A^\alpha d^4V = \int_{\partial\Omega} A^\alpha dS_\alpha, \tag{3}$$

with $\partial\Omega$ the boundary of the region Ω. In a chart (u, x^1, x^2, x^3) for which V is a surface of constant u, $dS_\alpha = \sqrt{|g|}\,\nabla_\alpha u d^3x$, and

$$\int_V A^\alpha dS_\alpha = \int_V A^u \sqrt{|g|} d^3x. \tag{4}$$

If V is nowhere null, one can define a unit normal,

$$\widehat{n}_\alpha = \frac{\nabla_\alpha u}{\left|\nabla_\beta u \nabla^\beta u\right|^{1/2}}, \tag{5}$$

and write

$$dS_\alpha = \widehat{n}_\alpha dV, \tag{6}$$

where

$$dV = \sqrt{|^3 g|}\, d^3x, \tag{7}$$

where ^3g is the determinant of the 3-metric induced on the surface V. But Gauss's theorem has the form (3) for any 3-surface S, bounding a 4-dimensional region $\mathcal{R}$, regardless of whether S is timelike, spacelike, or null.[1]

Similarly, if $F^{\alpha\beta}$ is an antisymmetric tensor, its integral over a 2-surface S of constant coordinates u and v is written

$$\int_S F^{\alpha\beta} dS_{\alpha\beta} = \int_S F^{uv} \sqrt{|g|} d^2x, \tag{8}$$

and a corresponding generalized Gauss's theorem has the form

$$\int_V \nabla_\beta F^{\alpha\beta} dS_\alpha = \int_{\partial V} F^{\alpha\beta} dS_{\alpha\beta}. \tag{9}$$

If n_α and $\widetilde{n}_\alpha$ are orthogonal unit normals to the surface S, for which $(\mathbf{n}, \widetilde{\mathbf{n}}, \boldsymbol{\partial}_2, \boldsymbol{\partial}_3)$ is positively oriented, then $dS_{\alpha\beta} = n_{[\alpha}\widetilde{n}_{\beta]}\sqrt{|^2g|}\, d^2x$.

Asymptotic notation: O and o

We will use the symbols $O(x)$ and $o(x)$ to describe asymptotic behavior of functions. For a function $f(x)$, $f = O(x)$ if there is a constant C for which $|f/x| < C$, for sufficiently small $|x|$; $f = o(x)$ if $\lim_{x\to 0} |f/x| = 0$. For example, if A is constant, $A/r = O(r^{-1})$, and $A/r^{3/2} = o(r^{-1})$.

[1] Note that in the text, n_α denotes the *future* pointing unit normal to a $t =$ constant hypersurface, $n_\alpha = -\nabla_\alpha t/|\nabla_\beta t \nabla^\beta t|^{1/2}$. In order that, for example, $\int \rho u^\alpha dS_\alpha$, be positive on a $t =$ constant surface, one must use $dS_\alpha = \nabla_\alpha t\sqrt{|g|}d^3x = \widehat{n}_\alpha dV = -n_\alpha dV$.

1

Stationary, axisymmetric equilibria

The Newtonian approximation describes to extraordinarily high precision the gravitational field of low-mass stars over the course of their evolution, from the instability to collapse that triggers their formation to their death as white dwarfs. In a high-mass star, however, when the nuclear reactions that halted its initial collapse ultimately die out, the core's renewed collapse leads either to a star above nuclear density or to a black hole at whose center is a speck that, at least momentarily, is vastly beyond any known density. In both of these final states of stellar evolution, general relativity plays a fundamental role. The relativistic stars of nature have a complex composition, spanning fifteen orders of magnitude in density. Thought to consist primarily of a gas of neutrons with a gradually varying density of free protons, electrons, and muons, they are surrounded by a crust of ordinary matter, and their cores may hold hyperons, pion or kaon condensates, or possibly free quarks. In fact, our uncertainty about the behavior of matter above nuclear density is (in 2012) great enough to allow what we call neutron stars to be strange-quark stars, collections of up, down, and strange quarks surrounded by a thin normal crust. In the conventional neutron-star model, a much thicker, 1-km crust surrounds an interior in which neutrons and protons form a two-component superfluid. High magnetic fields, whose strength in some cases appears to exceed 10^{14} G, are observed and thought to extend in quantized flux tubes through the superfluid interior. The angular velocities of observed millisecond pulsars range up to 716 Hz, and the vorticity of their velocity fields is similarly thought to be confined, in the neutron stars' interiors, to quantized tubes.

Departures from local isotropy are associated with the crust, with the vortex and magnetic flux tubes, and with heat flow and viscosity. Nevertheless, a neutron star *in equilibrium* is accurately approximated by a stationary self-gravitating perfect fluid, its structure determined by a balance among its intense gravity, the pressure of its degenerate particles, and its rotation. In particular, departures from perfect fluid equilibrium due to a solid crust are expected to be smaller than $\sim 10^{-3}$, corresponding to the maximum strain that an electromagnetic lattice can support [142]; this estimate is supported by observations of pulsar glitches, which are consistent with departures from a perfect fluid equilibrium of order 10^{-5} (see [235]).

Similarly, on scales of meters or larger, a single rotational velocity field u^α describes the averaged superfluid motion [58, 639, 424]. The error in computing the gravitational field is much smaller than errors in the fluid model, because the characteristic length over which a potential varies is much larger than the distance between vortices. Although the assumption of a perfect fluid is adequate for describing equilibrium configurations, studies of neutron-star dynamics – of formation, oscillations, and stability, and of the interaction of binaries during and just before merger – require a more detailed knowledge of the stars' microphysics.

1.1 Perfect fluids

The stress-energy tensor. A perfect fluid is a model for a large assembly of particles in which a continuous energy density ϵ can reasonably describe the macroscopic distribution of mass. One assumes that the microscopic particles collide frequently enough that their mean free path is short compared with the scale on which the density changes, so that the collisions enforce a local thermodynamic equilibrium. In particular, one assumes that a mean velocity field u^α and a mean stress-energy tensor $T^{\alpha\beta}$ can be defined in boxes – fluid elements – that are small compared to the macroscopic length scale but large compared to the mean free path. One also assumes that on scales large compared to the size of the fluid elements, the 4-velocity and thermodynamic quantities can be accurately described by continuous fields. An observer moving with the average velocity u^α of the fluid will see the collisions randomly distribute the nearby particle velocities so that the particle distribution will appear locally isotropic.

Because a comoving observer sees an isotropic distribution of particles, the components of the fluid's energy momentum tensor in her frame must have no preferred direction: $T^{\alpha\beta}u_\beta$ must be invariant under rotations that fix u^α. Denote by

$$q^{\alpha\beta} = g^{\alpha\beta} + u^\alpha u^\beta \tag{1.1}$$

the projection operator orthogonal to u^α. The momentum current

$$q^\alpha{}_\gamma T^{\gamma\beta} u_\beta$$

is a vector in the 3-dimensional subspace orthogonal to u^α, and it is therefore invariant under rotations of that subspace only if it vanishes. Similarly, the symmetric trace-free tensor ${}^3T^{\alpha\beta} - \frac{1}{3}q^{\alpha\beta}\,{}^3T \equiv q^\alpha{}_\gamma q^\beta{}_\delta T^{\gamma\delta} - \frac{1}{3}q^{\alpha\beta}q_{\gamma\delta}T^{\gamma\delta}$ belongs to a $j = 2$ representation of the rotation group and can be invariant only if it vanishes.

It follows that the only nonzero parts of $T^{\alpha\beta}$ are the rotational scalars

$$\epsilon := T^{\alpha\beta}u_\alpha u_\beta \tag{1.2}$$

and

$$p := \frac{1}{3}q_{\gamma\delta}T^{\gamma\delta}. \tag{1.3}$$

More concretely, in an orthonormal frame $\mathbf{e}_{\hat{\mu}}$, with $e_{\hat{0}}^{\alpha} = u^{\alpha}$, the components $T^{\hat{0}\hat{i}}$ and $T^{\hat{i}\hat{j}} - \frac{1}{3}\delta^{\hat{i}\hat{j}}T^{\hat{k}}{}_{\hat{k}}$ must vanish, implying that $T^{\alpha\beta}$ has components

$$\|T^{\hat{\mu}\hat{\nu}}\| = \begin{Vmatrix} \epsilon & & & \\ & p & & \\ & & p & \\ & & & p \end{Vmatrix}. \tag{1.4}$$

To summarize: The condition of local isotropy suffices to define a perfect fluid by enforcing a stress-energy tensor (synonymous with *energy-momentum tensor*) of the form

$$T^{\alpha\beta} = \epsilon u^{\alpha}u^{\beta} + pq^{\alpha\beta}. \tag{1.5}$$

The scalars ϵ and p are the *total energy density* (or, simply, energy density) and the *pressure*, as measured by a comoving observer (an observer with 4-velocity u^{α}).

Thermodynamics. We denote by n the baryon number density and assign a fixed rest mass m_B per baryon.[1] The rest-mass density (equivalently, baryon-mass density) is then

$$\rho := m_B n. \tag{1.6}$$

In general, the properties of matter in a compact object will depend on several parameters, including fluid and magnetic stresses, entropy gradients, composition, heat flow, and neutrino emission. Here, we restrict our attention to the case of a perfect fluid with equilibrium composition, where the energy density and pressure depend on two parameters that can be taken to be ρ and the *specific entropy* (entropy per unit rest mass) s,

$$\epsilon = \epsilon(\rho, s), \qquad p = p(\rho, s). \tag{1.7}$$

The thermodynamics of the fluid is described by the first law, which, in terms of ρ and s, takes the form

$$d\epsilon = \rho T ds + h d\rho, \tag{1.8}$$

where T is *temperature* and h is the *specific enthalpy* (enthalpy per unit rest mass)

$$h := \frac{\epsilon + p}{\rho}. \tag{1.9}$$

One can easily derive Eq. (1.8) from its more common form in terms of extensive quantities,

$$dE = TdS - p\,dV + \mu\,dN \equiv TdS - p\,dV + g\,dM_0, \tag{1.10}$$

[1] Assignment of a rest mass density is somewhat arbitrary, but the difference between choices is less than 0.1%. We follow earlier papers in using the mass per nucleon of ^{12}C, $m_B = 1.659 \times 10^{-24}$ g. This choice is equivalent, up to a constant factor, to assigning to a fluid the rest mass of the collection of free electrons and protons that would result from dispersing the fluid.

by introducing the energy E, entropy S, volume V, baryon number N, and rest mass $M_0 = m_B N$ of a fluid element as measured by a comoving observer. (Here $\mu = g m_B$, where the quantity g is to be defined below.) Then, writing

$$\epsilon = \frac{E}{V}, \quad s = \frac{S}{M_0}, \quad \rho = \frac{M_0}{V},$$

and using conservation of baryons in the form $dM_0 = 0$ to replace dV/V by $-d\rho/\rho$, we quickly obtain Eq. (1.8).

Note that, whether or not one adds baryons to the fluid, one can choose dV so that the number of baryons in the volume $V + dV$ is unchanged; Eq. (1.8) thus holds whether or not baryons are added to the fluid. On the other hand, if one simply enlarges the volume under consideration without changing the internal state of the fluid (keeping ρ and s constant), then $d\rho = ds = d\epsilon = 0$, $dM_0 = \rho dV$, and the extensive version (1.10) of the first law implies

$$\epsilon dV = d(\epsilon V) = T d(\rho s V) - p\, dV + g\, d(\rho V) = (\rho T s - p + \rho g) dV,$$

whence

$$g = \frac{\epsilon + p}{\rho} - Ts. \tag{1.11}$$

The quantity g is thus the *specific Gibbs free energy* (free energy per unit rest mass), and $\mu = g m_B = \dfrac{\epsilon + p - \rho T s}{n}$ is the *Gibbs free energy per baryon*, each defined for a comoving observer.

Defining the *specific internal energy* (internal energy per unit rest mass) e by the relation

$$\epsilon = \rho(1 + e), \tag{1.12}$$

one recovers the *Newtonian expression for the specific enthalpy*,

$$h_{\text{Newtonian}} = h - 1 = e + p/\rho. \tag{1.13}$$

Because the relativistic energy density ϵ includes the rest-mass density ρ, the relativistic enthalpy per unit rest mass differs from its Newtonian counterpart by the rest mass per unit rest mass, by $\rho/\rho = 1$.

Baroclinic (entropy nonconserving) flow. From the definition (1.9) and using Eq. (1.8), one finds

$$\begin{aligned} dh &= \frac{d\epsilon}{\rho} + \frac{dp}{\rho} - \frac{\epsilon + p}{\rho^2} d\rho = T ds + h\frac{d\rho}{\rho} + \frac{dp}{\rho} - h\frac{d\rho}{\rho} \\ &= T ds + \frac{dp}{\rho} \\ \Rightarrow \quad d \ln h &= \frac{T}{h} ds + \frac{dp}{\epsilon + p}, \end{aligned} \tag{1.14}$$

implying

$$\nabla_\alpha \ln h = \frac{T}{h}\nabla_\alpha s + \frac{1}{\epsilon + p}\nabla_\alpha p. \tag{1.15}$$

Taking the curl of (1.15), we arrive at

$$0 = \nabla_{[\alpha}\nabla_{\beta]} \ln h = \nabla_{[\alpha}\left(\frac{T}{h}\nabla_{\beta]} s\right) + \nabla_{[\alpha}\left(\frac{1}{\epsilon + p}\nabla_{\beta]} p\right). \tag{1.16}$$

The last relation implies that in the presence of entropy gradients ($\nabla s \neq 0$), surfaces of constant energy density (*isopycnic surfaces*) do not, in general, coincide with surfaces of constant pressure (*isobaric surfaces*). Such a flow is called *baroclinic*, and, for a rotating star, it implies the presence of *meridional circulation.*

Barotropic flow. One commonly uses a one-parameter equation of state to describe a compact star, because within a short time after formation, neutrino emission cools the star to 10^{10} K $\simeq 1$ MeV. This is much smaller than the Fermi energy of the interior, in which a density greater than nuclear ($0.16\,\mathrm{fm}^{-3}$) implies a Fermi energy greater than $E_F\,(0.16\,\mathrm{fm}^{-3}) \approx 60\,\mathrm{MeV}$. A neutron star is in this sense cold, and, because nuclear reaction times are shorter than the cooling time, one can use a zero-temperature *equation of state* (EOS) to describe the matter:

$$\epsilon = \epsilon(\rho), \quad p = p(\rho), \tag{1.17}$$

or, equivalently,

$$\epsilon = \epsilon(p). \tag{1.18}$$

In a stationary, one-component perfect fluid, a one-parameter equation of state of the form (1.18) holds, more generally, when the specific entropy is constant throughout the star ($\nabla s = 0$) – that is, for a *homentropic flow.* From Eq. (1.16) it is evident that in such a case, the isopycnic and isobaric surfaces coincide – that is, the homentropic flow of a one-component perfect fluid is *barotropic*, which is also implied by Eq. (1.18) itself.

In a homentropic star, the first law, Eq. (1.8), takes the form

$$d\epsilon = h d\rho, \tag{1.19}$$

and using Eq. (1.14), the specific enthalpy is also given by

$$h = \exp\left(\int_0^p \frac{dp}{\epsilon + p}\right), \tag{1.20}$$

with $\frac{\epsilon}{\rho} = 1$ at $p = 0$ (the gas is nonrelativistic at low densities).

Although a nonhomentropic star is, in general, a barocline, if one makes the assumption that $\epsilon = \epsilon(p)$, then Eq. (1.16) implies, for a one-component perfect fluid, that $s = s(\rho)$, and the star is still barotropic.[2] Similarly, if $s = s(\rho)$, the star is barotropic.

[2] Such models are sometimes called *pseudobarotropes* in the literature.

Fluid dynamics and conservation laws. For a two-parameter equation of state, five variables determine the state of a perfect fluid; they can be taken to be ϵ, p and three independent components of u^α. The dynamical evolution of the fluid is governed by the vanishing divergence of the stress-energy tensor

$$\nabla_\beta T^{\alpha\beta} = 0, \tag{1.21}$$

and by conservation of baryons,

$$\nabla_\alpha(\rho u^\alpha) = 0. \tag{1.22}$$

The projection of the equation $\nabla_\beta T^{\alpha\beta} = 0$ along u^α yields an energy conservation law, whereas the projection orthogonal to u^α is the relativistic Euler equation. For an intuitive understanding of these equations, it is helpful to look first at conservation of baryons.

Conservation of baryons. The baryon mass M_0 of a fluid element is conserved by the motion of the fluid. The proper volume of a fluid element is the volume V of a slice orthogonal to u^α through the history of the fluid element; conservation of baryons can be written in the form $0 = \Delta M_0 = \Delta(\rho V)$. The fractional change in V in a proper time $\Delta\tau$ is given by the 3-dimensional divergence of the velocity in the subspace orthogonal to u^α:

$$\frac{\Delta V}{V} = q^{\alpha\beta}\nabla_\alpha u_\beta \Delta\tau. \tag{1.23}$$

Because $u^\beta u_\beta = -1$, we have $u^\beta \nabla_\alpha u_\beta = \frac{1}{2}\nabla_\alpha(u_\beta u^\beta) = 0$, implying

$$q^{\alpha\beta}\nabla_\alpha u_\beta = \nabla_\beta u^\beta. \tag{1.24}$$

With $u^\alpha\nabla_\alpha\rho = \dfrac{d}{d\tau}\rho$, conservation of baryons takes the form

$$0 = \frac{\Delta(\rho V)}{V} = \Delta\rho + \rho\frac{\Delta V}{V} = (u^\alpha\nabla_\alpha\rho + \rho\nabla_\alpha u^\alpha)\Delta\tau, \tag{1.25}$$

or

$$\nabla_\alpha(\rho u^\alpha) = 0. \tag{1.26}$$

Conservation of energy. The projection $u_\alpha\nabla_\beta T^{\alpha\beta} = 0$ similarly expresses energy conservation for a fluid element:

$$\begin{aligned} 0 = u_\alpha\nabla_\beta T^{\alpha\beta} &= u_\alpha\nabla_\beta[\epsilon u^\alpha u^\beta + p q^{\alpha\beta}] \\ &= -\nabla_\beta(\epsilon u^\beta) + p u_\alpha\nabla_\beta(g^{\alpha\beta} + u^\alpha u^\beta) \\ &= -\nabla_\beta(\epsilon u^\beta) - p\nabla_\beta u^\beta, \end{aligned}$$

implying

$$\nabla_\beta(\epsilon u^\beta) = -p\nabla_\beta u^\beta. \tag{1.27}$$

The equation means that the total energy of a fluid element decreases by the work,

$$p\,dV = pV\,\nabla_\beta u^\beta\,d\tau, \tag{1.28}$$

that it does on its surroundings in proper time $d\tau$.

Relativistic Euler equation. The projection of the conservation of the stress-energy tensor orthogonal to u^α is

$$q^\alpha{}_\gamma \nabla_\beta T^{\beta\gamma} = 0, \tag{1.29}$$

so that

$$\begin{aligned} 0 &= q^\alpha{}_\gamma \nabla_\beta [\epsilon u^\beta u^\gamma + p q^{\beta\gamma}] \\ &= q^\alpha{}_\gamma \epsilon u^\beta \nabla_\beta u^\gamma + q^{\alpha\beta}\nabla_\beta p + q^\alpha{}_\gamma p \nabla_\beta (u^\beta u^\gamma) \\ &= \epsilon u^\beta \nabla_\beta u^\alpha + q^{\alpha\beta}\nabla_\beta p + p u^\beta \nabla_\beta u^\alpha, \end{aligned}$$

implying

$$(\epsilon + p) u^\beta \nabla_\beta u^\alpha = -q^{\alpha\beta}\nabla_\beta p. \tag{1.30}$$

For a barotropic fluid with constant entropy (a homentropic fluid), one can use Eq. (1.20) to write the relativistic Euler equation in the form

$$u^\beta \nabla_\beta u^\alpha = -q^{\alpha\beta}\nabla_\beta \ln h \tag{1.31}$$

or, equivalently,

$$u^\beta \nabla_{[\alpha}(h u_{\beta]}) = 0. \tag{1.32}$$

In this equation, the form $\omega_{\alpha\beta}$, defined by

$$\omega_{\alpha\beta} = \nabla_\alpha (h\,u_\beta) - \nabla_\beta (h\,u_\alpha), \tag{1.33}$$

is the *relativistic vorticity*.

Newtonian approximation. Let ε be a small parameter of order v/c or $v_{\rm sound}/c$, whichever is larger. In the Newtonian approximation, there are Cartesian coordinates for which the metric has the form

$$ds^2 = -(1 + 2\Phi)dt^2 + (dx^2 + dy^2 + dz^2)(1 + O(\varepsilon^2)), \tag{1.34}$$

and with off-diagonal terms of order $\varepsilon\Phi$. Here, Φ is the Newtonian potential, satisfying

$$\nabla^2\Phi = 4\pi\rho. \tag{1.35}$$

The components u^μ and scalars Φ, ϵ, p, ρ satisfy relations of the form $\partial_t\Phi = O(\varepsilon)\nabla_i\Phi$ and have the orders

$$u^t = 1 + O(\varepsilon^2), \tag{1.36}$$

$$u^i = v^i + O(\varepsilon^3), \tag{1.37}$$

$$p/\epsilon = O(\varepsilon^2), \tag{1.38}$$

$$\epsilon = \rho[1 + O(\varepsilon^2)]. \tag{1.39}$$

Conservation of baryons (1.26) then takes the form

$$\partial_t\rho + \partial_i(\rho v^i) = 0 + O(\rho\varepsilon^2). \tag{1.40}$$

The relativistic Euler equation becomes

$$\rho u^\mu \nabla_\mu u^i = -\nabla^i p;$$

and, after using the metric to compute $\Gamma^i_{tt} = \nabla_i\Phi[1 + O(\varepsilon^2)]$, we recover the Euler equation

$$\rho(\partial_t + v^j\nabla_j)v_i + \rho\partial_i\Phi = -\nabla_i p. \tag{1.41}$$

Conservation of energy (1.27) to the lowest nontrivial order, $O(\varepsilon^3)$, immediately reduces to conservation of baryons. To $O(\varepsilon^5)$, one obtains the energy conservation equation that arises from the Euler equation (1.41) by dotting it with v^i, but to deduce this relation from Eq. (1.27), one must keep subleading terms in the metric and 4-velocity.

Spacetime symmetries. A vector field ξ^α is a Killing vector if it Lie derives the metric

$$\mathcal{L}_{\boldsymbol{\xi}} g_{\alpha\beta} = \nabla_\alpha\xi_\beta + \nabla_\beta\xi_\alpha = 0. \tag{1.42}$$

A Killing vector generates a family of isometries, diffeomorphisms that leave the metric invariant. We will call ξ^α a *symmetry vector* of a perfect-fluid spacetime if ξ^α is a Killing vector that also Lie derives the fluid variables:

$$\mathcal{L}_{\boldsymbol{\xi}} u^\alpha = 0, \quad \mathcal{L}_{\boldsymbol{\xi}}\epsilon = 0, \quad \mathcal{L}_{\boldsymbol{\xi}} p = 0. \tag{1.43}$$

Associated with a symmetry ξ^α is a quantity $hu_\beta\,\xi^\beta$ that is conserved along the spacetime trajectories of the fluid,

$$\mathcal{L}_{\mathbf{u}}\left(h\,u_\beta\xi^\beta\right) = 0. \tag{1.44}$$

To derive this relation, we use the form (1.32) of the Euler equation to write

$$\begin{aligned} 0 &= u^\beta\left[\nabla_\beta(hu_\alpha) - \nabla_\alpha(hu_\beta)\right]\xi^\alpha \\ &= u^\beta\nabla_\beta(hu_\alpha\,\xi^\alpha) - u^\beta\,\mathcal{L}_\xi(hu_\beta) \\ &= u^\beta\nabla_\beta(hu_\alpha\,\xi^\alpha), \end{aligned}$$

as claimed.

Stationary flow. A stationary flow is described by a spacetime with an asymptotically timelike symmetry vector, t^α, the generator of time translations that leave the metric and the fluid variables fixed. The corresponding conservation law, (1.44),

$$\mathcal{L}_{\mathbf{u}}\left(h\, u_\beta t^\beta\right) = \mathcal{L}_{\mathbf{u}}\left(\frac{\epsilon+p}{\rho} u_\beta t^\beta\right) = 0, \tag{1.45}$$

is the relativistic form of Bernoulli's law, the conservation of *enthalpy per unit rest mass*, $-hu_t$, along the trajectories of a stationary flow.

To obtain the Newtonian approximation,

$$(\partial_t + \mathcal{L}_{\mathbf{v}})\left(h_{\text{Newtonian}} + \frac{1}{2}v^2 + \Phi\right) = 0,$$

one must use the $O(\varepsilon^2)$ form of u_t, namely $u_t = -(1+\Phi+\frac{1}{2}v^2)$, implied by Eqs. (1.34), (1.36), (1.37) and the normalization $u^\alpha u_\alpha = -1$. Note that the definition $h_{\text{Newtonian}} = h - 1$ coincides for isentropic flows with

$$h_{\text{Newtonian}} = \int_0^p \frac{dp}{\rho}, \tag{1.46}$$

as follows from Eq. (1.13) and the first law in the form $de = \dfrac{p}{\rho^2} d\rho$.

Axisymmetric flow. An axisymmetric flow is described by a spacetime with a rotational symmetry vector ϕ^α, a spacelike vector field whose orbits are circles, except on an axis of symmetry (a two-dimensional submanifold of the spacetime), where $\phi^\alpha = 0$. The corresponding conservation law, (1.44),

$$\mathcal{L}_{\mathbf{u}}(h\, u_\beta \phi^\beta) = 0, \tag{1.47}$$

expresses the conservation of a fluid element's *specific angular momentum*, $j := hu_\phi$, the angular momentum per unit rest mass about the axis of symmetry associated with ϕ^α. We will see in Section 1.8 that calling j the specific angular momentum is consistent with the integral expression for the total angular momentum J of the spacetime: $J = \int j dM_0$.

Isentropic flow. In the absence of shocks, the flow of a perfect fluid remains *isentropic* – that is, each fluid element conserves its specific entropy along its trajectory,

$$u^\alpha \nabla_\alpha s = 0. \tag{1.48}$$

Formally, the relation follows from conservation of baryons (1.26), conservation of energy (1.27), and from the first law (1.8) or equivalently the equation of state $\epsilon = \epsilon(\rho, s)$.

Conservation of vorticity and circulation in barotropic flows. The *relativistic vorticity* $\omega_{\alpha\beta}$ was defined by Eq.(1.33). From Eq. (1.32) and the Cartan identity (A.38), we have

$$0 = u^\beta [\nabla_\beta(hu_\alpha) - \nabla_\alpha(hu_\beta)] = \mathcal{L}_u(hu_\alpha) + \nabla_\alpha h. \tag{1.49}$$

From the fact that exterior derivatives and Lie derivatives commute, $\nabla_{[\alpha}\mathcal{L}_{\mathbf{u}} w_{\beta]} = \mathcal{L}_{\mathbf{u}}\nabla_{[\alpha} w_{\beta]}$, the variation of vorticity along fluid trajectories is given by

$$\mathcal{L}_{\mathbf{u}}\omega_{\alpha\beta} = -2\nabla_{[\alpha}\nabla_{\beta]}h = 0. \tag{1.50}$$

For a barotropic flow, it immediately follows that *vorticity is conserved along the fluid trajectories.*

The corresponding integral law is obtained as follows. Let c be a closed curve in the fluid, bounding a 2-surface S; and let c_τ be the curve obtained by moving each point of c a proper time τ along the fluid trajectory through that point. From the relation

$$\mathcal{L}_{\mathbf{u}}\omega_{\alpha\beta} = \nabla_\alpha \mathcal{L}_{\mathbf{u}}(h\,u_\beta) - \nabla_\beta \mathcal{L}_{\mathbf{u}}(h\,u_\alpha), \tag{1.51}$$

we have

$$\begin{aligned} 0 = \int_S \mathcal{L}_{\mathbf{u}}\omega_{\alpha\beta} dS^{\alpha\beta} &= \int_c \mathcal{L}_{\mathbf{u}}(h\,u_\alpha)dl^\alpha \\ &= \frac{d}{d\tau}\int_{c_\tau} h\,u_\alpha\, dl^\alpha, \end{aligned} \tag{1.52}$$

where Stokes's theorem was used to obtain the second equality and Eq. (A.83) of Appendix A was used in the last equality. That is, the line integral,

$$\int_{c_\tau} h\,u_\alpha\, dl^\alpha = \int_{c_\tau} \frac{\epsilon + p}{\rho}\, u_\alpha\, dl^\alpha \tag{1.53}$$

(the *circulation* of the flow along a closed curve), is independent of τ, conserved by the fluid flow.

Circular flow (absence of meridional circulation). Although newly born neutron stars are baroclinic, having meridional circulation and strong convection in the outer layers, as the star cools below the Fermi temperature for neutrons, its equation of state becomes essentially barotropic. The velocity field becomes circular (its only spatial velocity component is u^ϕ); viscosity and the magnetic field enforce uniform rotation.

1.2 The spacetime of a rotating star

A rotating star can be modeled by a stationary, axisymmetric, perfect-fluid spacetime, whose circular velocity field u^α can be written in terms of the two Killing vectors t^α and ϕ^α,

$$u^\alpha = u^t(t^\alpha + \Omega\phi^\alpha), \tag{1.54}$$

where the scalar

$$u^t := [-g_{\alpha\beta}(t^\alpha + \Omega\phi^\alpha)(t^\beta + \Omega\phi^\beta)]^{-1/2} \tag{1.55}$$

will be the t component of u^α, $u^t = u^\alpha \nabla_\alpha t$, with respect to natural coordinates t and ϕ that we describe in detail below. Here

$$\Omega \equiv \frac{u^\phi}{u^t} = \frac{d\phi}{dt} \tag{1.56}$$

is the angular velocity of the fluid as seen by an observer at rest at infinity (an asymptotic observer with 4-velocity along the asymptotically timelike Killing vector t^α). A star is *uniformly rotating* (as seen from infinity) if Ω is constant. Locally, the flow is shear-free if and only if rotation is uniform. Defining the shear tensor as

$$\sigma_{\alpha\beta} := q_\alpha{}^\gamma q_\beta{}^\delta \nabla_{(\gamma} u_{\delta)} - \frac{1}{3} q_{\alpha\beta} \nabla_\gamma u^\gamma, \tag{1.57}$$

Killing's equation implies for constant Ω

$$q_\alpha{}^\gamma q_\beta{}^\delta \nabla_{(\gamma} u_{\delta)} = 0. \tag{1.58}$$

The formalism here is general, allowing differentially rotating configurations.

Geometry. The spacetime geometry of an axisymmetric rotating star in equilibrium is described by a stationary metric $g_{\alpha\beta}$ of the form

$$ds^2 = -e^{2\nu} dt^2 + e^{2\psi}(d\phi - \omega dt)^2 + e^{2\mu}(dr^2 + r^2 d\theta^2), \tag{1.59}$$

where ν, ψ, ω, and μ are four *metric functions* that depend on the coordinates r and θ only. In the exterior vacuum, it is possible to reduce the number of metric functions to three. It is convenient to write e^ψ in the form [46]

$$e^\psi = r \sin\theta B e^{-\nu}, \tag{1.60}$$

where B is again a function of r and θ only.

A metric of the form (1.59) follows from the assumptions

1. The spacetime is asymptotically flat.
2. The spacetime is stationary and axisymmetric: There exist an asymptotically timelike symmetry vector t^α and a rotational symmetry vector ϕ^α.

   The spacetime is said to be *strictly* stationary if t^α is everywhere timelike. (Some rapidly rotating stellar models, as well as black-hole spacetimes, have *ergospheres*, regions in which t^α is spacelike.)
3. The Killing vectors commute,

$$[t, \phi] = 0, \tag{1.61}$$

   and there is an isometry of the spacetime that simultaneously reverses the direction of t^α and ϕ^α,

$$t^\alpha \to -t^\alpha, \quad \phi^\alpha \to -\phi^\alpha. \tag{1.62}$$

If the spacetime is strictly stationary, one does not need (1.61) as a separate assumption: A theorem by Carter [116] shows that $[t, \phi] = 0$.

The Frobenius theorem now implies the existence of scalars t and ϕ [377, 115] for which

$$t^\alpha \nabla_\alpha t = \phi^\alpha \nabla_\alpha \phi = 1, \tag{1.63a}$$

$$t^\alpha \nabla_\alpha \phi = \phi^\alpha \nabla_\alpha t = 0. \tag{1.63b}$$

That is, there is a family of 2-surfaces orthogonal to t^α and ϕ^α, the surfaces of constant t and ϕ, and it is natural to choose as coordinates $x^0 = t$ and $x^3 = \phi$. Requiring that the remaining coordinates x^1, x^2 that rule the surfaces of constant t and ϕ be Lie derived by t^α and ϕ^α, we have

$$t^\alpha = \boldsymbol{\partial}_t, \quad \phi^\alpha = \boldsymbol{\partial}_\phi. \tag{1.64}$$

With coordinates chosen in this way, the metric components are independent of t and ϕ.

Because time reversal inverts the direction of rotation, the fluid is not invariant under $t \to -t$ alone. For a spherical star, time-reversal symmetry $\mathcal{T} : t \to -t$ implies that t^α and ϕ^α are orthogonal: Noting that $t^\alpha \phi_\alpha = g_{t\phi}$, this follows from the fact that $g_{t\phi}$ is independent of t and ϕ and $\mathcal{T}$ maps $g_{t\phi}$ to $-g_{t\phi}$. That is,

$$g_{t\phi}(t, x^1, x^2, \phi) = (\mathcal{T}^* g)_{t\phi}(t, x^1, x^2, \phi) = -g_{t\phi}(-t, x^1, x^2, \phi) = -g_{t\phi}(t, x^1, x^2, \phi), \tag{1.65}$$

implying $g_{t\phi} = 0$. For a rotating star, the asymmetry under time reversal means that there is no family of surfaces orthogonal to t^α, implying a metric in which t^α and ϕ^α are not orthogonal to each other. The lack of orthogonality is measured by the potential ω that describes (as we detail below) the *dragging of inertial frames.*

In a fluid with meridional convective currents, one loses both time-reversal invariance and invariance under the simultaneous inversion $t \to -t, \phi \to -\phi$, because the inversion changes the direction of the circulation. In this case the asymmetry means that there will be no family of surfaces orthogonal to t^α and ϕ^α, and the spacetime metric (1.59) will have additional off-diagonal components.

A common choice for x^1 and x^2 are *quasi-isotropic coordinates*, for which $g_{r\theta} = 0$, $g_{\theta\theta} = r^2 g_{rr}$ (in spherical polar coordinates), or $g_{\varpi z} = 0$, $g_{zz} = r^2 g_{\varpi\varpi}$ (in cylindrical coordinates). In terms of the cylindrical coordinates ϖ and z, the metric reads

$$ds^2 = -e^{2\nu} dt^2 + e^{2\psi}(d\phi - \omega dt)^2 + e^{2\mu}(d\varpi^2 + dz^2). \tag{1.66}$$

In the slow-rotation formalism by Hartle [297], a different form of the metric is used, requiring $g_{\theta\theta} = g_{\phi\phi}/\sin^2\theta$.

The three metric functions ν, ψ, and ω are related to the norms of the Killing vectors, t^α and ϕ^α, and to their dot product by the relations

$$t_\alpha t^\alpha = g_{tt} = -e^{2\nu} + \omega^2 e^{2\psi}, \tag{1.67a}$$

$$\phi_\alpha \phi^\alpha = g_{\phi\phi} = e^{2\psi}, \tag{1.67b}$$

$$t_\alpha \phi^\alpha = g_{t\phi} = -\omega e^{2\psi}. \tag{1.67c}$$

The corresponding components of the contravariant metric are

$$g^{tt} = \nabla_\alpha t \nabla^\alpha t = -e^{-2\nu}, \tag{1.68a}$$

$$g^{\phi\phi} = \nabla_\alpha \phi \nabla^\alpha \phi = e^{-2\psi} - \omega^2 e^{-2\nu}, \tag{1.68b}$$

$$g^{t\phi} = \nabla_\alpha t \nabla^\alpha \phi = -\omega e^{-2\nu}. \tag{1.68c}$$

The geometry of the orthogonal 2-surfaces is determined by the conformal factor $e^{2\mu}$.

Physical interpretation. A striking difference between the description of a star in general relativity and that in the Newtonian approximation is the fact that particles dropped from infinity with zero angular momentum acquire a nonzero angular velocity in the direction of the star's rotation. Because inertial observers are freely falling observers, the dragging forward of free particles is called the dragging of inertial frames. Formally, associated with the rotational Killing vector ϕ^α is the conserved angular momentum $L = p_\alpha \phi^\alpha$ of a free particle with 4-momentum $p^\alpha = mu^\alpha$. Conservation of L follows from the geodesic equation, $u^\beta \nabla_\beta u^\alpha = 0$, and the definition (1.42) of a Killing vector:

$$u^\beta \nabla_\beta (m u_\alpha \phi^\alpha) = m u^\alpha u^\beta \nabla_\alpha \phi_\beta = m u^\alpha u^\beta \nabla_{(\alpha} \phi_{\beta)} = 0. \tag{1.69}$$

A particle with $L = 0$ satisfies $u_\phi = u_\alpha \phi^\alpha = 0$, or, using Eq. (1.59),

$$e^{2\psi}(u^\phi - \omega u^t) = 0,$$

implying that the angular velocity $d\phi/dt$ measured by an observer at infinity is

$$\frac{u^\phi}{u^t} = \omega. \tag{1.70}$$

Thus, an initially (at infinity) radially infalling particle (with zero angular momentum) will acquire an angular velocity as measured by an inertial observer at infinity

$$\Omega = \omega, \tag{1.71}$$

even though it maintains zero angular momentum along its trajectory.

In describing the fluid, it is helpful to introduce a family of zero-angular-momentum observers (ZAMOs) [45, 46], observers whose velocity has at each point the form (for circular, axisymmetric motion)

$$u^\alpha_{\text{ZAMO}} = u^t(t^\alpha + \omega\phi^\alpha) = e^{-\nu}(t^\alpha + \omega\phi^\alpha) \tag{1.72}$$

(the latter equality follows from Eqs. [1.80] and [1.83], discussed below). The worldlines of these observers are normal to the $t =$ const. hypersurfaces. Several properties of a fluid can be conveniently expressed with respect to ZAMOs (see, for example, the 3-velocity, defined below).

An interpretation of the metric function ν follows from Eq. (1.72): $e^{-\nu}$ is the *time dilation factor* relating the proper time of the local ZAMO to coordinate time t (proper time at infinity). Equivalently, the *redshift factor* z of a zero-angular-momentum photon sent to infinity by a ZAMO at point P is given by $z = 1 - e^{-\nu}|_P$.

Finally, the geometrical meaning of the metric function ψ is immediate from the form of the metric: $2\pi e^{\psi}$ is the *proper circumference* of a circle around the axis of symmetry, so that e^{ψ} is called the *proper circumferential radius*. For a nonrotating (spherical) star, e^{ψ} coincides with $R\sin\theta$, where R is the radial coordinate of the Schwarzschild metric (in Schwarzschild, not isotropic, coordinates).

Nonrotating limit. In the nonrotating limit, the metric (1.66) reduces to the metric of a spherical relativistic star in *isotropic coordinates*, so that, in the exterior vacuum region, the following relations hold:

$$e^{\nu} = \frac{1 - M/2r}{1 + M/2r}, \quad e^{\psi} = \varpi(1 + M/2r)^2, \quad e^{\mu} = (1 + M/2r)^2, \tag{1.73}$$

where M is the gravitational mass of the star (see [725]).

Ergospheres. In highly relativistic models, rapid rotation can lead to frame dragging extreme enough that all physical particles are dragged forward relative to an observer at infinity, or, equivalently, relative to the Killing vector t^{α}. A region in which this is true is called an *ergosphere*, whose definition (mentioned above) is a region in which the asymptotically timelike Killing vector t^{α} is spacelike. Because physical particles move along timelike or null lines, the definition implies that no physical particle can remain at rest relative to t^{α}.

At any point in the spacetime, the angular velocity of a physical particle is restricted in a way that looks asymmetric relative to infinity, but symmetric relative to a ZAMO. In Minkowski space, a particle can have a timelike or null trajectory only if $\varpi\Omega < 1$, implying $-1/\varpi < \Omega < 1/\varpi$. Here, the analogous requirement on $(u^t)^2$ for a particle with arbitrary velocity

$$\widetilde{u}^{\alpha} = \widetilde{u}^t(t^{\alpha} + \widetilde{\Omega}\phi^{\alpha} + \widetilde{v}^{\alpha}),$$

where $\widetilde{v}^{\alpha} \perp t^{\alpha},\ \phi^{\alpha}$, has the form

$$t^{\alpha}t_{\alpha} + 2\widetilde{\Omega}t^{\alpha}\phi_{\alpha} + \widetilde{\Omega}^2\phi^{\alpha}\phi_{\alpha} + \widetilde{v}^{\alpha}\widetilde{v}_{\alpha} < 0.$$

Then $\widetilde{v}^{\alpha}\widetilde{v}_{\alpha} \geq 0$ implies $\Omega_- < \widetilde{\Omega} < \Omega_+$. The bounds $\Omega_{\pm}$ take their extremal values for circular orbits in the equatorial plane (for $\widetilde{v}^{\alpha} = 0$):

$$\Omega_{\pm} = -\frac{\boldsymbol{t}\cdot\boldsymbol{\phi}}{\boldsymbol{\phi}\cdot\boldsymbol{\phi}} \pm \left[\left(\frac{\boldsymbol{t}\cdot\boldsymbol{\phi}}{\boldsymbol{\phi}\cdot\boldsymbol{\phi}}\right)^2 - \frac{\boldsymbol{t}\cdot\boldsymbol{t}}{\boldsymbol{\phi}\cdot\boldsymbol{\phi}}\right]^{1/2}$$

or

$$\Omega_{\pm} = \omega \pm \left(\omega^2 - \frac{\boldsymbol{t}\cdot\boldsymbol{t}}{\boldsymbol{\phi}\cdot\boldsymbol{\phi}}\right)^{1/2}. \tag{1.74}$$

For the metric (1.66),

$$\Omega_{\pm} = \omega \pm e^{\nu-\psi}. \tag{1.75}$$

One obtains the same expression (1.75) when requiring that $|v| < 1$, where v is the 3-velocity of circular motion in the equatorial plane, with respect to a ZAMO. At the boundary of the ergosphere (called the *stationary limit*), where $t^\alpha t_\alpha = 0$, we have $\Omega_- = 0$. Within the ergosphere, both Ω_+ and Ω_- are positive, implying that, seen from infinity, all particles must move in the direction of the star's rotation. In stellar models, ergospheres are toroidal and typically enclose the star's equator.

The conserved "energy" $E = -p_\alpha t^\alpha$ associated with t^α is, inside the ergosphere, a spatial component of momentum and can have either sign. This allows one to extract rotational energy from the star by the *Penrose process*: A particle with initial energy $E > 0$ decays in the ergosphere into two particles, one, with negative energy $E_1 < 0$, that remains trapped in the ergosphere, and one whose positive energy $E_2 = E - E_1$ is greater than that of the initial particle. The conserved energy of radiation fields on a background spacetime with an ergosphere (for example, of scalar and electromagnetic fields and solutions to the Einstein equations linearized about the background geometry) can similarly have negative values in the ergosphere. Because the negative energy part of the field is again trapped within the ergosphere, and because a time-dependent, nonaxisymmetric field radiates positive energy to infinity, fields of this kind are (as we show in a subsequent chapter) unstable in spacetimes with an ergosphere and no horizon [231]. The associated growth time, however, is comparable to the age of the universe [155].

Asymptotic behavior. The lowest-order asymptotic behavior of the metric functions ν and ω is

$$\nu \sim -\frac{M}{r} \tag{1.76}$$

$$\omega \sim \frac{2J}{r^3}, \tag{1.77}$$

where M and J are the gravitational mass and angular momentum, of the source of the gravitational field (see Section 1.8 for definitions and Section 3.2 for an expansion in terms of higher-order multiple moments.). The asymptotic expansion of the dragging potential ω shows that it decays rapidly far from the star, so that its effect will be significant mainly in the vicinity of the star.

In addition, the asymptotic relations

$$e^\psi = \varpi(e^{-\nu} + O(r^{-2})), \qquad e^\mu = e^{-\nu} + O(r^{-2}) \tag{1.78}$$

hold for the metric functions (1.73), and for the metric (1.66) as well, because any stationary, asymptotically flat spacetime agrees with the Schwarzschild geometry to order r^{-1}. If, following Bardeen and Wagoner [51], we write

$$\beta := \psi + \nu, \qquad \zeta := \mu + \nu, \qquad B := \frac{1}{\varpi}\, e^\beta, \tag{1.79}$$

then, asymptotically, β (or B) and ζ deviate by $O(r^{-2})$ from their values in the isotropic Schwarzschild metric.

4-Velocity. The 4-velocity for circular flow is written as in (1.54)

$$u^\alpha = u^t(t^\alpha + \Omega\phi^\alpha). \tag{1.80}$$

In the metric (1.66) the normalization $u_\alpha u^\alpha = -1$ determines u^t as

$$u^t = \frac{e^{-\nu}}{\sqrt{1-(\Omega-\omega)^2 e^{2(\psi-\nu)}}}. \tag{1.81}$$

Defining

$$v = (\Omega-\omega)e^{(\psi-\nu)}, \tag{1.82}$$

the contravariant and covariant components of the 4-velocity take the form

$$u^t = \frac{e^{-\nu}}{\sqrt{1-v^2}}, \qquad u^\phi = \Omega u^t, \tag{1.83}$$

$$u_t = -\frac{e^{\nu}}{\sqrt{1-v^2}}(1+e^{\psi-\nu}\omega v), \qquad u_\phi = \frac{e^{\psi}v}{\sqrt{1-v^2}}. \tag{1.84}$$

Written in this way, the denominator has the form of a Lorentz factor. Indeed, as shown later, v is identified as the 3-velocity measured in the frame of a ZAMO.

3-Velocity. Because the 3-velocity v is not a covariant quantity, one has to determine the physical frame with respect to which it is measured. This is most easily done by constructing an orthonormal tetrad, in which the metric has locally the form of the Minkowski metric

$$ds^2 = \eta_{\mu\nu}\boldsymbol{\omega}^{\hat\mu} \otimes \boldsymbol{\omega}^{\hat\nu}, \tag{1.85}$$

where $\boldsymbol{\omega}^{\hat\mu}$ are the basis covectors (the index denotes the different vectors, not components). A suitable example is the frame defined by the basis covectors

$$\boldsymbol{\omega}^{\hat 0} = e^{\nu}\boldsymbol{d}t, \quad \boldsymbol{\omega}^{\hat 1} = e^{\psi}(\boldsymbol{d}\phi - \omega\boldsymbol{d}t), \quad \boldsymbol{\omega}^{\hat 2} = e^{\mu}\boldsymbol{d}\varpi, \quad \boldsymbol{\omega}^{\hat 3} = e^{\mu}\boldsymbol{d}z, \tag{1.86}$$

with corresponding contravariant basis vectors

$$\boldsymbol{e}_{\hat 0} = e^{-\nu}(\boldsymbol{\partial}_t + \omega\boldsymbol{\partial}_\phi), \quad \boldsymbol{e}_{\hat 1} = e^{-\psi}\boldsymbol{\partial}_\phi, \quad \boldsymbol{e}_{\hat 2} = e^{-\mu}\boldsymbol{\partial}_\varpi, \quad \boldsymbol{e}_{\hat 3} = e^{-\mu}\boldsymbol{\partial}_z. \tag{1.87}$$

The nonzero components of the 4-velocity $u^{\hat\mu}$ along these frame vectors are written in terms of a fluid 3-velocity v as in Minkowski spacetime,

$$u^{\hat 0} = \frac{1}{\sqrt{1-v^2}}, \qquad u^{\hat 1} = \frac{v}{\sqrt{1-v^2}}. \tag{1.88}$$

Transforming these back to the coordinate frame, via $u^\nu = u^{\hat\mu}e^\nu_{\hat\mu}$, one obtains the components (1.83) only if

$$v = e^{\psi-\nu}(\Omega-\omega), \tag{1.89}$$

as in Eq. (1.82). Because $v = 0$ for $\Omega = \omega$ (for the ZAMO), the 3-velocity v is defined with respect to this observer, and the orthonormal tetrad (1.87) is the basis of the ZAMO frame.

1.3 Einstein's field equation

When an equation of state has been specified and if an equilibrium solution exists, the structure of the star is determined by solving four components of Einstein's gravitational field equation $G_{\alpha\beta} = 8\pi T_{\alpha\beta}$, or

$$R_{\alpha\beta} = 8\pi \left(T_{\alpha\beta} - \frac{1}{2} g_{\alpha\beta} T \right) \tag{1.90}$$

(where $R_{\alpha\beta}$ is the Ricci tensor and $T = T_\alpha{}^\alpha$), together with the equation of hydrostationary equilibrium (see next section). One approach for deriving the necessary equations is to select four components of the Einstein field equation, expressed in the tetrad frame of the ZAMO. In this frame, the stress-energy tensor becomes

$$T^{\hat{0}\hat{0}} = \frac{\epsilon + pv^2}{1 - v^2}, \qquad T^{\hat{0}\hat{1}} = (\epsilon + p)\frac{v}{1 - v^2}, \tag{1.91}$$

$$T^{\hat{1}\hat{1}} = \frac{\epsilon v^2 + p}{1 - v^2}, \qquad T^{\hat{2}\hat{2}} = T^{\hat{3}\hat{3}} = p. \tag{1.92}$$

With $\zeta = \mu + \nu$, one common choice for the components of the gravitational field equation is [46], [110]

$$\begin{aligned} \nabla \cdot (B \nabla \nu) = {} & \frac{1}{2} r^2 \sin^2 \theta B^3 e^{-4\nu} \nabla \omega \cdot \nabla \omega \\ & + 4\pi B e^{2\zeta - 2\nu} \left[\frac{(\epsilon + p)(1 + v^2)}{1 - v^2} + 2p \right], \end{aligned} \tag{1.93a}$$

$$\nabla \cdot (r^2 \sin^2 \theta B^3 e^{-4\nu} \nabla \omega) = -16\pi r \sin \theta B^2 e^{2\zeta - 4\nu} \frac{(\epsilon + p)v}{1 - v^2}, \tag{1.93b}$$

$$\nabla \cdot (r \sin \theta \nabla B) = 16\pi r \sin \theta B e^{2\zeta - 2\nu} p \tag{1.93c}$$

(these are, respectively the $R_{\hat{0}\hat{0}}, R_{\hat{0}\hat{1}}$, and $R_{\hat{0}\hat{0}} - R_{\hat{1}\hat{1}}$ field equations), supplemented by a first-order differential equation for ζ (see [110]), which comes from $e^{-\beta + 2\mu}(G^{\hat{3}\hat{3}} - G^{\hat{2}\hat{2}}) = e^{-\beta}(G_{zz} - G_{\varpi\varpi}) = 0$:

$$\begin{aligned} \frac{1}{\varpi} \zeta_{,\varpi} + \frac{1}{B}(B_{,\varpi} \zeta_{,\varpi} - B_{,z} \zeta_{,z}) = {} & \frac{1}{2\varpi^2 B}(\varpi^2 B_{,\varpi})_{,\varpi} - \frac{1}{2B} B_{,zz} + (\nu_{,\varpi})^2 \\ & - (\nu_{,z})^2 - \frac{1}{4} \varpi^2 B^2 e^{-4\nu} \left[(\omega_{,\varpi})^2 - (\omega_{,z})^2 \right]. \end{aligned} \tag{1.94}$$

In the first three equations above, ∇ is the ordinary *flat* 3-dimensional derivative operator, the derivative operator of the flat 3-metric

$$dl^2 = dr^2 + r^2(d\theta^2 + \sin^2 \theta d\phi^2) = d\varpi^2 + dz^2 + \varpi^2 d\phi^2.$$

Thus, three of the four components of the field equation are of elliptic type, whereas the fourth is a first-order partial-differential equation, relating only metric functions. The remaining non-zero components of the gravitational field equation yield two more elliptic equations and one first-order partial-differential equation, which are consistent with the above set of four equations.

1.4 Hydrostationary equilibrium equation

For a stationary, axisymmetric perfect fluid star, the equation of hydrostationary equilibrium can be written in several equivalent forms. Because scalars and components of vectors (in t, r, θ, ϕ coordinates) depend only on r and θ, the equation has only r and θ components: The t and ϕ components of each term vanish identically, and $q_\alpha{}^\beta \nabla_\beta p = \nabla_\alpha p$. The relativistic Euler equation (1.30) for a stationary, axisymmetric star then takes the form

$$\frac{\nabla_\alpha p}{(\epsilon + p)} = -u^\beta \nabla_\beta u_\alpha. \tag{1.95}$$

We note first that the equation can be written in terms of the scalars $u^t = u^\alpha \nabla_\alpha t$ and $u_\phi = u_\alpha \phi^\alpha$ as

$$\frac{\nabla_\alpha p}{(\epsilon + p)} = \nabla_\alpha \ln u^t - u^t u_\phi \nabla_\alpha \Omega. \tag{1.96}$$

To obtain (1.96), one uses the fact that, for a circular flow, the 4-velocity is a linear combination of two Killing vectors, $u^\alpha = u^t(t^\alpha + \Omega \phi^\alpha)$. For uniform rotation, $k^\alpha := u^\alpha / u^t$ is also a Killing vector, satisfying $\nabla_{(\alpha} k_{\beta)} = 0$. More generally, for differential rotation,

$$\nabla_{(\alpha} k_{\beta)} = \phi_{(\alpha} \nabla_{\beta)} \Omega, \tag{1.97}$$

which leads directly to

$$u^\beta \nabla_\beta u_\alpha = -\nabla_\alpha \ln u^t + u^t u_\phi \nabla_\alpha \Omega, \tag{1.98}$$

where we have used $u^\beta \nabla_\beta \ln u^t = 0 = u^\beta \nabla_\beta \Omega$ (from axisymmetry and stationarity) and $u^\beta \nabla_\alpha u_\beta = 0$ (from the normalization $u^\alpha u_\alpha = -1$).

Next, replacing u^t and u_ϕ in Eq. (1.96) by their expressions in Eqs. (1.83), (1.84), and using Eq. (1.89) for v, we have

$$\begin{aligned}\frac{\nabla p}{(\epsilon + p)} &= \nabla \ln \frac{e^{-\nu}}{\sqrt{1 - v^2}} - \frac{e^{\psi - \nu} v}{1 - v^2} \nabla \Omega, \\ &= -\frac{1}{1 - v^2} \left(\nabla \nu - v^2 \nabla \psi + e^{\psi - \nu} v \nabla \omega \right),\end{aligned} \tag{1.99}$$

Other, equivalent, forms of the hydrostationary equilibrium equation that can easily be obtained are

$$\frac{\nabla p}{(\epsilon+p)} = \nabla \ln u^t - u^t u_\phi \nabla\Omega, \tag{1.100a}$$

$$= \nabla \ln u^t - \frac{l}{1-\Omega l}\nabla\Omega, \tag{1.100b}$$

$$= -\nabla \ln(-u_t) + \frac{\Omega}{1-\Omega l}\nabla l, \tag{1.100c}$$

$$= -\nabla\nu + \frac{1}{1-v^2}\left(v\nabla v - \frac{v^2\nabla\Omega}{\Omega-\omega}\right), \tag{1.100d}$$

where $l := -u_\phi/u_t$ is conserved along fluid trajectories (since hu_t and hu_ϕ are conserved, so is their ratio and l is the angular momentum per unit energy). Note that (1.99) is explicitly independent of $\nabla\Omega$, with the $\nabla\Omega$ term in each equation of (1.100) canceling a term involving $\nabla\Omega$ in $\nabla \ln u^t$, $\nabla \ln(-u_t)$, or ∇v.

1.5 The Poincaré-Wavre theorem

For barotropes, one can prove a number of important properties. Because $\epsilon = \epsilon(p)$, one can define a function

$$H(p) := \int_0^p \frac{dp'}{\epsilon(p')+p'}, \tag{1.101}$$

satisfying $\nabla H = \nabla \ln h - \frac{T}{h}\nabla s$, so that (1.96) becomes

$$\nabla(H - \ln u^t) = -F\nabla\Omega, \tag{1.102}$$

where we have set $F := u^t u_\phi$. For homentropic stars, $H = \ln h$, and the equation of hydrostationary equilibrium takes the form

$$\nabla\left(\ln\frac{h}{u^t}\right) = -F\nabla\Omega. \tag{1.103}$$

Because scalars are independent of t and ϕ, we can regard ∇ in Eqs. (1.96) and (1.99) as the 2-dimensional gradient in the $r - \theta$ subspace. With A, B indices in that subspace, we have

$$\nabla_A(H - \ln u^t) = -F\nabla_A\Omega, \tag{1.104}$$

The curl of Eq. (1.104) has the form $\nabla_{[A}F\ \nabla_{B]}\Omega = 0$, implying either

$$\Omega = \text{constant} \tag{1.105}$$

(*uniform rotation*) or

$$F = F(\Omega) \tag{1.106}$$

in the case of *differential rotation*. In the latter case, Eq. (1.104) becomes $H - \ln u^t + \int_{\Omega_0}^{\Omega} F(\Omega')d\Omega' = \text{constant}$, or

$$H - \ln u^t + \int_{\Omega_{\text{pole}}}^{\Omega} F(\Omega')d\Omega' = \nu|_{\text{pole}}, \tag{1.107}$$

where the lower limit Ω_0 is chosen as the value of Ω at the pole, where H and v vanish. The above *global* first integral of the hydrostationary equilibrium equations is useful in constructing numerical models of rotating stars.[3]

Conversely, if, for a general equation of state, the velocity field is such that $F = F(\Omega)$, then (1.96) becomes

$$\nabla_\alpha \left(\ln u^t - \int_{\Omega_0}^{\Omega} F(\Omega')d\Omega' \right) = \frac{1}{\epsilon + p} \nabla_\alpha p, \tag{1.108}$$

which, by taking its curl, directly implies that the flow is barotropic, $\epsilon = \epsilon(p)$.

For a uniformly rotating star, (1.107) can be written as

$$H - \ln u^t = \nu|_{\text{pole}}, \tag{1.109}$$

which, in the case of a homentropic star, becomes

$$\frac{h}{u^t} = \mathcal{E}, \tag{1.110}$$

with $\mathcal{E} = e^\nu\,|_{\text{pole}}$ constant over the star. As we will see in Section 3.5, $\mathcal{E}$ has the meaning of the *injection energy*, the increase in a star's mass when a unit mass of baryons is injected at a point in the star. The constancy of $\mathcal{E}$ follows from the fact that an equilibrium configuration is an extremum of the mass for perturbations that move baryons from one place to another, with angular momentum and entropy fixed.

Another consequence of (1.107) is that *the effective gravity can be derived from a potential*, Φ_{eff}, as is clear from

$$\frac{\nabla_\alpha p}{\epsilon + p} = \nabla_\alpha \Phi_{\text{eff}} := \nabla_\alpha \left(\ln u^t - \int_{\Omega_0}^{\Omega} F(\Omega')d\Omega' \right). \tag{1.111}$$

Using (1.107), one finds

$$\nabla_\alpha \Phi_{\text{eff}} = \nabla_\alpha H, \tag{1.112}$$

and the surfaces of constant effective gravity (*level surfaces*) coincide with the surfaces of constant energy density ϵ (isopycnic surfaces).

[3] The global first integral (1.107) and its special case (1.110) are sometimes mistakenly referred to as Bernoulli's law. In Bernoulli's law (1.45), however, the conserved quantity is hu_t, and it is conserved only along each fluid trajectory; in the equation of hydrostationary equilibrium, the constant quantity is h/u^t (for a uniformly rotating, homentropic star), and it is constant throughout the star. The confusion may arise from the fact that, for a uniformly rotating star, the Newtonian form of the conserved quantity appearing in Bernoulli's law, $h_{\text{Newtonian}} + \frac{1}{2}v^2 + \Phi$, differs from the corresponding Newtonian first integral (1.115) only in the sign of the v^2 term.

In the Newtonian limit, because $e^{\psi} = \varpi + O(\varepsilon^2)$, $e^{\nu} = 1 + O(\varepsilon^2)$, we have, to Newtonian order,

$$u^t u_\phi = v\varpi = \varpi^2 \Omega, \tag{1.113}$$

and the functional dependence of Ω implied by Eq. (1.106) becomes the familiar requirement that, for a barotropic equation of state, Ω be stratified on cylinders,

$$\Omega = \Omega(\varpi). \tag{1.114}$$

The Newtonian limit of the integral of motion (1.110) is

$$h_{\text{Newtonian}} - \frac{1}{2}v^2 + \Phi = \text{constant}. \tag{1.115}$$

With the assumption that the topology of the star's surface is either spherical or toroidal, Abramowicz [2] shows in the relativistic context that the surfaces of constant Ω are topological cylinders.

From Eq. (1.102) and our subsequent discussion, the relativistic version of the classical Poincaré-Wavre theorem [676] follows: Consider a model of a rotating star, a stationary axisymmetric spacetime with a bounded connected perfect fluid having 4-velocity along $k^\alpha = t^\alpha + \Omega\phi^\alpha$.

Any one of the following statements implies the other three:

1. *$F := u^t u_\phi$ is a function of Ω only.*
2. *The effective gravity can be derived from a potential.*
3. *The effective gravity is normal to the surfaces of constant ϵ.*
4. *The surfaces of constant p and ϵ (isobaric and isopycnic surfaces) coincide.*

Generalizations to the stationary equilibrium of proto–neutron stars and magnetized relativistic stars are treated in Chapter 5.

1.6 Equation of state

Relativistic polytropes. Because old neutron stars have temperatures much smaller than the Fermi energy of their constituent particles, one can ignore entropy gradients and assume a uniform specific entropy s (i.e., a nearly uniform entropy per baryon). The increase in pressure and density toward the star's center are therefore adiabatic, if one neglects the slow change in composition. That is, they are related by the first law (1.8), with $ds = 0$,

$$d\epsilon = \frac{\epsilon + p}{\rho} d\rho, \tag{1.116}$$

with p given in terms of ρ by

$$\frac{\rho}{p}\frac{dp}{d\rho} = \frac{\epsilon + p}{p}\frac{dp}{d\epsilon} = \Gamma_1. \tag{1.117}$$

Here Γ_1 is the *adiabatic index*, the fractional change in pressure per fractional change in comoving volume, at constant entropy and composition:

$$\Gamma_1 := \frac{\partial \log p(\rho, s, Y_1, \dots, Y_N)}{\partial \log \rho} = \frac{\epsilon + p}{p} \frac{\partial p(\epsilon, s, Y_1, \dots, Y_N)}{\partial \epsilon}, \tag{1.118}$$

with Y_k the fractional number density of the kth species of constituent particle ($Y_k = n_k/n$, with n_k the number density of the kth species and n the total number density of baryons). In an ideal degenerate Fermi gas, in the nonrelativistic and ultrarelativistic regimes, Γ_1 has the constant values 5/3 and 4/3, respectively. The nonrelativistic value corresponds to Fermi momentum $p_F \sim \hbar n^{1/3}$ and hence to a pressure $p \sim p_F v n = \frac{p_F^2}{m} n \sim \frac{\hbar^2}{m} n^{5/3}$ (with m the particle mass), whereas in the relativistic case, v approaches 1, implying a degeneracy pressure of order $p \sim p_F n \sim \hbar n^{4/3}$.

Except in the outer crust, neutron-star matter is far from an ideal Fermi gas, but models often assume a constant effective adiabatic index, chosen to match an average stellar compressibility. An equation of state of the form

$$p = K\rho^\Gamma, \tag{1.119}$$

with K and Γ constants, is called *polytropic*; K and Γ are the *polytropic constant* and *polytropic exponent*, respectively. The corresponding relation between ϵ and p follows from Eq. (1.116), rewritten in the form $d\frac{\epsilon}{\rho} = \frac{p}{\rho^2} d\rho$. Requiring $\lim_{p \to 0} \frac{\epsilon}{\rho} = 1$, we have

$$\frac{\epsilon}{\rho} = 1 + \int_0^\rho K\rho^{\Gamma-2} d\rho = 1 + K\frac{\rho^{\Gamma-1}}{\Gamma - 1},$$

or

$$\epsilon = \rho + \frac{p}{\Gamma - 1}. \tag{1.120}$$

The polytropic exponent Γ is commonly replaced by a *polytropic index* N, given by

$$\Gamma = 1 + \frac{1}{N}. \tag{1.121}$$

For the above polytropic EOS, the quantity $c^{(\Gamma-2)/(\Gamma-1)} \sqrt{K^{1/(\Gamma-1)}/G}$ has units of length. In gravitational units ($c = G = 1$), one can thus use $K^{N/2}$ as a fundamental length scale to define dimensionless quantities. Equilibrium models are then characterized by the polytropic index N and their dimensionless central energy density. Equilibrium properties can be scaled to different dimensional values, using appropriate values for K. For $N < 1.0$ ($N > 1.0$) one obtains stiff (soft) models, whereas for $N \sim 0.5$–1.0, one obtains models whose masses and radii are roughly consistent with observed neutron-star masses and with the weak constraints on radius imposed by present observations and by candidate equations of state.

The definition (1.119), (1.120) of the relativistic polytropic EOS was introduced by Tooper [698], to allow a polytropic exponent Γ that coincides with the adiabatic index of a relativistic fluid with constant entropy per baryon (a homentropic fluid). Unfortunately, Tooper had previously used the form $p = K\epsilon^\Gamma$ [697]; because this equation of state does not satisfy Eq. (1.116), it is not consistent with the first law of thermodynamics for a fluid with uniform entropy. The literature, however, has not universally accepted one convention. In particular, two of the best introductions to the equation of state of compact objects, by Shapiro and Teukolsky [603] and by Glendenning [265], use the two different definitions, with Shapiro-Teukolsky adopting the choice, $p = K\rho^\Gamma$, that we regard as more natural and will use here. (Note, however, that Glendenning's use of the equation of state $p = K\epsilon^\Gamma$ is restricted to two cases where the definitions agree: nonrelativistic matter, for which $\epsilon = \rho$, and the ultrarelativistic Fermi gas, for which $\epsilon = 3p$.)

Hadronic equations of state. Cold matter below nuclear density, $\rho_0 = 2.7 \times 10^{14}$ g/cm^3 (or $n_0 = 0.16$ fm^{-3}), is thought to be well understood. A derivation of a sequence of equations of state at increasing densities, beginning with the semiempirical mass formula for nuclei, can be found in [603] (see also [287]). Another treatment, using experimental data on neutron-rich nuclei, was given in [286]. In a neutron star, matter below nuclear density forms a crust, whose outer part is a lattice of nuclei in a relativistic electron gas. At 4×10^{11} g/cm^3, the electron Fermi energy is high enough to induce *neutron drip*: Above this density, nucleons begin leaving their nuclei to become free neutrons. The inner crust is then a two-phase equilibrium of the lattice nuclei and electrons and a gas of free neutrons. The emergence of a free-neutron phase means that the equation of state softens immediately above neutron drip: Increasing the density leads to an increase in free neutrons and to a correspondingly smaller increase in pressure. Melting of the Coulomb lattice, marking the transition from crust to a liquid core of neutrons, protons, and electrons, occurs between 10^{14} g/cm^3 and ρ_0.

A review by Heiselberg and Pandharipande [312] describes the partly phenomenological construction of a primarily nonrelativistic many-body theory that gives the equation of state at and slightly below nuclear density. Two-nucleon interactions are matched to neutron-neutron scattering data and the experimentally determined structure of the deuteron. Parameters of the three-nucleon interaction are fixed by matching the observed energy levels of light nuclei.

Above nuclear density, however, the equation of state is still beset by substantial uncertainties. For a typical range of current candidate equations of state, values of the pressure differ by more than a factor of 5 at $2\rho_0 \sim 5 \times 10^{14}$ g/cm^3, and by at least that much at higher densities [284]. Although scattering experiments probe the interactions of nucleons (and quarks) at distances small compared to the radius of a nucleon, the many-body theory required to deduce the equation of state from the fundamental interactions is poorly understood. Heavy ion collisions do produce collections of nucleons at supranuclear densities, but here the unknown extrapolation is from the high temperature of the experiment to the low temperature of neutron-star matter.

Observations of neutron stars provide a few additional constraints. Of these astrophysical constraints, two are unambiguous and precise: The equation of state must allow a mass at least as large as $1.93M_\odot$, the largest accurately determined mass of a neutron star. (The observation by Demorest et al. [179] is of a $1.97 \pm 0.04M_\odot$

neutron star.) And it must allow a rotational period at least as small as 1.4 ms, the period of the fastest confirmed millisecond pulsar [313]. Observations of neutron star radii are much less precise, but a large number of observations of type I X-ray bursts or transient X-ray binaries may allow for the reconstruction of the neutron star equation of state [516, 515, 655].

The uncertainty in the equation of state above nuclear density is dramatically seen in the array of competing alternatives for the nature of matter in neutron star cores: Cores that are dominantly neutron matter may have sharply different equations of state, depending on the presence or absence of pion or kaon condensates, of hyperons, and of droplets of strange quark matter (described below). The inner core of the most massive neutron stars may be entirely strange quark matter. Other differences in candidate equations of state arise from constructions based on relativistic and on nonrelativistic many-body theory. A classic collection of early proposed EOSs was compiled by Arnett and Bowers [35], and reviews of many modern EOSs have been compiled by Haensel [284] and Lattimer and Prakash [391]. Detailed descriptions and tables of several modern EOSs, especially EOSs with phase transitions, can be found in Glendenning [265]; his treatment is particularly helpful in showing how one constructs an equation of state from a relativistic field theory. The Heiselberg-Pandharipande review [312], in contrast, presents a more phenomenological construction of equations of state that match experimental data. Detailed theoretical derivations of equations of state are presented in the book by Haensel, Potekhin, and Yakovlev [287].

Candidate EOSs are supplied in the form of an energy density versus pressure table, and intermediate values are interpolated. This results in some loss of accuracy because the usual interpolation methods do not preserve thermodynamic consistency. Swesty [673] devised a cubic Hermite interpolation scheme that does preserve thermodynamical consistency, and the scheme has been shown to indeed produce higher-accuracy neutron star models [498].

High-density equations of state with pion condensation were proposed in [472, 588] (see also [378]). Beyond nuclear density, the electron chemical potential could exceed the rest mass of π^- (139 MeV) by a margin large enough to overcome a pion-neutron repulsion and thus allow a condensate of zero-momentum pions. The critical density is thought to be $2\rho_0$ or higher, but the uncertainty is greater than a factor of 2; a condensate with both π^0 and π^- has also been suggested. Because the s-wave kaon-neutron interaction is attractive, kaon condensation may also occur, despite the higher kaon mass, a possibility suggested in [343] (for discussions with differing viewpoints see [101, 520, 312]). Pion and kaon condensates lead to significant softening of the equation of state.

As initially suggested in [15], when the Fermi energy of the degenerate neutrons exceeds the mass of a Λ or Σ, weak interactions convert neutrons to these hyperons: Examples are $2n \leftrightarrow p + \Sigma^-$, $n + p^+ \rightarrow p^+ + \Lambda$. Reviews and bibliographies can be found in [265, 42, 541], and more recent work, spurred by the r-mode instability (see Chapter 8), is reported in [427, 285, 381]. The critical density above which hyperons appear is estimated at 2 or 3 times nuclear density. Above that density, the presence of copious hyperons can significantly soften the equation of state. Because a softer core equation of state can support less mass against collapse, the larger the observed maximum mass, the less likely that neutron stars have cores with hyperons (or with pion or kaon condensates). In particular, a measured mass of $1.97 \pm 0.04 M_\odot$ for the pulsar PSR J16142230 with a white dwarf companion [179] limits the equation of state parameter space [558], ruling out several candidate equations of states with hyperons [517]. Whether a hyperon core is consistent with a mass this large remains an open question [668].

Strange quark matter. Before a density of about $6\rho_0$ is reached, lattice QCD calculations indicate a phase transition from quarks confined to nucleons (or hyperons) to a collection of free quarks (and gluons); heavy ion collisions at CERN and RHIC show evidence of the formation of such a quark-gluon plasma. A density for the phase transition higher than that needed for strange quarks in hyperons is similarly high enough to give a mixture of up, down, and strange quarks in quark matter, and the expected strangeness per unit baryon number is $\simeq -1$. If densities high enough for a phase transition to quark matter are reached, neutron-star cores may contain a transition region with a mixed phase of quark droplets in neutron matter [265].

Bodmer [76] and later Witten [732] pointed out that experimental data do not rule out the possibility that the ground state of matter at zero pressure and large baryon number is not iron but strange quark matter. If this is the case, all "neutron stars" may be strange quark stars, a lower-density version of the quark-gluon plasma, again with roughly equal numbers of up, down, and strange quarks, together with electrons to give overall charge neutrality [76, 213]. The first extensive study of strange quark star properties is due to Witten [732] (but see also [324, 96]), and hybrid stars that have a mixed-phase region of quark and hadronic matter have also been studied extensively (see, for example, Glendenning's review [265]).

The strange quark matter equation of state can be represented by the following linear relation between pressure and energy density:

$$p = a(\epsilon - \epsilon_0), \tag{1.122}$$

where ϵ_0 is the energy density at the surface of a bare strange star (neglecting a possible thin crust of normal matter). The MIT bag model of strange quark matter involves three parameters, the bag constant, $\mathcal{B} = \epsilon_0/4$, the mass of the strange quark, m_s, and the QCD coupling constant, α_c. The constant a in (1.122) is equal to $1/3$ if one neglects the mass of the strange quark, whereas it takes the value of $a = 0.289$ for $m_s = 250$ MeV. When measured in units of $\mathcal{B}_{60} = \mathcal{B}/(60 \text{ MeV fm}^{-3})$, the constant $\mathcal{B}$ is restricted to be in the range

$$0.9821 < \mathcal{B}_{60} < 1.525, \tag{1.123}$$

assuming $m_s = 0$. The lower limit is set by the requirement of stability of neutrons with respect to a spontaneous fusion into strangelets, the upper limit by the energy per baryon of ^{56}Fe at zero pressure (930.4 MeV). For other values of m_s, the above limits are modified somewhat (see also [185, 268] for other attempts to describe deconfined strange quark matter).

1.7 Rotation law

At birth, neutron stars are likely to be differentially rotating, with a rotation law that can depend on the rotation of their progenitors and on the way they are formed – the collapse of stellar cores [755, 435, 6], the accretion-induced collapse of white dwarfs [437], the merger of degenerate cores in double-degenerate systems, or the merger of neutron stars [556, 625, 501, 627].

Uniform angular velocity minimizes the total mass-energy of a configuration for a given baryon number and total angular momentum [95, 303]. As a result, apart from slight differential rotation following glitches, neutron stars, soon after birth, are expected to rotate uniformly. As the neutron star cools, several mechanisms act to enforce uniform rotation. Magnetic braking of differential rotation by Alfvén waves [602, 161, 438] could be the most effective damping mechanism, acting on a timescale of minutes or less. In contrast, kinematical shear viscosity acts on a timescale [219, 220, 170].

$$\tau \sim 18 \times \left(\frac{\rho}{10^{15}\,\mathrm{g\,cm^{-3}}}\right)^{-5/4} \left(\frac{T}{10^9\,\mathrm{K}}\right)^2 \left(\frac{R}{10^6\,\mathrm{cm}}\right) \quad \mathrm{yr}, \tag{1.124}$$

where ρ, T, and R are the central density, temperature and radius of the star. It has also been suggested that convective and turbulent motions may enforce uniform rotation on a timescale of the order of days [311]. Finally, differentially rotating stars may also be subject to a shear instability that tends to suppress differential rotation [745].

If the cooling to degeneracy, and hence to a nearly barotropic configuration, is faster than the time to establish uniform rotation, condition (1.106) is likely to govern the rotation law of nascent neutron stars. A simple choice of a differential rotation law that satisfies this condition is

$$F(\Omega) = A^2(\Omega_c - \Omega) = \frac{(\Omega - \omega)e^{2(\psi-\nu)}}{1 - (\Omega - \omega)^2 e^{2(\psi-\nu)}}, \tag{1.125}$$

where A is a constant with units of length that determines the length scale over which the angular velocity changes within the star [371, 372]. When $A \to \infty$, the above rotation law reduces to the uniform rotation case. In the Newtonian limit and when $A \to 0$, the rotation law becomes a so-called j-constant rotation law (where $j := hu_\phi$ is the specific angular momentum), a law that satisfies the Rayleigh criterion for local dynamical stability against axisymmetric disturbances: j should not decrease outward, $dj/d\Omega < 0$. The same criterion is also satisfied in the relativistic case (see Section 9.3.2). The above rotation law is a simple choice, and more physically plausible choices must be obtained through detailed numerical simulations of the formation of rotating relativistic stars.

When the temperature of a neutron star falls below about 10^9 K, its core is expected to become superfluid and superconducting (see [471, 177] and references therein), with pairing of neutrons and of protons analogous to the BCS pairing of electrons in a superconductor. One can define an amplitude $\psi(\mathbf{r}, t)$ for which $|\psi|^2$ measures the density of pairs.[4] Then, with a phase S defined modulo 2π by $\psi = |\psi|e^{iS}$, the superfluid velocity is given by $\mathbf{v} = \dfrac{\hbar}{m}\boldsymbol{\nabla} S$. It follows that $\boldsymbol{\nabla} \times \mathbf{v} = 0$ in the superfluid, and a nonzero average vorticity can be achieved only by a breakdown of neutron superfluidity. The matter can remain a superfluid everywhere except in thin *vortex tubes* if the vorticity vanishes outside the tubes, as long as ψ is well defined. This, in turn means that an integral of $\boldsymbol{\nabla} S$ along any closed loop c within the superfluid has the value $\int_c \boldsymbol{\nabla} S \cdot \mathbf{dl} = 2\pi n$ (where n is an integer). Then, the relation between $\mathbf{v}$ and S gives the Onsanger-Feynman quantization condition on the circulation about vortices,

$$\int_c \mathbf{v} \cdot \mathbf{dl} = \frac{2\pi\hbar}{m} n. \tag{1.126}$$

With each vortex having circulation $\frac{2\pi\hbar}{m}$, the intervortex spacing is

$$d_n \sim 5 \times 10^{-3} \Omega_2^{-1/2} \text{cm}, \tag{1.127}$$

where Ω_2 is the angular velocity of the star in 10^2 s^{-1}. On scales much larger than the intervortex spacing, for example, on the order of 1 cm, an averaged fluid velocity u^α can be defined, and it is this averaged velocity field that has the character of uniform rotation [639].

1.8 Equilibrium quantities

The mass and angular momentum of a stationary axisymmetric spacetime can each be defined as an integral at spatial infinity (their Komar form [370]), namely

$$M = -\frac{1}{4\pi} \int_{S_\infty} \nabla^\alpha t^\beta dS_{\alpha\beta}, \tag{1.128}$$

$$J = \frac{1}{8\pi} \int_{S_\infty} \nabla^\alpha \phi^\beta dS_{\alpha\beta}, \tag{1.129}$$

where $\int_{S_\infty} := \lim_{r\to\infty} \int_{S_r}$, and S_r is a sphere of constant radial coordinate r appropriate to the asymptotically flat metric. Stokes's theorem relates the surface integrals for M and J to integrals over the matter and implies that the surface integrals are independent of surface for any surface enclosing the matter. To see this, one uses the Killing-vector identity

$$\nabla_\beta \nabla^\alpha \xi^\beta = R^\alpha{}_\beta \xi^\beta, \tag{1.130}$$

[4] Readers can find a derivation of the (Landau-Ginsberg) description of a superfluid in terms of ψ in, for example, [178].

together with the field equation in the form

$$R^\alpha{}_\beta = 8\pi(T^\alpha{}_\beta - \delta^\alpha_\beta T), \tag{1.131}$$

to write

$$M = -2\int (T_\alpha{}^\beta - \frac{1}{2}\delta^\beta_\alpha T)\, t^\alpha dS_\beta, \tag{1.132}$$

$$J = \int T_\alpha{}^\beta \phi^\alpha dS_\beta. \tag{1.133}$$

In a time-dependent spacetime, the asymptotic forms above can be used to define the total mass and angular momentum associated with timelike and rotational symmetry vectors of a flat asymptotic metric, but they can no longer be written as integrals over the matter.

Associated with the differential baryon (and lepton) conservation law (1.26) is a conserved rest mass. That is, the integral of $\nabla_\alpha(\rho u^\alpha)$ over a 4-volume ${}^4V$ bounded by 3-surfaces V_1 and V_2 yields

$$\int_{{}^4V} \nabla_\alpha(\rho u^\alpha) d^4V = \left(\int_{V_2} - \int_{V_1}\right) \rho u^\alpha dS_\alpha,$$

implying that the total rest mass,

$$M_0 = \int_V \rho u^\alpha dS_\alpha, \tag{1.134}$$

is independent of the 3-surface V, for any 3-surface containing the fluid.

We can now make the connection between the specific angular momentum, $j = hu_\alpha\phi^\alpha$, introduced in Eq. (1.47), and the total angular momentum: If we denote by $dM_0 = \rho u^\alpha dS_\alpha$ the integrand in Eq. (1.134), then Eq. (1.133) for the total angular momentum has, when evaluated on an axisymmetric hypersurface, the form

$$J = \int j dM_0. \tag{1.135}$$

This follows from the fact that ϕ^α is tangent to an axisymmetric hypersurface, implying $\phi^\alpha dS_\alpha = 0$. Then

$$T^\alpha{}_\beta \phi^\beta dS_\alpha = (\epsilon + p) u_\beta \phi^\beta u^\alpha dS_\alpha = \frac{\epsilon + p}{\rho} u_\beta \phi^\beta \rho u^\alpha dS_\alpha = j dM_0.$$

The rotational kinetic energy is defined by

$$T = \frac{1}{2}\int \Omega dJ, \quad \text{where} \quad dJ = T_\alpha{}^\beta \phi^\alpha dS_\beta, \tag{1.136}$$

and it takes the form $T = \frac{1}{2}J\Omega$ for uniform rotation. A natural definition of the scalar moment of inertia for a uniformly rotating star is similarly

$$I = J/\Omega. \tag{1.137}$$

Table 1.1. *Equilibrium properties*

Gravitational mass	$M = -2\int (T_\alpha{}^\beta - \frac{1}{2}\delta_\alpha^\beta T)\, t^\alpha\, dS_\beta$
Rest mass	$M_0 = \int \rho u^\beta\, dS_\beta$
Internal energy	$U = \int u\; u^\beta\, dS_\beta$
Gravitational binding energy	$W = M - M_0 - T - U$
Angular momentum	$J = \int T_\alpha{}^\beta \phi^\alpha\, dS_\beta \equiv \int dJ$
Kinetic energy	$T = \frac{1}{2}\int \Omega dJ$
Moment of inertia (uniform rotation)	$I = J/\Omega$

Although J can be computed as an integral over the source, the scalar moment of inertia does not have an independent integral definition. As a result, it has no natural definition as a local integral for a star with arbitrary velocity field.[5]

The internal energy density, given by

$$u = \epsilon - \rho = \rho e, \tag{1.138}$$

can be used to define a total internal energy $U = \displaystyle\int u\; u^\alpha dS_\alpha$. Finally, with mass, rest mass, kinetic energy, and internal energy defined, one can define a gravitational binding energy W as the difference

$$W = M - M_0 - T - U. \tag{1.139}$$

(A *proper mass*, $M_P = M_0 + U$, is also commonly used, allowing one to write $W = M - M_P - T$.)

A summary of integral quantities that characterize stationary, axisymmetric models, together with the equations that define them, is displayed in Table 1.1.

The *mass-shedding* or *Kepler limit* along a sequence of rotating stellar models is reached when, at the equator, the angular velocity of the fluid, $\Omega(r_e, \pi/2)$, reaches the angular velocity of a free particle in circular orbit, Ω_K. The latter has the form

$$\Omega_K = \omega + \frac{\omega'}{2\psi'} + \left[e^{2\nu-2\psi}\frac{\nu'}{\psi'} + \left(\frac{\omega'}{2\psi'}\right)^2\right]^{1/2}, \tag{1.140}$$

where ($'$) means ∂_r. Equation (1.140) is obtained from the geodesic equation, $u^\beta\nabla_\beta u^\alpha = 0$, for a satellite in circular orbit, with $u^\alpha = u^t(t^\alpha + \Omega_K\phi^\alpha)$. As in Eq. (1.89), the normalization $u^\alpha u_\alpha = -1$ implies $u^t = \frac{e^{-\nu}}{\sqrt{1-v_K^2}}$, where $v_K = e^{\psi-\nu}(\Omega_K - \omega)$. Then, the geodesic equation gives

$$0 = (1 - v_K^2)\partial_r \ln u^t = -\nu' + v_K^2\psi' - v_K e^{\psi-\nu}\omega'. \tag{1.141}$$

Note that this relation agrees with (and could be deduced from) the hydrostationary equilibrium equation (1.99) when $\nabla p = 0$: At maximum rotation (when $\Omega = \Omega_K$,

[5] Multipole moments of the mass and current distributions are defined in terms of the asymptotic metric at spatial infinity, but these are symmetric, trace-free tensors and do not define I.

no pressure gradient contributes to the support of a ring of fluid at the equator, and the balance of forces on a fluid element is set by the equality of the gravitational[6] and centrifugal forces. A further increase in uniform rotation would lead to mass shedding at the equator. Solving the quadratic equation (1.141) for v_K, we have

$$v_K = e^{\psi-\nu}\frac{\omega'}{2\psi'} + \left[\frac{\nu'}{\psi'} + e^{2\psi-2\nu}\left(\frac{\omega'}{2\psi'}\right)^2\right]^{1/2}$$

and Eq. (1.140) immediately follows.

For a spherical star, Eq. (1.140) takes the Newtonian form

$$\Omega_K = \left(\frac{M}{R^3}\right)^{1/2}, \tag{1.142}$$

with R the Schwarzschild coordinate of the equator. As a spherical star spins up, it becomes oblate, and the Kepler frequency of a particle at larger equatorial radius is correspondingly smaller. By the time the star itself rotates at the Kepler frequency, Ω_K has typically fallen to about 60% of its value for the spherical configuration with the same baryon number.

Finally, we note that in the quasi-isotropic coordinates, $e^\psi = \sqrt{\phi^\alpha\phi_\alpha}$ is the radial coordinate for which $2\pi e^\psi$ is the circumference of the circular orbits of the symmetry vector ϕ^α. We will write $r_c(r) = e^\psi(r, \theta = \pi/2)$, noting that, for spherical stars, r_c coincides with the usual Schwarzschild radial coordinate. At the equator, r_c is the *circumferential equatorial radius* of the star, also denoted as R (coinciding with the Schwarzschild radial coordinate at the equator for spherical stars).

[6] More precisely, by the curvature of spacetime, including the effect of frame dragging.

2

3+1 split, action, Lagrangian, and Hamiltonian formalisms

2.1 The 3+1 split

The initial value problem in general relativity is associated with a slicing of spacetime by a family of spacelike hypersurfaces. For detailed accounts on the historical development of the subject, we refer to several books published in recent years [7, 79, 55, 271]. Here, we give a summary of the main definitions and equations of the 3+1 formalism.

A choice of time coordinate t gives a slicing by $t =$ constant surfaces Σ_t whose normal $\nabla_\alpha t$ is everywhere timelike. The spacetimes we consider have the form $\mathbb{R} \times \Sigma$, with each hypersurface Σ_t a copy $\{t\} \times \Sigma$ of Σ. Because the future is the direction in which t increases, the $-$+++ metric signature implies that the future pointing contravariant unit vector normal to each slice Σ_t is

$$n^\alpha = -\nabla^\alpha t \, (-\nabla_\beta t \nabla^\beta t)^{-1/2}. \tag{2.1}$$

The 3+1 formalism we present in this section leads to field equations derived in Section 2.2.5. The present section is otherwise essentially self-contained.

The Einstein equation (and the associated conservation of energy and momentum), the conservation of baryons, and the equation of state of the fluid determine the evolution of the geometry and the fluid variables from one hypersurface to another. The 3+1 decomposition of the spacetime is mirrored in a 3+1 decomposition of the equations and the tensors whose time-evolution they describe. We introduce the projection

$$\gamma^\alpha{}_\beta := \delta^\alpha_\beta + n^\alpha n_\beta \tag{2.2}$$

onto each Σ_t, analogous to the operator $q^\alpha{}_\beta$ of Eq. (1.1) that projects onto the spatial subspace normal to u^α: $\gamma^\alpha{}_\beta$ kills vectors $v^\alpha_\perp = v_\perp n^\alpha$ normal to Σ_t, because $\gamma^\alpha{}_\beta n^\beta = 0$, and it leaves invariant vectors $v^\alpha_\parallel$ tangent to Σ_t: $\gamma^\alpha{}_\beta v^\beta_\parallel = v^\alpha_\parallel$. The decomposition of any vector v^α (or any tensor index) into a part tangent and a part orthogonal to Σ_t is then given by

$$v^\alpha = \delta^\alpha_\beta v^\beta = (\gamma^\alpha{}_\beta - n^a n_\beta) v^\beta = \gamma^\alpha{}_\beta v^\beta - n_\beta v^\beta n^\alpha = v^\alpha_\parallel + v_\perp n^\alpha. \tag{2.3}$$

In the corresponding decomposition of the field equation $E^{\alpha\beta} := G^{\alpha\beta} - 8\pi T^{\alpha\beta} = 0$, namely

$$\begin{aligned} E^{\alpha\beta} &= E^{\alpha\beta}_\parallel \quad + \quad 2\, E^{(\alpha}_{\parallel\perp}\, n^{\beta)} \quad + \quad E_{\perp\perp}\, n^\alpha n^\beta \\ &= \gamma^\alpha{}_\gamma \gamma^\beta{}_\delta E^{\gamma\delta} - 2 n_\delta E^{\gamma\delta} \gamma_\gamma{}^{(\alpha} n^{\beta)} + n_\gamma n_\delta E^{\gamma\delta} n^\alpha n^\beta, \end{aligned} \tag{2.4}$$

we will see that the projection $E_{\|}^{\alpha\beta} = 0$ onto the hypersurface Σ_t yields dynamical equations for the metric, whereas the remaining parts, $G_{\alpha\beta}n^\beta = 8\pi T_{\alpha\beta}n^\beta$, of the field equation serve as constraints (equations involving only first time derivatives of the metric).

Writing the spacetime in the explicit product form $\mathbb{R} \times \Sigma$ picks out, in addition to n^α, a second direction, in which a point x of Σ is fixed and only t changes. If $\{x^i\}$ is any chart on Σ, that direction is along the t-coordinate line of the corresponding chart $\{t, x^i\}$ on spacetime, along the vector field $\boldsymbol{\partial}_t$ tangent to the line of fixed coordinates x^i. As in our description of stationary, rotating stars, we write t^α for $\boldsymbol{\partial}_t$. Then, in the chart $\{t, x^i\}$,

$$t^\mu = \delta^\mu_t, \qquad \nabla_\mu t = \delta^t_\mu. \tag{2.5}$$

Because of the dragging of frames, the timelike Killing vector t^α of a rotating star is not in general orthogonal to the surfaces Σ_t; not surprisingly, in the more general spacetimes that describe oscillating stars and binary systems, convenient choices of coordinates again have a time direction t^α that is not along the normal. We follow the usual numerical relativity notation in writing the 3+1 decomposition of t^α in the form

$$t^\alpha = \alpha n^\alpha + \beta^\alpha, \qquad \beta^\alpha n_\alpha = 0. \tag{2.6}$$

The component α of t^α along the unit normal is called the *lapse*, because $\alpha\, dt$ is the proper time elapsed in a normal direction between coordinate times t and $t + dt$. The spatial vector β^α is called the *shift*, because $\beta^\alpha\, dt$ is the spatial shift in position from a path of coordinate length dt along the normal n^α to a path of coordinate length dt along t^α. One regards the choice of the vector t^α as the direction in which the time-evolution proceeds, noting that the time derivative $\dot{T}^{\alpha\cdots\beta}{}_{\gamma\cdots\delta}$, defined by

$$\dot{T}^{\alpha\cdots\beta}{}_{\gamma\cdots\delta} := \mathcal{L}_{\boldsymbol{t}} T^{\alpha\cdots\beta}{}_{\gamma\cdots\delta}, \tag{2.7}$$

has components $\partial_t T^{\mu\cdots\nu}{}_{\sigma\cdots\tau}$ in a chart of the form $\{t, x^i\}$ (this follows from Eq. [2.5] and the definition [A.16] of *Lie derivative*).

The spacetime metric at a time t can be written in terms of the lapse, the shift, and the pullback γ_{ab} of the metric $g_{\alpha\beta}$ onto Σ_t. The Riemannian metric γ_{ab} is the *3-metric*, or *induced metric* on Σ_t, the metric for which the geometry of Σ_t is inherited from the geometry of the spacetime. We again use the symbol γ, because $\gamma_{\alpha\beta}$ is the unique 4-dimensional tensor that is orthogonal to the normal n^α and whose pullback to Σ_t is γ_{ab}. In a coordinate system $\{t, x^i\}$, the spatial components of γ_{ab}, $\gamma_{\alpha\beta}$ and $g_{\alpha\beta}$ coincide:

$$\gamma_{ij} = g_{ij}. \tag{2.8}$$

Similarly, because the shift vector β^α is tangent to Σ_t, we can associate it with a vector β^a on Σ_t (β_a is the pullback of β_α, β^α the push-forward of β^a, by the embedding map $\Sigma_t \to M$).

In the coordinates $\{t, x^i\}$, the 3+1 decomposition of the 4-metric, $g_{\alpha\beta} = \gamma_{\alpha\beta} - n_\alpha n_\beta$, has, from Eq. (2.6) and the relations $\beta^t = 0$ and $n_i = 0$, components

$$g_{tt} = t_\alpha t^\alpha = -\alpha^2 + \beta_i \beta^i, \qquad g_{ij} = \gamma_{ij}, \qquad g_{ti} = \gamma_{ti} = \beta_i, \tag{2.9}$$

implying

$$ds^2 = -\alpha^2 dt^2 + \gamma_{ij}(dx^i + \beta^i dt)(dx^j + \beta^j dt). \tag{2.10}$$

The corresponding decompositions of the volume element and the components of the inverse metric are given by

$$\sqrt{|g|} = \alpha\sqrt{\gamma}, \tag{2.11a}$$

$$g^{tt} = -\frac{1}{\alpha^2}, \qquad g^{ti} = \frac{\beta^i}{\alpha^2}, \tag{2.11b}$$

$$g^{ij} = \gamma^{ij} - \frac{\beta^i \beta^j}{\alpha^2}, \tag{2.11c}$$

where $\gamma = \det(\gamma_{ab})$, and the components of the unit normal by

$$n_\mu = -\alpha \nabla_\mu t = -\alpha\, \delta^t_\mu. \tag{2.12}$$

The choice of a 3+1 decomposition $\mathbb{R} \times \Sigma$ of the spacetime allows one to identify each hypersurface $\Sigma_t = \{t\} \times \Sigma$ with the fixed space Σ. We can then regard $\alpha(t)$, $\beta^a(t)$, $\gamma_{ab}(t)$, and the 3-dimensional projections of the fluid variables as time-dependent quantities on Σ. The correspondence between fields $\alpha(t), \beta^a(t)$, and $\gamma_{ab}(t)$ on Σ and α, β^α, and $\gamma_{\alpha\beta}$ on M extends to a correspondence between the time derivatives $\dot\alpha = \partial_t \alpha$, $\dot\beta^a = \partial_t \beta^a$, $\dot\gamma_{ab} = \partial_t \gamma_{ab}$ and the Lie derivatives $\dot\alpha = \mathcal{L}_{\boldsymbol{t}} \alpha$, $\dot\beta^\alpha = \mathcal{L}_{\boldsymbol{t}} \beta^\alpha$, $\dot\gamma_{\alpha\beta} = \mathcal{L}_{\boldsymbol{t}} \gamma_{\alpha\beta}$, immediate from the fact that the spatial components of corresponding tensors coincide (see the comment after Eq. [2.7]). In particular, the pullback to Σ of a covector σ_α can be written in the form $\sigma_a = \gamma^\alpha{}_a \sigma_\alpha$, where, in the charts $\{x^i\}$ for Σ and $\{t, x^i\}$ for spacetime, $\gamma^\alpha{}_a$ has components

$$\gamma^\mu{}_i = \delta^\mu_i. \tag{2.13}$$

Then $\gamma_{ab}(t) = \gamma^\alpha{}_a \gamma^\beta{}_b \gamma_{\alpha\beta}$, and

$$\dot\gamma_{ab} := \partial_t \gamma_{ab} = \gamma^\alpha{}_a \gamma^\beta{}_b \mathcal{L}_{\boldsymbol{t}} \gamma_{\alpha\beta} = \gamma^\alpha{}_a \gamma^\beta{}_b \dot\gamma_{\alpha\beta}. \tag{2.14}$$

More generally, each time-dependent tensor field $T^{a\cdots b}{}_{c\cdots d}(t)$ on Σ corresponds to a unique tensor field $T^{\alpha\cdots\beta}{}_{\gamma\cdots\delta}$ on M whose indices are all orthogonal to n_α, and $\mathcal{L}_{\boldsymbol{t}} T^{\alpha\cdots\beta}{}_{\gamma\cdots\delta}$ is the unique tensor field on M corresponding to $\partial_t T^{a\cdots b}{}_{c\cdots d}$ on Σ.

We need two final definitions to cast the field equation in the standard initial-value form. We denote by D_a the covariant derivative operator on Σ associated at a time t with the metric $\gamma_{ab}(t)$. Although one ordinarily chooses coordinates for which t^α is not normal to Σ_t, the time derivative normal to the surface naturally enters the projections of the field equation (2.23), (2.21), (2.22) below. This time derivative is (up to a factor $-1/2$) the *extrinsic curvature* K_{ab} of the surface Σ[1]:

$$K_{ab} := -\frac{1}{2}\gamma^\alpha{}_a\gamma^\beta{}_b\mathcal{L}_{\boldsymbol{n}}\gamma_{\alpha\beta}. \tag{2.15}$$

We denote by K its contraction

$$K = K_a{}^a = \gamma^{ab}K_{ab}. \tag{2.16}$$

Then, from Eqs. (2.15) and (2.6), we have

$$\begin{aligned} K_{ab} &= -\frac{1}{2}\gamma^\alpha{}_a\gamma^\beta{}_b\mathcal{L}_{\boldsymbol{n}}\gamma_{\alpha\beta} = -\frac{1}{2\alpha}[\partial_t\gamma_{ab} - \gamma^\alpha{}_a\gamma^\beta{}_b(\nabla_\alpha\beta_\beta + \nabla_\beta\beta_\alpha)] \\ &= -\frac{1}{2\alpha}(\partial_t\gamma_{ab} - D_a\beta_b - D_b\beta_a). \end{aligned} \tag{2.17}$$

Equation (2.17), in the form

$$\partial_t\gamma_{ab} = -2\alpha K_{ab} + D_a\beta_b + D_b\beta_a, \tag{2.18}$$

serves as an evolution equation for the 3-metric γ_{ab}.

The evolution equation for K_{ab} is the spatial projection of the Einstein equation in the form $R_{\alpha\beta} - 8\pi(T_{\alpha\beta} - \frac{1}{2}g_{\alpha\beta}T_\gamma{}^\gamma) = 0$, or, equivalently, its pullback to Σ. This equation and the constraint equations involve the decomposition of the stress-energy tensor $T_{\alpha\beta}$:

$$\rho_E := n^\alpha n^\beta T_{\alpha\beta} = (\epsilon + p)(\alpha u^t)^2 - p, \tag{2.19a}$$

$$j_a := -\gamma_a{}^\alpha n^\beta T_{\alpha\beta} = (\epsilon + p)\alpha u^t u_a, \tag{2.19b}$$

$$S_{ab} := \gamma_a{}^\alpha\gamma_b{}^\beta T_{\alpha\beta} = (\epsilon + p)u_a u_b + p\gamma_{ab}, \tag{2.19c}$$

where $u_a = \gamma_a{}^\alpha u_\alpha$. Here $\rho_{\rm E}$, j_a, and S_{ab} are, respectively, the energy density, the momentum density, and the stress tensor measured by an observer whose 4-velocity is n^α. We denote by R_{ab} the 3-dimensional Ricci tensor of the 3-metric γ_{ab} and by

$$R = \gamma^{ab}R_{ab} \tag{2.20}$$

the corresponding 3-dimensional Ricci scalar. The 4-dimensional Ricci scalar will be written ${}^4R$.

The projections $E_{\perp\perp} \equiv E^{\alpha\beta}n_\alpha n_\beta = 0$ and $E^a_{\parallel\perp} \equiv E^{\alpha\beta}\gamma^a{}_\alpha n_\beta = 0$ of the Einstein equation are, respectively, the *Hamiltonian constraint*

$$R + K^2 - K_{ab}K^{ab} - 16\pi\rho_{\rm E} = 0, \tag{2.21}$$

and the *momentum constraint*

$$D_b(K^{ab} - \gamma^{ab}K) - 8\pi j^a = 0. \tag{2.22}$$

[1] There is no consistent convention in the literature for the sign of the extrinsic curvature of a spacelike hypersurface. Our convention agrees with that of MTW [480] and disagrees with Wald's [715].

The remaining part of the Einstein equation, $\gamma_a{}^\alpha \gamma_b{}^\beta[R_{\alpha\beta} - 8\pi(T_{\alpha\beta} - \frac{1}{2}g_{\alpha\beta}T_\gamma{}^\gamma)] = 0$, has the form

$$\partial_t K_{ab} = -D_a D_b \alpha + \alpha(R_{ab} + KK_{ab} - 2K_{ac}K^c{}_b) + \mathcal{L}_{\boldsymbol{\beta}} K_{ab} - 8\pi\alpha\left[S_{ab} - \frac{\gamma_{ab}}{2}(S_c{}^c - \rho_{\mathrm{E}})\right]. \tag{2.23}$$

The system (2.18), (2.23) describes the evolution of the 3-dimensional tensor fields γ_{ab} and K_{ab} on Σ. Because Eqs. (2.18) and (2.23) do not contain time derivatives of the lapse function α or of the shift vector β^a, these metric functions are not dynamical variables. One can regard the four degrees of gauge freedom associated with the choice of coordinates (t, x^i) as the freedom to choose α and β^a. From this point of view, once α and β^a are prescribed, initial data for the geometry consist of initial values $\gamma_{ab}(0)$ and $K_{ab}(0)$ of the metric and extrinsic curvature satisfying the constraint equations (2.21) and (2.22).

We have so far presented only the equations governing the geometry. As we noted at the beginning of Chapter 1, an accurate description of neutron-star dynamics involves departures from a perfect fluid and from the effective one-parameter equation of state that describes an equilibrium star. In the present section introducing the initial-value problem, we consider the dynamical equations for the simplest case of a perfect fluid with a one-parameter equation of state and then present an extension of the formalism that is commonly used to model the entropy produced in shocks.

Where the 3+1 decomposition of the stress-energy tensor naturally involves the covariant components $u_i = \gamma_i{}^\mu u_\mu$ of the fluid velocity, it is the contravariant components u^i that naturally enter conservation laws and the Euler equation. Baryon conservation (1.26), for example, has the form $0 = \nabla_\alpha(\rho u^\alpha \sqrt{|g|}) = \partial_\mu(\rho u^\mu \sqrt{|g|})$. The upper and lower components are related by $u_i = \gamma_{i\mu}u^\mu = \gamma_{ij}u^j + \beta_i u^t$, or

$$u^i = \gamma^{ij}u_j - \beta^i u^t. \tag{2.24}$$

The difference is related to the relation between decompositions with n^α and with $t^\alpha = \alpha n^\alpha + \beta^\alpha$ as the timelike vector. To write this relation, we introduce the 3-velocity v_a measured by an observer with velocity n^α. Using the correspondence $v_a = \gamma_a{}^\alpha v_\alpha$ between v_a and a 4-vector v_α for which $v_\alpha n^\alpha = 0$, we have

$$u_\alpha = W(n_\alpha + v_\alpha), \tag{2.25}$$

where $u^\alpha u_\alpha = -1$ implies that W is the Lorentz factor

$$W = \frac{1}{\sqrt{1 - v^2}}, \tag{2.26}$$

with $v^2 = v^\alpha v_\alpha = \gamma_{ab}v^a v^b$. Using Eqs. (2.6) and (2.5), we can write the components of Eq. (2.25) as

$$W = \alpha u^t, \quad u^i = u^t\left(\alpha v^i - \beta^i\right), \tag{2.27}$$

and the decomposition of u^α with respect to t^α is

$$u^\alpha = u^t(t^\alpha + \alpha v^\alpha - \beta^\alpha). \tag{2.28}$$

The evolution of a perfect fluid is governed by conservation of baryons (the continuity equation), (1.26), the Euler equation, (1.30), and conservation of energy, (1.27), together with an equation of state. These first three equations have the 3+1 form

$$\partial_t(\rho u^t \alpha\sqrt{\gamma}) = -D_b[\rho u^t(\alpha v^b - \beta^b)\alpha\sqrt{\gamma}], \tag{2.29}$$

$$\partial_t j^a = \mathcal{L}_{\boldsymbol{\beta}} j^a + \alpha(2K^a{}_b + \delta^a_b K)j^b - D_b(\alpha S^{ab}) - \rho_E D^a\alpha. \tag{2.30}$$

$$\partial_t \rho_E = \mathcal{L}_{\boldsymbol{\beta}}\rho_E + \alpha K\rho_E - \frac{1}{\alpha}D_b(\alpha^2 j^b) + \alpha K_{ab}S^{ab}. \tag{2.31}$$

Equations (2.30) and (2.31) are, respectively the projections $\gamma^a{}_\alpha\nabla_\beta T^{\alpha\beta} = 0$ and $n_\alpha\nabla_\beta T^{\alpha\beta} = 0$. Detailed derivations are given by Gourgoulhon [271].

For a one-parameter equation of state, $\epsilon = \epsilon(\rho), p = p(\rho)$, satisfying the zero-entropy first law of thermodynamics, (1.116), the energy conservation equation, (2.31), is redundant, implied by baryon conservation and the equation of state. The evolution of the fluid is then given by Eqs. (2.29), (2.30), and the equation of state. Initial data for the fluid are the values of baryon mass density ρ and fluid 3-velocity u^a, with initial values of ϵ, p and u^t determined by the equation of state and the normalization $u_\alpha u^\alpha = -1$.

A solution $\gamma_{ab}(t), K_{ab}(t), \alpha(t), \beta^a(t), \epsilon(t), p(t), u^a(t)$ to the Einstein-Euler system, in the decomposed form (2.18), (2.23)–(2.30), then yields the 4-dimensional metric (2.10) whose source is a perfect fluid having 4-velocity $u^\alpha = u^t(t^\alpha + \gamma^\alpha{}_a u^a)$. In solving the system numerically, one specifies initial data satisfying the constraint equations at $t = 0$, and one essentially solves only the evolution equations, (2.18) and (2.23), for the metric and (2.29) and (2.30) for the fluid. The resulting evolution then preserves the constraints: Eqs. (2.21) and (2.22) are automatically satisfied. Preservation of the constraints is a consequence of the Bianchi identity, $\nabla_\beta G^{\alpha\beta} = 0$, and the fluid equation $\nabla_\beta T^{\alpha\beta} = 0$, because the equation $\nabla_\beta(G^{\alpha\beta} - 8\pi T^{\alpha\beta}) = 0$ expresses the time derivative of the constraint $(G^{\alpha\beta} - 8\pi T^{\alpha\beta})n_\alpha$ at a time t in terms of the field equation and its spatial derivatives at t.

The time evolution of a fluid with smooth initial data is, in general, smooth for a finite time (see below), after which shock waves appear. Within a perfect-fluid description, shocks are solutions in which the fluid variables are discontinuous on a characteristic hypersurface – the history of a two-surface that moves at the speed of sound. Formally maintaining a one-parameter equation of state can be appropriate for solutions without shocks or with small shocks, but it is not appropriate for strong shocks, because it ignores the heat generated by the shock. There is, however, a common approximation that roughly accounts for the heat generated without introducing the complications of neutrino and radiative transport. One adopts a two-parameter EOS of the form (1.7) and uses the continuity and energy conservation equations, (2.29) and (2.31), as independent evolution equations for

ρ and ϵ. With no shocks, as we have just seen, this evolution preserves the thermodynamic relation (first law) between ϵ and ρ that maintains a constant-entropy equation of state, $\epsilon = \epsilon(\rho)$. When there are shocks, the constant-entropy first law is violated by the time evolution. The evolution equations again determine ϵ and ρ, but ϵ is no longer a one-parameter function $\epsilon(\rho)$.

For both numerical evolution and for proving existence of solutions, it is important to note that the system as described is not strongly hyperbolic. To turn it into a strongly hyperbolic system, one can add to the dynamical equations linear combinations of the constraint equations or use harmonic coordinates, coordinates for which $\partial_\nu(g^{\mu\nu}\sqrt{|g|}) = 0$ (see also Section 10.1.2 for a generalized version). The harmonic gauge condition, in effect, replaces spatial derivatives in the constraint equations by time derivatives and leaves each component of the Ricci tensor in the manifestly hyperbolic form $R_{\mu\nu} = -\frac{1}{2}g^{\sigma\tau}\partial_\sigma\partial_\tau g_{\mu\nu} + F_{\mu\nu}(\partial g, g)$. With this form, local existence of solutions to the vacuum Einstein equation for analytic data is immediate from the Cauchy-Kovalevskaya theorem (see [164]) and was first proved for smooth data by Choquet-Bruhat [226]: For any smooth initial data $(\gamma_{ab}, \partial_t\gamma_{ab})$ on a hypersurface Σ, there is a solution $g_{\alpha\beta}$ to the vacuum Einstein equation in a neighborhood of $\{0\} \times \Sigma$ in $M = \mathbb{R} \times \Sigma$, and the solution is unique up to isometry. Local existence for perfect fluids without boundary was proved by Lichnerowicz [413], but for local evolution of stars, modeled as perfect fluids with boundary in an asymptotically flat spacetime, there is still no proof available (see [138, 563] for reviews of the Cauchy problem). Proof of existence of solutions with shocks (weak solutions to the Einstein-Euler system in which the fluid variables are discontinuous) is so far restricted to a few equations of state and to a few symmetric spacetimes [280, 396, 52].

2.2 Action for perfect-fluid spacetimes

2.2.1 Summary of results

We obtain in this section an action for the Einstein-Euler system governing a perfect-fluid spacetime, showing that it has the form

$$I = \int \left(\frac{1}{16\pi}R - \epsilon\right)\sqrt{|g|}d^4x. \tag{2.32}$$

At the same time, we begin our development of perturbation theory with a Lagrangian formalism for fluid perturbations that is used here in varying the action and in subsequent chapters to describe stellar oscillations and stability. This is one of the book's more technical chapters, and we begin with a brief summary for readers who do not need the formal derivations.

The change in a fluid flow that preserves the baryon number and entropy of each fluid element can be completely described by the change in its trajectories and by the accompanying change $h_{\alpha\beta}$ in the spacetime metric. A Lagrangian description of the fluid keeps track of the position of fluid elements, expressing the difference between the perturbed trajectory of a fluid element and its trajectory in the unperturbed flow by a connecting vector: A *Lagrangian displacement* ξ^α is a vector from the position of a fluid element in an unperturbed star to its position in the perturbed configuration. For a barotropic EOS or for a two-parameter EOS of the form $p = p(\rho, s), \epsilon = \epsilon(\rho, s)$, the change in the fluid variables can be written in terms of ξ^α and $h_{\alpha\beta}$ because, for a comoving observer, the change in s vanishes and the change in ρ is determined by the change in the volume of a fluid element measured by the comoving observer.

To make this precise, one defines the *Eulerian change* δs as the change in s at a fixed point of spacetime, whereas the *Lagrangian change*,

$$\Delta s := \delta s + \mathcal{L}_{\boldsymbol{\xi}} s = (\delta + \xi^\alpha \nabla_\alpha) s, \tag{2.33}$$

is the change in the value of s for the same fluid element. For adiabatic perturbations, we have

$$\Delta s = 0. \tag{2.34}$$

We can find the change $\Delta\rho$ heuristically in terms of the volume V of a fluid element. If we denote by ΔV the change in its volume (measured by a comoving observer), then conservation of baryons, $\Delta(\rho V) = 0$, implies $\dfrac{\Delta\rho}{\rho} = -\dfrac{\Delta V}{V}$, with the fractional change in volume given by the Lagrangian change in the metric orthogonal to u^α

$$\frac{\Delta\rho}{\rho} = -\frac{1}{2} q^{\alpha\beta} \Delta g_{\alpha\beta}, \tag{2.35}$$

where $\Delta g_{\alpha\beta} = (\delta + \mathcal{L}_{\boldsymbol{\xi}}) g_{\alpha\beta} = h_{\alpha\beta} + \nabla_\alpha \xi_\beta + \nabla_\beta \xi_\alpha$. When there is no change in the metric, the right side is just the spatial divergence $-q^{\alpha\beta}\nabla_\alpha \xi_\beta$. The Lagrangian change in the energy density and pressure is then given by the first law (1.8) and the definition (1.118) of the adiabatic index:

$$\frac{\Delta\epsilon}{\epsilon + p} = \frac{\Delta p}{\Gamma_1 p} = \frac{\Delta\rho}{\rho}. \tag{2.36}$$

Finally, the change in the 4-velocity u^α is obtained directly from the change in the trajectories and again involves only $\Delta g_{\alpha\beta}$, but with a projection along u^α:

$$\Delta u^\alpha = \frac{1}{2} u^\alpha u^\beta u^\gamma \Delta g_{\beta\gamma}. \tag{2.37}$$

Using these relations, we show that the variation of the action, for perturbations $h_{\alpha\beta}$, ξ^α vanishing outside a 4-dimensional region Ω and preserving entropy and baryon number, has the form

$$\delta I = \int_\Omega \left[-\frac{1}{16\pi} (G^{\alpha\beta} - 8\pi T^{\alpha\beta}) h_{\alpha\beta} - \xi_\alpha \nabla_\beta T^{\alpha\beta} \right] \sqrt{|g} \, d^4x, \tag{2.38}$$

yielding the Einstein-Euler system.

2.2.2 Lagrangian formalism and Lagrangian displacements

We begin with a more formal derivation of the expressions for the changes in the fluid variables ρ, s, and u^α in terms of the change of the fluid trajectories and metric. We consider a family of perfect-fluid spacetimes, labeled by a parameter s. The Lagrangian displacement ξ^α will be the infinitesimal version of a diffeo that relates trajectories of finitely different fluid flows. That is, we define a family of diffeos χ_s mapping the trajectories of an initial fluid flow to the corresponding trajectories of the flow labeled by s. For a fixed value of s, χ_s maps a point x to the point $\chi_\mathsf{s}(x)$; if $c(\tau)$ is a trajectory of the initial fluid parameterized by τ, then

$$\tilde{c}(\tau) := \chi_\mathsf{s}[c(\tau)] \tag{2.39}$$

is the corresponding trajectory in the history labeled by s.

Assume now that χ_0 is the identity map, $\chi_0(x) = x$. For a fixed value x_0 of x, $\chi_\mathsf{s}(x_0)$ is a curve through x_0 parametrized by s,

$$\mathsf{s} \mapsto \chi_\mathsf{s}(x_0).$$

Define $\xi^\alpha(x_0)$ as the tangent to this curve. In a chart $\{x^\mu\}$, a point of this curve has coordinates $\chi_\mathsf{s}^\mu(x_0)$, and ξ^α has components

$$\xi^\mu = \frac{d}{d\mathsf{s}}\chi_\mathsf{s}^\mu(x_0). \tag{2.40}$$

For small s, as shown in Fig. 2.1 the coordinates of the points x and $\chi_\mathsf{s}(x)$ differ by $\mathsf{s}\xi^\mu$:

$$\chi_\mathsf{s}^\mu(x) = x^\mu + \mathsf{s}\xi^\mu(x) + o(\mathsf{s}), \tag{2.41}$$

where $o(\mathsf{s})$ is a function that vanishes faster than s.

Introducing the shorthand

$$\mathcal{Q}(\mathsf{s}) := [g_{\alpha\beta}(\mathsf{s}), u^\alpha(\mathsf{s}), \rho(\mathsf{s}), s(\mathsf{s})], \tag{2.42}$$

for the quantities that describe a family of perfect fluid spacetimes, we define the Eulerian perturbation $\delta\mathcal{Q}$ in each quantity by[2]

$$\delta\mathcal{Q} := \left.\frac{d}{d\mathsf{s}}\mathcal{Q}(\mathsf{s})\right|_{\mathsf{s}=0}. \tag{2.43}$$

To compare a scalar or tensor at $\chi(x)$ to a tensor at x, one uses the pullback χ^* to map the tensor from $\chi(x)$ to x: The pullback of a scalar ρ is $\chi^*\rho(x) = \rho[\chi(x)]$; the pullback $\chi^* g_{\alpha\beta}$ of the covariant tensor $g_{\alpha\beta}$ has components in a chart given by

$$\chi^* g_{\mu\nu}(x) = \partial_\mu\chi^\sigma(x)\,\partial_\nu\chi^\tau(x)\,g_{\sigma\tau}[(\chi(x)], \tag{2.44}$$

[2] Note that in the formal treatment, we have replaced δQ and ξ^α of the heuristic introduction by $\mathsf{s}\delta Q$ and $\mathsf{s}\xi^\alpha$. The small parameter of the perturbation is now the dimensionless s.

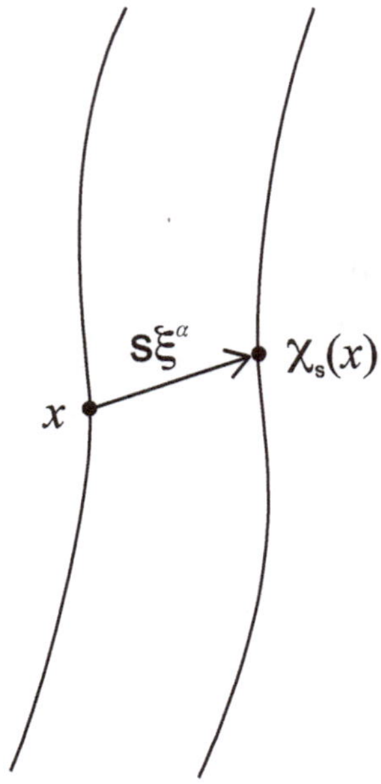

Figure 2.1. For small s, a Lagrangian displacement ξ^α can be regarded as a vector for which sξ^α joins the position x of a fluid element in an initial fluid flow to its position $\chi_{\mathsf{s}}(x)$ in the perturbed fluid flow.

and the pullback $\chi^* u^\alpha$ of the vector u^α has components

$$\chi^* u^\mu(x) = \partial_\nu \chi^{-1\,\mu}(x)\, u^\nu[\chi(x)], \tag{2.45}$$

where χ^{-1} means the inverse of the diffeo χ.

A *Lagrangian perturbation* $\Delta\mathcal{Q}$ is then defined by

$$\Delta\mathcal{Q} := \frac{d}{d\mathsf{s}}\chi_{\mathsf{s}}^*\mathcal{Q}(\mathsf{s})\bigg|_{\mathsf{s}=0} = (\delta + \mathcal{L}_{\boldsymbol{\xi}})\mathcal{Q}. \tag{2.46}$$

The second equality follows from the geometric definition of the Lie derivative of a tensor, $\frac{d}{d\mathsf{s}}\chi_{\mathsf{s}}^* T = \mathcal{L}_{\boldsymbol{\xi}} T$:

$$\frac{d}{d\mathsf{s}}\chi_{\mathsf{s}}^*\mathcal{Q}(\mathsf{s})\bigg|_{\mathsf{s}=0} = \chi_0^* \frac{d}{d\mathsf{s}}\mathcal{Q}(\mathsf{s})\bigg|_{\mathsf{s}=0} + \frac{d}{d\mathsf{s}}\chi_{\mathsf{s}}^*\mathcal{Q}(0)\bigg|_{\mathsf{s}=0} = (\delta + \mathcal{L}_{\boldsymbol{\xi}})\mathcal{Q}. \tag{2.47}$$

The local first law of thermodynamics for the fluid expresses the Lagrangian change in ϵ in terms of the Lagrangian changes in s and ρ:

$$\Delta\epsilon = \rho T \Delta s + h\Delta\rho, \tag{2.48}$$

where $h = (\epsilon + p)/\rho$. We will find that extremizing the action yields the Einstein-Euler system only for variations that locally conserve baryons and for which the entropy per baryon is unchanged – for which $\Delta s = 0$. We now obtain the first-order Lagrangian changes given in the introduction, expressing the variables $\mathcal{Q}$ in terms of the Lagrangian change $\Delta g_{\alpha\beta}$ in the metric. Again denoting the Eulerian change in the metric by $h_{\alpha\beta}$ and using the definition (2.46), we have

$$\Delta g_{\alpha\beta} = h_{\alpha\beta} + \nabla_\alpha \xi_\beta + \nabla_\beta \xi_\alpha. \tag{2.49}$$

We will also need the Lagrangian change in the density $\sqrt{|g|}$. To find it, we can use the matrix identity $(\log\det M)' = (\operatorname{tr}\log M)'$ or $(\det M)^{-1}(\det M)' = \operatorname{tr}(M^{-1}M')$, relating the derivative of the determinant of a symmetric matrix M to the trace of its derivative.[3] In particular,

$$\delta g = g g^{\alpha\beta}\delta g_{\alpha\beta}, \quad \mathcal{L}_{\boldsymbol{\xi}} g = g g^{\alpha\beta}\mathcal{L}_{\boldsymbol{\xi}} g_{\alpha\beta}, \quad \text{and} \quad \Delta g = g g^{\alpha\beta}\Delta g_{\alpha\beta}. \tag{2.50}$$

The Lagrangian change in $\sqrt{|g|}$ is then

$$\Delta\sqrt{|g|} = \frac{1}{2}\sqrt{|g|}\, g^{\alpha\beta}\Delta g_{\alpha\beta}. \tag{2.51}$$

To find the Lagrangian change in the 4-velocity, we use the relation (2.39) between the unperturbed trajectory c of a fluid element and its perturbed trajectory $\tilde{c}$. If c is parametrized by proper time, its tangent at a point x is $u^\alpha(x)$, and the (unnormalized) tangent $\tilde{w}^\alpha$ to $\tilde{c}$ at $\chi_{\mathsf{s}}(x)$ is the result of dragging u^α by χ_{s}. That is, $\tilde{w}^\alpha = \chi_{\mathsf{s}} u^\alpha$. The corresponding 4-velocity is the normalized tangent,

$$\tilde{u}^\alpha = \frac{\tilde{w}^\alpha}{\sqrt{-g_{\mathsf{s}\,\beta\gamma}\,\tilde{w}^\beta\tilde{w}^\gamma}}. \tag{2.52}$$

Then, by construction, dragging $\tilde{w}^\alpha$ back to x yields $u^\alpha(x)$:

$$\chi_{\mathsf{s}}^*\tilde{w}^\alpha = \chi_{\mathsf{s}}^{-1}\tilde{w}^\alpha = \chi_{\mathsf{s}}^{-1}\chi_{\mathsf{s}} u^\alpha = u^\alpha, \tag{2.53}$$

implying

$$\Delta w^\alpha = \frac{d}{d\mathsf{s}}\chi_{\mathsf{s}}^*\tilde{w}^\alpha\bigg|_{\mathsf{s}=0} = 0. \tag{2.54}$$

The Lagrangian change in u^α thus arises entirely from the Lagrangian change in the metric used to normalize it:

$$\begin{aligned}\Delta u^\alpha &= \Delta\frac{w^\alpha}{\sqrt{-g_{\beta\gamma}w^\beta w^\gamma}} = -\frac{1}{2}\frac{w^\alpha}{(-w_\delta w^\delta)^{3/2}}w^\beta w^\gamma\bigg|_{\mathsf{s}=0}(-\Delta g_{\beta\gamma})\\ \Delta u^\alpha &= \frac{1}{2}u^\alpha u^\beta u^\gamma\Delta g_{\beta\gamma},\end{aligned} \tag{2.55}$$

where we have used $u^\alpha = w^\alpha|_{\mathsf{s}=0}$.

We next obtain the Lagrangian change $\Delta\rho$ in baryon mass density. We first complete the heuristic derivation of the introduction, using the relation $\dfrac{\Delta\rho}{\rho} = -\dfrac{\Delta V}{V}$. The volume element is $\sqrt{q}$, with q the determinant of q_{ab}, the 3-metric on the subspace orthogonal to u^α. The 4-dimensional form of q_{ab} (the tensor orthogonal to u^α whose spatial components are q_{ij}) is the previously encountered tensor

[3] Here is a derivation, with the matrix denoted by $g_{\mu\nu}$: The coefficient of the component $g_{\mu\nu}$ in the determinant g is the minor $\Delta^{\mu\nu}$ of $g_{\mu\nu}$, and the inverse matrix is given by $g^{\mu\nu} = \frac{\Delta^{\mu\nu}}{g}$. Thus $\frac{\partial g}{\partial g_{\mu\nu}} = \Delta^{\mu\nu} = g g^{\mu\nu}$, and we have $g' = \frac{\partial g}{\partial g_{\mu\nu}} g'_{\mu\nu}$ or $\frac{g'}{g} = g^{\alpha\beta} g'_{\alpha\beta}$.

$q_{\alpha\beta} = g_{\alpha\beta} + u_\alpha u_\beta$. The fractional change in volume is given by the fractional change $\Delta\sqrt{q}/\sqrt{q}$ in $\sqrt{q}$, and the same calculation that gave Eq. (2.51) here gives

$$\frac{\Delta V}{V} = \frac{\Delta\sqrt{q}}{\sqrt{q}} = \frac{1}{2}q^{\alpha\beta}\Delta q_{\alpha\beta} = \frac{1}{2}q^{\alpha\beta}\Delta g_{\alpha\beta}, \tag{2.56}$$

whence

$$\Delta\rho = -\frac{1}{2}\rho q^{\alpha\beta}\Delta g_{\alpha\beta}. \tag{2.57}$$

The more formal version of the derivation obtains Eq. (2.35) as a first integral of the perturbed baryon conservation equation, written as the vanishing divergence of a vector density,

$$\nabla_\alpha(\rho u^\alpha\sqrt{|g|}) = 0. \tag{2.58}$$

To compute $\Delta\nabla_\alpha(\rho u^\alpha\sqrt{|g|})$ we use the vanishing of the commutator $[\mathcal{L}_{\boldsymbol{\xi}}, \nabla\cdot]$ acting on vector densities:

$$[\mathcal{L}_{\boldsymbol{\xi}}, \nabla_\alpha](A^\alpha\sqrt{|g|}) = 0, \tag{2.59}$$

(this is Eq. [A.49] of Appendix A.2.2, and it is equivalent to the dual relation $[\mathcal{L}_{\boldsymbol{\xi}}, d] = 0$ for forms). Then, because

$$[\delta, \nabla_\alpha](A^\alpha\sqrt{|g|}) = [\delta, \partial_\mu](A^\mu\sqrt{|g|}) = 0, \tag{2.60}$$

we have

$$[\Delta, \nabla_\alpha](\rho u^\alpha\sqrt{|g|}) = 0. \tag{2.61}$$

The perturbed conservation law now takes the form

$$\nabla_\alpha\Delta(\rho u^\alpha\sqrt{|g|}) = 0, \tag{2.62}$$

with first integral

$$\Delta(\rho u^\alpha\sqrt{|g|}) = 0. \tag{2.63}$$

Finally, using Eqs. (2.37) and (2.51), we recover Eq. (2.57).

We obtained in the summary the Lagrangian change in the energy density from the first law, equivalent to the requirement that, when $\Delta s = 0$, the change in energy $\Delta(\epsilon V)$ of a fluid element be the work $-p\Delta V$, or

$$\Delta\epsilon = -(\epsilon + p)\frac{\Delta V}{V} = -\frac{1}{2}(\epsilon + p)q^{\alpha\beta}\Delta g_{\alpha\beta}. \tag{2.64}$$

Formally, a derivation similar to that just given for $\Delta\rho$, using $\nabla_\alpha u^\alpha = 0$ for the unperturbed star, yields a first integral of the energy conservation equation (1.27),

$$0 = \Delta[\nabla_\beta(\epsilon u^\beta) + p\nabla_\beta u^\beta] = u^\gamma\nabla_\gamma\left[\Delta\epsilon + \frac{1}{2}(\epsilon + p)q^{\alpha\beta}\Delta g_{\alpha\beta}\right]. \tag{2.65}$$

Equations (2.51), and (2.64) express the changes in $\sqrt{|g|}$ and ϵ in terms of $h_{\alpha\beta}$ and ξ^α, in the combination $\Delta g_{\alpha\beta}$. From these expressions, it follows that the Lagrangian variation of the scalar density $\epsilon\sqrt{|g|}$ is given by

$$\begin{aligned}\frac{1}{\sqrt{|g|}}\Delta(\epsilon\sqrt{|g|}) &= -\frac{1}{2}(\epsilon+p)q^{\alpha\beta}\Delta g_{\alpha\beta} + \frac{1}{2}\epsilon g^{\alpha\beta}\Delta g_{\alpha\beta} \\ &= -\frac{1}{2}T^{\alpha\beta}\Delta g_{\alpha\beta},\end{aligned} \tag{2.66}$$

where $T^{\alpha\beta} = \epsilon u^\alpha u^\beta + pq^{\alpha\beta}$ is the stress-energy tensor of a perfect fluid. Then, by rewriting Eq. (2.66) in the form

$$\frac{1}{\sqrt{|g|}}\Delta(\epsilon\sqrt{|g|}) = -\frac{1}{2}T^{\alpha\beta}h_{\alpha\beta} + \xi_\alpha\nabla_\beta T^{\alpha\beta} - \nabla_\alpha(T^{\alpha\beta}\xi_\beta), \tag{2.67}$$

we conclude that the variation of the integral

$$I_M := -\int_\Omega \epsilon\, d^4V, \tag{2.68}$$

over a region Ω of spacetime, is given by

$$\delta I_M = \int_\Omega \left[\frac{1}{2}T^{\alpha\beta}h_{\alpha\beta} - \xi_\alpha\nabla_\beta T^{\alpha\beta}\right] d^4V. \tag{2.69}$$

That is, ignoring until the next section the fact that we have not formally defined the variables that are its arguments, we have shown that I_M is an action for the matter: Requiring $\delta I_M = 0$ for variations ξ^α in the fluid trajectories that vanish on $\partial\Omega$ yields the equations of motion of the fluid, $\nabla_\beta T^{\alpha\beta} = 0$, and the stress-energy tensor is given by

$$\frac{\delta I_M}{\delta g_{\alpha\beta}} = \frac{1}{2}T^{\alpha\beta}\sqrt{|g|}. \tag{2.70}$$

An action with s and ρ unconstrained. The first law of thermodynamics for a star describes the change in energy of an equilibrium configuration that results from local changes in the entropy, baryon number (baryon mass), and vorticity of the fluid. Our formalism already includes perturbations that change the fluid's vorticity; to allow the baryon number and entropy per baryon to change, one must allow the variations $\Delta \mathrm{j}^\alpha \equiv \Delta(\rho u^\alpha\sqrt{|g|})$ and Δs to be nonzero. By Eq. (2.37), Δu^α is proportional to u^α, implying that the change $\Delta \mathrm{j}^\alpha$ in the current is functionally proportional to j^α: Calling the function of proportionality $\Delta \ln j$, we have

$$\Delta(\rho u^\alpha\sqrt{|g|}) = (\Delta \ln j)\rho u^\alpha\sqrt{|g|}. \tag{2.71}$$

The equation has t component[4]

$$\Delta \ln j = \frac{\Delta(\rho\sqrt{|g|}\, u)^t}{\rho\sqrt{|g|}\, u^t}, \tag{2.72}$$

and Eq. (2.35) is replaced by $\dfrac{\Delta\rho}{\rho} = -\dfrac{1}{2}q^{\alpha\beta}\Delta g_{\alpha\beta} + \Delta \ln j$ or

$$\frac{\Delta\rho}{\rho} = -\frac{1}{2}q^{\alpha\beta}\Delta g_{\alpha\beta} + \frac{\Delta(\rho\sqrt{|g|}\, u)^t}{\rho\sqrt{|g|}\, u^t}. \tag{2.73}$$

Writing the local first law, Eq. (1.8), in the form $\Delta\epsilon = \rho T\Delta s + h\Delta\rho$ and using Eq. (2.73), we replace Eq. (2.66) by

$$\frac{1}{\sqrt{|g|}}\Delta(\epsilon\sqrt{|g|}) = \rho T\Delta s + \frac{h}{u^t\sqrt{|g|}}\Delta(\rho\sqrt{|g|}u)^t - \frac{1}{2}T^{\alpha\beta}h_{\alpha\beta} + \xi_\alpha\nabla_\beta T^{\alpha\beta} - \nabla_\alpha(T^{\alpha\beta}\xi_\beta). \tag{2.74}$$

Functional dependence of I_M. In the previous section, with the number of baryons and the entropy per baryon fixed, the functional I_M has two arguments. One is the metric, the tensor field $g_{\alpha\beta}$ on spacetime. The other is the diffeo χ that maps a fixed initial fluid configuration to any other configuration. That is, to define the action, one first fixes a fiducial fluid, specifying a set of fluid trajectories with unit tangent (4-velocity) u_0^α, functions s_0 and ρ_0, and metric $g_{0\alpha\beta}$.[5] The fiducial fluid trajectories are restricted to the star, to the support of ρ_0. Any other spacetime configuration of the fluid (with the same topology as that of the fiducial fluid) is then determined by the diffeo χ from the fixed fiducial fluid to the support of the new fluid configuration and by the metric $g_{\alpha\beta}$. As in Eq. (2.52), the 4-velocity is given in terms of χ by

$$u^\alpha = \frac{w^\alpha}{\sqrt{-g_{\beta\gamma}\, w^\beta\, w^\gamma}}, \quad \text{where } w^\alpha = \chi u_0^\alpha. \tag{2.75}$$

The rest-mass density (equivalently, baryon mass density) ρ is determined by conservation of baryons, $\mathrm{j}^\alpha = \chi \mathrm{j}_0^\alpha$, with $\mathrm{j}^\alpha = \rho u^\alpha\sqrt{|g|}$, and the energy density is then determined by the equation of state, $\epsilon = \epsilon(\rho)$.

Explicitly, with J the Jacobian det $\|\partial_\nu\chi^\mu\|$, the components of the baryon current satisfy

$$\rho u^\mu\sqrt{|g|}\Big|_P = J^{-1}\partial_\nu\chi^\mu\, \rho_0 u_0^\nu\sqrt{|g_0|}\Big|_{\chi^{-1}(P)}. \tag{2.76}$$

[4] The superscript t is outside the parentheses because the Lagrangian change in the t component $\mathrm{j}^t \equiv \mathrm{j}^\alpha\nabla_\alpha t$ of the vector density j^α is not in general the t component of the Lagrangian change in j^α: That is, unless $\xi^t = 0$, the Lagrangian change in the scalar t is nonzero, $\Delta t = \mathcal{L}_{\boldsymbol{\xi}} t = \xi^\alpha\nabla_\alpha t = \xi^t$, implying $(\Delta \mathrm{j})^t \equiv (\Delta \mathrm{j}^\alpha)\nabla_\alpha t \neq \Delta(\mathrm{j}^\alpha\nabla_\alpha t) \equiv \Delta(\mathrm{j}^t)$.

[5] One can specify the fiducial fluid more concisely by simply giving s, together with an initial baryon current density j_0^α, whose meaning, once a metric is specified, is $\rho\sqrt{|g|}u^\alpha$. The fiducial metric $g_{0\alpha\beta}$ is not needed: All fluid variables are determined by χ, $g_{\alpha\beta}$, s_0, and j_0^α. In particular, the configuration associated with $(\chi, g_{\alpha\beta})$ has baryon current density $\mathrm{j}^\alpha = \chi \mathrm{j}_0^\alpha$, current $j^\alpha = \mathrm{j}^\alpha/\sqrt{|g|}$, baryon mass density $\rho = \sqrt{-g_{\alpha\beta}j^\alpha j^\beta}$, and 4-velocity $u^\alpha = j^\alpha/\rho$. The usual form of the first law, however, is associated with a choice of ρ or, equivalently, n as a canonical variable.

Because u^μ is given by Eq. (2.75), with $w^\mu|_P = \partial_\nu \chi^\mu u_0^\nu|_{\chi^{-1}(P)}$, we have

$$\rho = \rho_0 J^{-1} \sqrt{\frac{g_0 \circ \chi^{-1}}{g}} (-g_{\mu\nu} \partial_\sigma \chi^\mu \partial_\tau \chi^\nu u_0^\sigma u_0^\tau \circ \chi^{-1})^{1/2}, \tag{2.77}$$

and ϵ is expressed as a function of $g_{\alpha\beta}$ and χ by

$$\epsilon = \epsilon(s_0, \rho). \tag{2.78}$$

In the notation of Eq. (2.46), perturbations of an arbitrary fluid configuration $\mathcal{Q}$ are described by a family of diffeos χ_{s} *of that configuration*, for which $\chi_{\mathsf{s}=0}$ is the identity. Because the fluid configuration $\mathcal{Q}$ is itself specified by a pair $g_{\alpha\beta}, \chi$ with χ a diffeo of the fiducial fluid configuration $\mathcal{Q}_0$, trajectories of the perturbed fluid are related to trajectories of the fiducial fluid $\mathcal{Q}_0$ by diffeos $\widetilde{\chi}_{\mathsf{s}} = \chi_{\mathsf{s}} \circ \chi$ for which $\widetilde{\chi}_{\mathsf{s}=0} = \chi$.

Because the coordinates of $\chi_{\mathsf{s}}(P)$ are $\chi^\mu_{\mathsf{s}} = x^\mu + \mathsf{s}\xi^\mu + o(\mathsf{s})$, we have $\frac{d}{d\mathsf{s}}\chi^\mu_{\mathsf{s}}(P)\big|_{\mathsf{s}=0} = \xi^\mu(P)$, or

$$\delta\chi^\alpha = \xi^\alpha. \tag{2.79}$$

We can then define the variational derivatives $\dfrac{\delta I_M}{\delta \chi^\alpha}$ and $\dfrac{\delta I_M}{\delta g_{\alpha\beta}}$ by

$$\begin{aligned}
\delta I_M \equiv \frac{d}{d\mathsf{s}} I_M \bigg|_{\mathsf{s}=0} &= \int_\Omega \left[\frac{\delta I_M}{\delta \chi^\alpha}\delta\chi^\alpha + \frac{\delta I_M}{\delta g_{\alpha\beta}}\delta g_{\alpha\beta}\right] d^4x \\
&= \int_\Omega \left[\frac{\delta I_M}{\delta \chi^\alpha}\xi^\alpha + \frac{\delta I_M}{\delta g_{\alpha\beta}} h_{\alpha\beta}\right] d^4x,
\end{aligned} \tag{2.80}$$

for displacements ξ^α that vanish at the intersection of the fluid with the boundary $\partial\Omega$ and for which $h_{\alpha\beta}$ and its first derivative vanish on $\partial\Omega$. In particular, Eq. (2.69) of the previous section implies

$$\frac{\delta I_M}{\delta \chi^\alpha} = -\nabla_\beta T^{\alpha\beta}\sqrt{|g|}. \tag{2.81}$$

Functional dependence of I_M with s and ρ unconstrained. If, instead of fixing s_0 and ρ_0, we promote them to independent arguments of I_M, then we can regard Eqs. (2.77) and (2.78) as expressions for ρ and ϵ in terms of the independent variables $g_{\alpha\beta}, \chi, s_0, \rho_0$.

2.2.3 Gravitational action

To complete this section, we recall the variation of the gravitational action

$$I_G := \frac{1}{16\pi}\int_\Omega {}^4R\, d^4V. \tag{2.82}$$

The formalism, involving the perturbed Riemann tensor, will be needed for the sections on perturbation theory as well as for the present section on the action for perfect-fluid spacetimes.

The Riemann tensor is defined by the commutator

$$R^\alpha{}_{\beta\gamma\delta} v^\beta = [\nabla_\gamma, \nabla_\delta] v^\alpha \tag{2.83}$$

for any vector v^α. For fixed v^α, we need to compute

$$\delta R^\alpha{}_{\beta\gamma\delta} v^\beta = \delta[\nabla_\gamma, \nabla_\delta] v^\alpha . \tag{2.84}$$

The perturbed covariant derivative operator, $\delta\nabla_\alpha$, acts as a tensor[6] $\delta\Gamma^\gamma{}_{\beta\alpha}$; on a vector v^α, its action is

$$\delta\nabla_\gamma v^\alpha = \delta\Gamma^\alpha{}_{\beta\gamma} v^\beta , \tag{2.85}$$

where

$$\delta\Gamma^\alpha{}_{\beta\gamma} = \frac{1}{2}\left(\nabla_\beta h^\alpha{}_\gamma + \nabla_\gamma h^\alpha{}_\beta - \nabla^\alpha h_{\beta\gamma}\right). \tag{2.86}$$

Equation (2.86) can be derived from the relation

$$0 = \delta(\nabla_\alpha g_{\beta\gamma}) = \nabla_\alpha h_{\beta\gamma} - \delta\Gamma^\delta{}_{\beta\alpha} g_{\delta\gamma} - \delta\Gamma^\delta{}_{\gamma\alpha} g_{\beta\delta} , \tag{2.87}$$

via the linear combination

$$0 = \frac{1}{2} g^{\alpha\delta} [\delta(\nabla_\beta g_{\gamma\delta}) + \delta(\nabla_\gamma g_{\beta\delta}) - \delta(\nabla_\delta g_{\beta\gamma})]. \tag{2.88}$$

To compute $\delta R_{\alpha\beta\gamma}{}^\delta$, we write

$$\begin{aligned} \delta(\nabla_\gamma \nabla_\delta v^\alpha) &= -\delta\Gamma^\beta{}_{\gamma\delta} \nabla_\beta v^\alpha + \delta\Gamma^\alpha{}_{\beta\gamma} \nabla_\delta v^\beta + \nabla_\gamma \delta(\nabla_\delta v^\alpha) \\ &= -\delta\Gamma^\beta{}_{\gamma\delta} \nabla_\beta v^\alpha + \delta\Gamma^\alpha{}_{\beta\gamma} \nabla_\delta v^\beta + \delta\Gamma^\alpha{}_{\beta\delta} \nabla_\gamma v^\beta \\ &\quad + \nabla_\gamma (\delta\Gamma^\alpha{}_{\beta\delta}) v^\beta . \end{aligned} \tag{2.89}$$

Because the first three terms on the right side are, together, symmetric in γ and δ, Eq. (2.84) implies, for all v^α,

$$\delta R^\alpha{}_{\beta\gamma\delta} v^\beta = (\nabla_\gamma \delta\Gamma^\alpha{}_{\beta\delta} - \nabla_\delta \delta\Gamma^\alpha{}_{\beta\gamma}) v^\beta ,$$

whence

$$\delta R^\alpha{}_{\beta\gamma\delta} = \nabla_\gamma \delta\Gamma^\alpha{}_{\beta\delta} - \nabla_\delta \delta\Gamma^\alpha{}_{\beta\gamma} . \tag{2.90}$$

The perturbation of $R_{\alpha\beta}$ follows immediately:

$$\delta R_{\alpha\beta} = \delta R^\gamma{}_{\alpha\gamma\beta} = \nabla_\gamma \delta\Gamma^\gamma{}_{\alpha\beta} - \nabla_\beta \delta\Gamma^\gamma{}_{\gamma\alpha} ; \tag{2.91}$$

and we have

$$\begin{aligned} \delta\, {}^4R = \delta(R_{\alpha\beta} g^{\alpha\beta}) &= -R_{\alpha\beta} h^{\alpha\beta} + g^{\alpha\beta} \delta R_{\alpha\beta} \\ &= -R_{\alpha\beta} h^{\alpha\beta} + \nabla_\alpha A^\alpha , \end{aligned} \tag{2.92}$$

[6] Although Christoffel symbols $\Gamma^\mu{}_{\nu\lambda}$ do not naturally constitute a tensor, the difference between the Christoffel symbols of two different covariant derivative operators do so. To linear order in the difference, this is implied by the explicit covariance of Eq. (2.86). Readers familiar with the definition of a tensor field as a map of vector fields that is linear under multiplication by scalar fields can note that the difference between two covariant derivative operators, $\tilde\nabla_\alpha - \nabla_\alpha$, is in fact linear in this sense: $(\tilde\nabla_\alpha - \nabla_\alpha)(f\xi^\beta) = f(\tilde\nabla_\alpha - \nabla_\alpha)\xi^\beta$.

where

$$A^\alpha = (g^{\beta\gamma}\delta\Gamma^\alpha{}_{\beta\gamma} - g^{\alpha\gamma}\delta\Gamma^\beta{}_{\beta\gamma}) = (g^{\alpha\gamma}g^{\beta\delta} - g^{\alpha\beta}g^{\gamma\delta})\nabla_\beta h_{\gamma\delta}. \tag{2.93}$$

Finally, Eqs. (2.50) and (2.92) imply

$$\delta({}^4R\sqrt{|g|}) = [-G_{\alpha\beta}h^{\alpha\beta} + \nabla_\alpha A^\alpha]\sqrt{|g|}. \tag{2.94}$$

I_G is thus the gravitational action for perturbations that vanish sufficiently rapidly at the boundary $\partial\Omega$:

$$\frac{\delta I_G}{\delta g_{\alpha\beta}} = -\frac{1}{16\pi}G_{\alpha\beta}\sqrt{|g|}. \tag{2.95}$$

2.2.4 Action for the Einstein-Euler system

From Eqs. (2.67) and (2.94), it follows that

$$\mathscr{L} := \left(\frac{1}{16\pi}{}^4R - \epsilon\right)\sqrt{|g|} \tag{2.96}$$

is a Lagrangian density for a perfect-fluid spacetime. That is,

$$\frac{1}{\sqrt{|g|}}(\delta\mathscr{L} - \nabla_\alpha\boldsymbol{\Theta}^\alpha) = -\frac{1}{16\pi}(G^{\alpha\beta} - 8\pi T^{\alpha\beta})h_{\alpha\beta} - \xi_\alpha\nabla_\beta T^{\alpha\beta}, \tag{2.97}$$

with

$$\frac{1}{\sqrt{|g|}}\boldsymbol{\Theta}^\alpha = (\epsilon + p)q^{\alpha\beta}\xi_\beta + \frac{1}{16\pi}(g^{\alpha\gamma}g^{\beta\delta} - g^{\alpha\beta}g^{\gamma\delta})\nabla_\beta h_{\gamma\delta}. \tag{2.98}$$

The corresponding action is

$$I := I_G + I_M = \int_\Omega \left(\frac{1}{16\pi}{}^4R - \epsilon\right)d^4V. \tag{2.99}$$

For perturbations that do not conserve entropy or baryon number, the variation of the Lagrangian is given by

$$\begin{aligned}\frac{1}{\sqrt{|g|}}(\delta\mathscr{L} - \nabla_\alpha\boldsymbol{\Theta}^\alpha) =& -\rho T\Delta s - \frac{h}{u^t\sqrt{|g|}}\Delta(\rho\sqrt{|g|}u)^t \\ & - \frac{1}{16\pi}(G^{\alpha\beta} - 8\pi T^{\alpha\beta})h_{\alpha\beta} - \xi_\alpha\nabla_\beta T^{\alpha\beta}. \end{aligned} \tag{2.100}$$

Restricting the variation to perturbations of compact support that locally preserve entropy and baryon number, we obtain Eq. (2.38).

2.2.5 Hamiltonian formalism

In parallel to our treatment of the Lagrangian formalism, we begin with the gravitational Hamiltonian and the corresponding decomposition of the vacuum field equation, turning in the next section to the Hamiltonian for the fluid.

Hamilton's equations will involve the time derivative of the 3-metric γ_{ab} and of its conjugate momentum, π^{ab}. To cast the gravitational action into Hamiltonian form, one adds to the Lagrangian density ${}^4R\sqrt{|g|}$ a total divergence that replaces the second time derivatives in ${}^4R$ by a kinetic term quadratic in first time derivatives. This is done using a form of the contracted Gauss-Codazzi equation

$$ {}^4R - 2\nabla_\alpha(n^\gamma\nabla_\gamma n^\alpha - n^\alpha\nabla_\gamma n^\gamma) = R + K_{ab}K^{ab} - K^2, \tag{2.101} $$

where $K = \gamma^{ab}K_{ab}$ (and we are equating the value of the scalar on the left at the point (t, x^i) of Σ_t with the value of the scalar on the right at the point x^i of Σ at time t). To avoid factors of 16π while retaining the standard definition of the momentum π^{ab} conjugate to γ_{ab}, we define a Lagrangian density $\tilde{\mathscr{L}}_G$ by

$$ \tilde{\mathscr{L}}_G = 16\pi\,\mathscr{L}_G = (K_{ab}K^{ab} - K^2 + R)\alpha\sqrt{\gamma}. \tag{2.102} $$

Because R involves only spatial derivatives of γ_{ab}, the time derivative $\dot{\gamma}_{ab}$ occurs only in the terms involving the extrinsic curvature K_{ab}, and the momentum conjugate to γ_{ab} with respect to $\tilde{\mathscr{L}}_G$ is given by

$$ \pi^{ab} := 8\pi\frac{\partial\tilde{\mathscr{L}}}{\partial(\dot{\gamma}_{ab})} = -(K^{ab} - \gamma^{ab}K)\sqrt{\gamma}. \tag{2.103} $$

In the canonical formulation of general relativity [480], $\gamma_{ab}, \pi^{ab}, \alpha$, and β^a are regarded as independent gravitational field variables, the arguments of the gravitational Hamiltonian $\tilde{H}_G := \int d^3x[\pi^{ab}\dot{\gamma}_{ab} - \tilde{\mathscr{L}}_G]$. To write the Hamiltonian as a function of these arguments, one as usual inverts Eq. (2.103) to write $\dot{\gamma}_{ab}$ in terms of π^{ab}: Using Eq. (2.18) to write $\dot{\gamma}_{ab}$ in terms of K_{ab},

$$ \dot{\gamma}_{ab} = -2\alpha K_{ab} + D_a\beta_b + D_b\beta_a, \tag{2.104} $$

and using Eq. (2.103) to write K_{ab} in terms of π_{ab},

$$ K_{ab} = -\gamma^{-1/2}\left(\pi_{ab} - \frac{1}{2}\gamma_{ab}\pi\right), \tag{2.105} $$

where $\pi := \gamma_{ab}\pi^{ab}$, we have

$$ \dot{\gamma}_{ab} = 2\alpha\gamma^{-1/2}\left(\pi_{ab} - \frac{1}{2}\gamma_{ab}\pi\right) + D_a\beta_b + D_b\beta_a. \tag{2.106} $$

Then

$$ \tilde{\mathscr{L}}_G = \alpha\left[\gamma^{1/2}R + \gamma^{-1/2}\left(\pi^{ab}\pi_{ab} - \frac{1}{2}\pi^2\right)\right], \tag{2.107} $$

and the Hamiltonian density takes the form

$$ \pi^{ab}\dot{\gamma}_{ab} - \tilde{\mathscr{L}}_G = \alpha\gamma^{-1/2}\left(\pi^{ab}\pi_{ab} - \frac{1}{2}\pi^2\right) - \alpha\gamma^{1/2}R + 2\pi^{ab}D_a\beta_b \tag{2.108} $$

$$ = \alpha\mathcal{H}_G + \beta^a\mathcal{H}_{G\,a} + D_a(2\pi^{ab}\beta_b), \tag{2.109} $$

where the scalar density

$$\mathcal{H}_G := \gamma^{-1/2}\left(\pi^{ab}\pi_{ab} - \frac{1}{2}\pi^2\right) - \gamma^{1/2}R, \tag{2.110}$$

and the vector density

$$\mathcal{H}_{G\,a} := -2D_b\pi^b{}_a, \tag{2.111}$$

are, up to a factor $\gamma^{1/2}$, the gravitational parts of the Hamiltonian and momentum constraints, Eqs. (2.21) and (2.22), written in terms of π_{ab}. The reason for the terminology *Hamiltonian constraint* and *momentum constraint* will be clear below, from their role in the dynamical evolution of γ_{ab}.

With a choice of lapse β^a for which $\pi^{ab}\beta_b$ vanishes at spatial infinity, the gravitational action $\tilde{I}_G = \int d^3x \tilde{\mathscr{L}}_G$ can now be written in the form

$$\tilde{I}_G = \int d^4x(\pi^{ab}\dot{\gamma}_{ab} - \alpha\mathcal{H}_G - \beta^a\mathcal{H}_{Ga}), \tag{2.112}$$

and the gravitational Hamiltonian is

$$\tilde{H}_G = \int d^3x[\alpha\mathcal{H}_G + \beta^a\mathcal{H}_{G\,a}]. \tag{2.113}$$

Because no time derivatives of β^a or α appear in the action $\tilde{I}_G$, the variational derivatives $\delta\tilde{I}_G/\delta\alpha$ and $\delta\tilde{I}_G/\delta\beta^a$ are constraint equations, the vacuum constraints

$$\mathcal{H}_G = 0, \qquad \mathcal{H}_{G\,a} = 0. \tag{2.114}$$

Equivalently, because no momenta conjugate to lapse and shift appear in H_G, the vacuum constraints are the parts of Hamilton's equations corresponding to these variables.

Dynamical equations. From the Hamiltonian $\tilde{H}_G$ we obtain Hamilton's equations for $\dot{\gamma}_{ab}$ and $\dot{\pi}^{ab}$ outside the star, namely

$$\dot{\gamma}_{ab} = \frac{\delta}{\delta\pi^{ab}}\int d^3x\ (\alpha\mathcal{H}_G + \beta^a\mathcal{H}_{G\,a}), \tag{2.115}$$

$$\dot{\pi}^{ab} = -\frac{\delta}{\delta\gamma_{ab}}\int d^3x\ (\alpha\mathcal{H}_G + \beta^a\mathcal{H}_{G\,a}). \tag{2.116}$$

When the shift is zero the spacetime evolution is along αn^α, perpendicular to Σ_t, and it is generated by $\int d^3x\ \alpha\mathcal{H}_G$. When, on the other hand, the lapse is zero, the evolution is purely spatial, along β^α, a vector that lies in Σ_t, and it is generated by $\int d^3x\ \beta^a\mathcal{H}_{G\,a}$. It is because $\alpha\mathcal{H}_G$ can in this sense be regarded as generating time evolution along αn^α, whereas $\beta^a\mathcal{H}_{G\,a}$ can be regarded as generating spatial evolution along β^α, that $\mathcal{H}_G$ and $\mathcal{H}_{G\,a}$ are referred to as the *Hamiltonian* and *momentum* constraints. (Recall that in ordinary Hamiltonian mechanics, the Hamiltonian is the generator of time translations, the momentum the generator of spatial translations.) From the definition of $\dot{T}^{a\cdots b}{}_{c\cdots d}$ in terms of $\mathcal{L}_t T^{a\cdots b}{}_{c\cdots d}$, we already know that the spatial part of the evolution (the evolution with $\alpha = 0$, $\beta^a \neq 0$) is given by

$$\dot{\gamma}_{ab} = \mathcal{L}_{\boldsymbol{\beta}}\gamma_{ab}, \qquad \dot{\pi}^{ab} = \mathcal{L}_{\boldsymbol{\beta}}\pi^{ab}. \tag{2.117}$$

We therefore only need the contributions to $\dot{\gamma}_{ab}$ and $\dot{\pi}^{ab}$ from $\alpha\mathcal{H}_G$, but for completeness, we include $\beta^a\mathcal{H}_{Ga}$ in the computation. We will use the form

$$\int \beta^a \mathcal{H}_{Ga} d^3x = \int \mathcal{L}_{\boldsymbol{\beta}}\gamma_{ab}\pi^{ab} d^3x = -\int \gamma_{ab}\mathcal{L}_{\boldsymbol{\beta}}\pi^{ab} d^3x. \tag{2.118}$$

We begin with $\dot{\gamma}_{ab}$, finding the change in $\mathcal{H}_G$ arising from a change in π^{ab}:

$$\delta\mathcal{H}_G = 2\gamma^{-1/2}\left(\pi_{ab} - \frac{1}{2}\pi\gamma_{ab}\right)\delta\pi^{ab}; \tag{2.119}$$

Then

$$\delta H = \int d^3x \left[2\alpha\gamma^{-1/2}\left(\pi_{ab} - \frac{1}{2}\pi\gamma_{ab}\right) + \mathcal{L}_{\boldsymbol{\beta}}\gamma_{ab}\right]\delta\pi^{ab}, \tag{2.120}$$

and we have

$$\dot{\gamma}_{ab} = \frac{\delta H}{\delta\pi^{ab}} = 2\alpha\gamma^{-1/2}\left(\pi_{ab} - \frac{1}{2}\pi\gamma_{ab}\right) + \mathcal{L}_{\boldsymbol{\beta}}\gamma_{ab}, \tag{2.121}$$

where the index symmetry of $\delta\pi^{ab}$ was used to replace $2D_a\beta_b$ by $\mathcal{L}_{\boldsymbol{\beta}}\gamma_{ab} = D_a\beta_b + D_b\beta_a$. Equation (2.105) implies that the dynamical equations (2.104) and (2.121) are equivalent.

Now turn to the dynamical equation for π^{ab}, finding the change in H_G arising from a change in γ_{ab}. Note first that $\delta(R\gamma^{1/2})$ has the same form as $\delta({}^4R\sqrt{|g|})$ given by Eqs. (2.93) and (2.94), with $\sqrt{|g|}$ replaced by $\delta\sqrt{\gamma}$ and $h_{\alpha\beta} := \delta g_{\alpha\beta}$ replaced by $\delta\gamma_{ab}$:

$$\delta(R\gamma^{1/2}) = \left(D^a D^b - \gamma^{ab} D^2 - R^{ab} + \frac{1}{2}\gamma^{ab}R\right)\delta\gamma_{ab}. \tag{2.122}$$

Then, from Eq. (2.109), we have

$$\begin{aligned}\delta\mathcal{H}_G = &-\frac{1}{2}\gamma^{-1/2}\left(\pi^{cd}\pi_{cd} - \frac{1}{2}\pi^2\right)\gamma^{ab}\,\delta\gamma_{ab} + 2\gamma^{-1/2}\left(\pi^{ac}\pi^b{}_c - \pi\pi^{ab}\right)\delta\,\gamma_{ab}\\ &+ \gamma^{1/2}\left(-D^a D^b + \gamma^{ab} D^2 + R^{ab} - \frac{1}{2}\gamma^{ab}R\right)\delta\gamma_{ab},\end{aligned} \tag{2.123}$$

and the change in $\int \mathcal{H}_G d^3x$ corresponding to $\delta\gamma_{ab}$ is given by

$$\begin{aligned}\delta\int \alpha\mathcal{H}_G d^3x = \int d^3x\,\delta\gamma_{ab}\Big\{&\gamma^{-1/2}\left[2\pi^{ac}\pi^b{}_c - \pi\pi^{ab} + \gamma^{ab}\left(-\frac{1}{2}\pi^{cd}\pi_{cd} + \frac{1}{4}\pi^2\right)\right]\\ &+ \gamma^{1/2}\left(-D^a D^b + \gamma^{ab}D^2 + R^{ab} - \frac{1}{2}\gamma^{ab}R\right)\Big\}\alpha.\end{aligned} \tag{2.124}$$

To find $\delta \int \mathcal{H}_{Ga}\beta^a d^3x$, we use the last form in Eq. (2.118) to write

$$\delta \int \mathcal{H}_{Ga}\beta^a d^3x = -\delta \int \gamma_{ab}\mathcal{L}_{\boldsymbol{\beta}}\pi^{ab} d^3x = -\int \delta\gamma_{ab}\pi^{ab} d^3x.$$

Finally, Eqs. (2.124) and (2.127) give the vacuum evolution equation $G^{ab} = 0$, namely

$$\begin{aligned}\dot{\pi}^{ab} = \dot{\pi}^{ab} - {}^4G^{ab} \equiv \gamma^{-1/2}\alpha \left[2\pi^{ac}\pi^b{}_c - \pi\pi^{ab} + \gamma^{ab}\left(-\frac{1}{2}\pi^{cd}\pi_{cd} + \frac{1}{4}\pi^2\right)\right] \\ + \gamma^{1/2}\left(R^{ab} - \frac{1}{2}\gamma^{ab} R - D^a D^b + \gamma^{ab} D^2\right)\alpha \\ + \mathcal{L}_{\boldsymbol{\beta}}\pi^{ab}. \qquad (2.125)\end{aligned}$$

Equation (2.125) is the equation ${}^4G_{ab} = 0$. To obtain the dynamical equation (2.23) for K_{ab} in its vacuum form ${}^4R_{ab} = 0$, one replaces π^{ab} by its expression in terms of K_{ab} and uses the linear combination ${}^4R_{ab} = {}^4G_{ab} + \frac{1}{2}\gamma_{ab}{}^4R$.

We next turn to the contribution of the fluid to the field equation. This can be quickly obtained from the action in the form

$$I = \frac{1}{16\pi}\int d^4x \left[\pi^{ab}\dot{\gamma}_{ab} - \alpha\mathcal{H}_G - \beta^a\mathcal{H}_{G\,a} - 16\pi\alpha\sqrt{\gamma}\epsilon\right]. \qquad (2.126)$$

We have already seen that the variation (2.69) of the matter action $I_M = -\int \epsilon\sqrt{|g|}$ with respect to the metric is proportional to the densitized stress-energy tensor, $T^{\alpha\beta}\sqrt{|g|}$. Its variations with respect to lapse, shift, and 3-metric comprise the 3+1 decomposition of the stress tensor

$$\frac{\delta I_M}{\delta\alpha} = -T_{\alpha\beta}n^\alpha n^\beta\sqrt{\gamma} = -\rho_E\sqrt{\gamma}, \qquad (2.127a)$$

$$\frac{\delta I_M}{\delta\beta^a} = -T_{\alpha\beta}\gamma^\alpha{}_a n^\beta\sqrt{\gamma} = j_a\sqrt{\gamma}, \qquad (2.127b)$$

$$\frac{\delta I_M}{\delta\gamma_{ab}} = \frac{1}{2}T^{\alpha\beta}\gamma_a{}^\alpha\gamma_b{}^\beta\sqrt{|g|} = \frac{1}{2}S^{ab}\alpha\sqrt{\gamma}. \qquad (2.127c)$$

The projections of the field equation then take the form

$$R - \frac{1}{\gamma}\left(\pi^{ab}\pi_{ab} - \frac{1}{2}\pi^2\right) = 16\pi\rho_E, \qquad (2.128a)$$

$$D_b\pi^{ab} = -8\pi j^a\sqrt{\gamma}, \qquad (2.128b)$$

$$G^{ab} = 8\pi S^{ab}, \qquad (2.128c)$$

with G^{ab} given by Eq. (2.125).

Hamiltonian form of the fluid action I_M. In Eq. (2.126) for the action, only the gravitational part of the action has been written in Hamiltonian form: The action is formally a function of $\pi^{ab}, \gamma_{ab}, \alpha, \beta^a$ and the diffeo χ that determines the configuration of the fluid in spacetime. Because we have already obtained the 3+1 form of the field equation and of the equation of motion for the fluid, the remainder of the text does not require the Hamiltonian form for the fluid action, but we give it here for completeness. A number of authors [677, 594, 483, 379, 316, 350, 151, 100] have obtained Hamiltonian formalisms for relativistic fluids, and several use 3-dimensional Lagrangian coordinates to describe the position of the fluid.

This is equivalent to using the gauge freedom in the diffeo χ to map the fiducial fluid at time t to the final fluid configuration at the same time t. For the remainder of this section, we adopt this gauge, and we also choose a static fiducial fluid, implying

$$\chi^t(t, \{x^i\}) = (t, \{\chi^i(t,x)\}), \qquad u_0^\alpha = u_0^t t^\alpha. \tag{2.129}$$

The baryon current $\mathrm{j}^\alpha := \rho u^\alpha \sqrt{|g|}$ of the fluid at a point P is given in terms of the baryon current $\mathrm{j}_0^\alpha := \rho_0 u_0^\alpha \sqrt{-g_0}$ of the fiducial fluid at a point $P_0 = \chi^{-1}(P)$ by $\mathrm{j}^\alpha = \chi j_0^\alpha$. From Eq. (2.76), we have

$$\mathrm{j}^\mu(P) = J^{-1}\, \partial_\nu \chi^\mu\, \mathrm{j}_0^\nu(P_0) = J^{-1}\, \dot{\chi}^\mu\, \mathrm{j}_0^t(P_0), \tag{2.130}$$

where $\dot{\chi}^\mu = \partial_t \chi^\mu$. Using $\dot{\chi}^t = 1$, we have

$$\mathrm{j}^t(P) = J^{-1}\, \mathrm{j}_0^t(P_0), \tag{2.131a}$$

$$\mathrm{j}^a(P) = J^{-1}\, \dot{\chi}^a\, \mathrm{j}_0^t(P_0), \text{ or} \tag{2.131b}$$

$$\mathrm{j}^a = \dot{\chi}^a\, \mathrm{j}^t. \tag{2.131c}$$

In the last equality, and in the equations below, explicit reference to the points P and P_0 is omitted for simplicity of notation.

The momentum conjugate to χ is then given at a point P by

$$\Pi_a = \frac{\partial\left(-\epsilon\sqrt{|g|}\right)}{\partial \dot{\chi}^a}, \tag{2.132}$$

where $\epsilon = \epsilon(\rho)$, $\rho = (g_{\alpha\beta}\mathrm{j}^\alpha \mathrm{j}^\beta / g)^{1/2}$, and $\mathrm{j}^a(\dot{\chi}, \chi, g_{ab})$ is given by Eq. (2.131b). Thus

$$\Pi_a = -\frac{\partial\epsilon}{\partial\rho}\frac{\partial\rho}{\partial \mathrm{j}^b}\frac{\partial \mathrm{j}^b}{\partial\dot{\chi}^a}\sqrt{|g|}, \tag{2.133}$$

with

$$\frac{\partial\epsilon}{\partial\rho} = \frac{\epsilon + p}{\rho} = h, \qquad \partial\rho/\partial \mathrm{j}^a = -u_a |g|^{-1/2}. \tag{2.134}$$

Now for our choice of gauge, $\partial_i \chi^t = 0, \partial_t \chi^t = 1$, implying that the Jacobian $J \equiv \det \|\partial_\mu \chi^\nu\|$ has the form $\det \|\partial_j \chi^i\|$, independent of $\dot{\chi}^a$. Then, writing the 3+1 decomposition of u^α in the form

$$u^\alpha = \alpha u^t (n^\alpha + \mathrm{v}^\alpha), \quad \text{with } \mathrm{v}^\alpha n_\alpha = 0, \tag{2.135}$$

we have $\partial\rho/\partial \mathrm{j}^a = -\dfrac{u^t}{\alpha\sqrt{|g|}}\mathrm{v}_a$; from the relation $\partial \mathrm{j}^b/\partial(\dot\chi^a) = \delta^b_a \mathrm{j}^t$, we obtain

$$\Pi_a = -h\left(-\frac{u^t}{\alpha\sqrt{|g|}}\mathrm{v}_a\right)(\delta^b_a \mathrm{j}^t)\sqrt{|g|}$$

$$= h\alpha^{-1}u^t\mathrm{v}_a\mathrm{j}^t, \qquad \text{or} \tag{2.136a}$$

$$\Pi_a = (\epsilon+p)(u^t)^2\mathrm{v}_a\sqrt{\gamma}. \tag{2.136b}$$

Finally, Eq. (2.131c) implies

$$\dot\chi^a = \mathrm{j}^a/\mathrm{j}^t = \mathrm{v}^a, \tag{2.137}$$

and the matter has Hamiltonian density $\mathcal{H}_M$ given by

$$\mathcal{H}_M = \Pi_a\dot\chi^a + \epsilon\sqrt{|g|} = \left[\epsilon\alpha + (\epsilon+p)(u^t\mathrm{v})^2\right]\sqrt{\gamma}, \tag{2.138}$$

with $\mathrm{v}^2 = \mathrm{v}_a\mathrm{v}^a$.

To write the Hamiltonian as a function of Π_a and χ, one must invert Eq. (2.136b) to find $\dot\chi^a$ as a function of Π_a. With $\dot\chi := \sqrt{\gamma_{bc}\dot\chi^b\dot\chi^c} = \sqrt{\gamma_{bc}mv^b mv^c}$, Eq. (2.136b) has the alternative form

$$\Pi_a = (\epsilon+p)\alpha^2\frac{\dot\chi^2}{1-\dot\chi^2}\dot\chi_a\sqrt{\gamma}, \tag{2.139}$$

implying

$$\Pi = (\epsilon+p)\alpha^2\frac{\dot\chi^2}{\sqrt{1-\dot\chi^2}}\sqrt{\gamma}, \tag{2.140}$$

where $\Pi := \sqrt{\Pi_a\Pi^a}$. Once Eq. (2.140) is solved for $\dot\chi(\Pi)$, Eq. (2.139) immediately gives $\dot\chi^a$ as a function of Π_a. In Eq. (2.140), however, ϵ and p are themselves functions of $\dot\chi$ via the relation $\rho = (g_{\alpha\beta}\mathrm{j}^\alpha\mathrm{j}^\beta/g)^{1/2}$, and Eq. (2.140) has no analytic inverse. Thus the functional dependence of $\dot\chi^a$ on Π_a is only given implicitly by Eq. (2.139), and the Hamiltonian density $\mathcal{H}$ is similarly implicit.

2.3 Gauge freedom and trivial displacements

Two kinds of gauge freedom. To obtain an unconstrained action, we introduced a diffeo χ that specifies fluid trajectories. In the action, χ plays the role of a potential in terms of which one can express the physical quantities ϵ, p, ρ, s and u^α. The price one pays for introducing such a potential is an additional gauge freedom [640, 119, 596]: The same physical configuration corresponds to more than one diffeo χ.

Diffeomorphism equivalence. Before we consider this new kind of gauge freedom, it is helpful to recall the usual gauge freedom one has in describing tensor fields in general relativity, the freedom to change every field by a diffeo, or, in its passive form, the freedom to make a coordinate transformation. As described in Appendix A.1, a diffeo ψ drags a tensor $T^\alpha{}_\beta$ at a point P to a tensor $\psi T^\alpha{}_\beta$ at $\psi(P)$ with components $\psi T^\mu{}_\nu|_{\psi(P)} = \frac{\partial \psi^{-1}{}^\mu}{\partial x^\sigma}\frac{\partial \psi^\tau}{\partial x^\nu}T^\sigma{}_\tau(P)$. In the passive version, one fixes the tensors and changes the coordinate system by the inverse transformation (see Eqs. [A.19] and [A.20]). A set of tensor fields, say $\mathcal{Q} = \{g_{\alpha\beta}, u^\alpha, \rho, s\}$, is physically equivalent to the set $\psi\mathcal{Q} = \{\psi g_{\alpha\beta}, \psi u^\alpha, \psi\rho, \psi s\}$, and any covariant equations constructed solely from this set of tensor fields (and their derivatives) are satisfied by the transformed fields.

This finite diffeomorphism invariance of the physics implies a corresponding gauge invariance of perturbed fields. Let $\mathcal{Q}(\mathsf{s}) := [g_{\alpha\beta}(\mathsf{s}), u^\alpha(\mathsf{s}), \rho(\mathsf{s}), s(\mathsf{s})]$ be a family of tensor fields, and let ψ_s be a family of diffeos with ψ_0 the identity. The set of perturbed fields $\delta\mathcal{Q} = \frac{d}{ds}\mathcal{Q}(\mathsf{s})|_{\mathsf{s}=0}$ is physically equivalent to the set $\widetilde{\delta}\mathcal{Q} = \frac{d}{ds}[\psi_\mathsf{s}\mathcal{Q}(\mathsf{s})] = \delta\mathcal{Q} - \mathcal{L}_\zeta\mathcal{Q}$, with ζ^α the vector field tangent to the trajectories of ψ_s. In particular, for any ζ^α, the set of perturbed fields $\{h_{\alpha\beta}, \delta u^\alpha, \delta\rho, \delta s\}$ is equivalent to the set $\{h_{\alpha\beta} - (\nabla_\alpha\zeta_\beta + \nabla_\beta\zeta_\alpha), \delta u^\alpha - \mathcal{L}_\zeta u^\alpha, \delta\rho - \boldsymbol{\zeta}\cdot\nabla\rho, \delta s - \boldsymbol{\zeta}\cdot\nabla s\}$.

Trivial displacements. The second kind of gauge invariance is associated with our description of a fluid configuration that expresses the fluid variables in terms of a diffeo χ of a fiducial fluid configuration. Here χ plays a different role. It does not act on the metric, and only s and the tangent vector to the fluid trajectories are dragged along by χ. The values of the fluid variables ρ and u^α are not given by $\chi\rho_0$ and χu_0^α; they have values (2.77) and (2.75) given by conservation of baryons and by normalizing the dragged-along tangent vector χw^α. Two diffeos χ and $\tilde{\chi}$ are equivalent only if they yield the *same* values of the fluid variables. (If χ and $\tilde{\chi}$ were to yield diffeomorphically related fluid variables, the fluid configurations would in general be physically different, because the metric is independent of χ.)

A corresponding infinitesimal symmetry is then associated with a family $\chi_\mathsf{s}\circ\chi$ of diffeos, all of which yield the same values of the fluid variables. A family of diffeos χ_s of a fluid configuration to itself leaves the fluid unchanged if the Eulerian changes in the fluid variables vanish:

$$\delta\mathcal{Q} := \frac{d}{ds}\mathcal{Q}\bigg|_{\mathsf{s}=0} = 0. \tag{2.141}$$

The Lagrangian displacement vector is the vector field ξ^α tangent to the trajectories of the family χ_s, and the Lagrangian changes in the fluid variables are expressed in terms of ξ^α by $\Delta s = 0$ and by Eqs. (2.37) and (2.35). The corresponding Eulerian changes are then

$$\delta s = -\xi^\alpha\nabla_\alpha s, \quad \delta\rho = -q^\alpha{}_\beta\nabla_\alpha(\rho\xi^\beta) - \rho\xi^\alpha u_\alpha\nabla_\beta u^\beta - \frac{1}{2}\rho q^{\alpha\beta}h_{\alpha\beta},$$
$$\delta u^\alpha = q^\alpha{}_\beta\mathcal{L}_{\boldsymbol{u}}\xi^\beta + \frac{1}{2}u^\alpha u^\beta u^\gamma h_{\beta\gamma}. \tag{2.142}$$

Any displacement of the form Gu^α satisfies these equations (for $h_{\alpha\beta} = 0$). In particular, conservation of entropy (1.52) in the flow of a perfect fluid implies $\delta s = \delta\rho = \delta u^\alpha = 0$. Equations (2.142) then involve only the projection $\xi_\perp^\alpha = q_\beta^\alpha\xi^\beta$

of ξ^α orthogonal to u^α:

$$\delta s = -\xi_\perp^\alpha \nabla_\alpha s, \quad \delta\rho = -q^\alpha{}_\beta \nabla_\alpha(\rho\xi_\perp^\beta) - \frac{1}{2}\rho q^{\alpha\beta} h_{\alpha\beta},$$

$$\delta u^\alpha = q^\alpha{}_\beta \mathcal{L}_{\boldsymbol{u}} \xi_\perp^\beta + \frac{1}{2} u^\alpha u^\beta u^\gamma h_{\beta\gamma}. \tag{2.143}$$

Our action assumes that the local state of the fluid is determined by s and ρ, via an equation of state $\epsilon = \epsilon(\rho, s)$, $p = p(\rho, s)$, so if $\delta s = 0$ and $\delta\rho = 0$, then the equation of state implies $\delta\epsilon$ and $\delta p = 0$.

Because all displacements of the form Gu^α are trivial, a general trivial displacement has the form $\eta^\alpha + Gu^\alpha$, where $\eta^\alpha = \eta_\perp^\alpha$ satisfies

$$\eta^\alpha \nabla_\alpha s = 0, \tag{2.144a}$$

$$q^\alpha{}_\beta \nabla_\alpha(\rho\eta^\beta) = 0, \tag{2.144b}$$

$$q^\alpha{}_\beta \mathcal{L}_{\boldsymbol{u}} \eta^\beta = 0. \tag{2.144c}$$

The meaning of the equations is straightforward. Equation (2.144a) implies that a trivial displacement η^α takes each fluid element to a fluid element with the same entropy per baryon; Eq. (2.144b) means that η^α leaves unchanged the volume orthogonal to the fluid trajectories; and Eq. (2.144c), in the form $q^\alpha{}_\beta \mathcal{L}_{\boldsymbol{\eta}} u^\beta = 0$, means that η^α takes fluid trajectories to fluid trajectories: the tangent u^α to a fluid trajectory changes only by a multiple of itself.

In characterizing the general solution to these equations it is helpful to introduce the antisymmetric tensor

$$\epsilon_{\alpha\beta\gamma} = \epsilon_{\alpha\beta\gamma\delta} u^\delta, \tag{2.145}$$

which can be regarded as the volume form in the subspace orthogonal to u^α.

Proposition. *Let $\mathcal{T}$ be the subspace spanned by vector fields of the form*

$$\eta^\alpha = \frac{1}{\rho}\epsilon^{\alpha\beta\gamma} \nabla_\beta F_1 \nabla_\gamma F_2, \tag{2.146}$$

where F_1 and F_2 are scalars that are constant along the fluid trajectories ($u^\alpha \nabla_\alpha F_1 = u^\alpha \nabla_\alpha F_2 = 0$) and where $\nabla_\alpha F_2 \propto \nabla_\alpha s$ when $\nabla_\alpha s \neq 0$. The space of trivial displacements is spanned by $\mathcal{T}$ together with displacements of the form

$$Gu^\alpha, \tag{2.147}$$

where G is an arbitrary scalar.

Proof. We begin by showing that any displacement η^α of the form (2.146) is trivial, that it satisfies Eqs. (2.144a)–(2.144c). Equation (2.144a) is immediate from the antisymmetry of $\epsilon^{\alpha\beta\gamma}$. To verify Eq. (2.144b), use the identity $\nabla_{[\alpha}\nabla_{\beta]} f = 0$ to write

$$q^\alpha{}_\beta \nabla_\alpha(\epsilon^{\beta\gamma\delta\epsilon} u_\epsilon \nabla_\gamma F_1 \nabla_\delta F_2) = q^\alpha{}_\beta \; \epsilon^{\beta\gamma\delta\epsilon}(\nabla_\alpha u_\epsilon) \nabla_\gamma F_1 \nabla_\delta F_2. \tag{2.148}$$

Then, because $u^\beta q^\alpha{}_\beta = 0$, $u^\gamma \nabla_\gamma F_1 = u^\delta \nabla_\delta F_2 = 0$, and $u^\epsilon \nabla_\alpha u_\epsilon = 0$, all four indices of $\epsilon^{\beta\gamma\delta\epsilon}$ are contracted with indices in the subspace orthogonal to u^α; since there are only three linearly independent vectors in that 3-dimensional subspace, $q^\alpha{}_\beta \epsilon^{\beta\gamma\delta\epsilon}(\nabla_\alpha u_\epsilon)\nabla_\gamma F_1 \nabla_\delta F_2 = 0$. To check Eq. (2.144c), note first that conservation of baryons, $\nabla_\alpha(\rho u^\alpha) = 0$, is equivalent to each of the relations

$$\mathcal{L}_{\boldsymbol{u}}(\rho\epsilon_{\alpha\beta\gamma}) = 0, \qquad \mathcal{L}_{\boldsymbol{u}}\left(\frac{1}{\rho}\epsilon^{\alpha\beta\gamma}\right) = 0. \tag{2.149}$$

Equation (2.144c) follows from the second of these and from $\mathcal{L}_{\boldsymbol{u}}\nabla_\alpha F_1 = \nabla_\alpha \mathcal{L}_{\boldsymbol{u}} F_1 = 0$, $\mathcal{L}_{\boldsymbol{u}}\nabla_\alpha F_2 = \nabla_\alpha \mathcal{L}_{\boldsymbol{u}} F_2 = 0$.

It remains to show that any trivial displacement η^α orthogonal to u^α is in $\mathcal{T}$. There is a 1-1 correspondence between tensors that are orthogonal to u^α and are Lie derived by u^α up to a multiple of u^α (for example, $q^\alpha{}_\beta \mathcal{L}_{\boldsymbol{u}} v^\beta = 0$), and tensors on the 3-dimensional manifold $\mathcal{M}$ of fluid trajectories [118]. Equations (2.144b) and (2.144c) have the meaning that $\rho\eta^\alpha$ is a divergence-free vector field on $\mathcal{M}$ and hence that it is a curl, having the form

$$\eta^\alpha = \frac{1}{\rho}\epsilon^{\alpha\beta\gamma}\nabla_\beta A_\gamma, \tag{2.150}$$

for some vector A_α with $A_\alpha u^\alpha = 0$, $q_\alpha{}^\beta \mathcal{L}_{\boldsymbol{u}} A_\beta = 0$.[7] Finally, any A_α with $A_\alpha u^\alpha = 0$ is a sum of terms of the form $F_1 \nabla_\alpha F_2$, with $u^\alpha \nabla_\alpha F_2 = 0$, and $q_\alpha{}^\beta \mathcal{L}_{\boldsymbol{u}} A_\beta = 0$ implies $u^\alpha \nabla_\alpha F_1 = 0$.

When $\nabla_\alpha s = 0$, there are no further restrictions on F_1 and F_2. When $\nabla_\alpha s \neq 0$, the vector $\rho\eta^\alpha$ again must have the form (2.150) and must be orthogonal to $\nabla_\alpha s$. Again it can be written as a sum of terms $\epsilon^{\alpha\beta\gamma}\nabla_\beta F_1 \nabla_\gamma F_2$, but now either ∇F_1 or ∇F_2 must be parallel to ∇s. In each case $\eta^\alpha \in \mathcal{T}$, as claimed.

2.4 Symmetry under trivial displacements implies conservation of circulation

We show in this section that the symmetry of the perfect-fluid Lagrangian associated with the class of trivial displacements is related by Noether's theorem to conservation of circulation in surfaces of constant entropy per baryon. A version in magnetohydrodynamics of isentropic, perfectly conducting fluids was discovered by Calkin [111]; the general relation for perfect fluids was found independently by Friedman and Schutz [241], and several subsequent authors have rediscovered that relation (see Salmon [585] for references). The general-relativistic version given here is based on [231].

[7] More formally, the identity

$$q^\alpha{}_\beta \nabla_\alpha(\rho\eta^\beta) = \frac{1}{2}\epsilon^{\alpha\beta\gamma}\nabla_\alpha(\rho\epsilon_{\beta\gamma\delta}\eta^\delta), \tag{2.151}$$

together with Eq. (2.149) implies that the 2-form $B_{\beta\gamma} := \rho\epsilon_{\beta\gamma\delta}\eta^\delta$, regarded as a tensor on $\mathcal{M}$, is closed: $dB = 0$. Then $B = dA$: $\rho\epsilon_{\beta\gamma\delta}\eta^\delta = 2\nabla_{[\beta}A_{\gamma]}$, for some vector A_α with $A_\alpha u^\alpha = 0$ and $q_\alpha{}^\beta \mathcal{L}_{\boldsymbol{u}} A_\beta = 0$. Contracting each side of the equation with $\epsilon^{\alpha\beta\gamma}$, we have $\eta^\alpha = \frac{1}{\rho}\epsilon^{\alpha\beta\gamma}\nabla_\beta A_\gamma$.

The Noether relation between symmetries and conserved quantities has the following general form for an unconstrained action (see, for example, descriptions by Trautman [700] and Sorkin [644, 641]). Let $\mathscr{L}$ be a Lagrangian density that depends on a set of fields $\{\phi^I\}$ and on their derivatives. An infinitesimal symmetry of the Lagrangian is a perturbation $\hat{\delta}\Phi^I$ for which the corresponding Eulerian change $\hat{\delta}\mathscr{L}$ in the Lagrangian is a divergence. That is, $\hat{\delta}\phi^I$ is an infinitesimal symmetry if

$$\hat{\delta}\mathscr{L} = \nabla_\alpha \mathcal{A}^\alpha, \tag{2.152}$$

where $\mathcal{A}^\alpha(x)$ is a vector density constructed from $\phi^I(x)$, $\hat{\delta}\Phi^I(x)$ and their derivatives at x.

The fact that $I = \int \mathscr{L} d^4x$ is an action for the field equations $E_I(\Phi) = 0$ means that, for *any* perturbation $\delta\Phi^I$, $\delta\mathscr{L} = E_I \delta\Phi^I + \nabla_\alpha \mathbf{\Theta}^\alpha(\Phi, \delta\Phi)$. Then, when the background fields ϕ^I satisfy the field equations, a symmetry $\hat{\delta}\Phi^I$ corresponds to a conserved current

$$\nabla_\alpha J^\alpha = \nabla_a \mathcal{J}^\alpha = 0, \qquad \text{with } J^\alpha\sqrt{|g|} = \mathcal{J}^\alpha := \mathbf{\Theta}^\alpha - \mathcal{A}^\alpha. \tag{2.153}$$

Trivial displacements are infinitesimal symmetries of the Einstein-Euler Lagrangian $\mathscr{L}$ for which $\delta\mathscr{L} = 0$. Equation (2.152) then implies that $\mathcal{A}^\alpha = 0$, and Eq. (2.98) for $\mathbf{\Theta}^\alpha$, with $h_{\alpha\beta} = 0$ and $\xi^\alpha = \eta^\alpha$, gives a conserved current of the form

$$J^\alpha = \mathbf{\Theta}^\alpha/\sqrt{|g|} = (\epsilon + p)q^{\alpha\beta}\eta_\beta. \tag{2.154}$$

The space of trivials is spanned by displacements η^α of the form (2.146) and trivials Gu^α. The latter have vanishing J^α, but for the former

$$J^\alpha = hu_\beta \epsilon^{\alpha\beta\gamma\delta}\nabla_\gamma F_1 \nabla_\delta F_2. \tag{2.155}$$

The associated conserved quantity is, as usual, obtained by integrating $\nabla_\alpha J^\alpha$ over a 4-dimensional region ${}^4V$ between two spacelike surfaces, which we can take to be two $t =$ constant surfaces, Σ_{t_1} and Σ_{t_2}, for some time coordinate t. We assume that Σ_{t_1} extends beyond the fluid, so that there is no timelike boundary on which J^α is nonzero. Then

$$0 = \int_{^4V} \nabla_\alpha J^\alpha = \int_{\Sigma_{t_2}} J^\alpha dS_\alpha - \int_{\Sigma_{t_1}} J^\alpha dS_\alpha, \tag{2.156}$$

implying that the integral $\int_{\Sigma_t} J^\alpha dS_\alpha$ is conserved.

The integral

$$\int_{\Sigma_t} J^\alpha dS_\alpha = \int_{\Sigma_t} hu_\beta \epsilon^{\alpha\beta\gamma\delta}\nabla_\gamma F_1 \nabla_\delta F_2 dS_\alpha = \int_{\Sigma_t} hu_\alpha \nabla_\beta F_1 \nabla_\gamma F_2 dS^{\alpha\beta\gamma} \tag{2.157}$$

sums the circulation $\int_{c_t} hu_\alpha dl^\alpha$ over curves c of constant F_1 and F_2, because the element of length dl^α along c is proportional to $dS^{\alpha\beta\gamma}\nabla_\beta F_1 \nabla_\gamma F_2$. In particular, if we choose coordinates r, θ, ϕ on Σ_1 and extend them through the fluid by requiring $u^\alpha\nabla_\alpha r = u^\alpha\nabla_\alpha\theta = u^\alpha\nabla_\alpha\phi = 0$, then with $F_1 = r$, $F_2 = \theta$, and with c_t the curve $\theta = \theta_0$, $r = r_0$ on Σ_{t_1}, the integral takes the form $\int \left(\oint_{c_t} hu_\phi d\phi\right) dr d\theta$. An integral over a single fluid ring c_t is picked out by choosing as F_1 and F_2 the step functions $F_1 = \Theta(r - r_0)$, $F_2 = \Theta(\theta - \theta_0)$. The conserved quantity is then

$$\int_{\Sigma_t} J^\alpha dS_\alpha = \int \left(\oint hu_\phi d\phi \right) \delta(r - r_0)\delta(\theta - \theta_0) dr d\theta = \int_{c_t} hu_\phi d\phi$$
$$= \int_{c_t} hu_\alpha dl^\alpha \equiv C[c_t], \tag{2.158}$$

the relativistic circulation along the curve c_t, as in Eq. (1.53). The requirement that F_1 and F_2 be Lie derived by u^α means that the curves c_t for different t are related by the fluid flow. If t is chosen so that $u^\alpha \nabla_\alpha t = 1$, then t measures proper time, and $c_{t+\tau}$ is obtained by dragging c_t along the fluid flow by a proper time τ, as in Eqs. (1.52) and (1.53). For general t, the curve $c_{t'}$ is the intersection of $\Sigma_{t'}$ with the history swept out by dragging c_t along the fluid flow.

When $\nabla_a s \neq 0$, the surfaces of constant F_2 must be chosen to be surfaces of constant specific entropy s, and the conservation law is restricted to conservation of circulation along curves c that lie in surfaces of constant s. The corresponding differential conservation law is the relativistic generalization of Ertel's theorem [211], conservation of the relativistic vorticity $\omega_{\alpha\beta} = \nabla_\alpha(hu_\beta) - \nabla_\beta(hu_\alpha)$ in surfaces of constant s. We can derive it directly from Euler's equation in the form (1.48) using Eq. (1.15), $\nabla_a h = T\nabla_a s + \nabla_a p/p$. The curl (exterior derivative) of (1.48) is then given by

$$\mathcal{L}_{\boldsymbol{u}} \omega_{\alpha\beta} = 2\nabla_{[\alpha} T \nabla_{\beta]} s, \tag{2.159}$$

immediately implying

$$\mathcal{L}_{\boldsymbol{u}} \omega_{[\alpha\beta} \nabla_{\gamma]} s = 0. \tag{2.160}$$

The Newtonian form of vorticity conservation as a vector equation,

$$(\partial_t + \mathcal{L}_{\boldsymbol{v}}) \left(\frac{1}{\rho} \boldsymbol{\nabla} \times \boldsymbol{v} \right) = 0, \tag{2.161}$$

also has a relativistic counterpart. Using Eq. (2.149), conservation of vorticity when $\nabla_\alpha s = 0$ can be written in terms of the vector $\omega^\alpha = \frac{1}{\rho}\epsilon^{\alpha\beta\gamma}\nabla_\beta(hu_\gamma)$ as

$$\mathcal{L}_{\boldsymbol{u}} \omega^\alpha = 0. \tag{2.162}$$

When $\nabla_\alpha s \neq 0$, Eq. (2.160) is equivalent to the scalar equation

$$\mathcal{L}_{\boldsymbol{u}} \widetilde{\omega} = 0, \tag{2.163}$$

where $\widetilde{\omega} = \frac{1}{\rho}\epsilon^{\alpha\beta\gamma}\nabla_\alpha(hu_\beta)\nabla_\gamma s$, whose Newtonian limit is

$$(\partial_t + \mathcal{L}_{\boldsymbol{v}}) \left(\frac{1}{\rho} \boldsymbol{\nabla} \times \boldsymbol{v} \cdot \boldsymbol{\nabla} s \right) = 0. \tag{2.164}$$

3

Asymptotics, virial identities, and nonaxisymmetric equilibria

3.1 ADM mass and angular momentum

Mass and angular momenta of an asymptotically flat spacetime are associated with symmetry vectors of a flat metric at spatial infinity and can be written as surface integrals that involve the initial data on a spacelike hypersurface, the spatial metric γ_{ab} and its conjugate momentum π^{ab} [36]. When expressed in this way, the quantities are called the ADM (Arnowitt-Deser-Misner) mass and angular momentum. We begin by presenting the surface integrals and giving asymptotic conditions for which they make sense. We then show how they arise in a Hamiltonian framework from the definition of mass and angular momentum as generators of time translations and rotations. Here we roughly follow work by Regge and Teitelboim [560, 295], but with a covariant formalism similar to that used in Noether-theorem derivations by Chruściel, Jezierski, and Kijowski [139, 140] and by Sorkin [641].

A spatial metric that is asymptotically flat at spatial infinity has the form

$$\gamma_{ab} = \eta_{ab} + h_{ab}, \tag{3.1}$$

where, as in Section 2.1, η_{ab} is the flat Euclidean 3-metric, the spatial part of $\eta_{\alpha\beta}$, having the form $dx_1^2 + dx_2^2 + dx_3^2 = dr^2 + r^2(d\theta^2 + \sin^2\theta d\phi^2)$, in natural coordinates for η_{ab}. We denote by $\boldsymbol{\nabla}$ or ∇_a the covariant derivative with respect to the flat metric η_{ab}, and we assume that the metric and conjugate momentum π^{ab} have asymptotic behavior

$$\begin{aligned} h_{ab} &= O(r^{-1}), \quad \nabla_c h_{ab} = O(r^{-2}), \quad \nabla_c\nabla_d h_{ab} = o(r^{-2}) \\ \pi^{ab} &= O(r^{-2}), \quad \nabla_c \pi^{ab} = o(r^{-2}). \end{aligned}$$

The ADM mass is given by

$$M_{\rm ADM} = \frac{1}{16\pi}\int_{S_\infty} (\eta^{ac}\eta^{bd} - \eta^{ab}\eta^{cd})\nabla_b h_{cd} dS_a, \tag{3.2}$$

and the ADM angular momentum associated with a rotational Killing vector ϕ^a of the flat metric η_{ab} has the form

$$J_{\rm ADM} = -\frac{1}{8\pi}\int_{S_\infty} \pi^a{}_b \phi^b d\sigma_a, \tag{3.3}$$

where $d\sigma_a = \gamma^{-1/2} dS_a = \nabla_a r d\theta d\phi$.

Here, as in Chapter 1, $\int_{S_\infty} = \lim_{r\to\infty} \int_{S_r}$, with S_r a sphere of constant coordinate r.

For a rotating star in coordinates for which the metric has the form (1.59), the metric's asymptotic behavior is given (to the order written in each component) by

$$ds^2 = -\left(1 - \frac{2M}{r}\right) dt^2 + \frac{2J}{r} \sin^2\theta dt d\phi + \left(1 + \frac{2M}{r}\right) (dr^2 + r^2 d\theta^2 + r^2 \sin^2\theta d\phi^2). \tag{3.4}$$

We can quickly check that Eqs. (3.2) and (3.3) correctly identify M and J for stationary axisymmetric stars. Using the fact that, to lowest nonvanishing order in $1/r$, we can raise indices with the flat metric η^{ab}, we have

$$\begin{aligned} M_{\text{ADM}} &= \frac{1}{16\pi} \lim_{r\to\infty} \int (\nabla_b h^{ab} - \nabla^a h^b_b) \nabla_a r r^2 d\Omega \\ &= \frac{1}{16\pi} \lim_{r\to\infty} \int \left(\nabla_b \frac{2M}{r} \eta^{ab} - \nabla^a \frac{6M}{r}\right) \nabla_a r r^2 d\Omega \\ &= M. \end{aligned} \tag{3.5}$$

To compute J, use $\beta^a = -\omega\phi^a$, $D_{(a}\phi_{b)} = 0$, to write

$$\begin{aligned} -K_{ab} &= D_{(a}\beta_{b)}[1 + O(1/r)] = D_{(a}\omega\phi_{b)}[1 + O(1/r)] = \nabla\left(a \frac{2J}{r^3}\phi_b\right)[1 + O(1/r)] \\ &= -\frac{6J}{r^4}\nabla_{(a} r\, \phi_{b)} + O(r^{-5}). \end{aligned} \tag{3.6}$$

Then

$$\begin{aligned} J_{\text{ADM}} &= \frac{1}{8\pi} \lim_{r\to\infty} \int \frac{6J}{r^4} \nabla_{(a} r\, \phi_{b)} \phi^b r^2 \nabla^a r d\Omega = \frac{1}{8\pi} \int 3J \sin^2\theta d\Omega \\ &= J. \end{aligned} \tag{3.7}$$

In the Hamiltonian framework, the mass and angular momentum can be defined as functions on phase space that generate time translations and rotations; in a Lagrangian framework, they arise as integrals of conserved currents associated by Noether's theorem with time translational and rotational symmetries of an action. Because most of the work involved in the Hamiltonian approach to mass and angular momentum is in deriving the Hamiltonian (2.110) for the metric, and we have already done that, the derivation of ADM mass and angular momentum will be quicker in this framework.

The second-order Hamiltonian (2.110), regarded as a function of π^{ab} and γ_{ab}, has two related problems. First, when π^{ab} and γ_{ab} satisfy the Hamiltonian and momentum constraints, H_G vanishes. This appears to contradict the identification of the generator of time translations with the total mass of a system; similarly, when $\alpha = 0$ and β^a is a rotational vector field ϕ^a, the vanishing of H_G appears to contradict the identification of the generator of rotations with angular momentum. Second, for general variations of γ_{ab} and π^{ab} satisfying the asymptotic conditions (3.2), δH_G is infinite (H_G is infinite for fields that do not satisfy the constraints); and the formal variation of H_G does not give the field equations, because the integration by parts leaves nonzero surface terms at infinity. It is this second difficulty that resolves the apparent contradiction: For a lapse α that does not vanish at infinity, it means that H_G is *not* the generator of time translations on a phase space large enough to include asymptotic time translations of asymptotically flat geometries, geometries with asymptotic behavior (3.2); also, for a shift β^a that agrees asymptotically with a rotational Killing vector ϕ^a of η_{ab}, H_G is not the generator of rotations on a phase space large enough to include asymptotic rotations of fields with asymptotic behavior (3.2).

Instead, we will see that adding a divergence to the second-order density $\alpha\mathcal{H}_G + \beta^{\mathcal{H}}_{Ga}$ yields a density that involves only first derivatives of the metric. The corresponding first-order Hamiltonian $H^{(1)}$ is finite; the surface terms arising from its variation vanish at spatial infinity; and for α and β^a having asymptotic behavior corresponding to asymptotic translations and rotations, $H^{(1)}$ thus generates translations and rotations of the metric and matter.

The surface terms arise from the integration by parts of the second derivatives of the spacetime metric, from second derivatives of γ_{ab} in the 3-dimensional Ricci scalar R and the divergence $D_b\pi^{ab}$ of π^{ab}. Introducing the covariant derivative operator ∇_a allows us to write the surface terms and the first-order Hamiltonian in a covariant way. We define a covariant version $C^a{}_{bc}$ of the Christoffel symbols by noting that the difference $D_a - \nabla_a$ between the two covariant derivative operators is a tensor $C^a{}_{bc}$:

$$(D_b - \nabla_b)\xi^a = C^a{}_{bc}\xi^c, \tag{3.8}$$

where

$$C^a{}_{bc} := \frac{1}{2}\gamma^{ad}\left(\nabla_b\gamma_{dc} + \nabla_c\gamma_{bd} - \nabla_d\gamma_{bc}\right). \tag{3.9}$$

The components $C^i{}_{jk}$ in a Cartesian chart $\{x^i\}$ for η_{ab} are then Christoffel symbols of D_a. The usual formula for the coordinate components of the 3-dimensional Riemann tensor in terms of the 3-dimensional Christoffel symbols,

$$R^k{}_{lij} = \partial_i\Gamma^k{}_{jl} - \partial_j\Gamma^k{}_{il} + \Gamma^k{}_{mi}\Gamma^m{}_{jl} - \Gamma^k{}_{mj}\Gamma^m{}_{il}, \tag{3.10}$$

thus has the covariant form

$$R^c{}_{dab} = \nabla_a C^c{}_{bd} - \nabla_b C^c{}_{ad} + C^c{}_{ea}C^e{}_{bd} - C^c{}_{eb}C^e{}_{ad}, \tag{3.11}$$

implying for the 3-dimensional Ricci scalar the corresponding form

$$R = \gamma^{bc}\nabla_a C^a{}_{bc} - \nabla^a C^b{}_{ba} + C^b{}_{bc}C^{ca}{}_a - C^b{}_{ac}C^{ac}{}_b. \tag{3.12}$$

Then $\alpha R\sqrt{\gamma}$ can be written as a sum of a first-order part and a divergence:

$$\begin{aligned}\alpha R\sqrt{\gamma} = &-\nabla_a(\alpha\sqrt{\gamma}\gamma^{bc})C^a{}_{bc} + \nabla_a(\alpha\sqrt{\gamma}\gamma^{ac})C^b{}_{bc} + (C^b{}_{bc}C^{ca}{}_a - C^b{}_{ac}C^{ac}{}_b)\alpha\sqrt{\gamma}\\ &+ \nabla_a[(\gamma^{bc}C^a{}_{bc} - \gamma^{ac}C^b{}_{bc})\alpha\sqrt{\gamma}].\end{aligned} \tag{3.13}$$

Using the relations $\nabla_a\gamma = \frac{1}{2}\gamma^{bc}\nabla_a\gamma_{bc}$ to replace derivatives of γ and γ^{bd} in the first-order part, and using Eq. (3.9) to replace $C^a{}_{bc}$ in the divergence, we obtain

$$\alpha R\sqrt{\gamma} = \nabla_a\alpha(C^b{}_b{}^a - C^{ab}{}_b)\sqrt{\gamma} + \alpha(C^b{}_{ac}C^{ac}{}_b - C^b{}_{bc}C^{ca}{}_a)\sqrt{\gamma}$$
$$+ \nabla_a[(\gamma^{ac}\gamma^{bd} - \gamma^{ab}\gamma^{cd})\nabla_b\gamma_{cd}\alpha\sqrt{\gamma}]. \tag{3.14}$$

Finally, from Eqs. (2.110), (2.111), (2.113), and the relation $-2\beta^a D_b\pi_a{}^b = 2D_b\beta^a\pi_a{}^b + D_b(2\beta^a\pi_a{}^b)$, a first-order Hamiltonian is given by

$$H^{(1)} := \frac{1}{16\pi}\int\Big[\gamma^{-1/2}\left(\pi^{ab}\pi_{ab} - \frac{1}{2}\pi^2\right) + \nabla_a\alpha(C^b{}_b{}^a - C^{ab}{}_b)\sqrt{\gamma} + 2\pi^{ab}D_a\beta_b$$
$$+ (C^b{}_{ac}C^{ac}{}_b - C^b{}_{bc}C^{ca}{}_a)\alpha\sqrt{\gamma} + 16\pi\mathcal{H}_M\alpha\sqrt{\gamma}\Big]d^3x \tag{3.15}$$

$$= H + \frac{1}{16\pi}\int_{S_\infty}[\alpha(\gamma^{ac}\gamma^{bd} - \gamma^{ab}\gamma^{cd})\nabla_b\gamma_{cd} - 2\beta^b\pi^a{}_b\gamma^{-1/2}]dS_a. \tag{3.16}$$

When the Hamiltonian and momentum constraints are satisfied, H vanishes, and the value of the first-order Hamiltonian $H^{(1)}$ is the surface integral

$$H^{(1)} = \frac{1}{16\pi}\int_{S_\infty}[\alpha(\eta^{ac}\eta^{bd} - \eta^{ab}\eta^{cd})\nabla_b\gamma_{cd} - 2\beta^b\pi^a{}_b\gamma^{-1/2}]dS_a, \tag{3.17}$$

where we have used the fact that only the leading term η^{ab} in γ^{ab} contributes to the surface integral when it is finite.

Finally, for $\alpha = 1 + O(r^{-1})$ and $\beta^a = O(r^{-1})$, $H^{(1)}$ now appropriately generates time translations, and its value when the constraints are satisfied is the ADM mass, Eq. (3.2),

$$H^{(1)}|_{\alpha=1,\beta^a=0} = \frac{1}{16\pi}\int_{S_\infty}(\eta^{ac}\eta^{bd} - \eta^{ab}\eta^{cd})\nabla_b\gamma_{cd}\,dS_a = M_{\rm ADM}. \tag{3.18}$$

Similarly, for $\alpha = O(r^{-1})$ and $\beta^a = \phi^a[1 + O(r^{-1})]$, $H^{(1)}$ generates rotations, and its value when the constraints are satisfied is the ADM angular momentum, Eq. (3.3),

$$H^{(1)}|_{\alpha=0,\beta^a=\phi^a} = -\frac{1}{8\pi}\int_{S_\infty}\pi^a{}_b\phi^b d\sigma_a = J_{\rm ADM}. \tag{3.19}$$

Note that, when the spacetime metric is known, the computation of π^{ab} and K_{ab} involves only $\mathcal{L}_{\mathbf{n}}\gamma_{\alpha\beta}$ and is in this sense independent of the values of α and β^a in Eqs. (3.18) and (3.19). In particular, when α vanishes, one cannot compute K_{ab} from the relation $2\alpha K_{ab} = -\dot{\gamma}_{ab} + 2D_{(a}\beta_{b)}$.

In the Noether-theorem approach, one again uses the first-order action, now written in terms of the 4-metric $g_{\alpha\beta}$ and its derivative $\nabla_\gamma g_{\alpha\beta}$ with respect to the flat background metric $\eta_{\alpha\beta}$. A detailed presentation is given by Chruściel [139].[1]

[1] In Chruściel's paper and in Sorkin's briefer derivation [641], the background metric is not required to be flat. The greater generality is needed for spacetimes whose topology is not $\mathbb{R}^4$.

The ADM mass and angular momentum are given in terms of asymptotically flat initial data γ_{ab}, π^{ab} on a hypersurface and require no symmetry of the spacetime. We have, however, already defined the mass and angular momentum of a stationary, axisymmetric spacetime in Section 1.8, and we need to check that the definitions agree. The earlier expressions of Eqs. (1.128) and (1.129) are commonly called the Komar mass and angular momentum, M_K and $J_{\rm K}$ [370]. The proof that, for stationary spacetimes, the Komar mass associated with the time-translational Killing vector t^α agrees with the ADM mass,

$$M_{\rm K} := -\frac{1}{4\pi}\int_{S_\infty} \nabla^\alpha t^\beta dS_{\alpha\beta} = \frac{1}{16\pi}\int_{S_\infty} (\eta^{ac}\eta^{bd} - \eta^{ab}\eta^{cd})\nabla_b h_{cd} dS_a =: M_{\rm ADM}, \tag{3.20}$$

is due to Beig and to Ashtekar and Magnon-Ashtekar and is given in Section 3.4. The proof that, for axisymmetric spacetimes, the Komar angular momentum associated with the rotational Killing vector ϕ^α agrees with the ADM angular momentum,

$$\begin{aligned} J_{\rm K} &= \frac{1}{8\pi}\int_{S_\infty} \nabla^\alpha \phi^\beta dS_{\alpha\beta} \\ &= -\frac{1}{8\pi}\int_{S_\infty} \pi^a{}_b \phi^b d\sigma_a, \\ &= J_{\rm ADM}, \end{aligned} \tag{3.21}$$

is much shorter and essentially follows from an expression for the extrinsic curvature in terms of the gradient $\nabla_\alpha n_\beta$, namely

$$\begin{aligned} K_{ab} &:= -\frac{1}{2}\gamma_a{}^\alpha \gamma_b{}^\beta \mathcal{L}_{\mathbf{n}}\gamma_{\alpha\beta} = -\frac{1}{2}\gamma_a{}^\alpha\gamma_b{}^\beta \mathcal{L}_{\mathbf{n}}(g_{\alpha\beta} + n_\alpha n_\beta) = -\frac{1}{2}\gamma_a{}^\alpha\gamma_b{}^\beta\mathcal{L}_{\mathbf{n}} g_{\alpha\beta} \\ &= -\gamma_a{}^\alpha\gamma_b{}^\beta\nabla_\alpha n_\beta, \end{aligned} \tag{3.22}$$

where in the last line the index symmetry of the right side follows from the relation $n_\alpha = -\alpha\nabla_\alpha t$.

With n_α and $\hat{r}_\alpha$ the unit normals to the $t =$ constant, $r =$ constant surface S_r, we have $\nabla^\alpha\phi^\beta\ dS_{\alpha\beta} = \nabla^\alpha\phi^\beta\ \hat{r}_{[\alpha}n_{\beta]}dS = \nabla^\alpha\phi^\beta\ \hat{r}_\alpha\, n_\beta\, dS$, using in the last equality the antisymmetry of $\nabla^\alpha\phi^\beta$. Then $\phi^\alpha n_\alpha = 0$ implies

$$\begin{aligned} \nabla^\alpha\phi^\beta n_\beta \hat{r}_\alpha &= -\phi^\beta\hat{r}^\alpha\nabla_\alpha n_\beta = -\phi^b\hat{r}^a\gamma_a{}^\alpha\gamma_b{}^\beta\nabla_\beta n_\alpha = K_{ab}\phi^b\hat{r}^a \\ &= -\pi^a{}_b\phi^b\hat{r}_a\gamma^{-1/2}, \end{aligned} \tag{3.23}$$

$$\int_{S_\infty}\nabla^\alpha\phi^\beta dS_{\alpha\beta} = -\int_{S_\infty}\pi^a{}_b\phi^b\hat{r}_a\gamma^{-1/2}\, dS = -\int_{S_\infty}\pi^a{}_b\phi^b d\sigma_a, \tag{3.24}$$

and Eq. (3.21) follows. (Given our previous formalism, there is a slightly shorter proof that uses Gauss's theorem to write each surface integral, [3.21] and [3.21], as a volume integral of $R^\alpha{}_\beta\phi^\beta n_\alpha$. That derivation, however, fails when the spacetime includes black holes and there is an interior boundary at the event horizon; it also obscures the fact that the identity depends only on the asymptotic behavior of the metric.)

3.2 Asymptotic behavior of equilibria

3.2.1 Asymptotic behavior of the metric

Like a source-free electromagnetic field in flat space, a stationary, axisymmetric, asymptotically flat vacuum geometry outside a rotating star is determined by a set of mass and current multipole moments. As in flat-space potential theory, successive moments of a stationary field arise at successively higher powers of r^{-1}. And as in electromagnetism, the asymptotic gravitational field is characterized by two sets of multipoles, mass multipoles and current multipoles, analogs of the electromagnetic charge multipoles and current multipoles. In contrast to the Newtonian and electromagnetic cases, however, only the lowest-order multipole moments of the gravitational field, the mass and angular momentum, can be written as integrals over the source; for all moments, the integrals involve the metric.

The dependence of metric components on the choice of coordinates leads to one final complication: In coordinate choices natural for a rotating star, including the quasi-isotropic coordinates we use, the asymptotic form of the metric includes information about the coordinates as well as about the multipole structure of the geometry. Because the metric potentials ν, ω, and ψ are scalars constructed locally from the metric and the symmetry vectors ϕ^α and t^α, their definition is in this sense coordinate-independent. The functional forms, $\nu(r,\theta)$, $\omega(r,\theta)$, $\psi(r,\theta)$, of course depend on r and θ, but it is not difficult to disentangle the physical mass and current moments from the coordinate contributions, and we do so at the end of the section.

We will restrict consideration here to the asymptotic behavior of the metric up to $O(r^{-3})$. At this order, the only contributing multipoles are the monopole and quadrupole mass moments and the $l = 1$ current moment. The definitions used in the literature of these lowest multipoles are unambiguous. In particular, two approaches to asymptotic multipoles of stationary systems, developed by Thorne (sections X and XI of [691]) and by Geroch [256] and Hansen [294], have become standard. They yield identical definitions for $l \leq 2$; higher multipoles in the two approaches are essentially the same, disagreeing only in the normalization chosen.[2] We will use Thorne's definition, but will also mention a coordinate-invariant characterization of the quadrupole moment due to Ryan [576], with a variant by Laarakkers and Poisson [380]. (These alternate definitions are accurate, but we point out an error in a subsequent paper by Ryan [576], repeated in [380], that misidentifies the quadrupole moment with the coefficient of $-P_2/r^3$ in the potential ν. The identification is correct for a class of asymptotic coordinate systems, but not for the quasi-isotropic coordinates that we and they use.)

[2] Gürsel [283] gives a clear summary of the relation between the Thorne and Geroch-Hansen multipoles and a proof of their equivalence up to normalization. Simon and Beig [634] provide a rigorous treatment of multipoles of stationary nonaxisymmetric systems in a framework similar to Thorne's and again make the connection to the Geroch-Hansen multipoles.

For a spherical star, the quasi-isotropic coordinates associated with our form (1.59) of the metric become the isotropic Schwarzschild coordinates, with metric components given by [480]

$$ds^2 = -\left(\frac{1 - M/2r}{1 + M/2r}\right)^2 dt^2 + \left(1 + \frac{M}{2r}\right)^4 (dr^2 + r^2 d\theta^2 + r^2 \sin^2\theta d\phi^2). \qquad (3.25)$$

We begin with the asymptotic forms of the Schwarzschild metric and the metric of a rotating star, and we then verify that the exhibited metric of a rotating star satisfies the asymptotic field equation, by checking its independent components, Eqs. (1.93) and (1.94). Denote by ν_S, μ_S, and ψ_S the metric potentials ν, μ, and ψ for the Schwarzschild metric in isotropic coordinates. For the asymptotic form of the Schwarzschild metric (3.25),

$$\begin{aligned} ds^2 &= -\left[1 - \frac{2M}{r} + 2\frac{M^2}{r^2} - \frac{1}{4}\frac{M^3}{r^3} + O(r^{-5})\right] dt^2 \\ &+ \left[1 + \frac{2M}{r} + \frac{3}{2}\frac{M^2}{r^2} + O(r^{-3})\right](r^2 \sin^2\theta d\phi^2 + dr^2 + r^2 d\theta^2), \end{aligned} \qquad (3.26)$$

the potentials have behavior

$$\nu_S = -\frac{M}{r} - \frac{1}{12}\frac{M^3}{r^3} + O(r^{-5}), \qquad (3.27a)$$

$$\mu_S = \frac{M}{r} - \frac{1}{4}\frac{M^2}{r^2} + \frac{1}{12}\frac{M^3}{r^3} + O(r^{-4}), \qquad (3.27b)$$

$$\psi_S = \log(r \sin\theta) + \mu_S. \qquad (3.27c)$$

The asymptotic metric of a rotating star differs from the Schwarzschild metric at $O(r^{-2})$, where a tensor is said to be $O(r^{-k})$ if each of its components along the asymptotically orthonormal frame,

$$\mathbf{e}_0 = \boldsymbol{\partial}_t, \ \mathbf{e}_1 = \boldsymbol{\partial}_r, \ \mathbf{e}_2 = \frac{1}{r}\boldsymbol{\partial}_\theta, \ \mathbf{e}_3 = \frac{1}{r\sin\theta}\boldsymbol{\partial}_\phi, \qquad (3.28)$$

is $O(r^{-k})$. Through $O(r^{-3})$ there are three corrections due to rotation:

1. the frame dragging potential $\omega \sim \frac{2J}{r^3}$, corresponding to an $l = 1$ current moment, the angular momentum J (this is a correction to the metric of order r^{-2} in g_{03});
2. a quadrupole correction to the diagonal metric coefficients at $O(r^{-3})$ associated with the mass quadrupole moment Q of the rotating star; and
3. coordinate-dependent monopole and quadrupole corrections to the diagonal metric coefficients, reflecting the asymptotic shape of the r- and θ-surfaces and associated with a dimensionless parameter a.

With the dimensionless form $q = Q/M^3$ of the quadrupole moment, the asymptotic form of the metric is given in terms of the parameters J, q, and a by

$$\nu = \nu_S + (a - 4aP_2 - qP_2)\frac{M^3}{r^3} + O(r^{-4}), \tag{3.29a}$$

$$\mu = \mu_S - (a - 4aP_2)\frac{M^2}{r^2} - (a - 4aP_2 - qP_2)\frac{M^3}{r^3} + O(r^{-4}), \tag{3.29b}$$

$$\psi = \log(r \sin\theta) + 3a\frac{M^2}{r^2} - (a - 4aP_2 - qP_2)\frac{M^3}{r^3} + O(r^{-4}), \tag{3.29c}$$

$$\omega = \frac{2J}{r^3} + O(r^{-4}), \tag{3.29d}$$

where $P_2 \equiv P_2(\cos\theta)$. In associating ω with an $l = 1$ current moment, we are referring to the fact that, to leading asymptotic order, the pair $(h_{t\theta}, h_{t\phi}) = (0, (2J/r)\sin^2\theta)$ comprises a vector proportional to an $l = 1$ vector spherical harmonic, $\mathbf{r} \times \nabla P_1(\cos\theta)$, whose components are $(0, \sin\theta\partial_\theta P_1) = (0, -\sin^2\theta)$.

To verify the asymptotic form (3.29), we must show that it satisfies the independent components, Eqs. (1.93) and (1.94), of the asymptotic field equation and that the coefficients of the r^{-2} and r^{-3} terms are related to one another in the manner implied by their expressions in terms of the parameters a and q. In addition, we obtain an expression for a as an integral over the star. The equations involve the auxiliary potentials $B = e^{\psi+\nu}/r\sin\theta$ and $\zeta = \mu + \nu$, which have, according to Eqs. (3.29), the asymptotic behavior

$$B = B_S + 3a\frac{M^2}{r^2} + O(r^{-4}), \quad \zeta = \zeta_S - (a - 4aP_2)\frac{M^2}{r^2} + O(r^{-4}), \tag{3.30}$$

with

$$B_S = 1 - \frac{1}{4}\frac{M^2}{r^2} + O(r^{-4}), \quad \zeta_S = -\frac{1}{4}\frac{M^2}{r^2} + O(r^{-4}). \tag{3.31}$$

We must verify the asymptotic behavior of B, ζ, ν, and ω. Our analysis follows that of Butterworth and Ipser [110], who exhibit the asymptotic expansion of B, ν, and ω to two additional orders in r^{-1}. Choosing the potentials B and ζ instead of ψ and μ simplifies the analysis, because in the vacuum B satisfies a decoupled linear equation, and $\zeta - \zeta_S$ can be quickly found to order r^{-3} from $B - B_S$ alone.

Outside the star, Eqs. (1.93) have the form

$$\nabla \cdot (B\nabla\nu) = \frac{1}{2}r^2\sin^2\theta B^3 e^{-4\nu}\nabla\omega \cdot \nabla\omega, \tag{3.32a}$$

$$r^{-2}\nabla \cdot (r^2\sin^2\theta B^3 e^{-4\nu}\nabla\omega) = 0, \tag{3.32b}$$

$$r^{-2}\nabla \cdot (r\sin\theta\nabla B) = 0. \tag{3.32c}$$

The left side of each equation has two spatial derivatives of a single potential, of ν, ω, and B, respectively, and it is helpful to refer to the equations as the ν-, ω-, and

B-equations. To keep terms up to order r^{-3} in the potentials, we must consistently keep all terms up to $O(r^{-5})$ in each equation, and we must verify that each equation is satisfied to that order.

We begin with Eq. (3.32a) for ν. The right side of the equation is $O(r^2\nabla\omega\cdot\nabla\omega) = O(r^{-6})$. Because the equation is satisfied by the Schwarzschild potentials ν_S and B_S, we have $\nabla\cdot(B_S\nabla\nu_S) = 0$. The left side is then given by

$$\begin{aligned}\nabla\cdot(B\nabla\nu) &= \nabla^2(\nu-\nu_S) + \nabla(B-B_S)\cdot\nabla\nu_S + O(r^{-6})\\ &= \nabla^2\left(a\frac{M^3}{r^3}\right) + \nabla\left(3a\frac{M^2}{r^2}\right)\cdot\nabla\left(-\frac{M}{r}\right) + O(r^{-6})\\ &= 0 + O(r^{-6}),\end{aligned} \tag{3.33}$$

where we have used $\nabla^2(P_2/r^3) = 0$ to write the second equality. Thus the equation is satisfied through $O(r^{-5})$, as claimed, and the check verifies the relation between the r^{-3} monopole contribution to ν and the r^{-2} monopole contribution to B.

The check of the ω-equation is similar and simpler: Because $B^3e^{-4\nu} = 1 + O(r^{-1})$ and $\nabla(B^3e^{-4\nu}) = O(r^{-2})$, the ω-equation has the asymptotic form

$$r^{-2}\nabla\cdot(r^2\sin^2\theta\nabla\omega) + O(r^{-6}) = 0.$$

Because $\nabla\cdot[r^2\sin^2\theta\nabla(2J/r^3)] = 0$, the equation is satisfied to $O(r^{-6})$. The identification of J with the star's angular momentum was discussed in detail in Section 3.1.

The B-equation involves no other potential and is exactly satisfied by B_S. The fact that the quadrupole contribution to B is $O(r^{-4})$ follows from the general solution to the B-equation: Following Butterworth and Ipser [110], we separate variables, writing $B = \mathcal{R}(r)\Theta(\theta)$ to obtain

$$\frac{1}{\mathcal{R}}\left(\partial_r^2 + \frac{3}{r}\right)\mathcal{R} + \frac{1}{\Theta}(\partial_\theta^2 + 2\cot\theta\partial_\theta)\Theta = 0. \tag{3.34}$$

The solutions to the eigenvalue equation $(\partial_\theta^2 + 2\cot\theta)\Theta = \lambda\Theta$ are the Gegenbauer polynomials $T_l^{1/2}(\cos\theta)$, with eigenvalues $\lambda = -l(l+2)$. They have the form

$$T_l^{1/2}(\cos\theta) = C_l\frac{1}{\sin\theta}\frac{d^l}{d\cos\theta^{\,l}}\sin^{2l+1}\theta, \tag{3.35}$$

with $T_0^{1/2} = \sqrt{\frac{2}{\pi}}$, $T_2^{1/2} = \frac{8}{3}\sqrt{\frac{2}{\pi}}(P_2 + \frac{1}{8})$, and they are orthonormal with respect to an inner product

$$\int T_l^{1/2}T_{l'}^{1/2}\sin^2\theta d\theta = \delta_{ll'}. \tag{3.36}$$

From Eq. (3.35), they have parity $(-1)^l$, implying only those with l even contribute to B. For each l, the asymptotically regular solution to the radial equation, $(\partial_r^2 + \frac{3}{r})\mathcal{R} = l(l+2)\mathcal{R}$, is $r^{-(l+2)}$. Thus B is spherically symmetric through $O(r^{-3})$ and has the claimed behavior

$$B = 1 + \left(3a - \frac{1}{4}\right)\frac{M^2}{r^2} + O(r^{-4}),$$

for some value of the constant $b := 3a - 1/4$.

Because the source of the B-equation (1.93c) is proportional to the pressure, the coefficient b vanishes for a pressureless rotating disk: When $p = 0$, Eq. (1.93c) has the unique asymptotically regular solution $B = 1$. Along a sequence of models with increasing angular velocity (and fixed baryon number), $|b|$ decreases from its initial value $1/4$ for a spherical star, as more of the support of the star is provided by rotation, less by pressure. We can obtain from the B-equation (1.93c) an expression for b as an integral over the star [110]: Using the orthogonality relation (3.36), we find

$$b = -\frac{8}{\pi M^2}\int pr\sin\theta\, e^{\nu}\, dV, \tag{3.37}$$

with $dV = \sqrt{^3g}\; dr d\theta d\phi = Be^{2\mu-\nu}r^2 \sin\theta dr d\theta d\phi$.[3] We again emphasize that, although the integral is over the star, $\sqrt{^3g}$ involves the potentials, and their determination by an iterative numerical computation involves finding them everywhere, not simply within the star. Thus, one can use (3.37) to compute b only after one has already found B everywhere in the spacetime. The integral may give a more accurate evaluation than the iterative procedure one has chosen to find the potentials, but in principle the iteration together with the asymptotic form of B independently determines b.

Finally, we check the ζ-equation, (1.94), which has, at order r^{-4}, the form

$$\frac{1}{\varpi}\partial_\varpi \zeta = \frac{1}{2\varpi^2}\partial_\varpi(\varpi^2\partial_\varpi B) - \frac{1}{2}\partial_z^2 B + (\partial_\varpi \nu)^2 - (\partial_z \nu)^2, \tag{3.38}$$

or

$$\frac{1}{r}\left(\partial_r + \frac{1}{r}\cot\theta\partial_\theta\right)(\zeta - \zeta_S) = -\left(\cos^2\theta\partial_r^2 + \frac{1}{r}\sin^2\theta\partial_r\right)(B - B_S) + O(r^{-6}). \tag{3.39}$$

From Eq. (3.30), left and right sides each have the value $-\frac{16aM^2}{r^4}(P_2 + \frac{1}{8})$, and it is this agreement that enforces the relation between the $O(r^{-2})$ monopole and quadrupole contributions to $\zeta - \zeta_S$, implying that they have coefficients $-a$ and $4a$, respectively.

Identification of the quadrupole moment. This concludes the verification of the asymptotic form (3.29) of the metric. What remains is to justify the identification

[3] For a spherical star, $b = -1/4$, and the relation becomes an identity $\frac{M^2}{32\pi} = \int pr^3 e^{\nu}(1 + M/2r)^6 dr$, whose commonly used Newtonian limit is $\frac{M^2}{32\pi} = \int pr^3 dr$.

of $Q = qM^3$ with the quadrupole moment of a stationary axisymmetric spacetime. Here we follow Thorne [691], restricting his treatment of stationary spacetimes to spacetimes with the symmetries of a rotating star: stationary, axisymmetric, and symmetric under $(t, \phi) \to (-t, -\phi)$. Because the potential ν has a coordinate-invariant definition in terms of the timelike symmetry vector t^α and has as its Newtonian limit the Newtonian potential Φ, it is natural to define Q as the coefficient of $-P_2/r^3$ in ν, agreeing in the Newtonian limit with the Newtonian quadrupole moment Q_N. The Newtonian multipole expansion

$$\Phi = -\frac{M}{r} - Q_N \frac{P_2}{r^3} + \cdots, \text{ with } Q_N = \int \rho P_2 r^2 dV, \tag{3.40}$$

would then have $\nu = -\frac{M}{r} - Q\frac{P_2}{r^3} + O(r^{-5})$ as its counterpart in the exact theory. A definition of this form, however, is not coordinate-invariant: In terms of a new radial coordinate $\bar{r}$ of the form $\bar{r} = r[1 + CP_2(\cos\theta)/r^2]$, for example, $\nu(\bar{r})$ acquires an additional P_2/r^3 contribution. The ambiguity is overcome by observing that coordinate transformations of this form lead to quadrupole terms in the metric of order r^{-2}, as is the case for the r-θ part of the metric in our quasi-isotropic coordinates. One obtains a gauge-invariant definition simply by restricting the definition to asymptotic coordinates for which the metric has only monopole and dipole contributions at $O(r^{-2})$.

Let $\{t, r, \theta, \phi\}$ be a chart for a stationary, axisymmetric, asymptotically flat spacetime, symmetric under $(t, \phi) \to (-t, -\phi)$, with t and ϕ the natural coordinates associated with the symmetry vectors, for which the metric components are independent of t and ϕ.

Definition. Suppose the metric components along the corresponding asymptotically orthonormal frame $\{\mathbf{e}_\mu\}$ of Eq. (3.28) satisfy

$$g_{tt} = -1 + \frac{2M}{r} + \frac{\text{0-pole}}{r^2} + \frac{2QP_2 + \text{0-pole}}{r^3} + O(r^{-4}), \tag{3.41}$$

$$g_{t3} = -\frac{2J}{r^2}\sin\theta + O(r^{-3}), \tag{3.42}$$

$$g_{ij} = \delta_{ij} + \frac{\text{0-pole}}{r} + \frac{\text{0-pole}}{r^2} + \frac{\text{2-pole} + \text{0-pole}}{r^3} + O(r^{-4}) \tag{3.43}$$

(and where $\boldsymbol{\nabla} O(r^{-k}) = O(r^{-(k+1)})$, $\nabla_i \nabla_j O(r^{-k}) = O(r^{-(k+2)})$).
Then Q is the *scalar quadrupole moment* of the spacetime.

This way of specifying a class of asymptotic charts conforms to Thorne's class of *asymptotically Cartesian and mass centered coordinates* (ACMC). That is, the Cartesian coordinates t, x, y, z associated with spherical coordinates t, r, θ, ϕ satisfying (3.43) are ACMC coordinates to order r^{-3}. We exhibit a simple choice of coordinates, $\bar{r}, \bar{\theta}$, satisfying the conditions (3.41)–(3.43) and use them to confirm the identification of Q. We then note that, having once decided on the meaning of

Q, one can easily find equivalent, *manifestly coordinate-invariant* definitions, and we give two examples.

Coordinates $\bar{r}$ and $\bar{\theta}$ that satisfy (3.43) – that is, for which the first quadrupole contribution to the metric is at order r^{-3} – are given by

$$r = \bar{r} + 4a\frac{M^2}{\bar{r}}P_2(\cos\bar{\theta}), \qquad \theta = \bar{\theta} + 2a\frac{M^2}{r^2}\partial_{\bar{\theta}}P_2(\cos\bar{\theta}). \tag{3.44}$$

Components of the metric in the t-ϕ subspace already satisfy the conditions (3.41–3.43), and the change of coordinates leaves them unaltered at $O(r^{-2})$. The metric components in the r-θ subspace are now

$$g_{22} = g_{\bar{r}\bar{r}} = \left(\frac{\partial r}{\partial \bar{r}}\right)^2 g_{rr} + \left(\frac{\partial \theta}{\partial \bar{r}}\right)^2 g_{\theta\theta}$$
$$= 1 + \frac{2M}{\bar{r}} + \left(\frac{3}{2} - 2a\right)\frac{M^2}{\bar{r}^2} + \left(\frac{1}{2} - 6a + 2qP_2\right)\frac{M^3}{\bar{r}^3}, \tag{3.45}$$

$$g_{33} = \bar{r}^{-2}g_{\bar{\theta}\bar{\theta}} = \bar{r}^{-2}\left[\left(\frac{\partial r}{\partial \bar{\theta}}\right)^2 g_{rr} + \left(\frac{\partial \theta}{\partial \bar{\theta}}\right)^2 g_{\theta\theta}\right]$$
$$= 1 + \frac{2M}{\bar{r}} + \left(\frac{3}{2} + 2a\right)\frac{M^2}{\bar{r}^2} + \left(\frac{1}{2} + 2a + 2qP_2\right)\frac{M^3}{\bar{r}^3}, \tag{3.46}$$

$$g_{23} = \bar{r}^{-1}g_{\bar{r}\bar{\theta}} = \bar{r}^{-1}\left(\frac{\partial r}{\partial \bar{r}}\frac{\partial r}{\partial \bar{\theta}} g_{rr} + \frac{\partial \theta}{\partial \bar{r}}\frac{\partial \theta}{\partial \bar{\theta}} g_{\theta\theta}\right) = O(\bar{r}^{-4}), \tag{3.47}$$

with no quadrupole contribution at $O(r^{-2})$.

From Eqs. (3.29a) and (3.44), the asymptotic behavior of ν is

$$\nu(\bar{r}) = -\frac{M}{r(\bar{r})} + \left(-\frac{1}{12} + a - 4aP_2 - qP_2\right)\frac{M^3}{\bar{r}^3} + O(r^{-4})$$
$$= -\frac{M}{\bar{r}} + \left(a - \frac{1}{12} - qP_2\right)\frac{M^3}{\bar{r}^3} + O(r^{-4}),$$

implying $Q = qM^3$, as claimed.

To summarize: *For the potential ν, in quasi-isotropic coordinates, the coefficient of $-P_2/r^3$ is $4aM^3 + Q$.* To obtain Q, one needs to extract a from the coefficient of $1/r^2$ in B or the coefficient of P_2/r^2 in μ.

Following (in part) Ryan [576] and Laarakkers and Poisson [380], we conclude the discussion by giving two definitions of the quadrupole moment that are manifestly coordinate invariant (the coordinate-invariant definition is correct as written in [576] and [380]). The idea here is that any scalar constructed locally from the metric and the symmetry vectors depends asymptotically only on the mass and current multipole moments, and its dominant asymptotic behavior will typically be determined by M, J, and Q. The functional form of a scalar will of course depend on the choice of r and θ, but if one chooses two independent scalars, say ν and e^{ψ}, and restricts each to the equatorial plane, then each scalar will be a function of r alone, and one can eliminate the dependence on r, writing one gauge-invariant scalar as a function of the other: $\nu = \nu(e^{\psi})$. This example is a natural one, because $e^\psi = \sqrt{\phi^\alpha \phi_\alpha}$ is the radial coordinate for which $2\pi e^\psi$ is the circumference of the circular orbits of the symmetry vector ϕ^α. As in Section 1.8, we will write $r_c(r) = e^\psi(r, \theta = \pi/2)$, noting that, for spherical stars, r_c is the usual Schwarzschild radial coordinate.

From the asymptotic form (3.29c) of ψ, $r_c(r)$ is given in the equatorial plane by

$$r_c = e^\psi\big|_{\theta=\pi/2} = r + M + \left(3a - 3a\frac{M^2}{r} - \frac{q}{2}\frac{M^3}{r^2}\right) + O(r^{-3}). \tag{3.48}$$

Inverting the series gives

$$r = r_c - M - \left(3a + \frac{1}{4}\right)\frac{M^2}{r_c} - \left(3a + \frac{1}{4} - \frac{q}{2}\right)\frac{M^3}{r_c^2} + O(r^{-3}), \tag{3.49}$$

and Eq. (3.29a) for ν becomes

$$\nu = -\frac{M}{r_c} - \frac{M^2}{r_c^2} + \left(\frac{q}{2} - \frac{4}{3}\right)\frac{M^3}{r_c^3} + O(r^{-3}), \tag{3.50}$$

valid in the equatorial plane. Thus, noting that $P_2(0) = -1/2$, we find that Q may again be defined as the coefficient of $-P_2/r_c^3$ in the expansion of ν as a function of $r_c \equiv e^\psi$. The dependence on a has now disappeared, as should be expected from the fact that a is an artifact of the isotropic coordinate choice, not algebraically expressible in terms of the mass and current multipoles.

The last definition (Ryan's in the form given by Laarakkers and Poisson) uses the fact that the dominant asymptotic behavior of geodesics is determined by M, J, and Q: A manifestly coordinate-invariant characterization of Q can be stated in terms of the expansion of the energy per unit mass, $E := -u_\alpha t^\alpha$, of circular geodesics [576] as a function of their angular velocity Ω_K, the Keplerian angular velocity of the orbits. A circular orbit in the equatorial plane is invariantly defined as an orbit tangent to the Killing vector ϕ^α, the spatial projection of a geodesic whose tangent at each point is in the t^α-ϕ^α plane. E is a function of Ω_K, and its asymptotic expansion for large orbits (small $M\Omega$), in powers of $M\Omega$, has the form

$$E(\Omega_K) = E_S(\Omega_K) - \frac{4}{3}\frac{J}{M^2}(M\Omega_K)^{5/3} - \frac{1}{2}\frac{Q}{M^2}(M\Omega_K)^2 + O[(M\Omega_K)^{7/3}]. \tag{3.51}$$

Then Q may be defined as the coefficient of $-\Omega_K^2/2$ in the expansion. One obtains the expansion $E(\Omega_K)$ in essentially the same way we found $\nu(r_c)$: Use the asymptotic expansions of the potentials and Eq. (1.140) to write Ω_K as a power series in r; invert the series to $r(\Omega_K)$, and replace r by $r(\Omega_K)$ in Eq. (1.84) for u_t.

3.3 Virial identities

For a stationary configuration, the Newtonian virial relation can be written in the form

$$0 = -\int_V x^a \nabla_b T_{\mathrm{N}a}{}^b\, dV = \int_V T_{\mathrm{N}b}{}^b\, dV = 2T + 3\Pi + W, \tag{3.52}$$

where x^a is the radial vector whose components in a Cartesian chart $\{x^i\}$ are just Cartesian coordinates x^i; $T_{\rm N}^{ab}$ is the Newtonian stress tensor,

$$T_{\rm N}^{ab} = \rho v^a v^b + p\delta^{ab} + \frac{1}{4\pi}(\nabla^a\Phi\nabla^b\Phi - \frac{1}{2}\delta^{ab}\nabla_c\Phi\nabla^c\Phi); \tag{3.53}$$

and T, Π, and W are, respectively, the kinetic energy, the integral of pressure, and the gravitational binding energy, given by

$$T = \int \frac{1}{2}\rho v^2 dV, \quad \Pi = \int p dV, \qquad W = -\frac{1}{8\pi}\int \nabla_b\Phi\nabla^b\Phi dV. \tag{3.54}$$

Notice that for an ideal gas with constant ratio of specific heats γ, the integral of pressure is related to the thermal internal energy U by $\Pi = (\gamma - 1)U$.

For a stationary solution in special relativity without gravity, the structure needed to write Eq. (3.52) is again available: One can use the Killing vector t^α that defines a stationary solution to fix a hypersurface V orthogonal to t^α. Then, with x^a a natural radial vector field of the flat metric on V, the steps in Eq. (3.52) go through as written, and we obtain a relation of the same form, $0 = \int_V T_b{}^b\, dV$.

In curved spacetime, however, there is no natural generalization of a radial vector field or of Cartesian coordinates that allow one to write the first equality in Eq. (3.52). In addition, because the general relativistic stress-energy tensor does not include a gravitational binding energy, the integral $\int_V T_b{}^b\, dV$ is nonzero. As a result, there is no entirely natural generalization of the virial relation. Analogous relations have, however, been constructed and are useful in testing the accuracy of numerical codes for rotating stars and for binary initial data and quasiequilibria. In the next two subsections, we present two general relativistic variants of the virial theorem, the first due to Gourgoulhon and Bonazzola [273]. The second is similar to a relation obtained by Shibata in [626]. A final subsection concludes with a 2-dimensional virial relation, due to Bonazzola and Gourgoulhon [84].

Earlier work by Chandrasekhar and Contopoulos [132] and by Chandrasekhar [127] gives a post-Newtonian extension of the virial theorem, and Vilain [711] gives a general relativistic generalization for spherically symmetric systems.

The form of asymptotic flatness at spatial infinity that we use is that of Section 3.1. With t^α as usual the asymptotically timelike Killing vector, there is a chart $t, \{x^i\}$ in a neighborhood of spatial infinity for which ∂_t is t^α and for which the metric components and their first derivatives satisfy

$$g_{\mu\nu} - \eta_{\mu\nu} = O(r^{-1}), \qquad \partial_i g_{\mu\nu} = O(r^{-2}), \tag{3.55}$$

with $r = \sum_i (x^i)^2$.

We also use the 3+1 notation of Section 2.1: Each $t =$ constant hypersurface Σ_t is identified with a fixed 3-manifold Σ, and γ_{ab} and η_{ab} are 3-metrics on Σ induced by the 4-metric $g_{\alpha\beta}$ and by the flat 4-metric $\eta_{\alpha\beta}$, respectively. Finally, D_a is the covariant derivative of γ_{ab} and ∇_b is the flat covariant derivative of η_{ab}. In the chart $t, \{x^i\}$,

$$\eta_{ij} = \delta_{ij}, \quad \gamma_{ij} - \eta_{ij} = O(r^{-1}), \quad \partial_i\gamma_{ij} = O(r^{-2}), \quad \partial_i\alpha = O(r^{-2}). \tag{3.56}$$

Introducing the flat metric η_{ab} allows us to write the virial identities in a covariant form.

3.3.1 Virial relation for stationary spacetimes

To obtain a virial relation, Gourgoulhon and Bonazzola [273] begin by replacing the Newtonian $T_a{}^a$ by $\gamma^{\alpha\beta}(T_{\alpha\beta} - \frac{1}{8\pi}G_{\alpha\beta})$, or $T_a{}^a - \dfrac{1}{8\pi}G_a{}^a$, where T_{ab} and G_{ab} are the pullbacks to Σ of $T_{\alpha\beta}$ and $G_{\alpha\beta}$. The form $\int_\Sigma (T_a{}^a - \frac{1}{8\pi}G_a{}^a)dV$ departs from the Newtonian pattern in two ways: The integrand involves second derivatives of the metric, and the field equation guarantees that it vanishes whether or not the spacetime is stationary. One obtains an expression whose Newtonian limit is Eq. (3.52), after an integration by parts that eliminates second derivatives of the metric. The resulting surface terms cancel only if the equality $M_{\rm K} = M_{\rm ADM}$ of the Komar and ADM mass holds.

The starting point of the derivation is the trace of Eq. (1.90), the contraction of the Einstein field equation with $\gamma^{\alpha\beta}$,

$$\gamma^{\alpha\beta}\left(R_{\alpha\beta} - \frac{1}{2}{}^4Rg_{\alpha\beta}\right) = 8\pi\gamma^{\alpha\beta}T_{\alpha\beta}. \tag{3.57}$$

For a stationary spacetime, the equation has the form

$$D^a D_a \ln\alpha - \frac{1}{4}R + D^a \ln\alpha D_a \ln\alpha - \frac{3}{4}\left(K_{ab}K^{ab} - K^2\right) + \nabla_\alpha(Kn^\alpha) = 4\pi S_a{}^a. \tag{3.58}$$

Integrating over Σ, we have

$$\begin{aligned}\int_\Sigma &\left[4\pi S_a{}^a - D^a \ln\alpha D_a \ln\alpha + \frac{3}{4}\left(K_{ab}K^{ab} - K^2\right)\right] dV \\ &= \int_\Sigma \nabla_\alpha\left(Kn^\alpha\right) dV + \int_\Sigma D^a D_a \ln\alpha dV - \frac{1}{4}\int_\Sigma R dV. \end{aligned}\tag{3.59}$$

The three terms on the right side of the above equation can be evaluated separately. Stationarity implies

$$\nabla_\alpha\left(Kn^\alpha\right) = -\frac{1}{\alpha}D_a\left(K\beta^a\right), \tag{3.60}$$

$$= -\alpha^{-1}K\beta^a D_a \ln\alpha - D_a\left(\alpha^{-1}K\beta^a\right), \tag{3.61}$$

so that the first term becomes

$$\int_\Sigma \nabla_\alpha\left(Kn^\alpha\right) dV = -\int_\Sigma \alpha^{-1}K\beta^a D_a \ln\alpha dV, \tag{3.62}$$

where we use Gauss's theorem to eliminate the term

$$\int_\Sigma D_a\left(\alpha^{-1}K\beta^a\right) dV = \int_{S_\infty} \alpha^{-1}K\beta^a dS_a = 0. \tag{3.63}$$

Because the asymptotic conditions imply $K = O(r^{-2})$, $\beta^a = O(r^{-1})$, the surface term vanishes. Again by Gauss's theorem, the second term becomes

$$\int_\Sigma D^a D_a \ln\alpha\, dV = \int_{S_\infty} D^a \ln\alpha\, dS_a = \int_{S_\infty} D^a \alpha\, dS_a = 4\pi M_{\rm K}, \tag{3.64}$$

where the form (1.128) of the Komar mass, $M_{\rm K}$ for a metric with asymptotic behavior (3.56), is used to write

$$\begin{aligned} M_K &= \frac{1}{8\pi}\int_{S_\infty}(\nabla^\beta t^\alpha - \nabla^\alpha t^\beta)dS_{\alpha\beta} = \frac{1}{8\pi}\int_{S_\infty}(-\nabla^t t^r + \nabla^r t^t)\epsilon_{tr\theta\phi}d\theta d\phi \\ &= \frac{1}{8\pi}\lim_{r\to\infty}\int \partial_r\alpha^2\, r^2 d\Omega \\ &= \frac{1}{4\pi}\int_{S_\infty} D^a \alpha dS_a, \end{aligned} \tag{3.65}$$

a form due to Misner [479].

From Eq. (3.14) with $\alpha = 1$, the Ricci scalar has the form

$$R = \gamma^{ab}\left(C^c{}_{ad}C^d{}_{bc} - C^c{}_{cd}C^d{}_{ab}\right) + D_a[(\gamma^{ac}\gamma^{bd} - \gamma^{ab}\gamma^{cd})\nabla_b\gamma_{cd}], \tag{3.66}$$

where $C^a{}_{bc}$ is the covariant version of the Christoffel symbols introduced in (3.9). The third term then becomes

$$\begin{aligned} -\frac{1}{4}\int_\Sigma R dV &= -\frac{1}{4}\int_{S_\infty}(\gamma^{ac}\gamma^{bd} - \gamma^{ab}\gamma^{cd})\nabla_b\gamma_{cd}dS_a \\ &\quad -\frac{1}{4}\int_\Sigma \gamma^{ab}\left(C^c{}_{ad}C^d{}_{bc} - C^c{}_{cd}C^d{}_{ab}\right)dV \\ &= -4\pi M_{\rm ADM} - \frac{1}{4}\int_\Sigma \gamma^{ab}\left(C^c{}_{ad}C^d{}_{bc} - C^c{}_{cd}C^d{}_{ab}\right)dV. \end{aligned} \tag{3.67}$$

Using the equality

$$M_{\rm K} = M_{\rm ADM}, \tag{3.68}$$

for a stationary, asymptotically flat spacetime (shown in Section 3.4), and collecting all terms, we obtain for the virial relation the form

$$\begin{aligned} \int_\Sigma \Bigg[& 4\pi S_a{}^a - D^a \ln\alpha D_a \ln\alpha + \frac{1}{4}\gamma^{ab}\left(C^c{}_{ad}C^d{}_{bc} - C^c{}_{cd}C^d{}_{ab}\right) \\ & + \frac{3}{4}\left(K_{ab}K^{ab} - K^2\right) + \alpha^{-1}K\beta^a D_a \ln\alpha \Bigg] dV = 0. \end{aligned} \tag{3.69}$$

A useful index of numerical accuracy of equilibrium initial data is the vanishing quantity

$$GRV3 = |1 - \lambda_3|, \tag{3.70}$$

where λ_3 is the ratio of the integrated first term, $\int_\Sigma 4\pi S_a{}^a dV$, to the remaining integral on the left side of Eq. (3.69). (The term *GRV3* is an abbreviation for *3-dimensional general relativistic virial index.*)

It is convenient to work in the tetrad (1.86), noting that $\boldsymbol{\omega}^{\hat{0}}$ is the normal $-\mathbf{n}$. Then $S_a{}^a$ is given by $S_a{}^a = T^{\hat{1}\hat{1}} + T^{\hat{2}\hat{2}} + T^{\hat{3}\hat{3}}$ and Eq. (1.91) implies

$$S_a{}^a = 3p + (\epsilon + p)\frac{v^2}{1 - v^2}. \tag{3.71}$$

With the remaining integrand on the left side of Eq. (3.69) evaluated for a metric of the form (1.59), λ_3 has the explicit form

$$\begin{aligned}
\lambda_3 \equiv 4\pi & \int_0^{+\infty}\int_0^{\pi}\left[3p + (\epsilon + p)\frac{v^2}{1 - v^2}\right] e^{2\mu+\psi}\, r\, dr\, d\theta \\
& \times\Bigg\{\int_0^{+\infty}\int_0^{\pi}\Bigg[\partial_r\nu\partial_r\nu + \frac{1}{r^2}\partial_\theta\nu\partial_\theta\nu - \frac{1}{2}\partial_r\mu\partial_r\psi - \frac{1}{2r^2}\partial_\theta\mu\partial_\theta\psi \\
& + \frac{e^{2(\mu-\psi)}}{2} r\sin^2\theta\left(\partial_r\mu + \frac{1}{r\tan\theta}\partial_\theta\mu\right) \\
& + \frac{1}{4r}\left(1 - e^{2(\mu-\psi)}r^2\sin^2\theta\right)\left(\partial_r\psi + \frac{1}{r\tan\theta}\partial_\theta\psi - \frac{1}{r\sin^2\theta}\right) \\
& - \frac{3}{8}e^{2(\psi-\nu)}\left(\partial_r\omega\partial_r\omega + \frac{1}{r^2}\partial_\theta\omega\partial_\theta\omega\right)\Bigg] e^{\psi}\, r\, dr\, d\theta\Bigg\}^{-1}.
\end{aligned} \tag{3.72}$$

In numerical applications, the above quantity has been shown to correlate well with the overall accuracy of stationary equilibrium models (see, for example, [498]).

3.3.2 Virial theorem associated with a pseudotensor

For each choice of asymptotically Cartesian coordinates, one can write an analog of the virial relation that replaces $T_{\rm N}^{ab}$ by the spatial part of a pseudotensor $\Theta^{\mu\nu}$, defined to satisfy $\partial_\nu\Theta^{\mu\nu} = 0$. As in Section 2.1, we introduce a flat metric to replace partial derivatives of components in the Cartesian chart by covariant derivatives of tensors with respect to the flat metric. This added structure is sufficient to replace the pseudotensor by a tensor, and we then obtain a corresponding covariant generalization of the Newtonian virial relation.

We present the first virial relation in terms of a covariant version of the Landau-Lifshitz pseudotensor $\Theta^{\mu\nu}$, but the Einstein pseudotensor would serve as well. To define the pseudotensor $\Theta^{\mu\nu}$ as a set of coordinate components, one introduces a quantity

$$H^{\mu\sigma\nu\tau} = |g|(g^{\mu\nu}g^{\sigma\tau} - g^{\mu\tau}g^{\nu\sigma}), \tag{3.73}$$

for which $\partial_\sigma\partial_\tau H^{\mu\sigma\nu\tau}$ is symmetric and $\partial_\nu\partial_\sigma\partial_\tau H^{\mu\sigma\nu\tau}$ vanishes identically by the index structure of $H^{\mu\sigma\nu\tau}$. When the field equation is satisfied,

$$\partial_\sigma\partial_\tau H^{\mu\sigma\nu\tau} = 16\pi|g|(T^{\mu\nu} + t^{\mu\nu}) := 16\pi\Theta^{\mu\nu}, \tag{3.74}$$

where $t^{\mu\nu}$ is quadratic in the first derivatives of the metric. The right side of this equation defines a symmetric pseudotensor $\Theta^{\mu\nu}$ that satisfies

$$\partial_\nu\Theta^{\mu\nu} = \partial_\nu\partial_\sigma\partial_\tau H^{\mu\sigma\nu\tau} = 0, \tag{3.75}$$

when the field equation is satisfied. For a stationary configuration, with a chart for which the asymptotically timelike Killing vector t^α is $\boldsymbol{\partial}_t$, $\Theta^{ij} = \partial_k\partial_l H^{ikjl}$.

We introduce the covariant version of the pseudotensor to state a virial relation for stationary spacetimes. Let Σ be an asymptotically flat spacelike hypersurface of the manifold M, and let η_{ab} be a flat metric that agrees asymptotically to $O(1/r)$ with the spatial metric γ_{ab}. Let x^a be a radial vector field of the flat Euclidean space. Then $\nabla_a x^b = \delta^b_a$. Define a *scalar* γ as the determinant of γ_{ab} with respect to the flat metric η_{ab} by

$$\gamma := \frac{1}{3!}\eta^{ad}\eta^{be}\eta^{cf}\epsilon_{abc}\epsilon_{def}, \tag{3.76}$$

with η^{ab} the *inverse* of η_{ab} – not the result of raising the indices with γ^{ab}. Then γ is the value of the determinant of γ_{ab} in a natural Cartesian chart $\{x^i\}$ of η_{ab}, and $\alpha\sqrt{\gamma}$ is the value of $\sqrt{|g|}$. We can now define tensors H^{acbd}, t^{ab}, and Θ^{ab} by

$$H^{acbd} := \alpha^2\gamma(\gamma^{ac}\gamma^{bd} - \gamma^{ab}\gamma^{cd}), \tag{3.77}$$

$$t^{ab} := \frac{1}{8\pi}\left(\frac{1}{2\alpha^2\gamma}\nabla_c\nabla_d H^{acbd} - G^{ab}\right), \tag{3.78}$$

$$\Theta^{ab} := \alpha^2\gamma(T^{ab} + t^{ab}). \tag{3.79}$$

When the field equation is satisfied,

$$\Theta^{ab} = 16\pi\nabla_c\nabla_d H^{acbd}, \tag{3.80}$$

whence

$$\nabla_b\Theta^{ab} = 0. \tag{3.81}$$

A generalization to stationary spacetimes of the Newtonian virial relation is then

$$0 = -\int_V x^a\eta_{ac}\nabla_b\Theta^{bc}\sqrt{^3\eta}\, d^3x = \int_V \eta_{ab}\Theta^{ab}\sqrt{^3\eta}\, d^3x. \tag{3.82}$$

The surface term vanishes because of the assumed $O(1/r^2)$ behavior of metric derivatives, together with the fact that, outside the fluid, Θ^{ab} is quadratic in the first derivatives of the metric.

3.3.3 2D virial identity

Another integral identity useful in checking the accuracy of numerical computations, derived by Bonazzola and Gourgoulhon [84], is known as the *2-dimensional virial identity* (see also [81] for a previous version). This identity holds for asymptotically flat spacetimes, without the assumption of any symmetry. In order to arrive at the identity, one introduces topologically spherical coordinates r, θ, ϕ on Σ and assumes a foliation of Σ by a family of 2-surfaces Σ_ϕ. Introducing a unit vector normal to Σ_ϕ, one can project the Einstein equations onto Σ_ϕ. The projection takes the form of a scalar Poisson-like equation, for which a Green function can be used to convert it into an integral equation. Using the condition of asymptotic flatness, the integral form simplifies and becomes the 2-dimensional virial identity.

By restricting the 2-dimensional virial identity to stationary, axisymmetric spacetimes, one obtains a useful index of numerical accuracy, namely the vanishing of the quantity

$$GRV2 := |1 - \lambda_2|, \tag{3.83}$$

where

$$\lambda_2 := \frac{8\pi \int_0^\infty \int_0^\pi \left[p + (\epsilon + p)\frac{v^2}{1-v^2}\right] e^{2\mu}\, r\, dr\, d\theta}{\int_0^\infty \int_0^\pi \left[\partial_r\nu\partial_r\nu + \frac{1}{r^2}\partial_\theta\nu\partial_\theta\nu - \frac{3}{4}e^{2(\psi-\nu)}\left(\partial_r\omega\partial_r\omega + \frac{1}{r^2}\partial_\theta\omega\partial_\theta\omega\right)\right] r\, dr\, d\theta}, \tag{3.84}$$

for a metric of the form of (1.59). Like GRV3, GRV2 correlates well with the accuracy of numerical models.

3.4 ADM mass = Komar mass

We introduced the Komar mass [370] of a stationary, asymptotically flat spacetime by Eq. (1.128), namely

$$M_{\rm K} = -\frac{1}{4\pi}\int_{S_\infty} \nabla^\alpha t^\beta dS_{\alpha\beta}. \tag{3.85}$$

For these spacetimes, Beig [59] and Ashtekar and Magnon-Ashtekar [39] prove the equality (3.20) of the Komar and ADM mass. We give here a version [626] of the proof patterned in part on Beig's work.

The Komar mass, as is clear from the form (3.65), involves only the leading term in $t^\alpha t_\alpha = g_{tt}$, whereas the ADM mass involves only the spatial metric. These different dependences correspond to different physical roles. Because the orbits of particles at large r are governed by g_{tt}, the Komar mass is measured by the Kepler relation between the radius and period of distant orbits. The ADM mass, on the other hand, arises as the conserved quantity associated with time translation. Because the two expressions involve different parts of the metric, the equality between them relies on the metric's asymptotically satisfying the vacuum field equation.

The theorem requires slightly stronger asymptotic conditions than those used in Section 3.1 to define mass and angular momentum. Suppose that the metric has the form (2.10), with

$$\gamma_{ab} = \eta_{ab} + h_{ab}, \tag{3.86}$$

$$h_{ab} = O(r^{-1}), \quad \nabla_c h_{ab} = o(r^{-3/2}), \quad \nabla_c \nabla_d h_{ab} = o(r^{-2}), \tag{3.87}$$

and suppose that the lapse, shift, and extrinsic curvature satisfy

$$\alpha = 1 - \frac{\alpha_1}{r} + o(r^{-1}), \quad \partial_r \alpha = \frac{\alpha_1}{r^2} + o(r^{-2}), \quad \partial_r^2 \alpha = -2\frac{\alpha_1}{r^3} + o(r^{-3}); \tag{3.88}$$

$$\beta^a = O(r^{-1}), \quad \nabla_b \beta^a = o(r^{-3/2}); \qquad K_{ab} = o(r^{-3/2}), \quad \nabla_c K_{ab} = o(r^{-2}). \tag{3.89}$$

Then $M_{\rm K} = M_{\rm ADM}$.[4]

To prove the theorem, we begin with the Komar mass in the form (3.65). To relate this form to the ADM mass, it useful to introduce $\delta\, {}^3G_{ab}$, the part of the 3-dimensional Einstein tensor, ${}^3G_{ab} := {}^3R_{ab} - \frac{1}{2}\gamma_{ab}{}^3R$, that is linear in h_{ab}. The idea is to show that if $\delta\, {}^3G_{ab}$ is the asymptotically dominant part of ${}^3G_{ab}$, then

$$\int_{S_\infty} {}^3G^a{}_b x^b \, dS_a = -8\pi M_{\rm ADM}. \tag{3.90}$$

One then uses the field equation for ${}^3G_{ab}$ to show that this integral can be written in the form (3.65) of the Komar mass, when K_{ab} and $\mathcal{L}_{\boldsymbol\beta} K_{ab}$ fall off rapidly enough at spatial infinity.

Formally,

$$\delta\, {}^3G_{ab} = \frac{1}{2}\Big(\nabla_a \nabla_c h_b{}^c + \nabla_b \nabla_c h_a{}^c - \nabla^2 h_{ab} - \nabla_a \nabla_b h^c{}_c + \eta_{ab} \nabla^2 h^c{}_c - \eta_{ab} \nabla_c \nabla_d h^{cd}\Big), \tag{3.91}$$

where the index of h_{ab} is raised by the flat metric η^{ab}. The asymptotic behavior of ${}^3G_{ab}$ is given by

$${}^3G_{ab} - \delta^3 G_{ab} = o(r^{-3}). \tag{3.92}$$

This is because each term in ${}^3G_{ab}$ involves either $\nabla_c \nabla_d h_{ab}$ or $\nabla_c h_{ab} \nabla_d h_{ef}$; then terms quadratic or higher order in h_{ab} fall off as rapidly as either $h_{ef} \nabla_c \nabla_d h_{ab}$ or $\nabla_c h_{ab} \nabla_d h_{ef}$.

From the linearized Bianchi identity, $\nabla_a \delta\, {}^3G^a{}_b = 0$, we have

$$\int_{S_\infty} {}^3G^a{}_b x^b dS_a = \int_{S_\infty} \delta\, {}^3G^a{}_b x^b dS_a = \int_V \nabla_a(\delta\, {}^3G^a{}_b\, x^b)\sqrt{\eta} d^3x$$
$$= \int_V \delta\, {}^3G_{ab}\, \eta^{ab} \sqrt{\eta} d^3x. \tag{3.93}$$

[4] The equality also follows with the slightly altered conditions, $h_{ab} = O(r^{-1-\epsilon})$, $\nabla_c \nabla_d h_{ab} = o(r^{-2+\epsilon})$, $\beta^a = O(r^{-1-\epsilon})$ and $\nabla_c K_{ab} = o(r^{-2+\epsilon})$.

Now $\delta\,^3G_{ab}\,\eta^{ab}\sqrt{\eta}$ is a simple divergence,

$$\eta^{ab}\delta\,^3G_{ab} = -\frac{1}{2}\nabla_a\nabla_b h^{ab} + \frac{1}{2}\nabla^2 h_b{}^b = \frac{1}{2}\nabla_a\left(\nabla^a h_b{}^b - \nabla_b h^{ab}\right), \tag{3.94}$$

and we return to a surface integral,

$$\int_{S_\infty} {}^3G^a{}_b x^b\, dS_a = -\frac{1}{2}\int_{S_\infty}\left(\nabla_b h^{ab} - \nabla_a h^c{}_c\right) dS_a =: -8\pi M_{\rm ADM}. \tag{3.95}$$

As written, the integrations by parts in Eqs. (3.93) and (3.95) appear to assume an interior with no boundary, a restriction that eliminates typical hypersurfaces of black-hole spacetimes; Beig explicitly takes $V = \mathbb{R}^3$. This restriction, however, is unnecessary. Because the steps from (3.93) to (3.95) do not use the field equation, one can replace h_{ab} by any smooth symmetric tensor field that agrees with h_{ab} for r greater than some radius R_2 and that vanishes for r less than some smaller radius R_1 outside all interior boundary points of the hypersurface V. The verification of the identity (3.95) is then valid as written. (An alternate way of going from Eqs. [3.93] to [3.95] that involves only surface integrals at infinity is given as a footnote in [626].)

Next, from Eqs. (2.23) and (2.21), ${}^3G_{ab}$ has, outside the matter, the form

$$\begin{aligned}
{}^3G_{ab} &= {}^3R_{ab} - \frac{1}{2}\gamma_{ab}{}^3R \\
&= \frac{1}{\alpha}D_a D_b\alpha - \frac{1}{2}\gamma_{ab}\frac{1}{\alpha}D^2\alpha + 2K_{ac}K_b{}^c - K_{ab}K \\
&\quad -\gamma_{ab}\left(K_{cd}K^{cd} - \frac{1}{2}K^2\right) + \frac{1}{\alpha}\mathcal{L}_{\boldsymbol\beta}K_{ab} - \frac{1}{2\alpha}\gamma_{ab}\gamma^{cd}\mathcal{L}_{\boldsymbol\beta}K_{cd}.
\end{aligned} \tag{3.96}$$

Because the terms involving K_{ab} are $o(r^{-3})$, we have

$$\int_{S_\infty} {}^3G^a{}_b x^b\, dS_a = \int_{S_\infty} \nabla_a\nabla_b\alpha\, \hat r^a \hat r^b r^3 d\Omega. \tag{3.97}$$

Now $\nabla_c\gamma_{ab} = o(r^{-3/2})$ and $\nabla_a\alpha = O(r^{-2})$ imply

$$D_r{}^2\alpha = \nabla_r{}^2\alpha + o(r^{-7/2}). \tag{3.98}$$

Finally, from Eq. (3.65) and the asymptotic form (3.88) of α, we have

$$M_{\rm K} = \frac{1}{4\pi}\int_{S_\infty}\alpha_1 d\Omega, \tag{3.99}$$

implying

$$\int_{S_\infty} {}^3G^a{}_b x^b dS_a = \int_{S_\infty}\nabla_r{}^2\alpha r^3 d\Omega = -8\pi M_{\rm K}, \tag{3.100}$$

whence

$$M_{\rm ADM} = M_{\rm K}. \tag{3.101}$$

3.5 First law of thermodynamics for relativistic stars

The first law of thermodynamics is a statement of energy conservation along a sequence of equilibrium configurations. For a rotating star, it relates the change in gravitational mass (i.e., total energy) to the change in its baryon number, entropy and angular momentum. We begin with a brief derivation for a local perturbation, in which the physics is transparent, and then proceed to a full derivation involving a general variation of the action.

Injection energy: first law for a group of baryons dropped from infinity. This section is drawn from previous discussions by Thorne [687], by Zel'dovich and Novikov [750] (with a section on injection energy by Thorne), and by Carter [117]. Consider a stationary, axisymmetric star with fluid velocity $u^\alpha = u^t(t^\alpha + \Omega\phi^\alpha)$. The contents of a box at infinity are to be injected into the star at a point P. The box holds a collection of baryons having the same composition as the matter at the point P, and the box, including its contents, has total energy δM_1, baryon number δN, entropy δS, and angular momentum δJ. For the matter to be injected in a thermodynamically reversible process, the specific entropy s of the box is to be that of the star at the point P, so that $\delta S = s\delta M_0$, where $\delta M_0 = m_B \delta N$ is the baryon mass of the δN baryons. Denoting by $p_{1\alpha}$ the initial 4-momentum of the box, we have

$$\delta M_1 = -p_{1\alpha} t^\alpha, \qquad \delta J = p_{1\alpha}\phi^\alpha. \tag{3.102}$$

Imagine the box falling freely to the point P of the star; when it reaches P, it has energy measured by an observer at rest with respect to the fluid

$$\begin{aligned} \delta E_1 &= -p_{1\alpha} u^\alpha = -u^t(p_{1\alpha}t^\alpha + \Omega p_{1\alpha}\phi^\alpha) \\ &= u^t(\delta M_1 - \Omega\delta J), \end{aligned} \tag{3.103}$$

where $u^\alpha = u^\alpha(P)$ is the fluid 4-velocity. Now, following Thorne, we suppose the observer catches the box at P and reversibly injects its contents into the fluid, imparting to the fluid an energy

$$\delta E = T\delta S + \mu\delta N = T\delta S + g\delta M_0, \tag{3.104}$$

where g is the specific Gibbs free energy, $g = (\epsilon + p)/\rho - Ts$, introduced in Eq. (1.11).

Because the initial specific entropy was already s, not all of the available energy is used: Our active observer uses the remaining energy to throw the empty box back up to infinity, on a trajectory with zero angular momentum, so that the angular momentum δJ is retained by the fluid. The returning box has energy $\delta E_2 = \delta E_1 - \delta E = p_{2\alpha}u^\alpha$, with $p_{2\alpha}$ the momentum with which it is thrown. Because its free trajectory conserves $p_{2\alpha}t^\alpha$, the box reaches infinity with redshifted energy $\delta M_2 = p_{2\alpha}t^\alpha = \dfrac{\delta E_2}{u^t}$. The change in mass of the star is then related to the change in baryon number, entropy, and angular momentum of the fluid by

$$\delta M = \delta M_1 - \delta M_2 = \frac{1}{u^t}(\delta E_1 - \delta E_2) + \Omega\delta J = \frac{\delta E}{u^t} + \Omega\delta J$$

$$= \frac{\mu}{u^t}\delta N + \frac{T}{u^t}\delta S + \Omega\delta J \quad (3.105a)$$

$$= \frac{g}{u^t}\delta M_0 + \frac{T}{u^t}\delta S + \Omega\delta J. \quad (3.105b)$$

The coefficient of δM_0 is the energy $\mathcal{E}$, the injection energy per unit rest mass of matter with zero initial entropy and angular momentum. Equation (3.105) is the first law of thermodynamics for relativistic stars (in two equivalent forms).

As we have seen in Section 1.5, $\mathcal{E}$ is constant in a star with constant entropy per baryon and constant angular velocity Ω. The first law provides the following interpretation of this Poincaré-Wavre theorem. In general relativity, an equilibrium configuration is an extremum of mass at fixed angular momentum, entropy, and baryon number: Small changes in the structure of the star leave the mass fixed to linear order in the perturbation. In particular, suppose one moves a ring of fluid from one location to another in a uniformly rotating white dwarf or neutron star. (Like neutron stars, white dwarfs can be modeled as zero-entropy fluids, because T is approximately zero; because white dwarfs are supported by degenerate electrons, the condition is $kT \ll E_F$, with E_F the Fermi energy of the electrons.) Using $g = h - Ts \approx h$, we have $\mathcal{E} = \frac{h}{u^t}$. Changing the location of the ring is equivalent to moving it out to infinity and back in to a new location in the star. According to Eq. (3.105b) with $T = 0$ and Ω constant,

$$\delta M = \left[\left(\frac{h}{u^t}\right)_2 - \left(\frac{h}{u^t}\right)_1\right]\delta M_0. \quad (3.106)$$

Requiring $\delta M = 0$ for small displacements of a fluid ring of a uniformly rotating star then implies

$$\left(\frac{h}{u^t}\right)_2 = \left(\frac{h}{u^t}\right)_1. \quad (3.107)$$

That is, the first law immediately implies that an equilibrium can be an extremum of the mass only if the injection energy is constant throughout the star.

Although the reversible injection of the matter in the box preserves the local thermodynamic equilibrium, the star itself is no longer in gravitational equilibrium after the process described in our heuristic derivation of the first law. That the ignored global corrections to the energy are of second order – that the first-order change in the star's mass is in fact given by the first law, Eq. (3.105) – is an implication of the formal derivation of the next section.

First law from a variation of the action. Most of the work to obtain the first law has already been done in evaluating the variation of the action, Eq. (2.100). We will integrate this equation over a spacelike hypersurface for a perturbation that joins nearby equilibria and will then directly evaluate the integral of the left side, showing that it has the form $\delta M - \int \Omega \Delta dJ$.

Consider a family of rotating stars, stationary, axisymmetric perfect-fluid spacetimes, all with angular momentum in the same direction. We can then choose a gauge for which the timelike and rotational Killing vectors, t^α and ϕ^α, are the same for each member of the family,

$$\delta t^\alpha = 0, \qquad \delta\phi^\alpha = 0. \tag{3.108}$$

Choosing to identify the Killing vectors is equivalent to choosing a gauge in which the metric components are independent of the same natural coordinates t and ϕ of Eqs. (1.63a) and (1.63b).[5] It is with respect to the coordinates t, ϕ that the metric takes the form (1.59) and the fluid has 4-velocity

$$u^\alpha = u^t k^\alpha, \tag{3.109}$$

where $k^\alpha = t^\alpha + \Omega\phi^\alpha$.

To compute the change in the mass M, it is then convenient to evaluate M on a $t =$ constant surface Σ_t. To do so for each member of the family of stars, we use the second kind of gauge freedom (2.147) in the choice of a Lagrangian displacement ξ^α, adding a trivial displacement of the form Gu^α to make $\xi^t = \xi^\alpha\nabla_\alpha t = 0$. Then ξ^α joins corresponding fluid elements in each $t =$ constant surface.

We are now ready to integrate Eq. (2.100) over the $t =$ constant hypersurface Σ_t, suppressing the subscript t. (Equivalently, one can regard Σ as the fixed hypersurface Σ identified with each Σ_t in the 3+1 discussion of Section 2.1.)

$$\int_\Sigma (-\delta\mathscr{L} + \nabla_\alpha \mathbf{\Theta}^\alpha) d^3x = \int_\sigma \Big[\rho T\Delta s + \frac{h}{u^t\sqrt{|g|}}\Delta(\rho u^t\sqrt{|g|}) \\ + \frac{1}{16\pi}(G^{\alpha\beta} - 8\pi T^{\alpha\beta})h_{\alpha\beta} + \xi_\alpha\nabla_\beta T^{\alpha\beta}\Big] t^\gamma dS_\gamma. \tag{3.110}$$

We have used $t^\alpha dS_\alpha = \sqrt{|g|}d^3x$ and have replaced $(\Delta u)^t$ by $\Delta(u^t)$, using $\Delta t = \delta t + \mathcal{L}_{\boldsymbol{\xi}} t = \xi^\alpha\nabla_\alpha t = 0$.

Writing

$$dM_0 = \rho u^\alpha dS_\alpha, \qquad dS = s\, dM_0, \tag{3.111}$$

we have

$$\Delta dM_0 = \Delta(\rho u^t\sqrt{|g|})d^3x, \qquad \Delta dS = \Delta s\, dM_0 + s\,\Delta dM_0. \tag{3.112}$$

[5] The diffeo relating a stationary axisymmetric spacetime with Killing vectors t^α and ϕ^α to one with Killing vectors $\tilde{t}^\alpha$ and $\tilde{\phi}^\alpha$ simply maps a point $\tilde{P}$ with coordinates $\tilde{t}, \tilde{\phi}, \tilde{r}, \tilde{\theta}$ to the point P for which $t = \tilde{t}, \phi = \tilde{\phi}, r = \tilde{r}, \theta = \tilde{\theta}$. One can in this way choose a diffeo for which two axisymmetric stars are described by spacetimes with the same rotational Killing vector.

Then Eq. (3.110) takes the form

$$\int_\Sigma (-\delta\mathscr{L} + \nabla_\alpha \boldsymbol{\Theta}^\alpha)\, d^3x = \int_\Sigma \left(\frac{T}{u^t}\Delta dS + \frac{h-\rho T s}{u^t}\Delta dM_0\right) + \int_\Sigma \left[\frac{1}{16\pi}(G^{\alpha\beta} - 8\pi T^{\alpha\beta})h_{\alpha\beta} + \xi_\alpha \nabla_\beta T^{\alpha\beta}\right] t^\gamma\, dS_\gamma. \quad (3.113)$$

What remains is a direct evaluation of the left side of this equation. We will find for each of the terms the simple expressions

$$-\int_\Sigma \delta\mathscr{L} d^3x = \frac{1}{2}\delta M - \int_\Sigma \Delta(\Omega dJ) \quad (3.114a)$$

$$\int_\Sigma \nabla_\alpha \boldsymbol{\Theta}^\alpha d^3x = \frac{1}{2}\delta M + \int_\Sigma (\Delta\Omega) dJ, \quad (3.114b)$$

where $dJ := T^\alpha{}_\beta \phi^\beta dS_\alpha$; Eqs. (3.113) and (3.114) immediately imply the first law.

We consider each term in turn. Recalling the definition (2.96), we write

$$-\int_\Sigma \mathscr{L} d^3x = -\int_\Sigma \left(\frac{1}{16\pi}R - \epsilon\right) t^\alpha dS_\alpha.$$

We replace the term involving ϵ using the relation

$$\begin{aligned} -T^\alpha{}_\beta t^\beta dS_\alpha &= -T^\alpha{}_\beta (t^\beta + \Omega\phi^\beta) dS_\alpha + T^\alpha{}_\beta \Omega \phi^\beta dS_\alpha \\ &= \epsilon t^\alpha dS_\alpha + \Omega\, dJ; \end{aligned} \quad (3.115)$$

and we replace $T^\alpha{}_\beta$ by $G^\alpha{}_\beta$, writing $T^\alpha{}_\beta = \frac{1}{8\pi}G^\alpha{}_\beta - \frac{1}{8\pi}(G^\alpha{}_\beta - 8\pi T^\alpha{}_\beta)$, where, for completeness, we keep the terms involving the field equations. We then have

$$\begin{aligned} -\int_\Sigma \mathscr{L} d^3x &= \int_\Sigma \left[-\frac{1}{8\pi}R^\alpha{}_\beta t^\beta dS_\alpha - \Omega dJ + \frac{1}{8\pi}(G^\alpha{}_\beta - 8\pi T^\alpha{}_\beta) t^\beta dS_\alpha\right] \\ &= \frac{1}{2}M - \int_\Sigma \left[\Omega dJ - \frac{1}{8\pi}(G^\alpha{}_\beta - 8\pi T^\alpha{}_\beta) t^\beta dS_\alpha\right], \end{aligned}$$

$$-\int_\Sigma \delta\mathscr{L} d^3x = \frac{1}{2}\delta M - \delta\int_\Sigma \left[\Omega dJ - \frac{1}{8\pi}(G^\alpha{}_\beta - 8\pi T^\alpha{}_\beta) t^\beta dS_\alpha\right]. \quad (3.116)$$

For a family of equilibria, the perturbed field equations are satisfied, and we obtain Eq. (3.114a).

Next we evaluate $\int_\Sigma \nabla_\alpha \boldsymbol{\Theta}^\alpha$. From the coordinate form $\partial_t \boldsymbol{\Theta}^t + \partial_i \boldsymbol{\Theta}^i$ of the divergence (or its covariant equivalent, $\mathcal{L}_{\boldsymbol{t}} \boldsymbol{\Theta}^\alpha \nabla_\alpha t + D_a \boldsymbol{\Theta}^a$), we have

$$\int_\Sigma \nabla_\alpha \boldsymbol{\Theta}^\alpha d^3x = \int_\Sigma \mathcal{L}_{\boldsymbol{t}} \boldsymbol{\Theta}^\alpha d\sigma_\alpha + \int_{S_\infty} \boldsymbol{\Theta}^a d\sigma_a. \quad (3.117)$$

Despite the fact that the perturbed spacetime is time independent, the first term does not vanish, because the Lagrangian displacement ξ^α depends on time: The separation between fluid elements with different angular velocities grows with time. From the definition (2.98) of $\boldsymbol{\Theta}^\alpha$, we have

$$\mathcal{L}_{\boldsymbol{t}} \boldsymbol{\Theta}^\alpha dS_\alpha = (\epsilon + p) q^\alpha{}_\beta\, \mathcal{L}_{\boldsymbol{t}} \xi^\beta dS_\alpha = (\epsilon + p) u^\alpha u_\beta\, \mathcal{L}_{\boldsymbol{t}} \xi^\beta dS_\alpha, \quad (3.118)$$

where we have used $\xi^\alpha dS_\alpha = 0$. From Eq. (3.109) for u^α and Eq. (2.37) for Δu^α, we have $\Delta u^\alpha = \frac{1}{2} u^t k^\alpha u^\beta u^\gamma \Delta g_{\beta\gamma} = (\Delta u^t) k^\alpha$. Because $\Delta u^\alpha = \Delta(u^t k^\alpha) = (\Delta u^t) k^\alpha + u^t \Delta k^\alpha$, we conclude $\Delta k^\alpha = 0$, whence

$$\mathcal{L}_\xi t^\alpha + \Omega \mathcal{L}_\xi \phi^\alpha + \Delta\Omega \phi^\alpha = 0 \tag{3.119}$$

$$\mathcal{L}_t \xi^\alpha = \Delta\Omega \phi^\alpha - \Omega \mathcal{L}_\phi \xi^\alpha . \tag{3.120}$$

Then, from Eqs. (3.118) and (3.120), we have

$$\begin{aligned}\int_\Sigma \mathcal{L}_{\boldsymbol{t}} \Theta^\alpha dS_\alpha &= \int_\Sigma \{\Delta\Omega(\epsilon + p) u^\alpha u_\beta \phi^\beta dS_\alpha - \mathcal{L}_{\boldsymbol{\phi}}[(\epsilon + p) u^\alpha u_\beta \xi^\beta dS_\alpha]\} \\ &= \int_\Sigma \Delta\Omega dJ,\end{aligned} \tag{3.121}$$

where the term involving $\mathcal{L}_{\boldsymbol{\phi}}$ vanishes because, for a scalar density $\mathfrak{f}$, $\mathcal{L}_{\boldsymbol{\phi}}\mathfrak{f} = \nabla_\alpha(\mathfrak{f}\phi^\alpha) = D_a(\mathfrak{f}\phi^a)$.

The second term in Eq. (3.117) is nearly identical to the expression (3.2) for the ADM mass, except that $\boldsymbol{\Theta}^\alpha$ involves 4-dimensional indices. The only changes in the evaluation (3.5) are then the replacement of M by δM and the replacement of of $h_b{}^b$ by the 4-dimensional trace $h_\beta{}^\beta$ whose asymptotic form in isotropic coordinates is $4\delta M/r$:

$$\int_{S_\infty} \boldsymbol{\Theta}^a d\sigma_a = \frac{1}{16\pi} \int_{S_\infty} (\nabla_b h^{ab} - \nabla^a h_\beta{}^\beta) \nabla_a r\, r^2 d\Omega = \frac{1}{2}\delta M. \tag{3.122}$$

Equations (3.121) and (3.122) give the form (3.114b) of $\int_\Sigma \nabla_\alpha \boldsymbol{\Theta}^\alpha d^3x$.

Finally, from Eq. (3.113), with left side given by the sum of Eqs. (3.116) and (3.114b), we obtain the desired expression for δM,

$$\begin{aligned}\delta M = &\int_\Sigma \left(\frac{T}{u^t} \Delta dS + \frac{g}{u^t} \Delta dM_0 + \Omega \Delta dJ \right) \\ &- \delta \int_\Sigma \frac{1}{8\pi} (G^\alpha{}_\beta - 8\pi T^\alpha{}_\beta) t^\beta \, dS_\alpha \\ &+ \int_\Sigma \left[\frac{1}{16\pi} (G^{\alpha\beta} - 8\pi T^{\alpha\beta}) h_{\alpha\beta} + \xi^\beta \nabla_\alpha T^\alpha{}_\beta \right] t^\gamma dS_\gamma,\end{aligned} \tag{3.123}$$

whence, for a family of equilibria, the first law is given by

$$\delta M = \int_\Sigma \left(\frac{T}{u^t} \Delta dS + \frac{g}{u^t} \Delta dM_0 + \Omega \Delta dJ \right). \tag{3.124}$$

Because neutron stars are effectively zero-temperature barotropic stars, the injection energy $\mathcal{E} = g/u^t = h/u^t$ is constant, and nearby uniformly rotating equilibria satisfy the first law in the form

$$\delta M = \bar{\mu} \delta N + \Omega \delta J = \mathcal{E} \delta M_0 + \Omega \delta J, \tag{3.125}$$

where $\bar{\mu} := \mu/u^t = \mathcal{E} m_B$ is the constant value of the injection energy per baryon.

If, on the other hand, one considers perturbations that fix the baryon number, entropy, and angular momentum of the fluid, and if the unperturbed field equation is satisfied, then one expects stationary variations of the Hamiltonian to yield the unperturbed field equations [596]. That is not quite the case here, because part of the perturbed constraint equations appears as well. The reason is that the mass that is varied coincides with the first-order Hamiltonian (as in Eq. (3.18)), only when the constraint equations are satisfied.[6] Finally, although Eq. (3.123) is restricted to stationary, axisymmetric perturbations, a slight rewriting of the term $\Omega\Delta dJ$ in that equation gives a form that holds for nonaxisymmetric perturbations: Writing $dJ = (\epsilon + p)u^\alpha u_\beta \phi^\beta dS_\alpha = \phi^\beta h u_\beta\, \rho u^\alpha dS_\alpha = \phi^\beta h u_\beta\, dM_0$, we have

$$\Omega\Delta dJ = \Omega\phi^\beta \Delta(hu_\beta)dM_0 + \Omega\phi^\beta hu_\beta \Delta dM_0, \tag{3.126}$$

and Eq. (3.123) becomes

$$\begin{aligned}\delta M = &\int_\Sigma \left[\frac{T}{u^t}\Delta dS + \left(\frac{g}{u^t} + \Omega\phi^\beta hu_\beta\right)\Delta dM_0 + \Omega\phi^\beta\Delta(hu_\beta)dM_0\right] \\ &-\delta\int_\Sigma \frac{1}{8\pi}(G^\alpha{}_\beta - 8\pi T^\alpha{}_\beta)t^\beta dS_\alpha \\ &+\int_\Sigma \left[\frac{1}{16\pi}(G^{\alpha\beta} - 8\pi T^{\alpha\beta})h_{\alpha\beta} + \xi^\beta\nabla_\alpha T^\alpha{}_\beta\right] t^\gamma dS_\gamma .\end{aligned} \tag{3.127}$$

3.6 Nonaxisymmetric equilibria

It seems intuitively clear that nonrotating stars must be spherical, but the result has only recently been proved for generic equations of state: A theorem by Masood-ul-Alam [465], assuming a one-parameter equation of state with p an increasing function of ϵ, shows that static solutions to the Einstein-Euler system are spherically symmetric. At least in a Newtonian context, however, the intuitive notion that rotating stars are axisymmetric is false, and analogous nonaxisymmetric equilibria are likely to exist as solutions to the Einstein-Euler system.

We will be concerned in this section with two kinds of nonaxisymmetric equilibria: nonaxisymmetric stellar models [129] that, like the Dedekind sequence of triaxial ellipsoids, are stationary in an *inertial frame*; and nonaxisymmetric stellar models that, like the Jacobi ellipsoids, are stationary in a *rotating frame*. In Section 6.2 we consider a third kind, binary systems stationary in a rotating frame. These equilibria play a fundamental role in the Newtonian description of a variety of astrophysical systems, including binaries, nonaxisymmetric modes of rotating stars, and spiral and elliptical galaxies. In relativistic systems, their role is diminished for two related reasons: Nonaxisymmetric systems compact enough to require a general relativistic description often evolve too quickly to be modeled with high precision as equilibria; and the Newtonian equilibria that are stationary in a rotating frame – binaries and nonaxisymmetric configurations like the Jacobi ellipsoids – have stationary

[6] The mass in Eq. (3.127) is the ADM mass: In Eq. (3.124), $\frac{1}{2}\delta M = \delta M_{\rm ADM} - \frac{1}{2}\delta M_K$, with h_{tt} contributing the Komar mass term: and this Komar term exactly cancels the Komar mass contribution from (3.122a).

counterparts in the full theory only as solutions with equal amounts of ingoing and outgoing gravitational radiation.

Nevertheless, until the final few orbits prior to merger, the radiation-reaction timescale for a compact binary is much longer than its period. The radiation-reaction time for an outgoing mode of a neutron star is also much longer than the mode's period (exceptions are the w-modes, modes that are nearly decoupled from the fluid). As a result, the orbits of compact binaries have been modeled by *quasiequilibrium sequences*, sequences of equilibrium models that approximate exact solutions. An estimate of the radiated energy is used to relate one equilibrium solution to another in the sequence. Equilibrium models that have been constructed include solutions to the post-Newtonian equations without radiation reaction (including only integer orders in the post-Newtonian expansion); solutions to the relativistic Euler equation together with the initial value equations; this last set of equations, together with the spatial trace of the Einstein equation, solved for a spatially conformally flat metric (the Isenberg-Wilson-Mathews [IWM] approximation); and solutions on an initial hypersurface to the full Einstein-Euler system, with the asymptotic behavior of a nonradiative system (the "waveless" ansatz). Such approximate solutions will be discussed in Chapter 6. Finally, solutions to the full Einstein-Euler system that are stationary in a rotating frame (and so have equal amounts of ingoing and outgoing radiation) have been constructed in a region whose radius is about one quadrupole wavelength ($R \sim \pi c/\Omega$). Outside the smaller radius $1/\Omega$, an observer would have to move faster than light to rotate with the star, so the phrase "stationary in a rotating frame" is misleading, replaced by *helically symmetric.*

The nonaxisymmetric configurations that can be genuinely stationary in general relativity are analogs of the triaxial Dedekind ellipsoids. The initial interest in these models came from the fact that rapidly rotating axisymmetric stars are unstable to a bar mode driven by gravitational radiation. Work by several authors suggested that the unstable stars would evolve to stable, Dedekind-like configurations. In the next section we describe the Dedekind sequence and argue that analogous nonaxisymmetric equilibria are likely to exist – that there are stationary, nonaxisymmetric perfect-fluid spacetimes. We review work on the possibility that they are temporary endpoints for the evolution of unstable stars. Finally, we present a result by Lindblom showing that, although perfect fluids may have nonaxisymmetric equilibria, dissipative fluids do not: The only stationary configurations of fluids with viscosity and heat flow are axisymmetric (and uniformly rotating).

3.6.1 Dedekind-like configurations

Stationary nonaxisymmetric models in general relativity?

The Dedekind sequence [129] comprises a set of uniform-density ellipsoids in the context of Newtonian gravity. Because they are ellipsoids, each configuration can be described by giving its three (unequal) principal axes $a_1 > a_2 > a_3$ and its velocity field. The fluid trajectories are a family of similar ellipses in the plane of the two longest principal axes, a_1 and a_2. The velocity field is linear in Cartesian coordinates, and its vorticity $\boldsymbol{\zeta} = \boldsymbol{\nabla} \times \mathbf{v}$ is thus constant. In coordinates x, y, z whose axes lie along the principal axes of the ellipsoid, the velocity field has the form

$$\mathbf{v} = \frac{\zeta}{a_1^2 + a_2^2}(-a_2^2\, y\, \hat{\mathbf{x}} + a_1^2\, x\, \hat{\mathbf{y}}), \tag{3.128}$$

where $\hat{\mathbf{x}}$ and $\hat{\mathbf{y}}$ are unit vectors.

The Dedekind sequence branches off the Maclaurin sequence of uniform density, uniformly rotating configurations where dimensionless measures of rotation have the values

$$a_3/a_1 = 0.58, \quad \Omega_0 = \frac{1}{2}\zeta_0 = 0.612\sqrt{\pi\rho}, \quad T/|W| = 0.1375, \tag{3.129}$$

where T is the kinetic energy, $T = \frac{1}{2}\Omega J$, and W the gravitational potential energy of the configuration. Along the Dedekind sequence the vorticity increases, the polar radius decreases, and the star approaches a cigar shape axisymmetric about the a_1 axis. That is, as $a_3/a_1 \to 0, \quad a_3/a_2 \to 1$.

Beyond the bifurcation point, the axisymmetric sequence is secularly unstable to a bar mode (an $l = m = 2$ instability) driven by gravitational waves [130]. (A secular instability is one that proceeds on a timescale long compared to a dynamical time and set by a dissipative mechanism such as viscosity, nuclear reactions, or gravitational radiation.) At the bifurcation point, the bar mode has zero frequency: It is a time-independent solution to the linearized Einstein-Euler system. The Dedekind sequence begins as this zero-frequency mode; that is, if λ is a parameter along the Dedekind sequence, with $\lambda = \lambda_0$ at the bifurcation point (for example, $\lambda = \zeta/\zeta_0$), the Eulerian changes, $\delta Q = \frac{d}{d\lambda}Q\big|_{\lambda_0}$, in the fluid variables and the potential satisfy the linearized equilibrium equations. Because nonaxisymmetric instabilities set in through zero-frequency modes in the full theory (and in the Newtonian approximation supplemented by a radiation-reaction term), a plausible conjecture is that a sequence of nonaxisymmetric equilibria, whose initial symmetry is that of a zero-frequency mode, branches at each instability point:

Conjecture. There are stationary, nonaxisymmetric stellar models – stationary, asymptotically flat solutions to the Einstein-relativistic Euler equations.

In fact, the investigations by Lyapunov and Poincaré of sequences of ellipsoids bifurcating in this way underlay the development of bifurcation theory; among its principal results are theorems showing for a variety of systems that, along a sequence of equilibria, zero-frequency modes (time-independent solutions to the linearized system) mark bifurcation points, where a second sequence of equilibria, agreeing to linear order with the mode, branches off. Ironically, however, for the system that spurred the theory – the case of self-gravitating perfect fluids – few results have been proved. Lyapunov's original work [412] was restricted to modes with zero frequency in the rotating frame of an equilibrium star. For these, he proved the existence of corresponding bifurcating sequences of uniformly rotating, nonaxisymmetric configurations, objects such as the Jacobi ellipsoid that we discuss in the next section. A paper by Lewis [409] does use bifurcation theory to prove existence of the Dedekind ellipsoids (and of the more general class of Riemann ellipsoids), but the work is restricted to these already known classical solutions. To our knowledge, there is no proof of existence of compressible analogs of the Dedekind ellipsoids or of other Newtonian sequences that are stationary in an inertial frame and are associated with modes belonging to angular harmonics other than that of the $m = 2$ bar mode. In a relativistic context, existence proofs are so far restricted to spherical stars [564].

Uniformly rotating Newtonian polytropes are stiff enough to allow a zero-frequency $m = 2$ mode when the polytropic index N is less than about 0.808 ($\Gamma > 2.238$) [340]. The bifurcation argument suggests that there should be corresponding Dedekind-like solutions for these polytropes, and their existence, at least for $N \leq 0.5$, has been numerically confirmed in work by Uryū and Eriguchi [704]. Using a radial coordinate adapted to the nonaxisymmetric surface, they construct models with meridional circulation that are stationary in an inertial frame and that have the reflection symmetries of the Dedekind ellipsoids, invariant under each of the three discrete symmetries $t \to -t, x \to -x;\quad t \to -t, y \to -y;$ and $t \to -t, z \to -z$. Following a conjecture by Ipser and Managan [330], Chambat [125] showed that compressible Dedekind-like equilibria could not have a velocity field that is both planar and divergence-free, and the Uryū-Eriguchi numerical work appears to imply the stronger result that the velocity field can be neither planar nor divergence-free.

Work by Chandrasekhar and Elbert [133] provides the strongest evidence for the existence of Dedekind-like configurations in general relativity. They find a sequence of explicit solutions to the Einstein-Euler equations for a uniform-density fluid at first post-Newtonian order. The post-Newtonian sequence differs from the Dedekind sequence by a "third-harmonic" deformation, a Lagrangian displacement cubic in the Cartesian coordinates. The Chandrasekhar-Elbert work restricted the post-Newtonian contributions to the velocity field to terms cubic in the Cartesian coordinates, and, as Gurlebeck and Petroff [282] point out, this restriction precludes an axisymmetric limit. By including post-Newtonian corrections linear in the coordinates, they appropriately extend the post-Newtonian Dedekind family to include post-Newtonian Maclaurin spheroids as axisymmetric members of the more general Dedekind sequences.

Two relativistic effects one might think of first as reasons why Dedekind-like solutions may not exist in general relativity are the departure from a Coulomb force law and dragging of inertial frames. The first of these precludes a stationary general relativistic analog of a stationary, elliptical ring of dust around a point mass in Newtonian gravity, a simplest, although singular, nonaxisymmetric equilibrium fluid. The second – frame dragging – seems intuitively capable of forbidding a non-axisymmetric shape when the fluid motion leads to frame dragging in a direction transverse to the star's boundary. The fact that both effects are already present at first post-Newtonian order effectively counters these initial intuitive arguments against Dedekind-like solutions. The existence of zero-frequency modes, stationary nonaxisymmetric solutions to the linearized Einstein-Euler equations to all post-Newtonian orders (Section 8.7), similarly adds weight to the conjecture that Dedekind-like solutions exist.

Astrophysical implications. We begin by reviewing the expectation that gravitational radiation will drive an unstable star to a Dedekind-like configuration; we then discuss more recent work that has made the astrophysical importance of Dedekind-like solutions less clear.

Maclaurin spheroids beyond the instability point, where the Dedekind sequence branches off, have higher energy $T + W$ than the corresponding Dedekind ellipsoid with the same mass and vorticity. Because relativistic vorticity is exactly conserved, the post-Newtonian radiation-reaction force conserves Newtonian vorticity $\boldsymbol{\zeta}$ and allows an unstable Maclaurin spheroid to evolve to the lower energy Dedekind ellipsoid. A first computation was done by Bonnie Miller [473], who found an evolution that begins at an unstable Maclaurin spheroid and ends at the expected Dedekind model with the same M and $\boldsymbol{\zeta}$. Because of the quadrupole character of the radiation reaction, the evolution proceeds through a set of ellipsoidal configurations for which the velocity field remains linear in the coordinates. The uniform density equilibria of this kind are the Riemann ellipsoids, and the evolutionary path is uniquely specified by the conditions

1. M and $\boldsymbol{\zeta}$ are constant;
2. $\boldsymbol{\Omega}$ is parallel to $\boldsymbol{\zeta}$ (because of the reflection symmetry in the equatorial plane of the perturbed configuration);
3. The first law, $dE = \Omega dJ$, relating neighboring equilibrium configuration implies that the energy and angular momentum radiated satisfy $dE/dt = \Omega dJ/dt$.

A plausible conclusion from Miller's computation is that, with no viscosity, rotating stars perturbed away from axisymmetry evolve to the Dedekind sequence, with the bar mode having, in effect, become the zero-frequency nonaxisymmetric shape of the Dedekind ellipsoid. This raises the question of whether a nascent neutron star rotating fast enough to be unstable would in fact evolve to a Dedekind-like configuration, with an equilibrium shape that reflects the angular behavior of the most unstable mode. That outcome now seems less likely, for two reasons. First, Lindblom [417] showed that any Dedekind-like configuration – any nonaxisymmetric perfect-fluid model stationary in an inertial frame – has a velocity field with nonzero shear. Any nonzero viscosity precludes a stationary nonaxisymmetric equilibrium, and the studies that followed Miller (see, for example, [183], [384], [613]) found that a star initially unstable to a bar mode ultimately evolves to an axisymmetric model, after having radiated enough angular momentum to reach a stable, axisymmetric, uniformly rotating configuration. For low viscosity, evolutions restricted to ellipsoidal configurations found that the evolving star first approaches a nearly stationary Dedekind-like configuration. Because the instability is driven by gravitational waves, and a stationary configuration does not radiate, viscosity then dominates radiation, ultimately reverses the direction of evolution, and returns the star to axisymmetry. A fully nonlinear evolution of a Newtonian model supplemented by a greatly enhanced radiation-reaction term, by Shibata and Karino [613], agrees with the first part of the picture; because the simulation did not include viscosity, there was no final evolution back to axisymmetry.

The second reason to expect that a Dedekind-like stage will not persist is a proof by Lebovitz and Lifshitz [397, 399] that *the entire sequence of Dedekind ellipsoids is dynamically unstable* to what is called an elliptical instability. For ellipsoids with highest vorticity (and shape that is most highly nonaxisymmetric), low-harmonic modes are unstable and the growth time is shortest. For ellipsoids near the Maclaurin sequence, only short wavelength modes with long growth times are unstable, and the instability is likely to be damped out by a realistic value of viscosity. A fully nonlinear study by Ou, Tohline, and Lindblom [511], for an $N = 1/2$ polytrope, similar to that of Shibata and Karino and also using a greatly enhanced radiation-reaction term, but with a longer run time, traces the evolution of an axisymmetric star with a bar-mode instability driven by gravitational radiation. The initial evolution to a Dedekind-like configuration closely matches the evolution described in the last paragraph; however, within ten dynamical times of reaching a nearly stationary configuration, short-wavelength perturbations grew rapidly. This is plausibly the nonlinear development of the predicted elliptical instability, but the authors emphasize their inability, on the basis of the simulation, to distinguish between that and a nonlinear decay of the bar mode through mode-mode coupling. Again, the study did not include viscosity.

The studies also ignored magnetic fields, and in nascent stars, fields generated by differential rotation are expected to enforce uniform rotation a shorter-than-viscous timescale. In addition, as we discuss in Chapter 8, the $l = m = 2$ r-mode (a perturbation with axial parity) and/or higher-order polar modes are more likely to be the dominant nonaxisymmetric instability of nascent neutron stars. It may still be the case in rapidly rotating nascent stars that the timescales for viscous and magnetic dissipation and for the elliptical instability are longer than the radiation-reaction timescale, and unstable neutron stars may initially evolve to a nonaxisymmetric configuration that remains stationary for a time longer than the radiation reaction time and shorter than the timescale for angular momentum transport set by viscosity and a magnetic field.

Stationary configurations of dissipative fluids are axisymmetric

We present here Lindblom's proof [417] that a stationary star, modeled as a dissipative fluid, must be axisymmetric. Dissipative fluids are viscous and heat-conducting, and the first part of the proof is a formal demonstration of a result that seems intuitively obvious: Because dissipation changes the state of a fluid, increasing its entropy, an equilibrium configuration of an imperfect fluid can be stationary only if it has no heat flow and only if the fluid velocity u^α is shear-free and divergence-free.

We begin by briefly introducing dissipative fluids in a relativistic context. As is commonly done, we use the Eckart formalism [203], a description of the fluid for which the equations of motion are first-order in quantities that describe the departure from local thermodynamic equilibrium. The energy dissipated is then quadratic in these quantities (in the heat flow, divergence, and shear of the fluid).[7]

The divergence θ of the velocity field u^α is the spatial divergence $q^{\alpha\beta}\nabla_\alpha u_\beta$ in the subspace orthogonal to u^α. Equivalently, one may define θ by requiring that in inertial coordinates of an observer comoving at P (with instantaneous 4-velocity $u^\alpha(P)$), it has the form $\partial_i u^i$. Because $u^\beta\nabla_\alpha u_\beta = 0$, we have

$$\theta := q^{\alpha\beta}\nabla_\alpha u_\beta = \nabla_\alpha u^\alpha. \tag{3.130}$$

Its definition implies that θ measures the fractional change in the comoving volume of a fluid element, as in Eq. (1.23). One can similarly define the shear tensor $\sigma_{\alpha\beta} := q_\alpha{}^\gamma q_\beta{}^\delta \nabla_{(\gamma} u_{\delta)} - \frac{1}{3} q_{\alpha\beta}\nabla_\gamma u^\gamma$ of Eq. (1.57) by requiring that, in our instantaneously comoving inertial frame, $\sigma_{\alpha\beta}$ has components $\sigma_{ij} = \partial_{(i}u_{j)} - \frac{1}{3}\delta_{ij}\partial_k u^k$, with $\sigma_{0\mu} = 0$.

The coefficients ζ and η of bulk and shear viscosity relate the divergence θ and shear $\sigma_{\alpha\beta}$ to the energy dissipated by the change in volume and by the shearing motion of the fluid, respectively. The corresponding contribution to the energy momentum tensor from bulk and shear viscosity is $-\zeta\theta q^{\alpha\beta} - 2\eta\sigma^{\alpha\beta}$. Finally, one defines a heat-flow vector q^α, with $q^\alpha u_\alpha = 0$, for which each spatial component q^i is the energy flux of heat across a surface perpendicular to $\boldsymbol{\partial}_i$. The contribution of heat flow to the stress-energy tensor is then $q^\alpha u^\beta + q^\beta u^\alpha$, and the stress-energy tensor of the dissipative fluid has the form

$$T^{\alpha\beta} = \epsilon u^\alpha u^\beta + (p - \zeta\theta)q^{\alpha\beta} - 2\eta\sigma^{\alpha\beta} + q^\alpha u^\beta + q^\beta u^\alpha, \tag{3.131}$$

with $\zeta > 0$, $\eta > 0$.

From the form of the stress-energy tensor, conservation of energy is given by

$$0 = u_\alpha \nabla_\beta T^{\alpha\beta} = -u^\alpha\nabla_\alpha\epsilon - (\epsilon + p - \zeta\theta)\theta - 2\eta u_\alpha\nabla_\beta\sigma^{\alpha\beta} - \nabla_\alpha q^\alpha + u_\alpha u^\beta\nabla_\beta q^\alpha. \tag{3.132}$$

Using the first law of thermodynamics,

$$u^\alpha\nabla_\alpha\epsilon = \rho T u^\alpha\nabla_\alpha s + h u^\alpha\nabla_\alpha\rho = \rho T u^\alpha\nabla_\alpha s - (\epsilon + p)\theta, \tag{3.133}$$

[7] See §15 of Landau and Lifshitz [386] for a Newtonian presentation and §22 of MTW [480] for a presentation similar to ours. Because the Eckart formalism includes heat flow but not particle diffusion, it does not describe the most general first-order dissipative fluid. The general formalism is given by Hiscock and Lindblom [315], and the proof we present is easily extended to this more general case by replacing Eq. (3.140) by Eq. (22) of [315]. Readers interested in dissipative fluids should be aware that first-order theories are not hyperbolic and so not causal. Strongly hyperbolic second-order theories can be constructed (see, for example, [488, 336, 337, 434, 258]) and are useful for ensuring regularity in time-evolution; but "the physical content of these theories is in most cases precisely the same" as that of the first-order theory [257]. Lindblom [419] shows "that the physical fluid states relax (on a time scale characteristic of the inter-particle interactions) to ones that are also well described by the simple relativistic Navier-Stokes theory."

and the identities

$$-2\eta u_\alpha \nabla_\beta \sigma^{\alpha\beta} = 2\eta\sigma_{\alpha\beta}\sigma^{\alpha\beta}, \qquad u_\alpha u^\beta \nabla_\beta q^\alpha = -u^\beta \nabla_\beta u_\alpha\, q^\alpha, \tag{3.134}$$

we can write Eq. (3.132) in the form

$$-\rho T u^\alpha \nabla_\alpha s + \zeta\theta^2 + 2\eta\sigma_{\alpha\beta}\sigma^{\alpha\beta} - \nabla_\alpha q^\alpha - u^\beta \nabla_\beta u_\alpha\, q^\alpha = 0. \tag{3.135}$$

The relativistic version of the Fourier law of heat conduction is

$$q^\alpha = -\kappa q^{\alpha\beta}(\nabla_\beta T + T u^\gamma \nabla_\gamma u_\beta), \text{ with } \kappa > 0. \tag{3.136}$$

The second term in parentheses here is a relativistic correction that accounts for the inertia of the heat flow. Finally, we need the contribution of q^α to the entropy. In the absence of heat flow, Eqs. (1.26) and (1.48) imply that the entropy current $s^\alpha := s\rho u^\alpha$ is conserved. A nonzero heat flow adds to s^α the term q^α/T,

$$s^\alpha = s\rho u^\alpha + \frac{q^\alpha}{T}, \tag{3.137}$$

and conservation of entropy is replaced by the condition that the entropy increase to the future: $\nabla_\alpha s^\alpha \geq 0$. To relate this differential law to an integral expressing the increase in entropy, let ${}^4V$ be a spacetime region whose spacelike boundary consists of spacelike hypersurfaces Σ_1 and Σ_2 with Σ_2 in the future of Σ_1; and assume that ${}^4V$ extends beyond the fluid (or simply that $s^\alpha dS_\alpha = 0$ on any timelike part of the boundary). Then the change in entropy from Σ_1 to Σ_2 is

$$\int_{\Sigma_2} s^\alpha dS_\alpha - \int_{\Sigma_1} s^\alpha dS_\alpha = \int_{{}^4V} \nabla_\alpha s^\alpha d^4V \geq 0. \tag{3.138}$$

We can now state Lindblom's result.

Proposition. *Consider a stationary, asymptotically flat, globally hyperbolic spacetime $(M, g_{\alpha\beta})$ that satisfies the Einstein equation for a dissipative fluid with stress-energy tensor of the form (3.131). Suppose that the fluid has compact spatial support and that the metric and fluid boundary satisfy suitable smoothness conditions.*[8] *Then the spacetime is axisymmetric.*

We begin by verifying the claim mentioned earlier, that a stationary dissipative fluid has vanishing heat flow, shear and divergence. Let t^α be the timelike Killing vector of our stationary spacetime. In Eq. (3.138), let Σ_2 be the result of time-translating Σ_1 forward by a time t (of dragging Σ_1 forward a parameter distance t along the integral curves of t^α). Stationarity then implies

$$\int_{\Sigma_2} s^\alpha dS_\alpha - \int_{\Sigma_1} s^\alpha dS_\alpha = 0. \tag{3.139}$$

[8] Technical smoothness conditions are used only to show that one can analytically extend a Killing vector from the interior of the star to the vacuum exterior: The conditions are that the metric is C^3 except on S; the timelike Killing vector t^α is C^4; in at least an open subset of the surface S, the metric can be analytically continued across S to some metric in the support of the fluid; and S in that subset is the level surface of an analytic function. Lindblom shows that these are sufficient conditions; it is likely that they can be relaxed.

We next show that $\nabla_\alpha s^\alpha$ is positive,

$$T\nabla_\alpha s^\alpha = \zeta\theta^2 + 2\eta\sigma_{\alpha\beta}\sigma^{\alpha\beta} + \frac{1}{\kappa T} q^\alpha q_\alpha, \tag{3.140}$$

and this, together with Eqs. (3.138) and (3.139) implies that q^α, $\sigma_{\alpha\beta}$, and θ vanish everywhere.

To obtain Eq. (3.140), we first use the definition (3.137) of the entropy current to compute $\nabla_\alpha s^\alpha$:

$$T\nabla_\alpha s^\alpha = \rho T u^\alpha \nabla_\alpha s + \nabla_\alpha q^\alpha - q^\alpha \nabla_\alpha \ln T. \tag{3.141}$$

From the energy conservation equation (3.135), the first two terms on the right side can be rewritten as

$$\rho T u^\alpha \nabla_\alpha s + \nabla_\alpha q^\alpha = \zeta\theta^2 + 2\eta\sigma_{\alpha\beta}\sigma^{\alpha\beta} - u^\beta \nabla_\beta u_\alpha\, q^\alpha. \tag{3.142}$$

Contracting the heat-flow equation (3.136) with q_α expresses $q_\alpha q^\alpha$ as a sum of the remaining term on the right side of Eq. (3.141) and the last term on the right side of (3.142),

$$\frac{1}{\kappa T} q_\alpha q^\alpha = -u^\beta \nabla_\beta u_\alpha\, q^\alpha - q^\alpha \nabla_\alpha \ln T; \tag{3.143}$$

and Eq. (3.140) now follows.

The remaining part of the proof shows the existence of a second Killing vector. At its heart is a short argument due to Pirani and Williams [536] showing that the 4-velocity of a stationary equilibrium of an imperfect fluid must be proportional to a Killing vector k^α: $u^\alpha = \Upsilon k^\alpha$ for some function Υ, and we present this part in detail. Note first that, because $q^\alpha = 0$, the heat-conduction equation implies that the acceleration is a gradient, $u^\beta \nabla_\beta u_\alpha = -\nabla_\alpha \ln T$. The Pirani-Williams result is:

Lemma. *Consider a dissipative fluid for which $\theta = 0$, $\sigma_{\alpha\beta} = 0$, and $u^\beta \nabla_\beta u_\alpha = -\nabla_\alpha \ln \Upsilon$, for some positive scalar Υ. Then $u^\alpha = \Upsilon k^\alpha$, with k^α a Killing vector.*

Proof of Lemma. From $\sigma_{\alpha\beta} = 0, \theta = 0$, we have $q_\alpha{}^\gamma q_\beta{}^\delta \nabla_{(\gamma} u_{\delta)} = 0$. Using $u^\beta \nabla_\alpha u_\beta = 0$, we have

$$\begin{aligned}\nabla_{(\alpha} u_{\beta)} &= \delta_\alpha^\gamma \delta_\beta^\delta \nabla_{(\gamma} u_{\delta)} = (q_\alpha{}^\gamma - u_\alpha u^\gamma)(q_\beta{}^\delta - u_\beta u^\delta)\nabla_{(\gamma} u_{\delta)} \\ &= -u^\gamma \nabla_\gamma u_{(\alpha}\, u_{\beta)} = \nabla_{(\alpha} \ln \Upsilon\, u_{\beta)}.\end{aligned} \tag{3.144}$$

Now define a vector field $k^\alpha := \Upsilon^{-1} u^\alpha$. Then

$$\nabla_{(\alpha} u_{\beta)} = \nabla_{(\alpha}(\Upsilon k_{\beta)}) = \nabla_{(\alpha}(\ln \Upsilon)\, u_{\beta)} + \Upsilon \nabla_{(\alpha} k_{\beta)}. \tag{3.145}$$

Equations (3.144) and (3.145) imply $\nabla_{(\alpha} k_{\beta)} = 0$. □

The final part of the proof of Lindblom's theorem is to show that k^α is a linear combination of the timelike Killing vector t^α and a rotational vector field ϕ^α, implying ϕ^α is a Killing vector of the spacetime – that the spacetime is axisymmetric. This relies on showing that k^α can be extended beyond the fluid to a Killing vector of the full spacetime. The argument here is more technical, and we will give only a rough outline. In vacuum, a Killing vector satisfies $\nabla_\beta \nabla^\beta k_\alpha = 0$. This equation, together with the value of k^α and $\nabla_\alpha k^\beta$ on S, is used to propagate k^α from S to the exterior. To show that k^α remains a Killing vector, one obtains from the relation $\nabla_\alpha \nabla_\gamma \nabla^\gamma k_\beta = 0$ an equation satisfied by the Killing form $\nabla_{(\alpha} k_{\beta)}$

$$0 = \nabla_\gamma \nabla^\gamma \nabla_{(\alpha} k_{\beta)} - [\nabla_\gamma \nabla^\gamma, \nabla_{(\alpha}] k_{\beta)} = \nabla_\gamma \nabla^\gamma \nabla_{(\alpha} k_{\beta)} - 2R_\alpha{}^\gamma{}_\beta{}^\delta \nabla_{(\gamma} k_{\delta)}. \quad (3.146)$$

Smoothness conditions on the metric and on k^α imply that both $\nabla_{(\alpha} k_{\beta)}$ and its gradient vanish on S, and Eq. (3.146) then implies that $\nabla_{(\alpha} k_{\beta)} = 0$ everywhere.

The spacetime thus has two Killing vectors t^α and k^α. At spatial infinity, each must coincide with a generator of the Poincaré group. Because no spacetime with nonzero mass is asymptotically symmetric under boosts or spatial translations, k^α is a linear combination of t^α and a rotational Killing vector ϕ^α, a Killing vector that asymptotically approaches a rotational symmetry vector of the Poincaré group at spatial infinity.

3.6.2 Jacobi-like configurations and helical symmetry

At the same bifurcation point along the Maclaurin sequence where the Dedekind ellipsoids branch off, a second sequence of ellipsoids of identical shape also begins. These are the Jacobi ellipsoids, congruent to their Dedekind counterparts, but with no internal motion. They are instead uniformly rotating. The Jacobi ellipsoid with angular velocity Ω has the same principal axes a_1, a_2, a_3 as the Dedekind ellipsoid with vorticity $\zeta = \left(\frac{a_1}{a_2} + \frac{a_2}{a_1}\right)\Omega$. It follows from Eq. (3.128) that the velocity fields of the corresponding Dedekind and Jacobi ellipsoids are given by

$$\mathbf{v}_{\text{Dedekind}} = \left(-\frac{a_2}{a_1}\, y\hat{\mathbf{x}} + \frac{a_1}{a_2}\, x\hat{\mathbf{y}}\right)\Omega, \quad (3.147)$$
$$\mathbf{v}_{\text{Jacobi}} = (-y\hat{\mathbf{x}} + x\hat{\mathbf{y}})\,\Omega.$$

There is an analog to the statement that the Dedekind sequence branches off the Maclaurin sequence where the rotation is fast enough that the frequency of an $m = 2$ mode vanishes in the *inertial* frame: The Jacobi sequence similarly branches off at the configuration for which the frequency of an $m = 2$ mode vanishes in the *rotating* frame. As we discuss in Chapter 8, in the Newtonian approximation, a mode with zero frequency in the rotating frame marks the onset of secular instability driven by shear viscosity. When viscosity is present, uniformly rotating axisymmetric Newtonian models are unstable for rotation faster than that at the instability point.

Because the Jacobi and Dedekind ellipsoids branch off at the same point, two bar ($m = 2$) modes of the Maclaurin sequence have, at that point, zero frequency in an inertial frame and in a rotating frame, respectively. The coincidence is not there for compressible models or in post-Newtonian models, but a bar mode has zero frequency in a rotating frame when the rotation, measured by $T/|W|$, has about the same value as in the Maclaurin sequence, $T/|W| \approx 0.14$.[9] Beyond the bifurcation point along the Maclaurin sequence, a Jacobi ellipsoid has lower energy than the Maclaurin spheroid with the same mass and angular momentum. As a result, with viscosity and no radiation reaction a perturbed unstable Maclaurin spheroid evolves to the Jacobi ellipsoid [544, 246].

Work by Lyapunov [412], using methods based on analyticity, proves that a mode of Maclaurin with zero frequency in a rotating frame corresponds to a family of exact solutions. The mode is the infinitesimal difference between the bifurcating sequence and the Maclaurin configuration at the bifurcation point: With λ a sequence parameter, the Eulerian changes, $\delta Q = \left.\frac{d}{d\lambda}Q\right|_{\lambda_0}$, in the fluid variables and the potential satisfy the linearized equilibrium equations in a rotating frame. Lyapunov, Lichtenstein [414], and subsequent authors in the 1930s extended these results to inhomogeneous models, still restricted to configurations stationary in a rotating frame (see Hagihara [293] for a review in English that outlines the arguments and gives a comprehensive bibliography of the work on rotating homogeneous stellar models prior to 1935).

Only stars with stiff equations of state can have uniform rotation fast enough for a Jacobi-like bifurcation point (roughly, for $T/|W|$ to reach the value 0.14). For Newtonian polytropes [340], only those whose polytropic index N is less than 0.808 ($\Gamma > 2.238$) have zero-frequency $m = 2$ (bar) modes in a rotating frame. For N smaller than this, Ipser and Managan [330] find numerically that the location of the zero-frequency mode along each sequence of axisymmetric polytropes is in fact a bifurcation point: They numerically construct the compressible analogs of the Jacobi ellipsoids, finding for $N = 0.5$ and 0.6 a sequence of uniformly rotating triaxial configurations that branches off from the sequence of uniformly rotating axisymmetric polytropes. The bifurcation points for compressible Newtonian stars have values of $T/|W|$ close to the 0.1375 value of the Maclaurin sequence: 0.1346 and 0.1364 for $N = 0.6$ and 0.5, respectively.

As discussed in Section 9.7, the branching of a Jacobi-like sequence (a zero-frequency bar mode) requires more rapid rotation in general relativity than in the Newtonian approximation [83]. Most candidate equations of state do not allow uniform rotation rapid enough for a bar mode to be dynamically unstable; however, differential rotation in stellar cores collapsing to form neutron stars, in the nascent neutron stars they form, and in hypermassive neutron stars formed by binary coalescence may be high enough, in some cases, to allow a nonaxisymmetric instability and a Jacobi-like configuration.

The symmetry vector of a Jacobi ellipsoid is a vector field that generates time translations in a rotating frame. A symmetry vector of spacetime of this kind is a *helical* Killing vector k^α. In the case of a uniformly rotating axisymmetric star, it

[9] This is the value of $T/|W|$ at which the bar mode is first unstable for uniformly rotating stars and for stars with moderate differential rotation. As discussed in Section 9.8, in stars with unusually large differential rotation a nonaxisymmetric instability can set in at small values of $T/|W|$.

has the form we have already encountered,

$$k^\alpha = t^\alpha + \Omega\phi^\alpha, \tag{3.148}$$

with t^α and ϕ^α asymptotically timelike and rotational. Note that at circumferential distance from the axis of symmetry greater than $1/\Omega$, a particle would have to move faster than light to rotate with angular velocity Ω; outside this light cylinder, k^α is spacelike. The spacetime of a uniformly rotating nonaxisymmetric system has neither a timelike nor a rotational Killing vector. Although its only symmetry vector is the helical Killing vector k^α, one can define helical symmetry in a similar way in terms of a timelike vector field and a vector field that is rotational in the sense of having circular orbits [86]:

Definition 1. A vector field k^α is *helical* if it can be written in the form

$$k^\alpha = t^\alpha + \Omega\phi^\alpha, \tag{3.149}$$

where ϕ^α is spacelike and has circular orbits with parameter length 2π, except where it vanishes; $\Omega \neq 0$ is a constant; and t^α is timelike outside some spatially compact region and is everywhere transverse to a family of Cauchy surfaces. (The compact-region caveat is needed because the Killing vector t^α is spacelike inside an ergosphere of a rotating axisymmetric black hole or of a rotating axisymmetric star compact enough to have an ergosphere. The transverse condition then prevents t^α from vanishing or generating closed timelike curves.)

The vector fields that appear in this definition are far from unique. The frequency Ω and the corresponding period $T = 2\pi/\Omega$, however, have an invariant meaning that defines the helical character of the orbits: Although k^α is spacelike outside the light cylinder, the helical orbit joins each point P to a timelike related point P' a parameter time $T = 2\pi/\Omega$ along the orbit. This allows a definition that avoids introducing the arbitrary fields t^α and ϕ^α [243].

Definition 2. A Killing vector field k^α is *helical* if it is transverse to a family of spacelike hypersurfaces that foliate the spacetime and if there is a smallest period $T > 0$ for which points P and $\chi_T(P)$ are timelike separated for every P outside a timelike surface.

The two definitions are equivalent in the spacetimes we consider (see [243], Proposition II.1).

In general relativity, nonaxisymmetric configurations radiate unless they are stationary in an inertial frame. A stationary analog of a Jacobi ellipsoid – a nonaxisymmetric solution to the Einstein-Euler equations that is stationary in a rotating frame – must have equal amounts of ingoing and outgoing radiation. The radiation field of such a stationary solution has infinite energy, because a constant flux of radiation is emitted at arbitrarily distant times in the past. That is, a spacelike hypersurface contains the energy from radiation emitted at all times in the past, and the total emitted energy is infinite. Thus spacetimes of nonaxisymmetric systems cannot be asymptotically flat if they are stationary in a rotating frame.[10] Instead, the asymptotic mass rises linearly with a naturally defined radial coordinate.

In the post-Newtonian approximation, however, the system is asymptotically flat through second post-Newtonian order (up to and including corrections of order v^4/c^4 to the Newtonian spacetime), because the radiation field is of order v^5/c^5. The first post-Newtonian sequence of uniformly rotating, uniform-density nonaxisymmetric models, corresponding to the Jacobi ellipsoids, was obtained by Chandrasekhar [128, 131]. As in the Dedekind case, the Newtonian and post-Newtonian sequences differ by a "third-harmonic" deformation, a Lagrangian displacement cubic in the Cartesian coordinates.

As Detweiler and Blackburn have emphasized [72, 181], even in exact solutions to the Einstein-Euler equations with helical symmetry, one can define an approximate asymptotic region in which the energy E emitted in gravitational waves in a dynamical time T is small compared to the mass of the system, $E \ll M$. The system will then be dominated by the mass of its stars up to distances of order $r = \frac{M}{E}T$, and in this "local wave zone," the geometry will describe gravitational waves propagating on a background of asymptotically Coulomb metric potentials.

The equation of hydrostatic equilibrium for Jacobi-like models – for helically symmetric configurations with uniform rotation – again has the first integral (1.110) for barotropic stars. That is, because the velocity u^α is along the Killing vector k^α, one can again define a scalar u^t by

$$u^\alpha = u^t k^\alpha. \tag{3.150}$$

Then Eq. (1.98) holds in the form $u^\beta \nabla_\beta u_\alpha = -\nabla_\alpha \ln u^t$, and the relativistic Euler equation takes the form

$$\nabla_\alpha \ln \frac{h}{u^t} = 0, \quad \text{or} \quad \frac{h}{u^t} = \mathcal{E}. \tag{3.151}$$

[10] A theorem that essentially implies this result uses the formal treatment of asymptotic flatness that adjoins boundaries at future null infinity $\mathcal{I}^\pm$ to the spacetime. A theorem by Gibbons and Stewart [261] shows that $\mathcal{I}^\pm$ cannot be periodic. Thus, no spacetime can be asymptotically flat if it is vacuum outside a compact region and has a helical Killing vector (and no other Killing vector).

4

Numerical schemes

Even in the Newtonian approximation, analytic models of rotating stars with specified equations of state are limited to the classical uniform-density ellipsoids, and analytic relativistic models are restricted to rotating disks [469]. But in the full theory, as in the Newtonian framework, accurate numerical models are readily obtained. All current codes involve an iteration in which one solves a set of nonlinear equations for the metric functions, together with the equation of hydrostationary equilibrium. In nearly all codes, one successively solves the equations for the metric functions for one function at a time, with the others set to their values from the last iteration; and the first integral (1.110) of the equation of hydrostationary equilibrium (with the updated metric functions) is then used to update the matter distribution. The methods fall into two classes, depending on whether each nonlinear equation is solved by a Newton-Raphson method or, in the case of elliptic equations for the metric functions, by inverting a flat-space elliptic operator with known Green's function. For uniform rotation and a specified barotropic equation of state, the family of equilibria is 2-dimensional (locally specified, for example, by central density and angular velocity Ω), and within each class of code, different choices are made of the two parameters that are held fixed to obtain a convergent iteration. Finally, the codes differ in whether spectral methods are used to represent radial and angular functions and to evaluate corresponding integrals and derivatives, and in the choice of independent components of the field equation that are to be solved.

The first such numerical solutions were found by Wilson [727], who generalized the Bardeen and Wagoner formalism for relativistic disks [50, 51] and by Bonazzola and Schneider [91], who applied the formalism presented in [90]. More extended numerical results were presented by Butterworth and Ipser (BI) [110], who used a Newton-Raphson linearization scheme. A robust, iterative numerical method, using an integral representation of components of the gravitational field equation, was applied by Komatsu, Eriguchi, and Hachisu (KEH) [371, 372] and extended by Cook, Shapiro, and Teukolsky (CST) [157, 159, 158]. A minimal-surface scheme was implemented by Neugebauer and Herold [494], whereas numerical schemes relying on pseudospectral methods were devised by Bonazzola et al. [89, 87] and, in full Newton-Raphson context, by Ansorg, Kleinwächrer, and Meinel (AKM) [33]. This last code is the one case where a full Newton-Raphson method is used, with a matrix inverted to simultaneously find the correction to the metric and fluid distribution in terms of their values in the previous iteration. The various schemes presented so far have their strengths and weaknesses. The most modern ones show a high degree of agreement on a large number of selected equilibrium models. Below, the KEH/CST scheme is presented in detail. A brief description is also given for other existing methods and numerical codes.

4.1 The KEH scheme

The KEH scheme [371, 372] is based on an integral representation of the independent field equation components (1.93), using appropriate Green's functions, and on an iteration until all metric equations and the first integral of the hydrostationary equilibrium equation are satisfied to a desired accuracy (starting from an initial guess of the solution). The scheme has similarities to the one introduced in [90, 91] but uses a different set of metric functions that are guaranteed to be asymptotically flat and also allow for the existence of ergospheres. Three of the four metric equations are used in an integral form, for which the boundary conditions are automatically satisfied at spatial infinity. For an isotropic stress-energy tensor, the fourth field equation is an ordinary first-order differential equation. During the iteration, the central density and ratio of polar to equatorial radii are held fixed until convergence to a desired accuracy is achieved. In [371, 372, 207], the original KEH code was used to construct uniformly and differentially rotating stars for both polytropic and realistic EOSs. Below, we summarize the main equations and numerical aspects of the KEH method.

4.1.1 Field equation

New metric variables are constructed so that associated components of the field equation can be written in a form suitable for inversion by the Green's function method. One defines

$$\gamma \equiv \ln B, \tag{4.1}$$

$$\rho \equiv 2\nu - \ln B, \tag{4.2}$$

so that the metric (1.59) is written as

$$ds^2 = -e^{\gamma+\rho}dt^2 + e^{\gamma-\rho}r^2\sin^2\theta(d\phi - \omega dt)^2 + e^{2\mu}(dr^2 + r^2 d\theta^2), \tag{4.3}$$

where, in this chapter, the metric function ρ should not be confused with the rest mass density.

Two field-equation components can be written in elliptic form, using the 3-dimensional, flat-space Laplacian in spherical polar coordinates

$$\nabla^2 = \partial_{rr} + \frac{2}{r}\partial_r + \frac{1}{r^2}\partial_{\theta\theta} + \frac{1}{r^2}\cot\theta\partial_\theta + \frac{1}{r^2\sin^2\theta}\partial_{\phi\phi}. \tag{4.4}$$

A third equation can be written in a form that involves the 2-dimensional, flat-space Laplacian in cylindrical coordinates. The field equations for ρ, ω, and γ are

$$\nabla^2 \left[\rho e^{\gamma/2}\right] = S_\rho(r,\mu), \tag{4.5}$$

$$\nabla^2 \left[r \sin\theta \cos\phi \omega e^{(\gamma-2\rho)/2}\right] = r \sin\theta \cos\phi S_\omega(r,\mu), \tag{4.6}$$

$$\left(\frac{\partial^2}{\partial \varpi^2} + \frac{\partial^2}{\partial z^2}\right)\left[\varpi \gamma e^{\gamma/2}\right] = \varpi S_\gamma(r,\mu), \tag{4.7}$$

where $\varpi = r\sin\theta$, $z = r\cos\theta$, and S_ρ, S_ω, and S_γ are the following source functions:

$$\begin{aligned} S_\rho(r,\theta) = e^{\gamma/2}\Bigg[& 8\pi e^{2\mu}(\epsilon+p)\frac{1+v^2}{1-v^2} + r^2 \sin^2\theta e^{-2\rho}\left[\omega_{,r}^2 + \frac{1}{r^2}\omega_{,\theta}^2\right] \\ & + \frac{1}{r}\gamma_{,r} + \frac{1}{r^2}\cot\theta\gamma_{,\theta} + \frac{\rho}{2}\Bigg\{16\pi e^{2\mu}p - \gamma_{,r}\left(\frac{1}{2}\gamma_{,r} + \frac{1}{r}\right) \\ & - \frac{1}{r^2}\gamma_{,\theta}\left[\frac{1}{2}\gamma_{,\theta} + \cot\theta\right]\Bigg\}\Bigg], \end{aligned} \tag{4.8}$$

$$\begin{aligned} S_\omega(r,\theta) = e^{(\gamma-2\rho)/2}\Bigg[& -16\pi e^{2\mu}\frac{(\Omega-\omega)(\epsilon+p)}{1-v^2} \\ & + \omega\Bigg\{-8\pi e^{2\mu}\frac{[(1+v^2)\epsilon + 2v^2 p]}{1-v^2} - \frac{1}{r}\left(2\rho_{,r} + \frac{1}{2}\gamma_{,r}\right) \\ & - \frac{1}{r^2}\cot\theta\left(2\rho_{,\theta} + \frac{1}{2}\gamma_{,\theta}\right) + \frac{1}{4}\left[\left(4\rho_{,r}^2 - \gamma_{,r}^2\right) + \frac{1}{r^2}\left(4\rho_{,\theta}^2 - \gamma_{,\theta}^2\right)\right] \\ & - r^2\sin^2\theta e^{-2\rho}\left[\omega_{,r}^2 + \frac{1}{r^2}\omega_{,\theta}^2\right]\Bigg\}\Bigg], \end{aligned} \tag{4.9}$$

$$S_\gamma(r,\theta) = e^{\gamma/2}\left\{16\pi e^{2\mu}p + \frac{\gamma}{2}\left[16\pi e^{2\mu}p - \frac{1}{2}\left(\gamma_{,r}^2 + \gamma_{,\theta}^2\right)\right]\right\}. \tag{4.10}$$

The field equation for the metric function μ can be written as

$$\begin{aligned} \mu_{,\theta} = & -\nu_{,\theta} + \sin\theta\{\sin^2\theta(1+rB^{-1}B_{,r})^2 + [\cos\theta + \sin\theta B^{-1}B_{,\theta}]^2\}^{-1} \\ & \times\Bigg\{\left[\nu_{,\theta}^2 - r^2\nu_{,r}^2 - \frac{1}{2}B^{-1}\left(r^2 B_{,rr} - B_{,\theta\theta} - 2\cot\theta B_{,\theta}\right)\right]\left[\cos\theta + \sin\theta B^{-1}B_{,\theta}\right] \\ & + rB^{-1}B_{,r}\left[\cos\theta\left(\frac{1}{2} + rB^{-1}B_{,r}\right) - \frac{1}{2}\sin\theta B^{-1}B_{,\theta}\right] \\ & + r\sin\theta\left[2\nu_{,r}\nu_{,\theta} + B^{-1}B_{,r\theta}\right]\left(1 + rB^{-1}B_{,r}\right) \\ & + \frac{1}{4}r^2\sin^3\theta B^2 e^{-4\nu}\left[\left(\cot\theta + B^{-1}B_{,\theta}\right)\left(r^2\omega_{,r}^2 - \omega_{,\theta}^2\right)\right. \\ & \left. - 2r\omega_{,r}\omega_{,\theta}\left(1 + rB^{-1}B_{,r}\right)\right]\Bigg\}, \end{aligned} \tag{4.11}$$

which is an ordinary differential equation for μ along any grid line of constant r. This can be integrated, from the pole to the equator, using the spatial flatness condition on the rotation axis, $\mu(\theta = 0) = (\gamma - \rho)/2$, as the starting value. Notice that the above equation for μ has been derived under the assumption of an *isotropic stress-energy tensor* and is thus not appropriate for computing models of, for example, magnetized stars.

4.1.2 Integral representation

The three elliptic field-equation components (4.5), (4.6), and (4.7) can be converted into integral equations, using the Green's functions of the flat-space, 3-dimensional Laplacian in spherical polar coordinates and of the flat-space, 2-dimensional Laplacian in cylindrical coordinates, respectively. In this way, one obtains the following implicit integral expressions:

$$\rho = -\frac{e^{-\gamma/2}}{4\pi} \int_0^{\infty} dr' \int_0^{\pi} d\theta' \sin\theta' \int_0^{2\pi} d\phi' r'^2 \frac{S_\rho(r', \theta')}{|\mathbf{r} - \mathbf{r}'|}, \tag{4.12}$$

$$\omega = -\frac{e^{(2\rho-\gamma)/2}}{4\pi r \sin\theta \cos\phi} \int_0^{\infty} dr' \int_0^{\pi} d\theta' \sin^2\theta' \int_0^{2\pi} d\phi' r'^3 \cos\phi' \frac{S_\omega(r', \theta')}{|\mathbf{r} - \mathbf{r}'|}, \tag{4.13}$$

$$\gamma = \frac{e^{-\gamma/2}}{2\pi r \sin\theta} \int_0^{\infty} dr' \int_0^{2\pi} d\theta' r'^2 \sin\theta' S_\gamma(r', \theta') \log|\mathbf{r} - \mathbf{r}'|, \tag{4.14}$$

where, in the last equation, the source function S_γ has been analytically continued to the range $\pi \le \theta \le 2\pi$ by assuming

$$S_\gamma(r, \theta) = S_\gamma(r, \theta - \pi). \tag{4.15}$$

The integrals in the above equations can be efficiently evaluated using the standard expansions of the Green's functions in angular harmonics, namely

$$\begin{aligned} \log|\mathbf{r} - \mathbf{r}'| &= \frac{1}{2} \log\left[r^2 + r'^2 - 2rr'\cos(\theta - \theta')\right] \\ &= -\sum_{n=1}^{\infty} \frac{1}{n} \frac{r_<^n}{r_>^n} (\cos n\theta \cos n\theta' + \sin\theta \sin n\theta') + \log(r_>), \end{aligned} \tag{4.16}$$

$$\begin{aligned} \frac{1}{|\mathbf{r} - \mathbf{r}'|} &= \sum_{n=0}^{\infty} \frac{r_<^n}{r_>^{n+1}} \Bigg[P_n(\cos\theta) P_n(\cos\theta') \\ &\quad + 2\sum_{m=1}^{n} \frac{(n-m)!}{(n+m)!} P_n^m(\cos\theta) P_n^m(\cos\theta') \cos m(\phi - \phi') \Bigg], \end{aligned} \tag{4.17}$$

where, as usual, $r_< = \min(r, r'), r_> = \max(r, r')$. An advantage of the integral representation is that the asymptotic flatness conditions, $\rho \sim O(1/r)$, $\gamma \sim O(1/r^2)$, and $\omega \sim O(1/r^3)$ for $r \to \infty$, are satisfied automatically.

4.1.3 Iterative procedure

A unique equilibrium model is characterized by at least two parameters, once the EOS has been chosen. One parameter is related to the compactness of the star, a second to the amount of rotation. For uniformly rotating models, a particular choice of these two parameters (which has proved to be robust in the KEH scheme) is the central energy density, ϵ_c, and the axis ratio, r_p/r_e, of polar to equatorial coordinate radii. In the case of differential rotation, one must also specify the rotation law, by choosing the function $F(\Omega)$ of Eq. (1.106). For example, for the rotation law (1.125), a single additional parameter, the constant A (which determines the length scale over which the angular velocity changes), is introduced. The particular choice of ϵ_c, r_p/r_e, and A then uniquely determines an equilibrium model satisfying the rotation law (1.125), except for cases where, because of a bifurcation, more than one equilibrium sequence exists (see [753, 754] for an example with both a spheroidal and a quasitoroidal sequence). Several different types of sequences of differentially rotating compact stars were found in [32].

Several slightly different versions of the detailed iterative procedure are implemented in KEH-type codes, but all adopt similar approaches. The essential features can be seen in the following summary of the iterative procedure used in recent versions of the `rns` numerical code [658], for rotation laws of the form (1.125):

1. The iterative procedure begins with a guess for the four metric functions $\rho, \gamma, \mu, \omega$; the energy density ϵ; and the angular velocity $\Omega(\varpi)$. This can be either a nonrotating Tolman, Oppenheimer, and Volkoff (TOV) solution of, for example, the same maximum energy density as the rotating model, or a previously computed, intermediate rotating model. During the iteration, the maximum energy density ϵ_{max}, the axis ratio r_p/r_e, and the rescaled constant $\hat{A} := A/r_e$ in rotation law (1.125) are held fixed, whereas the location of the energy density maximum is allowed to vary, to allow for quasitoroidal models.
2. The three metric potentials ρ, γ, and μ are rescaled by r_e^2, whereas the metric potential ω is rescaled by $1/r_e$.
3. A new value of the coordinate radius r_e is then obtained by equating the first integral of the hydrostationary equilibrium (1.107) at the location of the energy density maximum to its value at the pole. The value of Ω at the location of the energy density maximum is taken from the previous iteration step.
4. A new value for the angular velocity at the equator, Ω_e, is obtained by equating the first integral of the hydrostationary equilibrium (1.107) at the equator to its value at the pole. In this equation, the rotation law (1.125) is used to evaluate the integral of the function $F(\Omega)$ that specifies the rotation law; this results in a nonlinear equation for Ω_e that is solved by a root-finding method.
5. A new value for the angular velocity on the rotation axis, Ω_c, is obtained by solving the rotation law (1.125) at the equator, where a new value of Ω_e has just been computed in the previous step. This allows one to obtain a new value of $\Omega(\varpi)$ everywhere inside the star, again using a root-finding method for solving (1.125).

6. New distributions of the enthalpy, energy density, and pressure are obtained by solving the first integral of the hydrostationary equilibrium (1.107) everywhere inside the star (equating it to its value at the pole) and by using the chosen EOS.
7. New distributions of the metric functions ρ, γ, ω, and μ are obtained from (4.12), (4.13), (4.14), and (4.11), and the iteration is repeated from step 2 until convergence is achieved.

In the case of differentially rotating models for which the energy density maximum is at the center of the star, the above procedure can be simplified, because in step 3 no angular-velocity-dependent terms appear when the integral of hydrostationary equilibrium is applied at the center of the star. This is also true in case of uniform rotation, where the procedure is still simpler: Step 4 becomes algebraic for uniform angular velocity Ω, and step 5 is not needed.

A good choice of a variable for monitoring the convergence of the above iteration is the equatorial radius of the star. When the fractional change in the equatorial radius from one iteration to the next becomes less than a desired value, say 10^{-6}, the iterative procedure is terminated. In spite of the large number of equations, the above iterative procedure has proved to be, in most cases, robust and convergent. For some models, such as those based on a uniform-density EOS, the system of equations can become stiff, converging slowly or not at all. In such cases, under-relaxation (see [371]) can improve convergence significantly. Another case where the scheme may not converge is when the initial guess is too far away (in parameter space) from the desired rotating model. One can then compute a number of intermediate models before arriving at the final model. For example, it is usually safe to compute intermediate rotating models in steps of 0.2 in the axis ratio r_p/r_e.

4.1.4 The CST compactification

To improve the accuracy of integrals that extend to radial infinity, appearing in Eqs. (4.12), (4.13), and (4.14), one can compactify the radial variable. Infinity is then included in the range of integration, and the boundary conditions (vanishing of the metric potentials at infinity) are exactly enforced. Cook, Shapiro, and Teukolsky (CST) [157, 159, 158] introduce a new radial variable s, through the relation

$$r = r_e \frac{s}{1-s}, \tag{4.18}$$

where r_e is the equatorial coordinate radius, which maps the range $[0, \infty)$ to the finite segment $[0, 1]$. This substitution significantly increases the overall accuracy of the numerical solution, and it is used in the numerical code `rns` presented in [664] (SF). The `rns` code closely follows [157, 159, 158] and is available as a documented public-domain code [658].

4.1.5 Numerical issues

Two numerical problems with the KEH scheme, as it has been implemented in various codes, require attention. One problem is the presence of Gibbs errors at the surface of the star, due to steep gradients of physical variables. This problem could be solved using a surface-adapted coordinate system, as in [87, 274]. A second problem arises from the choice of $\cos\theta$ as the angular variable of the numerical grid. With this choice, some terms in the field equations are sensitive to division of small numbers as the symmetry axis is approached. As a result, the angular derivatives exhibit spurious oscillatory behavior near the axis, due to numerical error. In addition, one can notice an oscillatory behavior in the solution for the metric function μ, when the second-order accurate, central difference formula is used for the second-order radial derivative, as in the original KEH scheme. In [664] it was found that this behavior is smoothed out when twice the grid spacing is used to evaluate the second-order derivatives, resulting in an improvement in accuracy of up to 2% for some equilibrium properties of maximum-mass models. This improvement is implemented in the **rns** code [658] and is especially important for producing initial data for time-evolutions, with a low level of constraint violations.

Price, Markakis, and Friedman [549] analyze convergence (iteration stability) of the KEH method in a spherical Newtonian framework. The model is simple enough for a largely analytic understanding of code convergence, and its dependence on what quantities one holds fixed in an iteration is already present in this toy model.

4.2 Butterworth and Ipser (BI)

The BI scheme [110] solves the four field-equation components, following a Newton-Raphson–like linearization and iteration procedure, patterned after Stoeckly's Newton-Raphson code for Newtonian stellar models [667]. One starts with a non-rotating model and increases the angular velocity in small steps, treating a new rotating model as a linear perturbation of the previously computed rotating model. Because the AKM scheme outlined below also uses a Newton-Raphson method, and the iterative scheme in KEH is not simply a variant of the Newton-Raphson iteration (their convergence properties are fundamentally different), we outline here the Newton-Raphson iteration. To solve k nonlinear equations $f_i(\mathbf{X}) = 0$ for the k variables X_i, one updates the value $\mathbf{X}^{(n)}$ of $\mathbf{X}$ after the nth iteration to its new value $\mathbf{X}^{(n+1)}$ by writing $\mathbf{X}^{(n+1)} = \mathbf{X}^{(n)} + \delta\mathbf{X}$ and using the Taylor expansion $f_i(\mathbf{X}^{(n+1)}) = f(\mathbf{X}^{(n)}) + \partial_j f_i(\mathbf{X}^{(n)})\delta X_j$ to approximate $\delta\mathbf{X}$: That is, $\delta\mathbf{X}$ is the solution to the linear system $f(\mathbf{X}^{(n)}) + \partial_j f_i(\mathbf{X}^{(n)})\delta X_j = 0$, namely,

$$\delta X_i = -J^{-1}_{ij}\, f_j(\mathbf{X}^{(n)}), \tag{4.19}$$

where J is the Jacobian matrix $\|\partial_i f_j\|$.

For a uniformly rotating star, the updated metric potentials and fluid variable, $g_{\mu\nu}^{(n+1)} = g_{\mu\nu}^n + \delta g_{\mu\nu}$, $p^{(n+1)} = p^{(n)} + \delta p$, are then found by solving, for δp and the four independent components of $\delta g_{\mu\nu}$, the linearized equation of hydrostationary equilibrium (1.110), namely

$$\delta h/u^t - h/(u^t)^2\delta u^t - \delta\mathcal{E} = (\mathcal{E} - h/u^t)\big|_{g_{\mu\nu}^{(n)},p^{(n)}},$$

together with four analogously linearized field-equation components of the form

$$\delta G_{\mu\nu} - 8\pi\delta T_{\mu\nu} = (G_{\mu\nu} - 8\pi\delta T_{\mu\nu})^{(n)}.$$

Here $\delta G_{\mu\nu}$ and $\delta T_{\mu\nu}$ are the Einstein and stress-energy tensors, linearized about their values after the nth iteration. Each of these equations is discretized, and the resulting linear system is solved.

For Butterworth and Ipser, the memory and computer speed available in 1975 precluded storing and inverting the large Jacobian matrix needed for a full Newton-Raphson approach. Instead, each of the four field equations is solved separately to update a single metric potential, and the linearized equation of hydrostatic equilibrium is inverted separately to update p. Convergence is achieved with $\mathcal{E}$ held fixed in each iteration. Space is truncated at a finite distance from the star, and the boundary conditions there are imposed by expanding the metric potentials in powers of $1/r$. Angular derivatives are approximated by high-accuracy formulae, and models with density discontinuities are treated specially at the surface. An equilibrium model is specified by fixing its rest mass and angular velocity.

The original BI code was used to construct uniform density models and polytropic models [110, 109]. Friedman, Ipser, and Parker (FIP) [236, 237] extended the BI code to obtain a large number of rapidly rotating models based on a variety of realistic EOSs, whereas Lattimer et al. [395] used a code that was also based on the BI scheme to construct rotating stars using EOSs that include pion or kaon condensates and strange quark matter.

4.3 Bonazzola et al. and Lorene/rotstar

In the Bonazzola et al. (BGSM) scheme [89, 583, 584], the field equations are derived in the 3+1 formulation. All four chosen equations that describe the gravitational field are of elliptic type. This avoids the problem with the second-order radial derivative in the source term of the ordinary differential equation (ODE) used in BI and KEH. The equations are solved using a spectral method: That is, all functions are expanded in terms of a set of basis functions in both the angular and radial directions and a fast Fourier transform (FFT) is used to obtain coefficients. Outside the star a redefined radial variable is used, which maps infinity to a finite distance. Whereas the equations used in the BI and KEH schemes assume a perfect fluid, isotropic stress-energy tensor, the BGSM formulation makes no assumption about the isotropy of $T_{\alpha\beta}$. Thus, the BGSM code can easily be extended to compute stars with magnetic fields, a solid crust, or a solid interior, and it can also be used to construct rotating boson stars.[1]

The BGSM spectral method has been improved in [87, 274], allowing for several domains of integration. One of the domain boundaries is chosen to coincide with the surface of the star, and a regularization procedure is introduced for the divergent

[1] Similarly, the KEH scheme could also be modified to describe cases with nonisotropic stress-energy tensors, by replacing the ODE for μ by an elliptic-type partial differential equation derived from another component of the field equation.

derivatives at the surface (which appear in the density field when stiff equations of state are used). This allows models to be computed that are nearly free of Gibbs phenomena at the surface. The same method is also suitable for constructing quasistationary models of binary neutron stars.

4.4 Ansorg et al. (AKM)

Another multidomain spectral scheme was introduced in [33, 31]. The scheme can use several domains inside the star, one for each possible phase transition in the equation of state. Surface-adapted coordinates are used and approximated by a 2-dimensional Chebyshev expansion. Transition conditions are satisfied at the boundary of each domain, and the field and fluid equations are solved as a free boundary value problem by a full Newton-Raphson method, starting from an initial guess. The field-equation components are simplified by using a corotating reference frame. Applying this new method to the computation of rapidly rotating homogeneous relativistic stars, Ansorg et al. achieve near machine accuracy when about 24 expansion coefficients are used (see Section 4.5). For configurations near the mass-shedding limit, the relative error increases to about 10^{-5}, even with 24 expansion coefficients, because of the low differentiability of the solution at the surface. The AKM code has been used in systematic studies of uniformly rotating homogeneous stars [592] and differentially rotating polytropes [32]. A detailed description of the numerical method and a review of the results is given in [469].

4.5 Direct comparison of numerical codes

The accuracy of the above numerical codes can be estimated by comparing identically specified models constructed with different codes. The first such comparison of rapidly rotating models constructed with the FIP and SF codes was presented in [664]. Rapidly rotating models constructed with several EOSs agree to 0.1%–1.2% in the masses and radii and to better than 2% in any other quantity that was compared (angular velocity and momentum, central values of metric functions, etc.). This is a very satisfactory agreement, considering that the BI code was using relatively few grid points, because of limitations of computing power at the time of its implementation. In [664], it was also shown that a large discrepancy between certain rapidly rotating models constructed with the FIP and KEH codes, which was reported in [207], was only due to the fact that a different version of a tabulated EOS was used in [207] than by FIP.

Nozawa et al. [498] presented an extensive direct comparison of the BGSM, SF, and the original KEH codes, using a large number of models and equations of state. With a model specified by a choice of central density and ratio of polar to equatorial coordinate radius, more than twenty different quantities were compared for each model; the relative differences ranged from 10^{-3} to 10^{-4} or better, for smooth equations of state. The agreement is excellent for soft polytropes, which indicates

Table 4.1. *Detailed comparison of the accuracy of different numerical codes in computing a rapidly rotating, uniform-density model. The absolute value of the relative error in each quantity, compared to the AKM code, is shown for the numerical codes Lorene/rotstar, SF, BGSM, and KEH (see text). The resolution for the SF code is (angular × radial) grid points. See [498] for the definition of various equilibrium quantities. Table adapted from [660].*

	AKM	Lorene/ rotstar	SF (260 × 400)	BGSM	KEH
$\bar{p}_c$	1				
r_p/r_e	0.7			1e-3	
$\bar{\Omega}$	1.41170848318	9e-6	3e-4	1e-2	1e-2
$\bar{M}$	0.135798178809	2e-4	2e-5	9e-3	2e-2
$\bar{M}_0$	0.186338658186	2e-4	2e-4	1e-2	2e-3
$\bar{R}_{circ}$	0.345476187602	5e-5	3e-5	3e-3	1e-3
$\bar{J}$	0.0140585992949	2e-5	4e-4	2e-2	2e-2
Z_p	1.70735395213	1e-5	4e-5	2e-2	6e-2
Z^f_{eq}	−0.162534082217	2e-4	2e-3	4e-2	2e-2
Z^b_{eq}	11.3539142587	7e-6	7e-5	8e-2	2e-1
|GRV3|	$4e-13$	3e-6	3e-5	4e-3	1e-1

that all three codes compute the desired models to an accuracy that depends on the number of grid points used to represent the spacetime.[2]

For the extreme case of uniform density, the agreement drops to the level of 10^{-2}. In the BGSM code, this is because the spectral expansion in terms of trigonometric functions cannot accurately represent functions with discontinuous first-order derivatives at the surface of the star. In the KEH and SF codes, the three-point finite-difference formulae cannot accurately represent derivatives across the discontinuous surface of the star.

The accuracy of the three codes is also estimated by the use of the two virial identities GRV3 and GRV2. Overall, the BGSM and SF codes show a better and more consistent agreement than the KEH code with BGSM or SF. This is largely due to the difference in accuracy with which metric functions computed in the BGSM and SF codes, with integrals that formally extend to a point at infinity, compared to the KEH code, in which the domain of integration extends only to a finite radius.

Another direct comparison of different codes was presented by Ansorg et al. [33]. Their multidomain spectral code was compared to the BGSM, KEH, and SF codes for a particular uniform-density model of a rapidly rotating relativistic star. An extension of the detailed comparison in [33], which includes results obtained by the Lorene/rotstar code in [269] and by the SF code with higher resolution than the resolution used in [498], is shown in Table 4.1. The comparison confirms that

[2] Up to resolutions where Gibbs phenomena at the surface of the star become important.

the virial identity GRV3 is a good indicator for the accuracy of each code. For the particular model in Table 4.1, the AKM code achieves near double-precision machine accuracy, whereas the Lorene/rotstar code has typical relative accuracy of 2×10^{-4} to 7×10^{-6} in various quantities. The SF code at high resolution comes close to the accuracy of the Lorene/rotstar code for this model. Lower accuracies are obtained, as expected, with the BGSM and KEH codes at the low resolutions used in [498].

5

Equilibrium models

5.1 Models in uniform rotation

5.1.1 Bulk properties

Neutron star models constructed with different proposed EOSs have strikingly different bulk properties, due to the large uncertainties in the equation of state above nuclear density. The overall relation between the equation of state and the structure of the star is easy to see: Because more compressible (soft) EOSs require a higher density to supply the same pressure, they yield stellar models with higher central density and smaller radius for a given mass than models based on less compressible (stiff) EOSs. Because less compact models have less gravitational binding energy for the same mass, models based on stiff EOSs can support more mass against gravitational collapse. Thus the maximum mass for a given EOS increases as the EOS becomes less compressible, attaining its largest value for models based on an incompressible EOS – for uniform-density models. Because the moment of inertia increases with increasing mass and (quadratically) with increasing radius, it shows the most rapid rate of increase with increasing stiffness of the EOS. On the other hand, because more compact stars can spin faster before reaching their maximum (Keplerian) rotation, the maximum spin for a fixed baryon mass increases as the compressibility increases.

The sensitivity of the maximum mass to the compressibility of the neutron-star core is responsible for the strongest astrophysical constraint on the equation of state of cold matter above nuclear density. With the mass measurement of $1.97 \pm 0.04 M_\odot$ for PSR J16142230 [179], several candidate EOSs that yielded models with maximum masses of nonrotating stars below this limit were abandoned, but the remaining range of candidate EOSs is still large, yielding compact objects with substantially different properties. Despite these differences, one can typically express bulk properties in dimensionless forms whose values are approximately independent of the EOS. For example, because the moment of inertia is CMR^2, with C a constant of order unity, it is natural to express I in the dimensionless form $I/(MR^2)$; although there are sharp differences in the density profiles associated with different EOSs, the ratio $I/(MR^2)$ is constant to an accuracy of about 10%. As discussed in Section 5.1.3, the maximum angular velocity Ω_K of a star of given baryon mass scales with much higher accuracy as $\sqrt{M/R^3}$, with M and R the mass and radius of the corresponding spherical model, and $\sqrt{M/R^3}$, therefore, the Keplerian angular

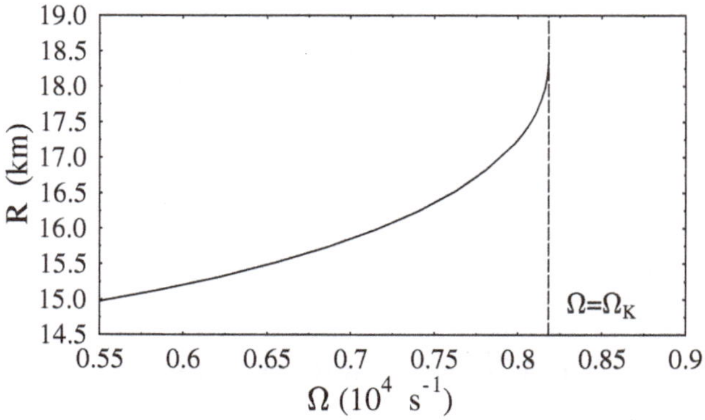

Figure 5.1. The radius R of a star increases sharply as the Kepler (mass-shedding) limit ($\Omega = \Omega_K$) is approached. The particular sequence of models shown here has a constant central energy density of $\epsilon_c = 1.21 \times 10^{15}$ g cm^{-3} and was constructed with EOS L. (Figure from [664]. Reproduced by permission of the AAS.)

velocity of a satellite orbiting the equator of the spherical model. In this case the ratio $\Omega_K/\sqrt{M/R^3}$ is constant to within a few percent. Similarly, the ratio $\Omega_{\max}/\sqrt{M_{\max}/R^3}$, where $\Omega_{\max}$ is the maximum angular velocity among all models based on a given EOS and $M_{\max}$ the maximum mass among all *spherical* models, is nearly independent of the EOS. Other dimensionless measures of maximum rotation are the ratio of rotational to gravitational energy, $T/|W|$, with range 0.11 ± 0.02; a dimensionless angular momentum, cJ/GM^2, with range 0.64 ± 0.06; and an eccentricity with range 0.66 ± 0.04 [235]. One exception to the uniformity of such ratios should be noted: Stellar models based on strange quark EOSs (with a possible crust of normal matter) have smaller radii for the same mass than a star composed entirely of normal matter. Because of this, the ratio $T/|W|$ for the maximum-mass model can become more than 60% larger than for hadronic EOSs (see [748]).

Rotation increases the equatorial radius of a star and also increases the mass that can be sustained at a given central energy density. As a result, the mass of the maximum mass rotating model is roughly 15% to 20% higher than the mass of the maximum mass nonrotating model, for hadronic EOSs. The corresponding increase in radius is between 30% and 40%. Figure 5.1 shows an example of a sequence of equilibrium models with constant central energy density of[1] $\epsilon_c = 1.21 \times 10^{15}$ g cm^{-3}, constructed with EOS L [523, 521]. Near the Kepler (mass-shedding) limit ($\Omega = \Omega_K$), the radius increases sharply. This leads to the appearance of a cusp in the equatorial plane. The effect of rotation in increasing the mass and radius becomes more pronounced in the case of strange quark EOSs (see Section 5.1.8).

Laarakkers and Poisson [380] computed the scalar quadrupole moment for several equations of state and found that, along a sequence of fixed gravitational mass, M,

[1] Following the standard convention, we report numerical values of ϵ_c as ϵ_c/c^2.

Q is given by the empirical formula

$$Q = -\kappa \frac{J^2}{Mc^2}, \tag{5.1}$$

where J is the angular momentum of the star and κ is a dimensionless quantity that depends on the equation of state. The above quadratic fit reproduces Q with remarkable accuracy. (Note, however, that their quadrupole moment is accurate only to about 3% as a result of the error referred to in Section 3.2.1.) The quantity κ varies between a few times unity for very soft EOSs and several times unity for very stiff EOSs, for neutron star models with $M = 1.4M_\odot$. Even for the most relativistic models, with the softest EOSs, the value of $a = J/M^2$ is considerably larger than for a Kerr black hole, for which $a = 1$ [691]. The significant difference in the structure of the exterior spacetime, between rotating neutron stars and Kerr black holes, could, in principle, be used to observationally distinguish the two classes of spacetimes.

5.1.2 Sequences of equilibrium models

1-Dimensional sequences. For a given EOS, nonrotating models form a one-parameter sequence that can be characterized by, for example, the central energy density ϵ_c. The parameter space of uniformly rotating models is 2-dimensional, and one can construct different sequences of rotating models, depending on which quantity is held fixed along the sequence. Examples are sequences of constant central energy density, constant angular momentum, or constant rest mass. Numerical codes usually construct sequences with a fixed central energy density. In diagnosing stability against collapse, however, one typically considers sequences with fixed baryon number or fixed angular momentum; sequences of equilibria that model the accretion-powered spin-up of neutron stars, or neutron-star spin-down by magnetic dipole radiation, by synchrotron radiation from spiraling particles, or by gravitational waves all involve sequences of equilibria in which the ratio dJ/dM is determined by the mechanism that changes the spin. To obtain these more general sequences, one constructs a large number of nearby models with the same central density (but different axis ratio) and uses a root-finding method to determine the axis ratio of the model for which the constraint that determines the sequence is satisfied with a desired accuracy. Figure 5.2 displays a representative sample of fixed rest mass sequences for EOS WFF3 [731] in a graph of mass versus central energy density, where the sequence of nonrotating models and the sequence of models at the mass-shedding limit are also shown.[2] The rest mass of the maximum-mass nonrotating model is $2.17M_\odot$. Below this value, all fixed rest mass sequences have a nonrotating member. Along such a sequence, the gravitational mass increases somewhat, because it also includes the rotational kinetic energy. Above $M_0 = 2.17M_\odot$,

[2] Notice that, although this particular EOS does not satisfy the current observational constraint of a $2M_\odot$ maximum-mass nonrotating model, the qualitative features of all sequences of models discussed here are generic for practically all EOSs.

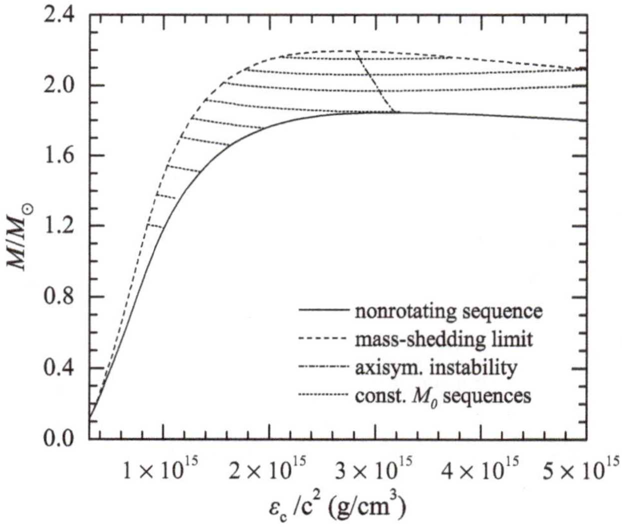

Figure 5.2. Representative sequences of rotating stars with fixed baryon mass, for EOS WFF3 [731]. Above a rest mass of $M_0 = 2.17M_\odot$, only supramassive stars exist, which reach the axisymmetric instability limit when spun down. The onset of axisymmetric instability approximately coincides with the minima of the constant rest mass sequences.

none of the fixed rest mass sequences have a nonrotating member. Instead, the sequences terminate at the *axisymmetric instability limit* (see Section 9.2). The onset of the instability occurs just prior to the minimum of each fixed rest mass sequence, and models to the right of the instability line are unstable. Models with $M_0 > 2.17M_\odot$ have masses larger than the maximum mass nonrotating model and are called *supramassive* [157]. A millisecond pulsar spun up by accretion can become supramassive, in which case it would subsequently spin down along a sequence with approximately fixed rest mass, finally reaching the axisymmetric instability limit and collapsing to a black hole. Some relativistic stars could also be born supramassive or become so as the result of a binary merger; here, however, the star would be initially differentially rotating, and collapse would be triggered by a combination of spin-down and by viscosity (or magnetic field wind-up) driving the star to uniform rotation.

Figure 5.3 displays the same fixed rest mass sequences as in Fig. 5.2, but in a mass versus radius graph, which clearly shows the large increase in radius for rapidly rotating models. Notice that there exists a region where models that are axisymmetrically unstable overlap with models that are stable, but which have lower central energy density.

The same fixed rest mass sequences as in Fig. 5.2, but in an angular velocity versus angular momentum graph, are shown in Fig. 5.4. For normal sequences, angular velocity increases monotonically with angular momentum. However, supramassive sequences show a more complex dependence. Near the mass-shedding limit a star that evolves along a fixed rest mass sequence, losing angular momentum, encounters

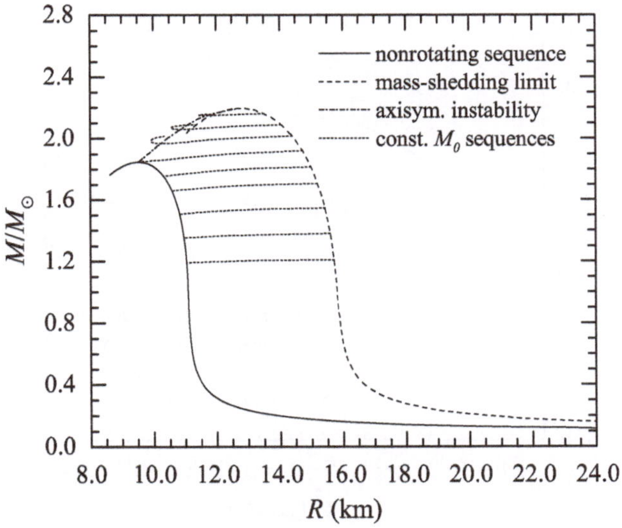

Figure 5.3. Mass versus radius dependence for the same constant rest mass sequences as in Fig. 5.2. A small overlap of models is seen in the unstable region (to the left of the line of turning points that approximates the axisymmetric instability curve).

a region where it spins up, even though its angular momentum is decreasing [157, 159, 158]. For sequences near the maximum mass, this tendency is dominant. Such high-mass stars would only spin up while losing angular momentum. A similar (but smaller) effect exists for stars approaching the mass-shedding limit, if the equation of state is extremely stiff or extremely soft. If the EOS features a phase transition

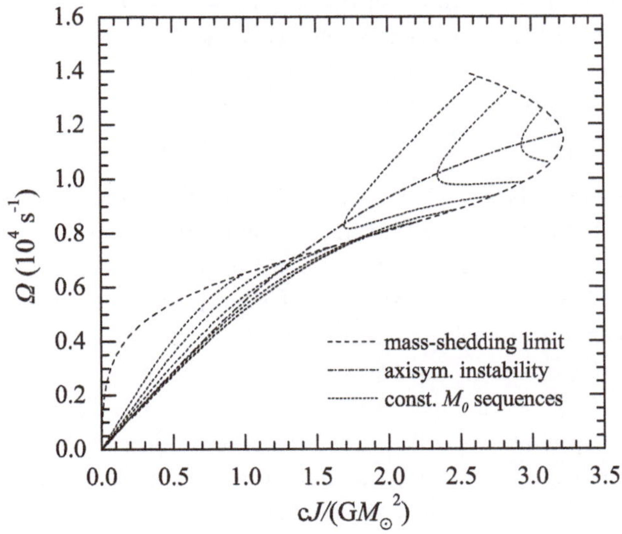

Figure 5.4. Angular velocity versus angular momentum dependence for the same constant rest mass sequences as in Fig. 5.2. Supramassive stars near the mass-shedding limit spin up with loss of angular momentum.

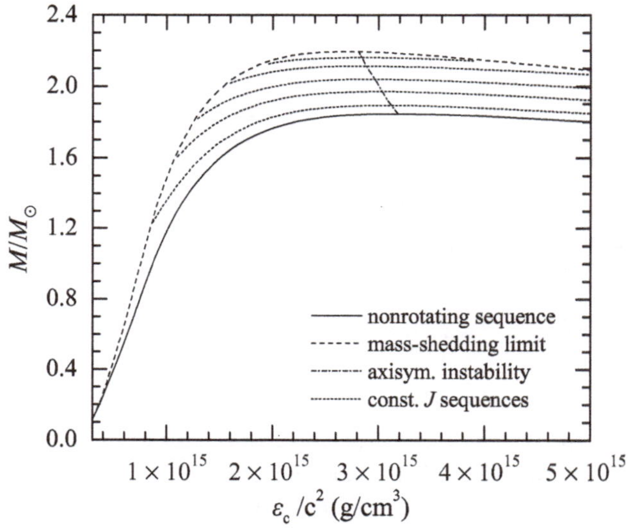

Figure 5.5. Representative sequences of rotating stars with fixed angular momentum, for EOS WFF3 [731]. The onset of axisymmetric instability approximately coincides with the maxima of the constant angular momentum sequences.

to, for example, quark matter, the spin-up region can be very large, and most millisecond pulsars (if supramassive) would need to be spinning up – the absence of spin-up in known millisecond pulsars indicates that either large phase transitions do not occur, or that the equation of state is sufficiently stiff, so that millisecond pulsars are not supramassive [650].

Sequences of fixed angular momentum (for the same EOS) are shown in Fig. 5.5. The maximum in a mass versus central energy density graph along such a sequence also nearly coincides with the onset of axisymmetric instability. Fixed angular momentum sequences terminate at the mass-shedding limit. The maximum angular momentum is attained near the maximum mass rotating model.

Maximum mass and rotation. Among all stable rotating models for a given EOS, the model with maximum mass is ordinarily close to the models with maximum angular velocity Ω, angular momentum J, and rest mass M_0. This coincidence is indicated in Fig. 5.6. The coincidence is not exact, and there is a (usually) small difference between the models of maximum M and Ω [157]. In addition, models with maximum values of J and M_0 also fail to coincide exactly with the maximum Ω model [664], as is illustrated in Fig. 5.6. This additional lack of coincidence can be explained as follows: Given that models with maximum Ω and maximum M among all equilibria do not coincide, either $\epsilon_c(\Omega_{\rm max}) < \epsilon_c(M_{\rm max})$ or $\epsilon_c(\Omega_{\rm max}) > \epsilon_c(M_{\rm max})$. In the first case, stable models with maximum M, Ω, J and M_0 coincide. In the second case, $(\partial M/\partial\epsilon_c)_J > 0$ for any $\epsilon_c < \epsilon_c(\Omega_{\rm max})$, and $J(M_{\rm max}) > J(\Omega_{\rm max})$. Furthermore, $\partial M/\partial\epsilon_c = 0$ at the maximum mass model. Hence, there exists $\epsilon_c < \epsilon_c(M_{\rm max})$ with $J > J(M_{\rm max})$. A similar argument, using $(\partial M/\partial\epsilon_c)_{M_0}$, yields $\epsilon_c(M_{0{\rm max}}) \neq \epsilon_c(M_{\rm max})$.

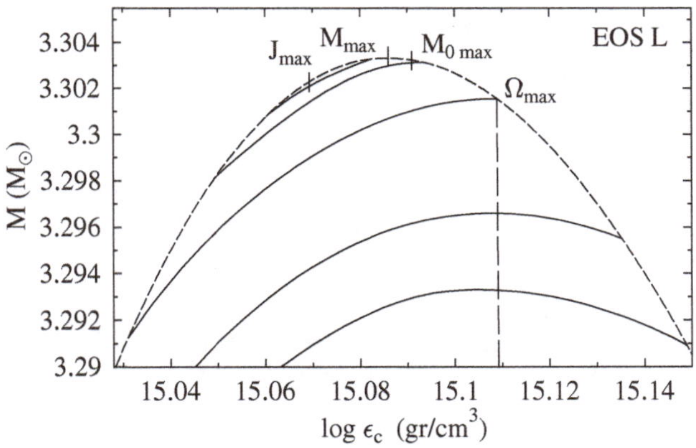

Figure 5.6. Detail near $M_{\rm max}$ for rotating models constructed with EOS L [523, 521]. Four different stable models are associated with maximum values of M, M_0, J, and Ω. Solid lines are constant J sequences, the short-dashed line is the mass-shedding limit, and the long-dashed line approximates the axisymmetric instability limit. (Figure from [664]. Reproduced by permission of the AAS.)

Depending on the central density of the fastest rotating stable model, one can distinguish two classes of equations of state: If the fastest rotating stable model is at a lower density than the maximum mass equilibrium model, the equilibrium models with maximum M, M_0, and J are all unstable (for example, EOS FP [230]); otherwise they are all stable (for example, EOS L; see Fig. 5.6). We note that in the first case, the fastest rotating *stable* model is also the configuration with maximum M, M_0, and J, whereas in both cases the four different equilibrium models are all distinct. Even when the coincidence of stable models is not precise, the models with maximum M, Ω, J, and M_0 are generally very close in central density and can be considered to coincide for practical purposes. A notable exception are EOSs with a large phase transition near the central density of the maximum mass model, such as EOS M [522], for which the distinction between the above-mentioned maximal models becomes significant [158].

Restricting the discussion of maximum mass to uniform rotation is appropriate for the neutron stars that have been observed. Proto–neutron stars, however, as well as the short-lived remnants formed in the coalescence of double neutron-star systems, will be differentially rotating, and the maximum mass that can be temporarily supported is significantly higher. Baumgarte, Shapiro, and Shibata [56] and their subsequent collaborators find upper limits on the mass of differentially rotating equilibria, using numerical evolutions to determine the stability against collapse of the initial configurations. For polytropes [449] and for candidate neutron-star EOSs [484], the increase δM in maximum mass over that of spherical stars with the same EOS can be as high as $\delta M/M_{\rm max,spherical} = 0.65$.[3]

[3] If one considers all differentially rotating equilibria, allowing toroidal configurations and not restricting consideration to models that are dynamically stable to nonaxisymmetric perturbations, there is no upper mass limit: Thin torii can have arbitrarily large masses (see [32] and references therein).

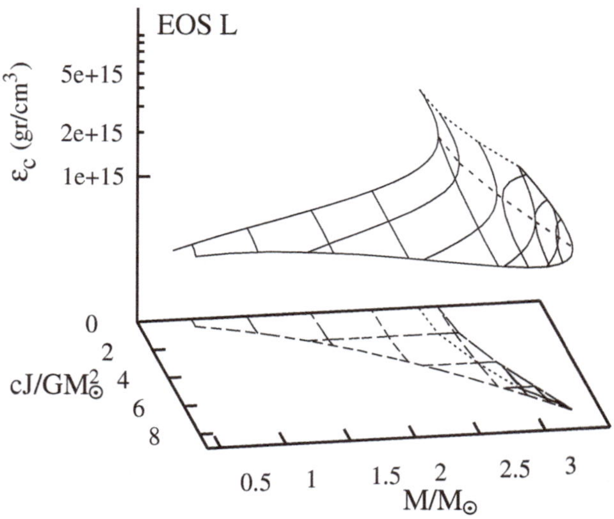

Figure 5.7. Two-dimensional surface of equilibrium models for EOS L [523, 521]. The surface is bounded by the nonrotating sequence ($J = 0$) and the mass-shedding limit and formed by constant J and constant M_0 sequences (solid lines). The projections of these sequences in the J-M plane are shown as dashed lines. Also shown is the axisymmetric instability sequence (short-dashed line), approximated by the line connecting the turning points. The projection of the 2-dimensional surface in the J-M plane shows an overlapping of models (see dotted lines). (Figure from [664]. Reproduced by permission of the AAS.)

Two-dimensional surface of equilibrium models. For a given zero-temperature EOS, the uniformly rotating equilibrium models form a two-dimensional surface in the three-dimensional space of central energy density, gravitational mass and angular momentum [664], as shown in Fig. 5.7 for EOS L. The corresponding surfaces for other EOSs are strikingly similar. The surface is limited by the nonrotating models ($J = 0$) and by the models rotating at the mass-shedding (Kepler) limit. The ridge of maximum mass at fixed J (or minimum mass at fixed baryon mass M_0), indicated by a dashed line, approximates the onset of axisymmetric instability (instability to collapse). Stars stable against collapse lie on the low-density part bounded by the dashed line. The surface folds over at the ridge; the projection of this smooth boundary onto the J-M plane has an apparent cusp at the maximum mass configuration.

5.1.3 Empirical relations for the mass-shedding limit

In the Newtonian limit the maximum angular velocity for uniformly rotating polytropic stars is approximately $\Omega_K \simeq (2/3)^{3/2}(GM/R^3)^{1/2}$ (this is derived using the Roche model; see [603]). For relativistic stars, the empirical formula [291, 237, 232, 289]

$$\Omega_K = 0.67\sqrt{\frac{GM^{\max}_{\rm sph}}{(R^{\max}_{\rm sph})^3}}, \tag{5.2}$$

gives the maximum angular velocity in terms of the mass and radius of the maximum mass *nonrotating* (spherical) model with an accuracy of 5% to 7%, without actually requiring construction of rotating models. Expressed in terms of the minimum period $P_{\rm min} = 2\pi/\Omega_K$, the empirical formula reads

$$P_{\rm min} \simeq 0.82 \left(\frac{M_\odot}{M_{\rm sph}^{\rm max}}\right)^{1/2} \left(\frac{R_{\rm sph}^{\rm max}}{10{\rm km}}\right)^{3/2} {\rm ms}. \tag{5.3}$$

The empirical formula results from universal proportionality relations that exist between the mass and radius of the maximum mass rotating model and those of the maximum mass nonrotating model for the same EOS. Lasota, Haensel, and Abramowicz [389] found that, for most EOSs, the numerical coefficient in the empirical formula is an almost linear function of the parameter

$$\chi_s = \frac{2GM_{\rm sph}^{\rm max}}{R_{\rm sph}^{\rm max} c^2}. \tag{5.4}$$

The Lasota et al. empirical formula

$$\Omega_K = (0.468 + 0.378\chi_s)\sqrt{\frac{GM_{\rm sph}^{\rm max}}{(R_{\rm sph}^{\rm max})^3}}, \tag{5.5}$$

reproduces the exact values with a relative error of only 1.5%. The corresponding formula for $P_{\rm min}$ is

$$P_{\rm min} \simeq \frac{0.187}{(\chi_s)^{3/2}(1 + 0.808\chi_s)} \left(\frac{M_\odot}{M_{\rm sph}^{\rm max}}\right) {\rm ms}. \tag{5.6}$$

The above empirical relations are specifically constructed for the most rapidly rotating model for a given EOS. Lattimer and Prakash [390] suggest the empirical relation

$$P_{\rm min} \simeq 0.96 \left(\frac{M_\odot}{M}\right)^{1/2} \left(\frac{R_{\rm sph}}{10{\rm km}}\right)^{3/2} {\rm ms} \tag{5.7}$$

for any neutron star model with mass M and radius $R_{\rm sph}$ of the nonrotating model with same mass, as long as its mass is not close to the maximum mass allowed by the EOS. Using the above relation, one can set an approximate constraint on the radius of a nonrotating star with mass M, given the minimum observed rotational period of pulsars.

5.1.4 Upper limits on mass and rotation: Theory versus observation

Maximum mass. Candidate EOSs for high-density matter predict vastly different maximum masses for nonrotating models. One of the stiffest proposed EOSs (EOS L) has a nonrotating maximum mass of $3.3M_\odot$. Some core-collapse simulations suggest a bimodal mass distribution of the remnant, with peaks at about

$1.3M_\odot$ and $1.7M_\odot$ [694] (the second peak depends on the assumption for the high-density EOS – if a soft EOS is assumed, then black hole formation of this mass is implied).

Observationally, the masses of a large number of compact objects have been determined, but, in most cases, the observational error bars are still large. In 2012, the most accurately determined maximum mass was the mass of PSR J16142230, with $M = 1.97 \pm 0.04M_\odot$ [179], but there are strong indications of higher masses (see [287] for a detailed account). Masses of compact objects have been measured in different types of binary systems: double neutron-star binaries, neutron star–white dwarf binaries, X-ray binaries, and binaries composed of a compact object around a main sequence star. For most double neutron-star binaries, masses have already been determined with good precision and are restricted to a narrow range of about 1.2 to $1.4M_\odot$ [693]. This narrow range of relatively small masses is probably related to the conditions of stability when double neutron-star systems form. Masses determined for compact stars in X-ray binaries still have large error bars but are consistently higher than $1.4M_\odot$, which is probably the result of mass accretion. In particular, the mass of Vela X-1 could be at least $M = 1.53M_\odot$ [554]. Other good candidates of X-ray binaries containing massive neutron stars are Cyg X-2 and 4U 1700-37. Several pulsar–white dwarf binaries contain possibly massive neutron stars.

Minimum period. When magnetic-field effects are ignored, conservation of angular momentum can yield very rapidly rotating neutron stars at birth. Simulations of the rotational core collapse of evolved rotating progenitors [309, 245] have demonstrated that rotational core collapse could result in the creation of neutron stars with rotational periods of the order of 1 ms (and similar initial rotation periods have been estimated for neutron stars created in the accretion-induced collapse of a white dwarf [437]). However, magnetic fields may complicate this picture. Spruit and Phinney [649] have presented a model in which a strong internal magnetic field couples the angular velocity between core and surface during most evolutionary phases. The core rotation decouples from the rotation of the surface only after central carbon depletion takes place. Neutron stars born in this way would have very small initial rotation rates, even smaller than the ones that have been observed in pulsars associated with supernova remnants. In this model, an additional mechanism is required to spin up the neutron star to observed periods. On the other hand, Livio and Pringle [440] argue for a much weaker rotational coupling between core and surface by a magnetic field, allowing for the production of more rapidly rotating neutron stars than in [649]. In [310], intermediate initial rotation rates were obtained. Clearly, more detailed studies of the role of magnetic fields are needed to resolve this important question.

Independently of their initial rotation rate, compact stars in binary systems are spun up by accretion, reaching high rotation rates. In principle, accretion could drive a compact star to its mass-shedding limit. However, there are a number of different processes that could limit the maximum spin to lower values. In one model, the minimum rotational period of pulsars could be set by the occurrence of the r-mode instability in accreting neutron stars in low-mass X-ray binaries (LMXBs) [68, 23], during which gravitational waves carry away angular momentum. Other models are based on the standard magnetospheric model for accretion-induced spin-up [726], or on the idea that the spin-up torque is balanced by gravitational radiation produced by an accretion-induced quadrupole deformation of the deep crust [68, 707], by deformations induced by a very strong toroidal field [169], or by magnetically confined "mountains" [470, 710]. With the maximum observed pulsar spin frequency (in 2012) at 716 Hz [313] and a few more pulsars at somewhat lower rotation rates [124], it is likely that one of the above mechanisms ultimately dominates over the accretion-induced spin-up, setting an upper limit that may be somewhat dependent on the final mass, the magnetic field, or the spin-up history of the star.

For a wide range of candidates for the neutron-star EOS, the mass-shedding limit sets a minimum period of about 0.5 to 0.9 ms [233]. However, even pulsar surveys that were in principle sensitive down to a few tenths of a millisecond have not been able to detect a submillisecond pulsar [103, 173, 166, 204].[4] This is consistent with the presence of a spin-limiting mechanism, or they could be present only in binary systems with small orbital period ($P_{orb} \sim 1$ hr), for which the current pulsar surveys are still lacking the required sensitivity. In wide systems, on the other hand, accretion could be halted by the turning on of the accreting compact stars as pulsars.

One can systematize the observational constraints on the neutron-star EOS by introducing a parameterized EOS above nuclear density with a set of parameters large enough to encompass the wide range of candidate EOSs and small enough that the number of parameters is smaller than the number of relevant observations. Read et al. [558] found that one can match a representative set of EOSs to within about 3% rms accuracy with a 4-parameter EOS based on piecewise polytropes.

The EOS has the polytropic form (1.119) on each of the three density intervals $\rho_0 \le \rho \le \rho_1$, $\rho_1 \le \rho \le \rho_2$, and $\rho > \rho_2$, namely,

$$p(\rho) = K_i \rho^{\Gamma_i}, \tag{5.8}$$

where the adiabatic index Γ_i is constant in each interval $[\rho_{i-1}, \rho_i]$, and the polytropic constant K_i is fixed by continuity. As in the case of a simple polytrope, the first law in the form $d\dfrac{\epsilon}{\rho} = -pd\dfrac{1}{\rho}$ determines the energy density. The dividing densities are $\rho_1 = 10^{14.7}$ g/cm^3 $= 1.85\rho_{\rm nuc}$, $\rho_2 = 10^{15.0}$ g/cm^3. Following Lattimer and Prakash's observation that neutron-star radii are closely tied to the pressure somewhat above nuclear density, the value $p_1 = p(\rho_1)$ of the pressure is taken as one parameter, and the remaining parameters can be chosen as the slopes Γ_1, Γ_2, Γ_3 or as pressures p_i (with an additional pressure needed above ρ_2).

The model-independent constraints are causality, the largest observed mass, and the largest observed spin. Because the maximum spin for nearly all candidate EOSs

[4] A claim for the detection of a submillisecond pulsar [342] remained unconfirmed as of 2012.

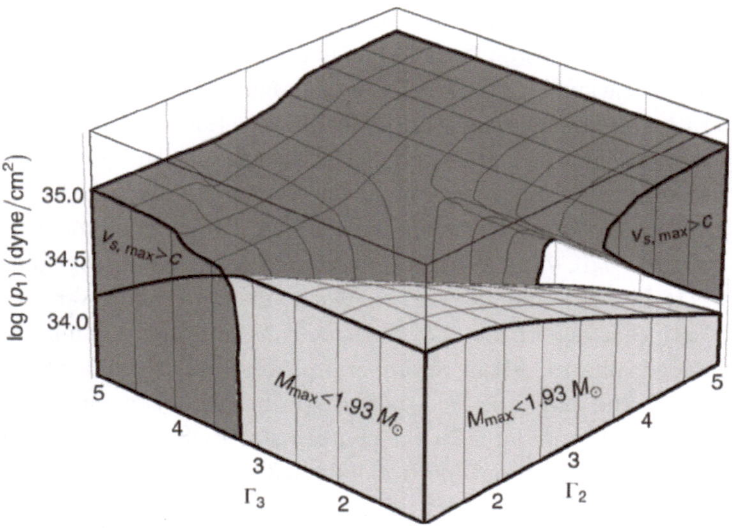

Figure 5.8. Constraints imposed by causality and a maximum mass above $1.93 M_\odot$ restrict the parameters p_1, Γ_2, and Γ_3 of a parameterized space of equations of state to the part of parameter space lying outside the shaded regions.

is higher than the maximum observed spin of 716 Hz, only the first two constraints significantly restrict the neutron-star EOS. The region in the EOS parameter space that they allow is shown in Fig. 5.8, due to B. Lackey, where the darker shaded region is forbidden by causality, and the lighter shaded region by the requirement that the maximum mass exceed $1.93 M_\odot$. In particular, most equations of state with hyperon or quark cores are ruled out. Causality is taken to be the requirement that the speed of sound be less than the speed of light for densities below the central density of the maximum-mass configuration. Parts of the parameter space above the darker shaded region represent EOSs that violate causality but can be altered to be causal without changing the set of stable stellar models.

Using spectral modeling to simultaneously estimate the radius and mass of a set of neutron stars in transient low-mass X-ray binaries, Özel, Baym, and Güver [515] and Steiner, Lattimer, and Brown [655] find more stringent constraints. They adopt equivalent piecewise-polytropic parametrizations, with pressures as the EOS parameters to find the more restricted region of the EOS space. Future gravitational-wave observations of inspiraling neutron-star binaries [559] may yield comparably accurate constraints without the model dependence of the current electromagnetic studies.

5.1.5 *Maximum mass set by causality*

If one is interested in obtaining an upper limit on the mass, independent of the current uncertainty in the high-density part of the EOS for compact stars, one can construct a schematic EOS that satisfies only a *minimal set of physical constraints* and which yields a model of absolute maximum mass. The minimal set of constraints are

(0) *A relativistic star is described as a self-gravitating, uniformly rotating perfect fluid with a 1-parameter EOS*, an assumption that is satisfied to high accuracy by cold neutron stars.
(1) *Matter at high densities satisfies the causality constraint* $c_s \equiv \sqrt{dp/d\epsilon} < 1$, where c_s is the sound speed. Relativistic fluids are governed by hyperbolic equations whose characteristics lie inside the light cone (consistent with the requirement of causality) only if $c_s < 1$ [258].
(2) *The EOS is known at low densities.* One assumes that the EOS describing the crust of cold relativistic stars is accurately known up to a matching energy density ϵ_m.

For *nonrotating stars*, Rhoades and Ruffini [568] showed that the EOS that satisfies the above constraints and yields the maximum mass consists of a high density region at the causal limit, $dp/d\epsilon = 1$ (as stiff as possible), that matches directly to the assumed low-density EOS at $\epsilon = \epsilon_m$,

$$p(\epsilon) = \begin{cases} p_{\text{crust}}(\epsilon) & \epsilon < \epsilon_m , \\ p_m + \epsilon - \epsilon_m & \epsilon > \epsilon_m , \end{cases} \tag{5.9}$$

where $p_m = p_{\text{crust}}(\epsilon_m)$. For this *maximum mass EOS* and a specific value of the matching density, they computed a maximum mass of $3.2M_\odot$. However, this is not the theoretically maximum possible mass of nonrotating neutron stars, as is sometimes quoted in the literature. Hartle and Sabbadini [302] pointed out that $M_{\max}$ is sensitive to ϵ_m, and Hartle [299] computed $M_{\max}$ as

$$M_{\max} = 4.8 \left(\frac{2 \times 10^{14} \text{g/cm}^3}{\epsilon_m / c^2} \right)^{1/2} M_\odot . \tag{5.10}$$

In the case of *uniformly rotating stars*, Friedman and Ipser [234] assumed that the *absolute maximum mass*, $M_{\max}^{\text{rot}}$, is obtained by the same EOS (5.9) as in the nonrotating case and computed $M_{\max}^{\text{rot}}$ as a function of ϵ_m, assuming that the BPS EOS holds at low densities. A more recent computation [373] uses the FPS EOS at low densities, arriving at a result similar to that in [234],

$$M_{\max}^{\text{rot}} = 6.1 \left(\frac{2 \times 10^{14} \text{g/cm}^3}{\epsilon_m / c^2} \right)^{1/2} M_\odot , \tag{5.11}$$

where $2 \times 10^{14} \text{g/cm}^3$ is roughly nuclear saturation density for the FPS EOS. Thus, as long as there is still uncertainty in the high-density EOS above nuclear saturation density, uniformly rotating compact stars with masses as high as $6M_\odot$ are not excluded by causality. This absolute maximum mass is expected to be lowered in the future, according to (5.11), when the EOS of compact stars can be trusted at higher densities.

5.1.6 Minimum period set by causality

A rigorous limit on the minimum period of uniformly rotating, gravitationally bound stars allowed by causality has been obtained in [373] (hereafter KSF), extending previous results by Glendenning [264]. The same three minimal constraints (0), (1), and (2) of Section 5.1.5, as in the case of the maximum mass allowed by causality, yield the minimum period. However, the *minimum-period EOS* is different from the maximum mass EOS (5.9). KSF found that just the two constraints (0), (1) (without matching to a known low-density part) suffice to yield a simpler, *absolute minimum-period EOS* and an absolute lower bound on the minimum period.

Absolute minimum period, without matching to low-density EOS. Considering only assumptions (0) and (1), so that the EOS is constrained only by causality, the minimum-period EOS is simply

$$p(\epsilon) = \begin{cases} 0 & \epsilon \leq \epsilon_C , \\ \\ \epsilon - \epsilon_C & \epsilon \geq \epsilon_C , \end{cases} \tag{5.12}$$

describing a star entirely at the causal limit $dp/d\epsilon = 1$, with surface energy density ϵ_C. This is not too surprising. A soft EOS yields stellar models with dense central cores and thus small rotational periods. Soft EOSs, however, cannot support massive stars. This suggests that the model with minimum period arises from an EOS that is *maximally stiff* ($dp/d\epsilon = 1$) at high density, allowing stiff cores to support against collapse, but *maximally soft* at low density ($dp/d\epsilon = 0$), allowing small radii and thus fast rotation, in agreement with (5.12). The minimum-period EOS is depicted in Fig. 5.9 and yields an absolute lower bound on the period of uniformly rotating stars obeying the causality constraint, independent of any specific knowledge about the EOS for the matter composing the star. Choosing different values for ϵ_C, one constructs EOSs with different $M_{\rm sph}^{\rm max}$. All properties of stars constructed with EOS (5.12) scale according to their dimensions in gravitational units, and thus the following relations hold between different maximally rotating stars computed from minimum-period EOSs with different ϵ_C:

$$P_{\rm min} \propto M_{\rm sph}^{\rm max} \propto R_{\rm sph}^{\rm max} , \tag{5.13}$$

$$\epsilon_{\rm sph}^{\rm max} \propto \frac{1}{\left(M_{\rm sph}^{\rm max}\right)^2} , \tag{5.14}$$

$$M_{\rm rot}^{\rm max} \propto M_{\rm sph}^{\rm max} , \tag{5.15}$$

$$R_{\rm rot}^{\rm max} \propto R_{\rm sph}^{\rm max} , \tag{5.16}$$

$$\epsilon_{\rm rot}^{\rm max} \propto \epsilon_{\rm sph}^{\rm max} . \tag{5.17}$$

A fit to the numerical results, yields the following relation for the absolute minimum period:

$$\frac{P_{\rm min}}{\rm ms} = 0.196 \left(\frac{M_{\rm sph}^{\rm max}}{M_\odot} \right) . \tag{5.18}$$

Thus, for $M_{\rm sph}^{\rm max} = 2M_\odot$, the absolute minimum period is 0.39 ms.

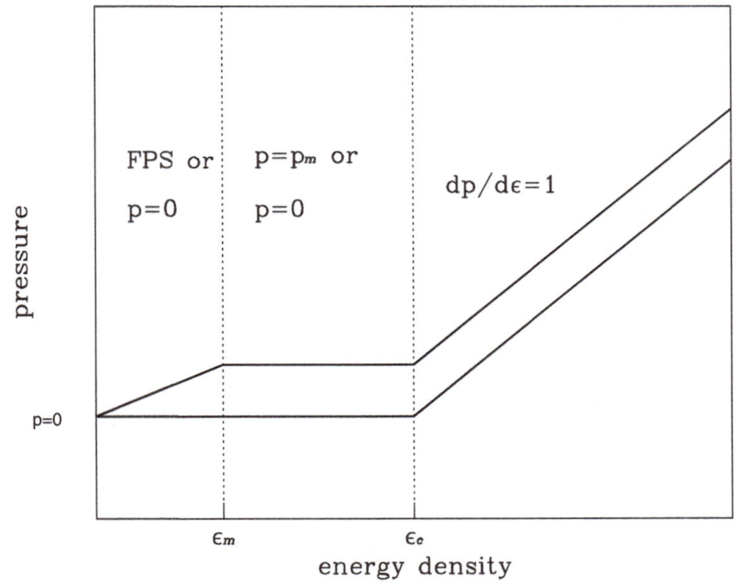

Figure 5.9. Schematic representations of the minimum-period EOSs (5.12) and (5.19). For the minimum-period EOS (5.12), the pressure vanishes for $\epsilon < \epsilon_C$. The minimum-period EOS (5.19) matches the FPS EOS to a constant-pressure region at an energy density ϵ_m. For $\epsilon > \epsilon_C$, both EOSs are at the causal limit with $dp/d\epsilon = 1$. (Figure from [373]. Reproduced by permission of the AAS.)

Minimum period when low-density EOS is known. Assuming all three constraints (0), (1), and (2) of Section 5.1.5 (so that the EOS matches to a known EOS at low density), the minimum-period EOS is

$$p(\epsilon) = \begin{cases} p_{\rm crust}(\epsilon) & \epsilon \leq \epsilon_m, \\ p_m & \epsilon_m \leq \epsilon \leq \epsilon_C, \\ p_m + \epsilon - \epsilon_C & \epsilon \geq \epsilon_C. \end{cases} \tag{5.19}$$

Between ϵ_m and ϵ_C, the minimum-period EOS has a constant-pressure region (a first-order phase transition) and is maximally soft, whereas above ϵ_C the EOS is maximally stiff; see Fig. 5.9. For a matching number density of $n_m = 0.1$ fm^{-3} to the FPS EOS, the minimum period allowed by causality is shown as a function of $M_{\rm sph}^{\rm max}$ in Fig. 5.10. A quite accurate linear fit of the numerical results is

$$\frac{P_{\rm min}}{\rm ms} = 0.295 + 0.203 \left(\frac{M_{\rm sph}^{\rm max}}{M_\odot} - 1.442 \right). \tag{5.20}$$

Thus, if $M_{\rm sph}^{\rm max} = 2M_\odot$, the minimum period is $P_{\rm min} = 0.41$ ms. This result is rather insensitive to n_m, for $n_m < 0.2$ fm^{-3}, but starts to depend significantly on n_m for larger matching densities.

Comparing (5.20) to (5.18), it is evident that the currently trusted part of the nuclear EOS plays a negligible role in determining the minimum period due to

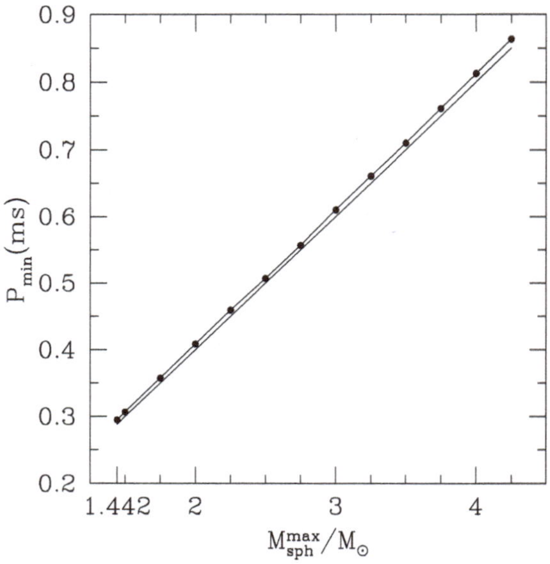

Figure 5.10. Minimum period $P_{\rm min}$ allowed by causality for uniformly rotating, relativistic stars as a function of the mass $M_{\rm sph}^{\rm max}$ of the maximum-mass nonrotating model. The lower curve is constructed using the absolute minimum-period EOS (5.12), which does not match at low densities to a known EOS. The upper curve is constructed using the minimum-period EOS (5.19), which matches at low densities to the FPS EOS. The region below the curves is inaccessible to stars, because of the causality constraint. (Figure from [373]. Reproduced by permission of the AAS.)

causality. In addition, because matching to a known low-density EOS *raises* $P_{\rm min}$, (5.18) represents an *absolute minimum period.*

Searching for the minimum-period EOS. In order to search for the *minimum-period EOS* satisfying constraints (0), (1), and (2) or just (0) and (1), KSF used Glendenning's ansatz [264] for splitting the EOS into a known low-density regime and an intermediate 2-parameter analytic regime (designed to mimic EOSs of various degrees of stiffness) followed by a regime at the causal limit. The EOS has the form

$$\epsilon(n) = \begin{cases} \epsilon_{\rm FPS}(n) & n \le n_m, \\ p_m \left\{ \frac{\kappa}{\gamma - 1} \left[\left(\frac{n}{n_m} \right)^\gamma - \frac{n}{n_m} \right] + \frac{n}{n_m} \frac{\epsilon_m}{p_m} \right. \\ \qquad \left. + \left(1 - \frac{n}{n_m} \right) (\kappa - 1) \right\} & n_m \le n \le n_C, \\ \epsilon_C - p_C + p(n) & n \ge n_C, \end{cases} \tag{5.21}$$

$$p(n) = \begin{cases} p_{\rm FPS}(n) & n \le n_m, \\ p_m \left\{ \kappa \left[\left(\frac{n}{n_m} \right)^\gamma \right] + 1 \right\} & n_m \le n \le n_C, \\ \frac{1}{2} \left\{ p_C - \epsilon_C + (p_C + \epsilon_C) \left(\frac{n}{n_C} \right)^2 \right\} & n \ge n_C, \end{cases} \tag{5.22}$$

where $\epsilon_{\rm FPS}$ and $p_{\rm FPS}$ are the energy density and pressure of the FPS EOS, below some matching number density n_m. The two dimensionless parameters κ and γ parameterize the EOS in the intermediate regime, up to the number density n_C (and energy density ϵ_C, pressure p_C) where the intermediate EOS reaches the causal limit of $\sqrt{dp/d\epsilon} = 1$. Above $n = n_C$, the EOS remains at the causal limit.[5]

Assumptions (0) and (1) of Section 5.1.5 require

$$\gamma > \quad 1, \tag{5.23}$$

$$\kappa > \quad 0, \tag{5.24}$$

$$\kappa\gamma \leq \frac{\epsilon_m}{p_m} + 1. \tag{5.25}$$

This restricted parameter space was searched for the values of κ and γ that yield the minimum-period EOS, by first constructing curves $\gamma(\kappa)$ in the parameter space for fixed values of maximum mass for nonrotating models $M_{\rm sph}^{\rm max}$. For any chosen $M_{\rm sph}^{\rm max}$, the minimum-period EOS was then numerically found to correspond to the limiting case $\kappa \to 0$, in which the EOS becomes independent of both κ and γ and depends only on the constants ϵ_m and ϵ_C. For given ϵ_m, the value of ϵ_C determines $M_{\rm sph}^{\rm max}$. Small changes in the parameters of the minimum-period EOS always increase the period.

5.1.7 Moment of inertia and ellipticity

The scalar moment of inertia of a neutron star, defined in Eq. (1.137) as the ratio $I = J/\Omega$, has been computed using the codes described in Chapter 4 for polytropes and for a wide variety of candidate equations of state (see, for example, [666, 158, 159, 236]). For a given equation of state the maximum value of the moment of inertia typically exceeds its maximum value for a spherical star by a factor of $1.5 - 1.6$. For spherical models, Bejger, Bulik, and Haensel [60] obtain an analytic approximation to the maximum value of I for a given EOS in terms of the maximum mass for that EOS and the radius of that maximum-mass configuration,

$$I_{\rm max,\Omega=0} \approx 0.97 \times 10^{45}\,{\rm g\,cm^2}\,\frac{M_{\rm max}}{M_\odot}\left(\frac{R_{M_{\rm max}}}{10\,{\rm km}}\right)^2, \tag{5.26}$$

with greater accuracy (rms error of about 2%) obtained if one replaces 0.97 by a term, $(-0.368 + 4.822 M_{\rm max}/R_{M_{\rm max}})$, involving the star's dimensionless compactness $M_{\rm max}/R_{M_{\rm max}}$, expressed in gravitational units.

Neutron-star moments of inertia can in principle be measured by observing the periastron advance of a binary pulsar [174]. Because the mass of each star can be found to high accuracy, this would allow a simultaneous measurement of two properties of a single neutron star [485, 393, 60, 558]. In the context of a parameterized EOS, like that of Eq. (5.8), a measurement of this kind would essentially restrict the

[5] For some combinations of κ and γ, the EOS never reaches the causal limit.

parameter space to the intersection of the two surfaces corresponding to the measured values of mass and moment of inertia (more precisely, to the error ellipsoid associated with the measurements).

Our discussion of equilibrium configurations has been restricted to perfect fluids, but neutron-star crusts are expected to be strong enough to support mountains a few centimeters high – high enough that the gravitational waves emitted by rotating, bumpy neutron stars may be observable by interferometric detectors. The size of the departure from axisymmetry is commonly expressed in terms of the ellipticity ε, defined in a Newtonian context by

$$\varepsilon := \frac{I_{xx} - I_{yy}}{I} = \sqrt{\frac{8\pi}{15}} \frac{Q_{22}}{I}, \tag{5.27}$$

where $I = I_{zz}$ is the moment of inertia about the star's rotation axis, and the $m = 2$ part of a neutron-star's quadrupole moment is given by

$$Q_{22} := \mathrm{Re} \int \rho Y_{22} r^2 dV. \tag{5.28}$$

Following Ushomirsky, Cutler, and Bildsten [707], Owen [512] finds for the maximum value of a neutron star's ellipticity the expression

$$\begin{aligned} \varepsilon_{\max} = 3.3 \times 10^{-7} \left(\frac{\sigma_{\max}}{10^{-2}}\right) \left(\frac{1.4 M_\odot}{M}\right)^{2.2} \left(\frac{R}{10\,\mathrm{km}}\right)^{4.26} \\ \times \left[1 + 0.7 \left(\frac{M}{1.4 M_\odot}\right) \left(\frac{10\,\mathrm{km}}{R}\right)\right]^{-1}. \end{aligned} \tag{5.29}$$

Here $\sigma_{\max}$ is the breaking strain of the crust, with an estimated value of order 10^{-2} for crusts below 10^8 K [142].

5.1.8 Rotating strange quark stars

Most rotational properties of strange quark stars differ considerably from the properties of rotating stars constructed with hadronic EOSs. First models of rapidly rotating strange quark stars were computed by Friedman, Ipser, and Parker [237] and by Lattimer et al. [395]. Nonrotating strange stars obey relations that scale with the constant $\mathcal{B}$ in the MIT bag model of the strange quark matter EOS. In [274], scaling relations for the model with maximum rotation rate were also found. The maximum angular velocity scales as

$$\Omega_{\max} = 9.92 \times 10^3 \sqrt{\mathcal{B}_{60}}\ \mathrm{s}^{-1}, \tag{5.30}$$

whereas the allowed range of $\mathcal{B}$ implies an allowed range of 0.513 ms $< P_{\min} <$ 0.640 ms. The empirical formula (5.2) also holds for rotating strange stars with an accuracy of better than 2%. Rotation increases the mass and radius of the maximum mass model by 44% and 54%, correspondingly, significantly more than for hadronic EOSs.

Accreting strange stars in LMXBs will follow different evolutionary paths in a mass versus central energy density diagram than accreting hadronic stars [749]. When (and if) strange stars reach the mass-shedding limit, the innermost stable circular orbit (ISCO) still exists [666] (whereas it disappears for most hadronic EOSs). In [666] it was shown that the radius and location of the ISCO for the sequence of mass-shedding models also scales as $\mathcal{B}^{-1/2}$, whereas the angular velocity of particles in circular orbit at the ISCO scales as $\mathcal{B}^{1/2}$. Additional scalings with the constant a in the strange quark EOS (that were proposed in [395]) were found to hold within an accuracy of better than ~9% for the mass-shedding sequence:

$$M \propto a^{1/2}, \qquad R \propto a^{1/4}, \qquad \Omega \propto a^{-1/8}. \tag{5.31}$$

In addition, it was found that models at the mass-shedding limit can have $T/|W|$ as large as 0.28 for $M = 1.34\ M_{\odot}$.

If strange stars have a solid normal crust, then the density at the bottom of the crust is the neutron drip density $\epsilon_{\rm ND} \simeq 4.1 \times 10^{11}$ g cm^{-3}, as neutrons are absorbed by strange quark matter. A strong electric field separates the nuclei of the crust from the quark plasma. In general, the mass of the crust that a strange star can support is very small, of the order of $10^{-5} M_{\odot}$. Rapid rotation increases by a few times the mass of the crust, and the thickness at the equator becomes much larger than the thickness at the poles [748]. The mass $M_{\rm crust}$ and thickness $t_{\rm crust}$ of the crust can be expanded in powers of the spin frequency $\nu_3 = \nu/(10^3$ Hz) as

$$M_{\rm crust} = M_{\rm crust,0}(1 + 0.24\nu_3^2 + 0.16\nu_3^8), \tag{5.32}$$

$$t_{\rm crust} = t_{\rm crust,0}(1 + 0.4\nu_3^2 + 0.3\nu_3^6), \tag{5.33}$$

where a subscript "0" denotes nonrotating values [748]. For $\nu \leq 500$ Hz, the above expansion agrees well with a quadratic expansion derived previously in [266]. The presence of the crust reduces the maximum angular momentum and ratio of $T/|W|$ by about 20%, compared to corresponding bare strange star models.

5.2 Proto–neutron-star models

Following the gravitational collapse of a massive stellar core, a proto–neutron star (PNS) is born. Initially it has a large temperature of order 50 MeV and a correspondingly large radius of up to 100 km. In most cases, the PNS may be born with a relatively low angular momentum (leading to a slowly rotating pulsar at birth). Then, rotational effects will not dominate, and one can study the evolution of the PNS assuming spherical symmetry (see, for example [106, 105, 77, 347, 348, 541, 540, 539, 542, 669]). Up to a time of about 100 ms after core bounce, the PNS is lepton rich and consists of an *unshocked core* at densities $n > 0.1$ fm^{-3}, with entropy per baryon $s \sim 1$, surrounded by a *transition region* and a low-density but high-entropy *shocked envelope* with $s \sim 4$ to 10, which extends to large radii. The lepton number Y_l is roughly 0.4, and neutrinos in the core and in the transition region are trapped – that is, the PNS is still opaque to neutrinos, whereas at densities less than $n \sim 6 \times 10^{-4}\ \mathrm{fm}^{-3}$, the outer envelope becomes transparent to neutrinos.

Within about 0.5 s, the outer envelope cools and contracts, accreting onto the PNS. The entropy per baryon is now roughly $s \sim 2$ throughout the star, whereas the lepton number in the outer envelope drops to $Y_l \sim 0.3$. Further cooling results in a fully deleptonized, hot neutron star at several tens of seconds after core bounce, with a roughly constant entropy per baryon of $s \sim 1$ to 2 throughout the star. Only after several minutes, when the neutron star has cooled to $T < 1$ MeV, are thermal effects negligibly small in the bulk of the star so that a zero-temperature EOS can be used to describe its main properties.

The structure of PNSs is described by finite-temperature EOSs. A few such EOSs have been constructed and are in use (see, for example [394, 670, 540, 606]). These candidate EOSs differ in several respects, such as in thermal pressure at high densities, so that PNS models also show significant differences among various EOSs. Because the extension of a zero-temperature nuclear EOS to a finite-temperature EOS is not a trivial task, the sample of cold EOSs that has been extended is so far very limited and does not correspond to the wide range of possibilities allowed by current observational constraints. Therefore, PNS models that have been constructed cover only a small region of the allowed parameter space. The general trend, however, is that PNSs have initially a hot core supported by thermal pressure and a large radius that limits their rotation rates to values smaller than for a corresponding cold star.

Understanding the detailed evolution of a PNS is significant, as the star could undergo transformations that could be associated with direct or indirect observational evidence. For example, if a large-amplitude phase transition takes place that softens the cold EOS while the PNS cools, then a delayed collapse to a black hole could take place [101, 57]. This could happen, because the maximum mass for cold neutron stars would be lower than the maximum mass of a hot, differentially rotating PNS, in which case the PNS is called *hypermassive* [56].

In the less common scenario that the PNS is born very rapidly rotating, its evolution will sensitively depend on the rotation rate. A complete numerical simulation of a rotating PNS, including neutrino transport, magnetic field effects, and so forth, on a long timescale of tens of seconds is a formidable task. Some partial understanding, however, has emerged by studying quasiequilibrium sequences of rotating models [305, 669, 672, 712]. The main difficulty in constructing quasiequilibrium sequences comes from the fact that exact equilibria exist only under certain limited assumptions that allow for a first integral of the form (1.107). Exact equilibria can be found in the case that the model is considered to be barotropic, all thermodynamical quantities (energy density, pressure, entropy, temperature) depending on baryon number density only. Special cases, such as homentropic or isothermal stars, can also be considered. Because the equation of state for hot matter depends on the neutrino fraction, one can construct temperature and entropy profiles assuming a certain neutrino fraction. More realistic profiles have been considered in [712], where a barotropic EOS was constructed by rescaling temperature, entropy, and lepton number profiles that were obtained from detailed, 1-dimensional simulations of PNS evolutions, whereas the rotational properties of the models were taken from 2-dimensional core-collapse simulations.

Depending on how the temperature and entropy profiles are chosen that characterize a given model of a PNS, it may be useful to consider the first integral of the hydrostationary equilibrium equation (1.107) in a slightly different form. Using (1.14), one finds

$$\ln h - \ln u^t - \int_{s_0}^{s} \frac{T}{h} ds + \int_{\Omega_0}^{\Omega} F(\Omega')d\Omega' = \text{const.}, \tag{5.34}$$

which holds when the model is assumed to be barotropic, that is, when $T = T(n, s(n))$ and $h = h(n, s(n))$.

The main conclusion from the studies of sequences of quasiequilibrium models is that PNSs that are born with moderate rotation will contract and spin up during the cooling phase. This could lead to a PNS rotating at a large enough rate that secular or dynamical instabilities become interesting. Surprisingly, stars belonging to a supramassive sequence exhibit the opposite behavior. If one assumes that a PNS evolves without losing angular momentum or accreting mass, then a cold neutron star produced by the cooling of a hot PNS has a smaller angular velocity than its progenitor. This effect was pointed out in [305] and confirmed in [276]. It should be noted here that a small amount of differential rotation significantly affects the mass-shedding limit, allowing more massive stars to exist than uniform rotation allows. Taking differential rotation into account, Goussard, Haensel, and Zdunik [277] suggest that PNSs created in a gravitational collapse cannot have periods shorter than 1.7 ms. A similar result has been obtained in [669]. It is not clear, however, whether the quasistationary approximation is valid when the stars reach the mass-shedding limit, as, on further thermal contraction, the outer envelope could actually be shed from the star, resulting in an equatorial stellar wind.

Studies of PNSs are being extended to include additional effects, such as entropy- and lepton-driven convective instabilities and hydromagnetic instabilities [205, 439, 106, 104, 475, 476, 477, 180], meridional flows [208], local and mean-field magnetic dynamos [684, 734, 80, 562, 492], magnetic breaking and viscous damping of differential rotation [602, 438, 196, 685, 195], and the magnetorotational (MRI) and Tayler instabilities [5, 374, 685, 34, 464, 616, 121, 463, 656, 71, 648, 352, 499]. These effects will be important for the evolution of PNSs formed after core collapse, as well as for hypermassive neutron stars possibly formed after a binary neutron star merger. The actual evolution of a rapidly rotating PNS is still uncertain, and a better picture will emerge only after more detailed and consistent dynamical computations become feasible.

5.3 Magnetized equilibrium models

The presence of a magnetic field has been ignored in the models of rapidly rotating relativistic stars that were considered in the previous sections. The reason is that the observed surface dipole magnetic field strength of pulsars ranges between 10^8 G and

2×10^{13} G. These values of the magnetic field strength imply a magnetic field energy density that is too small, compared to the energy density of the fluid, to significantly affect the structure of a neutron star. However, there exists observational evidence for another class of compact objects with much stronger magnetic fields than normal pulsars – *magnetars*, which could have fields up to the order of 10^{16} G [199], possibly born initially with high spin (but quickly spinning down to rotational periods of a few seconds). In addition, even though moderate magnetic field strengths do not alter the bulk properties of neutron stars, they may have an effect on the damping or growth rate of various oscillations of an equilibrium star, affecting its stability. For these reasons, a fully relativistic description of magnetized neutron stars is desirable.

Electromagnetism and MHD. The equations of electromagnetism and magnetohydrodynamics (MHD) in general relativity have been discussed in a number of works; see, for example [413, 480, 61, 30, 275] and references therein.

In general (as in special) relativity, the electromagnetic (E/M) field is described by a *vector potential* A_α, from which one constructs the antisymmetric *Faraday tensor*

$$F_{\alpha\beta} = \nabla_\alpha A_\beta - \nabla_\beta A_\alpha, \tag{5.35}$$

satisfying *Maxwell's equations*

$$\nabla_\beta {}^*F^{\alpha\beta} = 0, \tag{5.36}$$

$$\nabla_\beta F^{\alpha\beta} = 4\pi J^\alpha, \tag{5.37}$$

where ${}^*F_{\alpha\beta} = \frac{1}{2}\epsilon_{\alpha\beta\gamma\delta}F^{\gamma\delta}$, with $\epsilon_{\alpha\beta\gamma\delta}$ the totally antisymmetric Levi-Civita tensor. In Eq. (5.37), J^α is the *4-current* creating the E/M field, whereas Eq. (5.36) has the equivalent forms $\nabla_{[\gamma}F_{\alpha\beta]} = 0$ and, in terms of coordinate components, $F_{\mu\nu,\lambda} + F_{\nu\lambda,\mu} + F_{\lambda\mu,\nu} = 0$. The Faraday tensor can be decomposed in terms of an *electric 4-vector* $E_\alpha = F_{\alpha\beta}u^\beta$ and a *magnetic 4-vector* $B_\alpha = {}^*F_{\beta\alpha}u^\beta$ as

$$F_{\alpha\beta} = u_\alpha E_\beta - u_\beta E_\alpha + \epsilon_{\alpha\beta\gamma\delta}u^\gamma B^\delta. \tag{5.38}$$

Because $F_{\alpha\beta}$ is antisymmetric, the electric and magnetic 4-vectors satisfy the conditions $E_\alpha u^\alpha = B_\alpha u^\alpha = 0$.

The stress-energy tensor of the E/M field is

$$T^{(\mathrm{em})}_{\alpha\beta} = \frac{1}{4\pi}\left(F_{\alpha\gamma}F_\beta{}^\gamma - \frac{1}{4}F^{\gamma\delta}F_{\gamma\delta}g_{\alpha\beta}\right), \tag{5.39}$$

which is to be added linearly to the stress-energy tensor of the fluid, $T^{(\mathrm{f})}_{\alpha\beta}$, so that the total stress-energy tensor is $T_{\alpha\beta} = T^{(\mathrm{f})}_{\alpha\beta} + T^{(\mathrm{em})}_{\alpha\beta}$. Using Maxwell's equations, it is easy to show that

$$\nabla_\beta T^{(\mathrm{em})\alpha\beta} = -F^\alpha{}_\beta J^\beta. \tag{5.40}$$

Requiring that the total stress-energy tensor satisfies $\nabla_\beta T^{\alpha\beta} = 0$, one then obtains $\nabla_\beta T^{(\mathrm{f})\alpha\beta} = F^\alpha{}_\beta J^\beta$. Using $q_{\alpha\beta} := g_{\alpha\beta} + u_\alpha u_\beta$ to project this equation normal to the 4-velocity, one obtains the Euler equation in magnetohydrodynamics

$$(\epsilon+p)u^{\beta}\nabla_{\beta}u^{\alpha} = -q^{\alpha\beta}\nabla_{\beta}p + q^{\alpha}{}_{\delta}F^{\delta}{}_{\gamma}J^{\gamma}, \tag{5.41}$$

where the last term represents the *Lorentz force density* (force per unit volume), $f_{\alpha} := q_{\alpha}{}^{\delta}F_{\delta\gamma}J^{\gamma}$.

The 4-current J^{α} can be decomposed along and normal to the 4-velocity as

$$J^{\alpha} = qu^{\alpha} + j^{\alpha}, \tag{5.42}$$

where $q = -J^{\alpha}u_{\alpha}$ is the proper charge density, qu^{α} is the *convection current*, and j^{α} is the *conduction current*. Assuming a linear relation between j^{α} and E^{α}, *Ohm's law* is written as

$$j^{\alpha} = \sigma^{\alpha\beta}E_{\beta}, \tag{5.43}$$

where $\sigma^{\alpha\beta}$ is the *conductivity tensor*. This relation is simplified [61] when $eB\tau/m_e \ll 1$, where e is the electron charge, m_e is the electron mass, B is the strength of the magnetic field, and τ is the mean collision time in the plasma. Then $\sigma_{\alpha\beta} = \sigma g_{\alpha\beta}$, where $\sigma \sim n_e e^2\tau/m_e$ is the *scalar conductivity* (where n_e is the electron number density) and Ohm's law becomes simply

$$j^{\alpha} = \sigma E^{\alpha}. \tag{5.44}$$

The generalization of the energy conservation equation to MHD becomes

$$u_{\alpha}\nabla_{\beta}T^{(\mathrm{pf})\alpha\beta} = -u_{\alpha}\nabla_{b}T^{(\mathrm{em})\alpha\beta} = \sigma E_{\gamma}E^{\gamma}, \tag{5.45}$$

representing the Joule heating of the plasma by the E/M field.

The ideal MHD approximation. When $n_e e^2\tau/m_e \gg 1$ everywhere, one can consider the conductivity to be practically infinite, $\sigma \to \infty$, which constitutes the *ideal MHD approximation* in general relativity.[6] Then, in order for the current to be finite, Ohm's law implies $E^{\alpha} = 0$. A consequence of the last relation is that in a frame comoving with the fluid, there is no electric field. In addition, Eq. (5.45) implies that there is no conversion of E/M energy into heat, and in this sense the magnetic field is *frozen into the fluid*. The Faraday tensor becomes

$$F^{\alpha\beta} = \epsilon^{\gamma\delta\alpha\beta}B_{\gamma}u_{\delta}, \tag{5.46}$$

whereas the first set of Maxwell's equations, (5.36), becomes

$$\nabla_{\beta}(u^{\alpha}B^{\beta} - u^{\beta}B^{\alpha}) = 0. \tag{5.47}$$

The second set, (5.37), relates the magnetic field B^{α} to the 4-current J^{α}, whereas the E/M stress-energy tensor simplifies to

$$T^{(\mathrm{em})}_{\alpha\beta} = \frac{1}{4\pi}\left[B^{\gamma}B_{\gamma}\left(u_{\alpha}u_{\beta} + \frac{1}{2}g_{\alpha\beta}\right) - B_{\alpha}B_{\beta}\right]. \tag{5.48}$$

[6] Notice that the *nonrelativistic* ideal MHD approximation makes additional assumptions; see [338].

The MHD Euler equation takes the form

$$\left(\epsilon + p + \frac{B_\gamma B^\gamma}{4\pi}\right) u^\beta \nabla_\beta u_\alpha = -q_\alpha{}^\beta \left[\nabla_\beta \left(p + \frac{B_\gamma B^\gamma}{8\pi}\right) - \frac{1}{4\pi}\nabla_\gamma (B_\beta B^\gamma)\right], \quad (5.49)$$

whereas the Lorentz force density becomes simply $f_\alpha = \epsilon_{\alpha\beta\gamma\delta} u^\gamma B^\delta j^\beta$, where only the conduction current j^α contributes, because the convection current, being along u^α, drops out.

Circularity conditions. In general, a magnetized compact object will possess a magnetic field with both poloidal and toroidal components. Then its velocity field may include *noncircular* flows that give rise to the toroidal component. In such a case, the spacetime metric cannot be of the form (1.59) but will include additional nonvanishing components. The general formalism describing such a spacetime has been presented by [272], but no numerical solutions of equilibrium models have been constructed so far. Instead, either one can *assume* that the spacetime has (approximately) the form (1.59), or one can look for special cases where the velocity field is circular.

A theorem by [117] allows one to arrive at one such special case. A stationary, axisymmetric spacetime is *circular* if the following conditions hold:

$$t^\alpha R_{\delta[\alpha} t_\beta \phi_{\gamma]} = 0, \quad (5.50)$$

$$\phi^\alpha R_{\delta[\alpha} t_\beta \phi_{\gamma]} = 0, \quad (5.51)$$

where t^α and ϕ^α are the timelike and spacelike Killing vectors, respectively. A flow with 4-velocity u^α is circular, if

$$u^{[\alpha} t^\beta \phi^{\gamma]} = 0. \quad (5.52)$$

For a perfect fluid, it follows that

$$t_\alpha T^{(\mathrm{pf})\alpha[\beta} t^\gamma \phi^{\delta]} = 0, \quad (5.53)$$

$$\phi_\alpha T^{(\mathrm{pf})\alpha[\beta} t^\gamma \phi^{\delta]} = 0. \quad (5.54)$$

From the circularity conditions for the spacetime and using the Einstein equation, it immediately follows that a similar relation holds for the total stress-energy tensor. Thus, a similar relation must hold separately for the E/M part of the stress-energy tensor:

$$t_\alpha T^{(\mathrm{em})\alpha[\beta} t^\gamma \phi^{\delta]} = 0, \quad (5.55)$$

$$\phi_\alpha T^{(\mathrm{em})\alpha[\beta} t^\gamma \phi^{\delta]} = 0. \quad (5.56)$$

The latter equation is satisfied if the 4-current J^α is circular, that is, if

$$J_{[\alpha} t_\beta \phi_{\gamma]} = 0. \quad (5.57)$$

In other words, J_α has components $(J_t, 0, 0, J_\phi)$. Thus, for a stationary, axisymmetric spacetime with a general E/M field, one can find equilibrium solutions with circular velocity flows for which the current is purely toroidal, producing a purely *poloidal magnetic field.*

In ideal MHD, $T^{(\mathrm{em})}_{\alpha\beta}$ has the simpler form (5.48). Then the circularity conditions (5.55), (5.56) become [504]

$$B_t B_{[\alpha} t_\gamma \phi_{\delta]} = 0, \tag{5.58}$$

$$B_\phi B_{[\alpha} t_\gamma \phi_{\delta]} = 0. \tag{5.59}$$

For $B_t = B_\phi = 0$, we recover the purely poloidal solution. But, the above equations also allow for a purely *toroidal magnetic field*, $(B_t, 0, 0, B_\phi)$, generated by a current of the form $(0, J_r, J_\theta, 0)$.

More generally, in ideal MHD, a magnetized fluid with only a circular flow generates a circular spacetime if, and only if, the magnetic field is either purely poloidal or purely toroidal [504].

Equilibria with purely toroidal or purely poloidal magnetic fields, however, are unstable [733, 679], whereas some mixed equilibria appear to be stable (see, for example, Duez, Braithwaite, and Mathis [198], who infer stability of mixed equilibria in a Newtonian context from numerical evolutions). For relativistic stars, Gourgoulhon et al. [275] find a general form of stationary axisymmetric magnetic fields for noncircular equilibria.

Equilibrium models. A realistic magnetized star is never in exact equilibrium, because of magnetic braking of differential rotation and due to magnetic diffusion. However, assuming uniform rotation, circularity of spacetime, and ideal MHD, equilibrium solutions can be found. From the Euler equation (5.49), we obtain the equation of hydrostationary equilibrium in ideal MHD (expressed in terms of the vector potential A_α)

$$\nabla_\alpha (H - \ln u^t) - \frac{1}{\epsilon + p} (\nabla_\alpha A_\beta + \nabla_\beta A_\alpha) j^\beta = 0 \tag{5.60}$$

(with α corresponding only to indexes describing the 2-dimensional subspace normal to u^α, i.e., r and θ).

For a *purely poloidal* magnetic field, the ideal MHD condition leads directly to the following algebraic relation between the nonvanishing components of the vector potential:

$$\nabla A_t = -\Omega \nabla A_\phi . \tag{5.61}$$

Taking the curl of this equation then leads to the conclusion that Ω must be either a constant or a function of A_ϕ. Because there is no physical justification for the latter case, magnetized stars with a purely poloidal magnetic field must be *uniformly rotating* (a differentially rotating star is thus not stationary and evolves towards uniform rotation due to magnetic braking). Equation (5.61) then has the first integral

$$A_t = -\Omega A_\phi + C, \tag{5.62}$$

where C is a constant related to the total electric charge. The Euler equation becomes

$$\nabla(H - \ln u^t) - \frac{1}{\epsilon + p}(j^\phi - \Omega j^t)\nabla A_\phi = 0. \tag{5.63}$$

and has a first integral in three different cases. These are (a) $(j^\phi - \Omega j^t) = 0$, (b) $(\epsilon + p)^{-1}(j^\phi - \Omega j^t) = \text{const.}$, and (c) $(\epsilon + p)^{-1}(j^\phi - \Omega j^t) = f(A_\phi)$. The first case corresponds to a vanishing Lorentz force and has been considered in [62, 504] (force-free field). The second case is difficult to realize but has been considered as an approximation in, for example, [148]. The third case is more general and was first considered in [89, 74]. The force-free case (a) is self-consistent but implies the absence of a purely poloidal field for nonrotating stars – under this assumption, the field is generated only by convective currents in rotating charged stars. The more general choice (c) describes a field that is generated by the sum of convective and conduction currents, but requires the additional assumption of the existence of a conduction current in the nonrotating limit (or the existence of slow rotation).

After making a choice for the current (more precisely, for $(\epsilon + p)^{-1}(j^\phi - \Omega j^t)$) and for the total charge, the system consisting of the Einstein equations, the hydrostationary equilibrium equation and Maxwell's equations can be solved for the spacetime metric, the hydrodynamical variables, and the vector-potential components A_t and A_ϕ, from which the magnetic and electric fields in various observer frames are obtained.

For a *purely toroidal magnetic field*, the only nonvanishing component of the Faraday tensor is $F_{r\theta}$. Then, the ideal MHD condition does not lead to a restriction on the angular velocity of the star. It is easy to show that for uniformly rotating stars, the Euler equation can be written as [354, 275]

$$\nabla(H - \ln u^t) + \frac{1}{4\pi(\epsilon + p)g_2}\sqrt{\frac{g_2}{g_1}}F_{r\theta}\nabla\left(\sqrt{\frac{g_2}{g_1}}F_{r\theta}\right) = 0, \tag{5.64}$$

where $g_1 = g_{rr}g_{\theta\theta} - (g_{r\theta})^2$, $g_2 = -g_{tt}g_{\phi\phi} + (g_{t\phi})^2$, which implies the existence of solutions for which $\sqrt{\frac{g_2}{g_1}}F_{r\theta}$ is a function of $(\epsilon + p)g_2$ (see [354] for representative numerical solutions).

6

Approximation methods for equilibria

6.1 Slow-rotation approximation

To order $O(\Omega^2)$ the structure of a star changes only by quadrupole terms and the equilibrium equations become a set of ordinary differential equations. In addition to these $O(\Omega^2)$ changes in pressure and density, there is an $O(\Omega)$ change in the metric that describes the dragging of inertial frames. Hartle's [297, 304] method computes rotating stars in this *slow-rotation approximation.* A review of slowly rotating models can be found in [175]. Weber et al. [722, 724] also implement Hartle's formalism to explore the rotational properties of four more recent EOSs.

Weber and Glendenning [723] improved on Hartle's formalism in order to obtain a more accurate estimate of the angular velocity at the mass-shedding limit, but their models still showed large discrepancies compared to corresponding models computed without the slow-rotation approximation [583]. Thus, the slow-rotation approximation is appropriate for constructing models of typical pulsars and even millisecond pulsars, but it is not suitable for computing models of rapidly rotating relativistic stars near the mass-shedding limit. A detailed comparison of the numerical solutions obtained in the slow-rotation approximation to solutions obtained without making this assumption was presented by Berti et al. [65].

6.1.1 The nonrotating limit

We begin the discussion of the slow-rotation approximation by first reviewing the well-known nonrotating limit. In Schwarzschild coordinates (t, r, θ, ϕ), the metric takes the form

$$ds^2 = -e^{2\nu(r)}dt^2 + \left[1 - \frac{2m(r)}{r}\right]^{-1} dr^2 + r^2(d\theta^2 + \sin^2\theta d\phi^2), \tag{6.1}$$

where $\nu(r)$ and $m(r)$, as well as the pressure $p(r)$, are determined by the Tolman-Oppenheimer-Volkoff (TOV) equations

$$\frac{dm}{dr} = 4\pi r^2 \epsilon, \tag{6.2}$$

$$\frac{d\nu}{dr} = -\frac{1}{\epsilon + p}\frac{dp}{dr}, \tag{6.3}$$

$$\frac{dp}{dr} = -\frac{(\epsilon + p)(m + 4\pi r^3 p)}{r(r - 2m)}, \tag{6.4}$$

by integrating the above system from the center to the surface, $r = R_{\rm sph}$, with conditions $m(0) = 0$, $p(0) = p_c$, and $\nu(0)$ arbitrary. Here, p_c is the chosen value of central pressure. The arbitrariness in the initial value for $\nu(r)$ is removed by matching the solution at the surface of the star to the analytic exterior solution

$$e^{2\nu(r)} = 1 - \frac{2M_{\rm sph}}{r}. \tag{6.5}$$

where $M_{\rm sph} := m(R_{\rm sph})$ is the gravitational mass of the nonrotating (spherical) star.

6.1.2 Stationary axisymmetric spacetime in quasi-Schwarzschild coordinates

In Chapter 1 the metric of a stationary, axisymmetric spacetime was described in quasi-isotropic coordinates, as in Eq. (1.59). However, for the slow-rotation expansion, Hartle [297] described the metric in terms of *quasi-Schwarzschild* coordinates (t, r, θ, ϕ) [303][1]

$$ds^2 = -H(r,\theta)^2 dt^2 + Q(r,\theta)^2 dr^2 + r^2 K(r,\theta)^2 \left[d\theta^2 + \sin^2\theta(d\phi - \omega(r,\theta)dt)^2\right]. \tag{6.6}$$

As described in Section 1.2, the metric function $\omega(r, \theta)$ is the dragging of inertial frames, the angular velocity of a zero–angular-momentum particle falling from infinity to a coordinate location (r, θ). In the following, we will assume that the star is rotating with a *uniform angular velocity* Ω. Differential rotation is treated in, for example, [298, 654].

6.1.3 Slow-rotation expansion to $O(\Omega)$

Because of the assumed stationarity, axisymmetry, and symmetry with respect to the equatorial plane of a rotating star, the diagonal metric functions $\nu(r,\theta)$, $Q(r,\theta)$, $K(r,\theta)$ contain only even powers in an expansion in terms of Ω. Thus, at $O(\Omega)$ these metric functions remain unaffected by rotation. In contrast, the expansion of the off-diagonal metric function $\omega(r, \theta)$ contains only odd powers in Ω, that is;

$$\omega(r,\theta) = \omega_1(r,\theta) + O(\Omega^3), \tag{6.7}$$

where $\omega_1(r, \theta)$ is the dragging of inertial frames at $O(\Omega)$.

[1] Notice first that in this section we will use r for the radial coordinate of the slow-rotation metric, which coincides with the circumferential radius to $O(\Omega)$, whereas in other chapters the same symbol is used for the radial coordinate of the quasi-isotropic metric (1.59), which differs from the circumferential radius at any rotation rate. Note also that to translate our notation to Hartle's, make the following replacements (Hartle's on the right): $\nu \to \nu/2$, $\omega \to L$, $\omega_1 \to \omega$.

The $O(\Omega)$ expansion of the t-ϕ component of the field equation admits a regular and asymptotically flat solution only if $\omega_1 = \omega_1(r)$. Defining

$$\bar{\omega} := \Omega - \omega(r), \tag{6.8}$$

$$j(r) := e^{-\nu(r)}\sqrt{1 - \frac{2m(r)}{r}} \equiv e^{-[\nu(r)+\lambda(r)]}, \tag{6.9}$$

the t-ϕ component of the field equation yields, up to powers of $O(\Omega)$,

$$\frac{1}{r^4}\frac{d}{dr}\left(r^4 j\frac{d\bar{\omega}}{dr}\right) + \frac{4}{r}\frac{dj}{dr}\bar{\omega} = 0. \tag{6.10}$$

Starting with an arbitrary value for $\bar{\omega}$ and with $d\bar{\omega}/dr = 0$ at $r = 0$, one can integrate the above differential equation and then match it at the surface, $r = R_{\rm sph}$, to its analytic exterior solution

$$\bar{\omega}(r) = \Omega - \frac{2J}{r^3}, \tag{6.11}$$

where J is the total angular momentum of the star, given by

$$J = \frac{R_{\rm sph}^4}{6}\left(\frac{d\bar{\omega}}{dr}\right)_{r=R_{\rm sph}}. \tag{6.12}$$

Because (6.10) is linear in $\bar{\omega}$, which is proportional to Ω, the $O(\Omega)$ equations of structure need only be integrated once for an arbitrary value of $\bar{\omega}(r=0)$. The obtained solution $\bar{\omega}(r)$ can then be easily scaled to any value of Ω.

Summarizing, the metric of a uniformly rotating star to $O(\Omega)$ is

$$ds^2 = -e^{2\nu(r)}dt^2 + \left[1 - \frac{2m(r)}{r}\right]^{-1} dr^2 + r^2\{d\theta^2 + \sin^2\theta[d\phi - \omega(r)dt]^2\} + O(\Omega^2), \tag{6.13}$$

where the functions $\nu(r)$ and $m(r)$ are solutions of the TOV equations for a non-rotating star of same central density, whereas the dragging of inertial frames, $\omega(r)$, is obtained from Eq. (6.10). In the exterior, the metric is given analytically by

$$\begin{aligned} ds^2 = &- \left(1 - \frac{2M_{\rm sph}}{r}\right)dt^2 + \left(1 - \frac{2M_{\rm sph}}{r}\right)^{-1} dr^2 \\ &+ r^2\left[d\theta^2 + \sin^2\theta\left(d\phi - \frac{2J}{r^3}dt\right)^2\right] + O(\Omega^2), \end{aligned} \tag{6.14}$$

where $M_{\rm sph} = m(R_{\rm sph})$ is the gravitational mass and J the angular momentum of the star, given by Eq. (6.12).

6.1.4 Slow-rotation expansion to $O(\Omega^2)$

When the slow-rotation approximation is extended to $O(\Omega^2)$, all scalar fluid variables (pressure, density etc.), as well as the diagonal metric components, are altered by rotation. In contrast, the fluid's velocity, angular momentum, and the dragging of inertial frames associated with the off-diagonal metric component $g_{t\phi}$ are not affected by the $O(\Omega^2)$ terms in the expansion, because these quantities are functions of only odd powers in Ω. The $O(\Omega^2)$ terms in the expansion's fluid and metric variables are 2-dimensional functions of (r, θ), the angular dependence of which can be expanded in terms of Legendre polynomials, $P_l(\cos\theta)$. One can then show that only the monopole ($l = 0$) and quadrupole ($l = 2$) terms in this expansion are nonvanishing. Furthermore, because transformations of $r \to f(r)$ do not change the form of the Schwarzschild metric, the monopole deformation in the $r^2(d\theta^2 + \sin^2\theta d\phi^2)$ part of the line element can be absorbed in a redefinition of the radial coordinate r. One is left with monopole deformations in g_{tt} and g_{rr} and with quadrupole deformations in all diagonal components of the metric. In this way, Hartle and Thorne [304] write the metric to $O(\Omega^2)$ as

$$\begin{aligned} ds^2 &= -e^{2\nu(r)}\{1 + 2[h_0(r) + h_2(r)P_2(\cos\theta)]\}\,dt^2 \\ &\quad + \left[1 - \frac{2m(r)}{r}\right]^{-1}\left\{1 + \frac{2[m_0(r) + m_2(r)P_2(\cos\theta)]}{r - 2m(r)}\right\}dr^2 \\ &\quad + r^2\{1 + 2[v_2(r) - h_2(r)]P_2(\cos\theta)\}\left\{d\theta^2 + \sin^2\theta\,[d\phi - \omega_1(r)dt]^2\right\} \\ &\quad + O(\Omega^3), \end{aligned} \tag{6.15}$$

where $h_0(r)$ and $m_0(r)$ are functions related to monopole deformations and $h_2(r)$, $m_2(r)$, and $v_2(r)$ represent the radial dependence of the quadrupole deformations.

Similarly, the pressure is expanded as

$$p(r, \theta) = p(r) + (\epsilon + p)\,[p_0^*(r) + p_2^*(r)P_2(\cos\theta)] + O(\Omega^4), \tag{6.16}$$

where $p(r)$ is the pressure distribution of a nonrotating model of same central density and $p_0^*(r)$ and $p_2^*(r)$ are related to the $O(\Omega^2)$ monopole and quadrupole deformations of the pressure distribution, respectively. The corresponding expansion of the energy density distribution is

$$\epsilon(r, \theta) = \epsilon(r) + (\epsilon + p)\frac{d\epsilon}{dp}\,[p_0^*(r) + p_2^*(r)P_2(\cos\theta)] + O(\Omega^4), \tag{6.17}$$

where $d\epsilon/dp$ is obtained from the equation of state.

With the above definitions, the Einstein field equation for a stationary, axisymmetric configuration can be separated into a set of equations for the nonrotating background (which are satisfied by the TOV solution), a set of equations for the $l = 0$ part, and a set of equations for the $l = 2$ part. From these, one can obtain first-order ordinary differential equations for the functions $m_0(r)$, $p_0^*(r)$, $v_2(r)$, and $h_2(r)$ and an algebraic equation for $m_2(r)$. Expanding the hydrostationary equilibrium equation and the 4-velocity to order $O(\Omega^2)$ and separating the resulting equation into nonrotating, monopole, and quadrupole parts, one can obtain algebraic equations for the remaining unknown functions $h_0(r)$ and $p_2^*(r)$.

Monopole part of deformations. From the $l = 0$ part of the $G_t{}^t = 8\pi T_t{}^t$ equation, one obtains a first-order differential equation for $m_0(r)$,

$$\frac{dm_0}{dr} = 4\pi r^2 \frac{d\epsilon}{dp}(\epsilon + p)p_0^* + \frac{1}{12} j^2 r^4 \left(\frac{d\bar{\omega}}{dr}\right)^2 - \frac{1}{3} r^3 \frac{dj^2}{dr} \bar{\omega}^2, \tag{6.18}$$

and combining the $l = 0$ parts of $G_r{}^r = 8\pi T_r{}^r$ and of the hydrostationary equilibrium equation, one obtains a first-order differential equation for $p_0^*(r)$,

$$\frac{dp_0^*}{dr} = -\frac{m_0(1 + 8\pi r^2 p)}{(r - 2m)^2} - \frac{4\pi(\epsilon + p)r^2}{(r - 2m)} p_0^* + \frac{r^4 j^2}{12(r - 2m)} \left(\frac{d\bar{\omega}}{dr}\right)^2 + \frac{1}{3}\frac{d}{dr}\left(\frac{r^3 j^2 \bar{\omega}^2}{r - 2m}\right). \tag{6.19}$$

The above two equations are integrated outward from the center of the star, where the assumption that the rotating star has the same central density as the nonrotating model leads to the boundary conditions $m_0(r = 0) = 0$ and $p_0^*(r = 0) = 0$.

The last monopole deformation, h_0, is obtained from the $l = 0$ part of the hydrostationary equilibrium equation, through the algebraic relation

$$h_0 = -p_0^* + \frac{1}{3} r^2 e^{-2\nu} \bar{\omega}^2 + h_{0c}, \tag{6.20}$$

where h_{0c} is an integration constant that is determined by matching to the exterior solution at the surface.

The exterior solution for m_0 and h_0 is obtained analytically by solving Eq. (6.18) and the first-order differential equation obtained from the $l = 0$ part of $G_r{}^r = 8\pi T_r{}^r$,

$$\frac{dh_0}{dr} = \frac{m_0(1 + 8\pi r^2 p)}{(r - 2m)^2} + \frac{4\pi(\epsilon + p)r^2}{(r - 2m)} p_0^* - \frac{r^4 j^2}{12(r - 2m)} \left(\frac{d\bar{\omega}}{dr}\right)^2. \tag{6.21}$$

The analytic solution is

$$m_0(r) = \delta M - \frac{J^2}{r^3}, \tag{6.22}$$

$$h_0(r) = -\frac{\delta M}{r - 2M_{\rm sph}} + \frac{J^2}{r^3(r - 2M_{\rm sph})}, \tag{6.23}$$

where δM is a constant, that is, by inspection of the line element (6.15), identified with the *change in the gravitational mass of the star* due to rotation

$$\delta M = m_0(R_{\rm sph}) + \frac{J^2}{R_{\rm sph}^3}, \tag{6.24}$$

where the comparison is between two stars with the same central density.

Quadrupole part of deformations. Two first-order differential equations for $v_2(r)$ and $h_2(r)$ are obtained from the $l=2$ parts of the $R_r{}^\theta = 0$ and $G_r{}^r = 8\pi T_r{}^r$ field-equation components. With the help of the $l=2$ parts of the hydrostationary equilibrium equation and of the relation $R_\theta{}^\theta - R_\phi{}^\phi = 8\pi(T_\theta{}^\theta - T_\phi{}^\phi)$, one obtains

$$\frac{dv_2}{dr} = -2\frac{d\nu}{dr}h_2 + \left(\frac{1}{r}+\frac{d\nu}{dr}\right)\left[-\frac{1}{3}r^3\frac{dj^2}{dr}\bar{\omega}^2 + \frac{1}{6}j^2r^4\left(\frac{d\bar{\omega}}{dr}\right)^2\right], \tag{6.25}$$

$$\begin{aligned}\frac{dh_2}{dr} = &\left\{-2\frac{d\nu}{dr} + \frac{r}{2(r-2m)}\left(\frac{d\nu}{dr}\right)^{-1}\left[8\pi(\epsilon+p) - \frac{4m}{r^3}\right]\right\}h_2\\
&-\frac{2v_2}{r(r-2m)}\left(\frac{d\nu}{dr}\right)^{-1} + \frac{1}{6}\left[\frac{d\nu}{dr}r - \frac{1}{2(r-2m)}\left(\frac{d\nu}{dr}\right)^{-1}\right]r^3j^2\left(\frac{d\bar{\omega}}{dr}\right)^2\\
&-\frac{1}{3}\left[\frac{d\nu}{dr}r + \frac{1}{2(r-2m)}\left(\frac{d\nu}{dr}\right)^{-1}\right]r^2\frac{dj^2}{dr}\bar{\omega}^2.\end{aligned} \tag{6.26}$$

The above system of equations is integrated starting with the boundary conditions $h_2 = v_2 = 0$ at $r=0$. In the exterior, one can find an analytic solution, as a sum of a particular solution and a solution to the homogeneous system. Of the two possible solutions of the homogeneous system, the physical one is the solution that vanishes at infinity. A particular solution can be obtained by making the ansatz that v_2 and h_2 behave as $ar^{-3}+br^{-4}+cr^{-5}$ (with a, b, and c different constants in each case). In this way, the exterior solution for $v_2(r)$ and $h_2(r)$ is obtained as

$$v_2 = -\frac{J^2}{r^4} + K\frac{2M_{\rm sph}}{[r(r-2M_{\rm sph})]^{1/2}}Q_2^1\left(\frac{r}{M_{\rm sph}}-1\right), \tag{6.27}$$

$$h_2 = J^2\left(\frac{1}{M_{\rm sph}r^3}+\frac{1}{r^4}\right) + KQ_2^2\left(\frac{r}{M_{\rm sph}}-1\right), \tag{6.28}$$

where Q_n^m are the associated Legendre functions of the second kind and K is a constant, determined by matching the interior and exterior solutions at the surface.

The quadrupole deformation $m_2(r)$ is obtained directly as an algebraic equation from the $l=2$ part of the relation $R_\theta{}^\theta - R_\phi{}^\phi = 8\pi(T_\theta{}^\theta - T_\phi{}^\phi)$, which vanishes identically in the nonrotating limit

$$m_2 = (r-2M_{\rm sph})\left[-h_2 - \frac{1}{3}r^3\left(dj^2/dr\right)\bar{\omega}^2 + \frac{1}{6}r^4j^2\left(\frac{d\bar{\omega}}{dr}\right)^2\right], \tag{6.29}$$

and the quadrupole deformation $p_2^*(r)$ is also obtained algebraically from the $l=2$ part of the hydrostationary equilibrium equation

$$p_2^* = -h_2 - \frac{1}{3}r^2e^{-2\nu}\bar{\omega}^2. \tag{6.30}$$

This completes the determination of the fluid and spacetime properties of a rotating star to $O(\Omega^2)$.

6.1.5 Physical properties in the slow-rotation approximation

Gravitational mass. To $O(\Omega^2)$, the gravitational mass of a star with angular momentum J is

$$M = M_{\rm sph} + m_0(R_{\rm sph}) + J^2/R_{\rm sph}^3 + O(\Omega^4), \tag{6.31}$$

(cf. Eq. [6.24]), where $M_{\rm sph}$ and $R_{\rm sph}$ are the gravitational mass and radius of a nonrotating star of same central density, respectively.

Radius, surface deformation, and eccentricity. Because of the $O(\Omega^2)$ rotational corrections, the shape of the star is deformed from a sphere to a spheroid. Each surface of constant density at a radial distance $r_{\rm sph}$ in the nonrotating star is displaced in the rotating configuration to an θ-dependent radial distance

$$r = r_{\rm sph} + \xi_0(r_{\rm sph}) + \xi_2(r_{\rm sph})P_2(\cos\theta) + O(\Omega^4), \tag{6.32}$$

where

$$\xi_0 = -p_0^*(\epsilon + p)\left(\frac{dp}{dr}\right)^{-1}, \tag{6.33}$$

$$\xi_2 = -p_2^*(\epsilon + p)\left(\frac{dp}{dr}\right)^{-1}. \tag{6.34}$$

In order to define an invariant parameterization of a surface of constant density, one can construct a surface in a 3-dimensional flat space with polar coordinates (r^*, θ^*, ϕ^*) with the same intrinsic geometry. To $O(\Omega^2)$, the surface in flat space is the spheroid defined by

$$r^*(\theta^*) = r_{\rm sph} + \xi_0(r_{\rm sph}) + \{\xi_2(r_{\rm sph}) + r_{\rm sph}[v_2(r_{\rm sph}) - h_2(r_{\rm sph})]\}P_2(\cos\theta^*). \tag{6.35}$$

Its eccentricity is

$$e := [(r^*_{\rm equator}/r^*_{\rm pole})^2 - 1]^{1/2}, \tag{6.36}$$

$$= \{-3[v_2(r_{\rm sph}) - h_2(r_{\rm sph}) + \xi_2(r_{\rm sph})/r_{\rm sph}]\}^{1/2}. \tag{6.37}$$

The equatorial circumferential radius of the star is obtained from $\sqrt{g_{\phi\phi}}$, for which the analytic exterior solutions for $v_2(r)$ and $h_2(r)$ can be used, evaluated at the surface.

Moment of inertia. A scalar moment of inertia can be defined as the ratio

$$I = J/\Omega = \frac{R_{\rm sph}^4}{6\Omega}\left(\frac{d\bar\omega}{dr}\right)_{r=R_{\rm sph}} + O(\Omega^2). \tag{6.38}$$

One can also obtain an integral relation involving only metric functions, by integrating (6.10) from $r = 0$ to $r = R_{\rm sph}$, by parts,

$$I = -\frac{2}{3}\int_0^{R_{\rm sph}} dr r^3 \left(\frac{dj}{dr}\right)\left(\frac{\bar\omega}{\Omega}\right) + O(\Omega^2), \tag{6.39}$$

which can be reduced to the usual integral expression in the Newtonian limit. Alternatively, one can use the TOV equations describing the equilibrium of the nonrotating star, in order to write Eq. (6.39) as [288]

$$I = \frac{8\pi}{3} \int_0^{R_{\rm sph}} dr r^4 \, (\epsilon + p) \, e^{\lambda - \gamma} \left(\frac{\bar{\omega}}{\Omega} \right) + O(\Omega^2). \tag{6.40}$$

Obtaining a higher-order expression for I would require $O(\Omega^3)$ corrections to J and Ω.

Analytic exterior solution to $O(\Omega^2)$. Comparing the analytic expressions for various metric components obtained above to their Newtonian counterparts, one can identify the scalar quadrupole moment Q (the coefficient of the $r^{-3} P_2(\cos\theta)$ term in the Newtonian gravitational potential) as

$$Q = \frac{J^2}{M_{\rm sph}} + \frac{8}{5} K M_{\rm sph}^3, \tag{6.41}$$

where K is the constant in Eqs. (6.27) and (6.28). In this way, the exterior metric can be written in the following analytic form:

$$\begin{aligned}
ds^2 = & -\left(1 - \frac{2M}{r} + \frac{2J^2}{r^4}\right) \left\{ 1 + 2\left[\frac{J^2}{Mr^3}\left(1 + \frac{M}{r}\right) \right.\right. \\
& \left.\left. + \frac{5}{8} \frac{Q - J^2/M}{M^3} Q_2^2\left(\frac{r}{M} - 1\right) \right] P_2(\cos\theta) \right\} dt^2 \\
& + \left(1 - \frac{2M}{r} + \frac{2J^2}{r^4}\right)^{-1} \left\{ 1 - 2\left[\frac{J^2}{Mr^3}\left(1 - \frac{5M}{r}\right) \right.\right. \\
& \left.\left. + \frac{5}{8} \frac{Q - J^2/M}{M^3} Q_2^2\left(\frac{r}{M} - 1\right) \right] P_2(\cos\theta) \right\} dr^2 \\
& + r^2 \left[1 + 2\left\langle -\frac{J^2}{Mr^3}\left(1 + \frac{2M}{r}\right) + \frac{5}{8} \frac{Q - J^2/M}{M^3} \left\{ \frac{2M}{[r(r-2M)]^{1/2}} \right.\right.\right. \\
& \left.\left.\left. \times Q_2^1\left(\frac{r}{M} - 1\right) - Q_2^2\left(\frac{r}{M} - 1\right) \right\} \right\rangle P_2(\cos\theta) \right] \\
& \times \left\{ d\theta^2 + \sin^2\theta \left[d\phi - \left(\frac{2J}{r^3}\right) dt \right]^2 \right\} + O(\Omega^4),
\end{aligned} \tag{6.42}$$

where the gravitational mass, M, of the rotating configuration has been used in place of $M_{\rm sph}$, because their difference affects the above metric only at $O(\Omega^4)$. The above analytic exterior solution is characterized entirely by the three invariant multipole moments M, J, and Q.

6.2 Spatial conformal flatness and quasiequilibrium approximations

We discuss in this section related approximations that are widely used in finding initial data for the inspiral of compact binary systems; in quasiequilibrium models of binary inspiral; and in describing pulsating and collapsing stars. At first post-Newtonian order or for spherical stars, the spatial metric is conformally flat. Up to second post-Newtonian order, a binary system can be modeled as an equilibrium configuration, stationary in a rotating frame. In the approximations we discuss, one retains one or both of these assumptions – spatial conformal flatness and stationarity in a rotating frame – and finds solutions to components of the exact Einstein-Euler system.

We use in this and the following sections the 3+1 decomposition of Section (2.1), with the metric written in the form (2.10),

$$ds^2 = -\alpha^2 dt^2 + \gamma_{ij}(dx^i + \beta^i dt)(dx^j + \beta^j dt);$$

with the projections S_{ab}, j^a and ρ_E of the stress-energy tensor given by Eqs. (2.19); and with the time derivative of the 3-metric, $\partial_t \gamma_{ab}$, related to the extrinsic curvature K_{ab} by Eq. (2.18),

$$\partial_t \gamma_{ab} = -2\alpha K_{ab} + D_a \beta_b + D_b \beta_a.$$

6.2.1 Spatial conformal flatness: The IWM-CFC approximation

In the classical Newtonian development of the theory of stellar equilibria, binary systems are modeled as fluid configurations stationary in a frame that rotates with the orbital angular velocity. In a spacetime picture, the binary is *helically symmetric*, with a symmetry vector of the form $k^\alpha = t^\alpha + \Omega\phi^\alpha$ discussed in Section 3.6.2. Although helical symmetry requires equal amounts of ingoing and outgoing radiation and is inconsistent with asymptotic flatness, one can construct helically symmetric initial data sets that are asymptotically flat by solving a truncated set of field equations for a metric in which the radiative degrees of freedom are eliminated. This is most commonly done by assuming that the spatial part of the metric is conformally flat. The way one solves the truncated Einstein equation, together with the relativistic Euler equation, is closely patterned on the KEH method for constructing rotating stars.

Assuming spatial conformal flatness reduces the number of metric functions to five, resulting in a simpler system. The idea was proposed by Isenberg [334] and by Wilson and Mathews [729, 728, 730] (IWM) as a scheme that could be applied for the evolution of initial data in numerical relativity, such as binary neutron star systems (by constructing a series of equilibrium solutions), without solving the full dynamical Einstein field equations. For single rotating stars, the approximation scheme was tested by Cook, Shapiro, and Teukolsky [160].

In the IWM-CFC approximation, the 3-metric is *conformally flat*,

$$\gamma_{ab} = \psi^4 \eta_{ab}, \tag{6.43}$$

where ψ is a conformal factor and η_{ab} is the flat 3-metric. The acronym *CFC* abbreviates the *conformal flatness condition.* In this approximation, the metric is given by the conformal factor ψ, the shift vector β^a, and the lapse function α. These two scalars and one vector are determined by two scalar and one vector equation: the Hamiltonian constraint, the momentum constraint (the vector equation), and the spatial trace of the Einstein equation.

The metric variables can be determined at any instant (for given sources) by the Hamiltonian and momentum constraints and by choosing *maximal slicing*,

$$K = 0, \tag{6.44}$$

as the gauge condition for the lapse. One writes the Hamiltonian constraint, Eq. (2.21), in terms of ψ by using the form of the Ricci scalar of a conformally flat metric (see, e.g., Eq. [D.9] of Wald [715]),

$$R = -8\psi^{-5}\nabla^2\psi, \tag{6.45}$$

where ∇^2 is the flat-space Laplacian. The constraint (2.21) then takes the form

$$\nabla^2\psi = -\psi^5\left(\frac{1}{8}K_{ab}K^{ab} + 2\pi\rho_E\right). \tag{6.46}$$

The momentum constraint becomes an equation for the shift β^a,

$$\begin{aligned}\nabla^2\beta^a + \frac{1}{3}\nabla^a\nabla_c\beta^c = {} & \nabla_b \ln\left(\psi^{-6}|\alpha|\right) \\ & \times\left(\nabla^b\beta^a + \nabla^a\beta^b - \frac{2}{3}\eta^{ab}\nabla_c\beta^c\right) \\ & + 16\pi\alpha\psi^4 j^a,\end{aligned} \tag{6.47}$$

where ∇_a is the flat-space derivative operator – the derivative operator of η_{ab} – and $\nabla^a := \eta^{ab}\nabla_b$. This form of the constraint is derived as follows. Because $K = 0$, K_{ab} is obtained from the trace-free part of Eq. (2.18):

$$K_{ab} = K_{ab}^{\mathrm{TF}} = \frac{1}{2\alpha}(\mathcal{L}_{\boldsymbol{\beta}}\gamma_{ab} - \partial_t\gamma_{ab})^{\mathrm{TF}} = \frac{1}{2\alpha}(\mathcal{L}_{\boldsymbol{\beta}}\gamma_{ab})^{\mathrm{TF}}, \tag{6.48}$$

where we have used $\gamma_{ab}^{\mathrm{TF}} = 0$ and $\partial_t\gamma_{ab} = \gamma_{ab}\partial_t \ln\psi^4$ to infer $\partial_t\gamma_{ab}^{\mathrm{TF}} = 0$. Explicitly,

$$K_{ab} = \frac{1}{2\alpha}\left(\mathcal{L}_{\boldsymbol{\beta}}\gamma_{ab} - \frac{1}{3}\gamma_{ab}\gamma^{cd}\mathcal{L}_{\boldsymbol{\beta}}\gamma_{cd}\right) =: \frac{1}{2\alpha}L_{\boldsymbol{\beta}}\gamma_{ab}. \tag{6.49}$$

The trace-free part $L_{\boldsymbol{\beta}}\gamma_{ab}$ of $\mathcal{L}_{\boldsymbol{\beta}}\gamma_{ab}$ is called the conformal Killing form, and it is easy to check that it changes only by scaling under a conformal transformation:

$$L_{\boldsymbol{\beta}}\gamma_{ab} = \psi^4 L_{\boldsymbol{\beta}}\eta_{ab} = \psi^4\left(\nabla_a\tilde{\beta}_b + \nabla_b\tilde{\beta}_a - \frac{2}{3}\eta_{ab}\nabla_c\beta^c\right), \tag{6.50}$$

where $\tilde{\beta}_a := \eta_{ab}\beta^b$. Then

$$K_a{}^b = \frac{1}{2\alpha}\left(\nabla_a\beta^b + \nabla^b\tilde{\beta}_a - \frac{2}{3}\delta_b^a\nabla_c\beta^c\right). \tag{6.51}$$

Again using $K = 0$ and the conformally flat form of the metric, we obtain

$$D_b K_a{}^b = \psi^{-6} \nabla_b (\psi^6 K_a{}^b), \tag{6.52}$$

and the momentum constraint takes the form

$$\nabla^2 \tilde{\beta}_b + \frac{1}{3} \nabla_b \nabla_c \beta^c + 2\alpha K_b{}^c \nabla_c \ln \frac{\psi^6}{\alpha} = 16\pi j_b. \tag{6.53}$$

Finally, contracting with $\eta^{ab} = \psi^4 \gamma^{ab}$ to raise the index b, we obtain the form (6.47).

The remaining equation used to determine the metric is the spatial trace of the field equation, $\gamma^{\alpha\beta}(G_{\alpha\beta} - 8\pi T_{\alpha\beta}) = 0$. Imposing the condition $\partial_t K = 0$ turns the spatial trace of the Einstein equation into an elliptic equation for $\alpha\psi$,

$$\nabla^2(\alpha\psi) = \alpha\psi^5 \left[\frac{7}{8} K_{ab} K^{ab} + 2\pi(\rho_E + 2S_c{}^c)\right], \tag{6.54}$$

where $S_c{}^c := \gamma_{ab} S^{ab}$. To obtain (6.54), contract (2.23) with γ^{ab}, use (6.45) to eliminate R, and subtract $5\times$ (6.46).

Equations (6.46), (6.47), and (6.54) are solved for ψ, β^a, and $\alpha\psi$, respectively. The second derivatives in the two scalar equations occur only in the flat-space Laplacian, and they are solved using the flat-space Green's function as in the KEH iteration. Several alternative methods have been used in numerically solving the momentum constraint for β^a. The simplest is just to regard all terms except $\nabla^2 \beta^a$ as an effective source, and again follow the KEH iteration. The alternatives avoid a source with second derivatives of the shift vector by decomposing β^a in ways that yield coupled elliptic equations for its parts [94, 502] (see Grandclément et al. [278] for a comparison of methods).

For a single, axisymmetric star, the shift vector in spherical polar coordinates has only one nonvanishing component, β^ϕ. Then,

$$K_{ab} K^{ab} = \frac{\sin^2\theta}{2\alpha^2} \left[(r\partial_r \beta^\phi)^2 + (\partial_\theta \beta^\phi)^2 \right]. \tag{6.55}$$

The system of equations for determining ψ, α, and β^ϕ, for given sources ρ_E and j_ϕ, is

$$\nabla^2 \psi = -\psi^5 \left\{ \frac{\sin^2\theta}{16\alpha^2} \left[(r\partial_r \beta^\phi)^2 + (\partial_\theta \beta^\phi)^2 \right] + 2\pi\rho_E \right\}, \tag{6.56}$$

$$\nabla^2(\alpha\psi) = \alpha\psi^5 \left\{ \frac{7\sin^2\theta}{16\alpha^2} \left[(r\partial_r \beta^\phi)^2 + (\partial_\theta \beta^\phi)^2 \right] + 2\pi(\rho_E + 2S) \right\}, \tag{6.57}$$

and

$$\begin{aligned} \left(\nabla^2 + \frac{2}{r}\partial_r + \frac{2\cot\theta}{r^2}\partial_\theta\right) \beta^\phi = {} & \partial_r \ln(\psi^{-6}|\alpha|) \partial_r \beta^\phi \\ & + \frac{1}{r^2} \partial_\theta \ln(\psi^{-6}|\alpha|) \partial_\theta \beta^\phi \\ & + \frac{16\pi\alpha}{r^2 \sin^2\theta} j_\phi. \end{aligned} \tag{6.58}$$

The above equations have the same elliptic operators as the exact equations in the KEH formulation and can thus be solved by exactly the same Green's function approach.

Spatial conformal flatness is exact in the first post-Newtonian approximation or for spherical stars described in isotropic coordinates. As a first check on the size of its error in a more general context, Cook, Shapiro, and Teukolsky [160] compared numerical models of rotating stars obtained from the exact equations and in the IWM-CFC approximation. Here the translation between the IWM-CFC metric functions and those of the exact metric is

$$\alpha = e^{\nu}, \tag{6.59}$$

$$\psi^4 = e^{2\mu}, \tag{6.60}$$

$$\beta^{\phi} = -\omega. \tag{6.61}$$

Comparison to models constructed in the exact formulation showed that the IWM-CFC approximation introduces errors that are at most of order a few percent for the most rapidly rotating cases (assuming uniform rotation) and much smaller than those for moderate rotation rates. Such models are therefore useful in studying rotating relativistic stars, when a moderate accuracy level is acceptable.

As mentioned above, this formalism has been used in dynamical studies of spacetimes describing pulsating and collapsing relativistic stars and inspiraling compact binary systems; see, for example, [189, 190, 501, 577, 578, 193, 500]. In certain cases, however, the above formulation suffers from nonuniqueness of solutions. This problem is circumvented by a slight reformulation, introducing an auxiliary variable [577, 162].

Although Wilson and Matthews initially used the approximation for numerical simulations of inspiral, its most common subsequent use for binary systems has been in constructing initial data sets and quasiequilibrium sequences. In addition to the system (6.46), (6.47), (6.54), one solves the Euler and energy conservation equations, as described in the next subsection.

6.2.2 Irrotational flow and helical symmetry

The small size of neutron stars means that the tidal forces that typically lock close binaries into corotation are much too weak to do so for neutron stars [355, 69]. As a result, the spin of neutron stars near the end of inspiral is essentially that of the early inspiral, at large separation. Because the spin of neutron stars in observed double neutron-star systems is slow ($\Omega \ll \Omega_K$), neutron stars in compact binary systems are commonly modeled as *irrotational* [88, 682, 607], as fluids whose velocity fields have zero vorticity:

$$\nabla_\alpha(hu_\beta) - \nabla_\beta(hu_\alpha) = 0. \tag{6.62}$$

This is a convenient assumption, because any irrotational velocity field satisfies the Euler equation, $q^\alpha{}_\gamma \nabla_\beta T^{\beta\gamma} = 0$: This is immediate from the form (1.32) valid for a homentropic model,

$$u^\beta [\nabla_\beta(hu_\alpha) - \nabla_\alpha(hu_\beta)] = 0. \tag{6.63}$$

In this, its Lichnerowicz form [413], the Euler equation is manifestly satisfied by any irrotational flow.

Vanishing vorticity implies the existence of a velocity potential Ψ for which

$$hu_\alpha = \nabla_\alpha \Psi; \tag{6.64}$$

and one solves the Euler equation by choosing a Ψ. Finally, with Ψ and the EOS specified, one obtains initial data by solving the elliptic equations (6.56)–(6.58).

Note, however, that energy conservation (1.27), the component $u_\alpha \nabla_\beta T^{\alpha\beta} = 0$, is not automatically satisfied. For a barotropic star, conservation of energy (1.27) and of baryons (1.22) are equivalent (because, for example, of the relation $d\epsilon/(\epsilon + p) = d\rho/\rho$) and have the form

$$0 = \nabla_\beta(\rho u^\beta) = \nabla_\beta \left(\frac{\rho}{h} \nabla^\beta \Psi\right). \tag{6.65}$$

For a generic Ψ, this gives the time evolution of ρ.

When one is setting initial data, however, one further restricts Ψ, and the equation plays a different role. In particular, for initial data describing matter that is stationary in a rotating frame, one restricts Ψ by assuming that $hu_\alpha = \nabla_\alpha \Psi$ is helically symmetric, Lie derived by the vector field $k^\alpha = t^\alpha + \Omega\phi^\alpha$, with Ω the angular velocity of the binary system and with t^α and ϕ^α denoting as usual the vectors $\boldsymbol{\partial}_t$ and $\boldsymbol{\partial}_\phi$:

$$\mathcal{L}_{\mathbf{k}} \nabla_\alpha \Psi = 0. \tag{6.66}$$

Using the identity $[\mathcal{L}_{\mathbf{k}}, d] = 0$, we have

$$0 = \nabla_\alpha(k^\beta \nabla_\beta \Psi) = 0 \Longrightarrow k^\beta \nabla_\beta \Psi = -C, \tag{6.67}$$

for some constant C.

The remaining presentation is essentially that of Uryū and Eriguchi [705]. Equation (6.67) has a form analogous to the first integral (1.110) of the equation of hydrostationary equilibrium, if one decomposes u^α in the manner

$$u^\alpha = u^t(k^\alpha + v^\alpha), \quad \text{with } v^\alpha n_\alpha = 0. \tag{6.68}$$

Then, with $\nabla_\alpha \Psi$ replaced by hu_α and with Eq. (6.68) used to write k_α in terms of u^α, we have

$$\frac{h}{u^t} + hu_\alpha v^\alpha = C, \tag{6.69}$$

where we have used $u_\alpha u^\alpha = -1$. In the numerical iteration, one uses Eqs. (6.69) and $u^\alpha u_\alpha = -1$ to find h and u^t algebraically in terms of $\nabla_a \Psi$; Eq. (6.65) then takes the form

$$D_a D^a \Psi - (\beta^a + \Omega\phi^a) D_a \Lambda + [D^a \Psi - \Lambda(\beta^a + \Omega\phi^a)]\, D_a \ln \frac{\alpha\rho}{h} = 0, \tag{6.70}$$

where $\Lambda := [C + (\beta^a + \Omega\phi^a)\nabla_a\Psi]/\alpha^2$. For fixed h and Λ, and with the EOS used to find ρ from h, this equation is elliptic, and it is solved for Ψ with the boundary condition that, measured by a corotating observer, the fluid flow be tangent to the surface of the star:

$$[\nabla^a\Psi - \Lambda(\beta^a + \Omega\phi^a)]\nabla_a\rho|_S = 0. \tag{6.71}$$

To summarize: In the IWM-CFC approximation, one finds helically symmetric initial data for an irrotational binary system in close analogy with the KEH method of finding the exact solution for a rotating star. In each iteration:

(1) Using a flat-space Green's function, one sequentially solves Eqs. (6.46), (6.47), and (6.54) for ψ, β^a, and $\alpha\psi$, respectively.
(2) Given the metric potentials, one updates the matter distribution and velocity field: One finds h and u^t algebraically from $u^\alpha u_\alpha = -1$ and Eq. (6.69), and one then solves the elliptic equation (6.70) for Ψ.

6.2.3 Waveless formulation for binary systems: Beyond conformal flatness

We have seen that the assumption of spatial conformal flatness allows one to find initial data for a binary system by solving five elliptic equations for five metric potentials. The (intertwined) errors associated with this ansatz are due to (1) the absence of radial motion – the requirement of a circular orbit; (2) the absence of the remaining metric degrees of freedom; and (3) a formalism in which the radiation content of the field is unrelated to the actual radiation field. The first of these errors has been addressed by using either a point-particle post-Newtonian approximation or a short time evolution (or both) to estimate the radial motion. In addressing the second error, the goal has been to add the remaining parts of the metric in a way that is accurate near the neutron stars. To do this, one uses the fact that solutions to the wave equation are dominated near the source by their Coulomb field if the size of the source is small compared to a wavelength of the field.

In the dynamical evolution of the binary system, the remaining parts of the metric are associated asymptotically with the gravitational wave degrees of freedom: In the far zone, at distances large compared to a wavelength, the radiative part of the field dominates the Coulomb contribution, falling off for each multipole as $1/r$ instead of $1/r^{l+1}$. In the near zone, however, within a wavelength of the source, their Coulomb field dominates.

Extensions of of the IWM-CFC approximation have been presented in [334], called *waveless approximation theories*, and in [706, 589, 626, 85]. They restore the remaining metric components but impose conditions on initial data that remove second time derivatives from the dynamical equations. This allows one to solve the full Einstein equation on an initial hypersurface by solving an extended system of elliptic equations. The resulting solution captures the Coulomb behavior of the additional metric potentials and lacks the wavelike asymptotic behavior of the physical metric. We describe here the waveless formalism of Uryū et al. [703, 706, 626].

One again introduces a conformal decomposition of the spatial metric,

$$\gamma_{ab} = \psi^4 \tilde{\gamma}_{ab}, \tag{6.72}$$

where ψ is defined by requiring that $\tilde{\gamma}_{ab}$ have determinant $\tilde{\gamma} = {}^3\eta$, with ${}^3\eta$ the determinant of a flat metric η_{ab}, ordinarily associated with a choice of coordinates. In Cartesian coordinates, for example, η_{ab} is the metric $dx^2 + dy^2 + dz^2$, and ${}^3\eta = 1$. To obtain an elliptic system that includes $\tilde{\gamma}_{ab}$, one chooses a Dirac gauge for the conformal metric, writing

$$\nabla_b \tilde{\gamma}^{ab} = 0, \tag{6.73}$$

where ∇_a is again the derivative operator of η_{ab}. One again imposes the slicing conditions

$$K = 0, \quad \partial_t K = 0. \tag{6.74}$$

The Hamiltonian and momentum constraints of Eqs. (6.46) and (6.47) then have the slightly modified forms

$$\tilde{\Delta}\psi := \tilde{\gamma}^{ab} \tilde{D}_a \tilde{D}_b \psi = -\psi^5 \left(\frac{1}{8} K_{ab} K^{ab} + 2\pi \rho_E \right) + \frac{1}{8} \psi \tilde{R}, \tag{6.75}$$

$$\tilde{D}_b (\psi^6 K_a{}^b) = 8\pi j_a \psi^6, \tag{6.76}$$

where $\tilde{R}$ and $\tilde{D}_a$ are the Ricci scalar and the covariant derivative of the metric $\tilde{\gamma}_{ab}$. The spatial trace of the Einstein equation similarly retains its elliptic form as an equation for $\alpha\psi$,

$$\tilde{\Delta}(\alpha\psi) = \alpha\psi^5 \left[\frac{7}{8} K_{ab} K^{ab} + 2\pi(\rho_E + 2 S_c{}^c) \right] + \frac{1}{8} \alpha\psi \tilde{R}. \tag{6.77}$$

Finally, the remaining part $\tilde{\gamma}_{ab}$ of the metric is found from the remaining part of the Einstein equation, the trace-free part of Eq. (2.23), or $E_{ab}^{\mathrm{TF}} := \gamma_a{}^\alpha \gamma_b{}^\beta E_{\alpha\beta} - \frac{1}{3}\gamma_{ab}\gamma^{\gamma\delta} E_{\gamma\delta} = 0$. The equation involves $\partial_t \gamma_{ab}$ and $\partial_t K_{ab}$ – first and second time derivatives of γ_{ab}. If we were seeking a solution that was exactly helically symmetric – exactly stationary in a rotating frame – we would use helical symmetry to replace these time derivatives by ϕ derivatives, writing

$$\mathcal{L}_{\mathbf{k}} \gamma_{ab} = 0, \tag{6.78a}$$

$$\mathcal{L}_{\mathbf{k}} K_{ab} = 0. \tag{6.78b}$$

Instead, to avoid the hyperbolic operator and the asymptotic standing waves of exact helical symmetry, one imposes the initial condition

$$\partial_t \tilde{\gamma}_{ab} = 0. \tag{6.79}$$

This condition, together with the slicing conditions $K = \partial_t K = 0$, is not consistent with helical symmetry of K_{ab}; one imposes only the trace-free part of the symmetry condition, namely,

$$(\mathcal{L}_{\mathbf{k}} K_{ab})^{TF} = \partial_t K_{ab} + (\mathcal{L}_{\Omega\boldsymbol{\phi}} K_{ab})^{TF} = 0, \tag{6.80}$$

to replace $\partial_t K_{ab}$ by $-(\mathcal{L}_{\Omega\boldsymbol{\phi}} K_{ab})^{TF}$ in the trace-free part of Eq. (2.23).

Because the trace-free part, $E_{ab}^{TF} = 0$, of (2.23) eliminates the term in the source proportional to γ_{ab}, we can write

$$0 = E_{ab}^{\rm TF} = \mathcal{E}_{ab}^{\rm TF}, \tag{6.81}$$

where

$$\mathcal{E}_{ab} = R_{ab} - \alpha^{-1} D_a D_b \alpha - 2K_{ac} K_b{}^c + \alpha^{-1} \mathcal{L}_{\boldsymbol{\beta}+\Omega\boldsymbol{\phi}} K_{ab} - 8\pi S_{ab}. \tag{6.82}$$

The 3-dimensional Ricci tensor is a sum,

$$R_{ab} = \tilde{R}_{ab} + R^{\psi}_{ab}, \tag{6.83}$$

of the Ricci tensor of the conformal metric $\tilde{\gamma}_{ab}$ and a part that involves only derivatives of the conformal factor ψ; in the Dirac gauge, $\tilde{R}_{ab}$ is elliptic in $\tilde{\gamma}_{ab}$:

$$R_{ab} = -\frac{1}{2}\tilde{\gamma}^{cd} \nabla_c \nabla_d \tilde{\gamma}_{ab} + \tilde{R}_{ab}^{\text{first-order}} + R^{\psi}_{ab}, \tag{6.84}$$

where $\tilde{R}_{ab}^{\text{first-order}}$ involves only first-order derivatives of $\tilde{\gamma}_{ab}$. Explicitly,

$$\tilde{R}_{ab}^{\text{first-order}} := -\frac{1}{2}(\nabla_b \tilde{\gamma}^{cd} \nabla_c \tilde{\gamma}_{ad} + \nabla_a \tilde{\gamma}^{cd} \nabla_c \tilde{\gamma}_{bd}) - C^d{}_{ac} C^c{}_{bd}, \tag{6.85}$$

where

$$C^c{}_{ab} := \frac{1}{2}\tilde{\gamma}^{cd}\Big(\nabla_a \tilde{\gamma}_{bd} + \nabla_b \tilde{\gamma}_{ad} - \nabla_d \tilde{\gamma}_{ab}\Big) \tag{6.86}$$

and

$$R^{\psi}_{ab} = -\frac{2}{\psi}\tilde{D}_a \tilde{D}_b \psi - \frac{2}{\psi}\tilde{\gamma}_{ab}\tilde{\Delta}\psi + \frac{6}{\psi^2}\tilde{D}_a \psi \tilde{D}_b \psi - \frac{2}{\psi^2}\tilde{\gamma}_{ab}\tilde{D}_c \psi \tilde{D}^c \psi. \tag{6.87}$$

Equation (6.82), together with the three equations (6.75)–(6.77), determine initial data on an initial hypersurface.

Because the departure of $\tilde{\gamma}_{ab}$ from the flat metric η_{ab} is small, one obtains greater numerical accuracy by solving Eq. (6.82) for the difference

$$h_{ab} := \tilde{\gamma}_{ab} - \eta_{ab} \tag{6.88}$$

than for $\tilde{\gamma}_{ab}$ itself. Rewritten in terms of h_{ab}, the equation has the form

$$\nabla^2 h_{ab} = 2\mathcal{S}_{ab}^{TF} - \frac{1}{3}\tilde{\gamma}_{ab} \nabla^e (\tilde{\gamma}^{cd} - \eta^{cd}) \nabla_e h_{cd}, \tag{6.89}$$

where

$$\mathcal{S}_{ab} = R^{\psi}_{ab} + \tilde{R}^{\rm NL}_{ab} - \alpha^{-1} D_a D_b \alpha - 2K_{ac} K_b{}^c + \alpha^{-1} \mathcal{L}_{\boldsymbol{\beta}+\Omega\boldsymbol{\phi}} K_{ab} - 8\pi S_{ab} \tag{6.90}$$

and $\tilde{R}^{\rm NL}_{ab} := \tilde{R}_{ab}^{\text{first-order}} - \frac{1}{2}(\tilde{\gamma}^{cd} - \eta^{cd}) \nabla_c \nabla_d h_{ab}$. Here, in accord with the notation of [626, 706, 703], "NL" abbreviates "nonlinear" and refers to the fact that $R^{\rm NL}_{ab}$ is of quadratic order in the small departure h_{ab}. That is, because $\nabla_c \tilde{\gamma}_{ab} = \nabla_c h_{ab}$, $R^{\rm NL}_{ab}$ is $O(h_{..}{}^2)$.

To summarize: The Einstein equation is solved on a spacelike hypersurface Σ for the conformal factor ψ, the shift β^a, the lapse α, and the conformal metric $\tilde{\gamma}_{ab}$, subject to the Dirac gauge condition $\mathcal{D}_b\tilde{\gamma}^{ab} = 0$, the slicing conditions $K = \partial_t K = 0$, and the additional conditions

$$\partial_t\tilde{\gamma}_{ab} = 0, \quad (\mathcal{L}_{\mathbf{k}} K_{ab})^{TF} = 0. \tag{6.91}$$

Explicitly, the set of equations (6.75)–(6.77) and (6.89) are solved for ψ, β^a, $\alpha\psi$, and $h_{ab} = \tilde{\gamma}_{ab} - \eta_{ab}$ [626].

6.3 Exact vacuum solutions

As we have seen in Section 6.1.3, to lowest order in rotation, the exterior metric (6.14) of a rotating star depends only on its mass and total angular momentum: For fixed J and M, rotating stars with different equations of state, as well as Kerr black holes, have metrics that coincide at $O(\Omega)$. At higher order, the multipole moments of rapidly rotating stellar models based on different equations of state differ from one another, and all differ from the multipole moments of the Kerr spacetime. There have been many attempts to find exact solutions to the Einstein equations in the stationary, axisymmetric case that could (to some approximation) describe the exterior of a rapidly rotating neutron star. Some relevant solutions will be described here.

In the vacuum region surrounding a stationary and axisymmetric star, the spacetime depends on only three metric functions (whereas four metric functions are needed for the interior). The most general form of the metric was given by Papapetrou [528]:

$$ds^2 = -f(dt - \omega d\phi)^2 + f^{-1}\left\{e^{2\gamma}(d\tilde{\varpi}^2 + d\tilde{z}^2) + \tilde{\varpi}^2 d\phi^2\right\}. \tag{6.92}$$

Here f, ω and γ are functions of the quasicylindrical Weyl-Lewis-Papapetrou coordinates $(\tilde{\varpi},\ \tilde{z})$. Because the metric agrees with the form (1.66) up to a (t, ϕ)-independent change of coordinates $(\varpi, z) \to (\tilde{\varpi},\ \tilde{z})$, the potential ω is the usual dragging of inertial frames, whereas $f = e^{2\nu}$, and both f and ω are defined in terms of the Killing vectors t^α and ϕ^α by Eqs. (1.67).

Starting from this metric, one can write the vacuum Einstein-Maxwell equations as two equations for two complex potentials $\mathcal{E}$ and Φ, following a procedure due to Ernst [209, 210]. The equations are

$$(Re\{\mathcal{E}\} + |\Phi|^2)\nabla^2\mathcal{E} = (\nabla\mathcal{E} + 2\Phi^*\nabla\Phi)\cdot\nabla\mathcal{E}, \tag{6.93}$$

$$(Re\{\mathcal{E}\} + |\Phi|^2)\nabla^2\Phi = (\nabla\mathcal{E} + 2\Phi^*\nabla\Phi)\cdot\nabla\Phi. \tag{6.94}$$

Once the potentials are known, the metric can be reconstructed. Sibgatullin [631] devised a powerful procedure for reducing the solution of the Ernst equations to simple integral equations. The exact solutions are generated by choosing the values of the Ernst potentials on the symmetry axis

$$e(\tilde{z}) \equiv \mathcal{E}(\tilde{\varpi} = 0, \tilde{z}), \qquad f(\tilde{z}) \equiv \Phi(\tilde{\varpi} = 0, \tilde{z}). \tag{6.95}$$

Defining two new functions, ξ and q, through

$$\mathcal{E} = \frac{1-\xi}{1+\xi}, \tag{6.96}$$

$$\Phi = \frac{q}{1+\xi}, \tag{6.97}$$

their values on the axis are given as a series expansion in terms of the physical multipole moments of the spacetime,

$$\xi(\tilde{\varpi}=0) = \sum_{n=0}^{\infty} m_n \tilde{z}^{-(n+1)}, \tag{6.98}$$

$$q(\tilde{\varpi}=0) = \sum_{n=0}^{\infty} q_n \tilde{z}^{-(n+1)}, \tag{6.99}$$

where the real and imaginary parts of m_n are the mass and current multipoles, respectively (see also Section 7.6), whereas the real and imaginary parts of q_n are the electric and magnetic multipoles. In [221] it was shown that the function ξ everywhere in the exterior spacetime can be given as a double series expansion, the coefficients of which can be obtained by a recursive relation, involving the physical multipole moments. This expansion was used in [617] (truncated at a certain order) to derive approximate analytic expressions for the location of the innermost stable circular orbit around a rotating relativistic star.

6.3.1 The 3-parameter Manko et al. solution

An interesting exact vacuum solution, given in a closed, algebraic form, was found by Manko et al. [458, 459]. For nonmagnetized sources of zero net charge, it reduces to a 3-parameter solution, involving the gravitational mass, M, the specific angular momentum, $a = J/M$, and a third parameter, b, related to the quadrupole moment of the source. The Ernst potential $\mathcal{E}$ on the symmetry axis is

$$e(z) = \frac{(z-M-ia)(z+ib)+d-\delta-ab}{(z+M-ia)(z+ib)+d-\delta-ab}, \tag{6.100}$$

where

$$\delta = \frac{-M^2b^2}{M^2-(a-b)^2}, \tag{6.101}$$

$$d = \frac{1}{4}\left[M^2-(a-b)^2\right]. \tag{6.102}$$

To be able to write the metric in rational form, one must introduce generalized spheroidal coordinates

$$x = \frac{r_+ + r_-}{2k}, \qquad y = \frac{r_+ - r_-}{2k}, \tag{6.103}$$

where $r_{\pm} = \sqrt{\tilde{\varpi}^2 + (\tilde{z} \pm k)^2}$ and $k = \sqrt{d + \delta}$. The inverse transformation is

$$\tilde{\varpi} = k(1 - y^2)^{1/2}(x^2 - 1)^{1/2}, \qquad \tilde{z} = kxy. \tag{6.104}$$

The metric is then written as

$$ds^2 = -f(dt - \omega d\phi)^2 + \frac{k^2}{f}\left[e^{2\gamma}(x^2 - y^2)\left(\frac{dx^2}{x^2 - 1} + \frac{dy^2}{1 - y^2}\right) + (x^2 - 1)(1 - y^2)d\phi^2\right], \tag{6.105}$$

with

$$f = \frac{E}{D}, \tag{6.106}$$

$$e^{2\gamma} = \frac{E}{16k^8(x^2 - y^2)^4}, \tag{6.107}$$

$$\omega = \frac{-(1 - y^2)F}{E}, \tag{6.108}$$

and

$$\begin{aligned} D = {} & \{4(k^2x^2 - \delta y^2)^2 + 2kmx[2k^2(x^2 - 1) \\ & + (2\delta + ab - b^2)(1 - y^2)] \\ & + (a - b)[(a - b)(d - \delta) - m^2 b](y^4 - 1) - 4d^2\}^2 \\ & + 4y^2\{2k^2(x^2 - 1)[kx(a - b) - mb] - 2mb\delta(1 - y^2) \\ & + [(a - b)(d - \delta) - m^2 b](2kx + m)(1 - y^2)\}^2, \end{aligned} \tag{6.109}$$

$$\begin{aligned} E = {} & \{4[k^2(x^2 - 1) + \delta(1 - y^2)]^2 \\ & + (a - b)[(a - b)(d - \delta) - m^2 b](1 - y^2)^2\}^2 \\ & - 16k^2(x^2 - 1)(1 - y^2)\{(a - b)[k^2(x^2 - y^2) + 2\delta y^2] \\ & + m^2 b y^2\}^2, \end{aligned} \tag{6.110}$$

$$\begin{aligned} F = {} & 8k^2(x^2 - 1)\{(a - b)[k^2(x^2 - y^2) + 2\delta y^2] + y^2 m^2 b\} \\ & \times \{kmx[(2kx + m)^2 - 2y^2(2\delta + ab - b^2) - a^2 + b^2] \\ & - 2y^2(4\delta d - m^2 b^2)\} + \{4[k^2(x^2 - 1) + \delta(1 - y^2)]^2 \\ & + (a - b)[(a - b)(d - \delta) - m^2 b](1 - y^2)^2\} \\ & \times (4(2kmbx + 2m^2 b)[k^2(x^2 - 1) + \delta(1 - y^2)] \\ & + (1 - y^2)\{(a - b)(m^2 b^2 - 4\delta d) \\ & - (4kmx + 2m^2)[(a - b)(d - \delta) - m^2 b]\}). \end{aligned} \tag{6.111}$$

In order for the solution to satisfy the requirements of axisymmetry, stationarity, and reflection symmetry in the equatorial plane, all three parameters M, a, and b must be real.

Multipolar structure. The three lowest-order multipole moments of the solution are the gravitational mass $Re\{m_0\} := M$, the angular momentum $Im\{m_1\} := J = aM$, and the quadrupole moment $Re\{m_2\} := Q$, related nonlinearly to the third parameter of the solution, b, by

$$Q = -M(d - \delta - ab + a^2). \tag{6.112}$$

Because the solution has three independent parameters, one can only set the above three multipole moments at will. The next multipole moment, the current octupole $Im\{m_3\} := S_3$, is given in terms of the three parameters of the solution as

$$S_3 = -M\left\{a^3 - 2a^2b + a\left[b^2 + 2(d-\delta)\right] - b(d-\delta)\right\}. \tag{6.113}$$

Because a and b are independent parameters, setting a equal to zero does not automatically imply a vanishing Q and S_3, as would be the case for a realistic solution of a nonrotating perfect fluid star. Instead, the nonrotating solution ($a = 0$) has a quadrupole moment equal to

$$Q(a=0) = -\frac{M}{4}\frac{\left(M^2+b^2\right)^2}{\left(M^2-b^2\right)} \tag{6.114}$$

and a current octupole moment equal to

$$S_3(a=0) = -bQ(a=0). \tag{6.115}$$

It is obvious that there exists no real value of the parameter b for which the quadrupole moment vanishes for a nonrotating star. For $|b| < M$, the solution is oblate ($Q < 0$) with a minimum quadrupole deformation obtained for $b = 0$,

$$|Q|_{\min}(a=0) = M^3/4. \tag{6.116}$$

At $b = \pm M$, the nonrotating multipole moments Q and S_3 diverge, whereas for $|b| > M$, the nonrotating solution is prolate ($Q > 0$) with a minimum quadrupole deformation of

$$Q_{\min}(a=0) = 2M^3, \tag{6.117}$$

at $b = \pm\sqrt{3}M$. Hence, the 3-parameter solution by Manko et al. does not reduce continuously to the Schwarzschild solution as the rotation vanishes. It can only reduce to other forms of nonrotating vacuum solutions (for example, the well-known Weyl solutions) that could be matched to other interior solutions, such as nonrotating stars with nonisotropic stresses, inducing nonvanishing quadrupole deformations. It is interesting that the Kerr solution can still be obtained algebraically, if one assumes the following imaginary form for the parameter b:

$$b = i\sqrt{M^2 - a^2}, \tag{6.118}$$

with $a \leq M$, and one recovers the correct expressions $Q = -a^2 M$ and $S_3 = -a^3 M$.

Matching to interior numerical solutions. Although the Manko et al. 3-parameter solution is not suitable for describing slowly rotating stars, it is still relevant for rapidly rotating models, when the quadrupole deformation induced by rotation roughly exceeds the minimum nonvanishing oblate quadrupole deformation of the solution in the absence of rotation, that is, roughly when

$$|Q| > M^3/4. \tag{6.119}$$

Because the quadrupole moment is roughly proportional to $a^2 M$, one can expect that the exact solution could be relevant for rotation rates of roughly $j > 0.5$, where $j \equiv J/M^2$ is a dimensionless measure of the angular momentum of the star.

A matching of the vacuum exterior solution to numerically constructed interior solutions of rapidly rotating stars (by identifying three multipole moments) was presented by Berti and Stergioulas in [64]. The first two parameters of the solution can be set directly by the values of the mass and angular momentum of the numerical solution. The third parameter, b, is determined by finding the roots of the equation

$$Q - Q_{\rm N} = 0, \tag{6.120}$$

where $Q_{\rm N}$ is the value of the quadrupole moment for the numerical interior solution. Above a critical rotation rate, $j > j_{\rm crit}$, along a sequence of rotating models, two possible real solutions for b exist: a solution that is usually negative, b_-, and a solution that is always positive, b_+. Thus, for each set of physical parameters M, a, and Q, there exist two different branches of solutions, with parameters $(M,\ a,\ b_-)$ and $(M,\ a,\ b_+)$, respectively, corresponding to very different spacetimes. We will refer to these as the *negative solution* $(-)$ and the *positive solution* $(+)$. Comparing the current octupole moment, S_3, obtained for the exact vacuum solutions and the numerical interior solution, it was shown in [64] that the negative solution branch is a better match. Below the critical rotation rate, $j < j_{\rm crit}$, no real solutions exist, and the Manko et al. 3-parameter solution is not relevant for rotating relativistic stars. In [458], the quadrupole moment was also used for matching the exact vacuum solution to numerical interior solutions, but only the positive branch of solutions was studied, which is not a good approximation to rotating stars.

For a wide range of candidate EOSs, the critical rotation rate $\Omega_{\rm crit}/\Omega_{\rm K}$ above which the Manko et al. 3-parameter solution is relevant ranges from ~0.4 to ~0.7 for sequences of models with $M = 1.4 M_\odot$, with the lower ratio corresponding to the stiffest EOS. For the maximum-mass sequence, the ratio is ~0.9, nearly independent of the EOS. Hence, the Manko et al. 3-parameter exterior solution could be useful for studying rapidly rotating neutron stars and even accreting neutron stars in LMXBs, with rotational periods of a few milliseconds, if the EOS is sufficiently stiff (see also [632, 633] and [457] for an exact vacuum solution that was not represented in terms of rational functions).

6.3.2 Other exact vacuum solutions

Another exact exterior vacuum solution found by Manko, Martín, and Ruiz [456] involves (in the case of vanishing charge and magnetic field) four parameters, which can be directly related to the four lowest-order multipole moments of a source. The solution is obtained by assuming that the Ernst potential on the symmetry axis takes the form

$$e(z) = \frac{(z - M - ia)(z + ib) - k}{(z + M - ia)(z + ib) - k}, \tag{6.121}$$

where M, a, k, b are the four parameters of the solution. In terms of these, the four lowest-order multipole moments are

$$Re\{m_0\} := M, \tag{6.122}$$

$$Im\{m_1\} := J = aM, \tag{6.123}$$

$$Re\{m_2\} := Q = -(a^2 - k)M, \tag{6.124}$$

$$Im\{m_3\} := S_3 = -[a^3 - (2a - b)k]M. \tag{6.125}$$

The explicit relations for the metric functions f, $e^{2\gamma}$ and ω are given in [456].

The advantage of the above solution is that its four parameters are introduced linearly in the four lowest-order multipole moments. For this reason, one can always match the exact solution to a numerical solution by identifying the four lowest-order multipole moments. Therefore, the 4-parameter Manko, Martín, and Ruiz [456] solution is relevant for studying rotating relativistic stars at any rotation rate. It is easily seen that the 3-parameter Manko et al. [458, 459] solution discussed in the previous section is, in fact, a special case of the 4-parameter solution. In [529], the two solutions are compared to numerical solutions of rapidly rotating relativistic stars, finding good agreement.

An exact vacuum solution (constructed via Bäcklund transformations) that can be matched to numerically constructed solutions with an arbitrary number of constants was presented in [680], where very good agreement with numerical solutions was found even for a small number of parameters.

7

Perturbation theory of relativistic fluids

7.1 The perturbed Einstein-Euler equations

Perturbations of rotating stars have been discussed in the context of general relativity by a number of authors. The present discussion of perfect fluids is essentially the Lagrangian formalism of Friedman and Schutz [240, 231], and it is closely related to work by Taub [677, 678], Schutz [594, 598], Chandrasekhar and Friedman [136, 137], Carter [118], and Schutz and Sorkin [596]. The treatment of imperfect fluids is based on that given by Lindblom and Hiscock [423, 315].

In discussing stellar oscillations, one is interested in the time evolution of a star close to a given equilibrium with the same baryon number and the same total entropy. The configuration of the perturbed star is regarded as a deformation of the original equilibrium, with each fluid element of the equilibrium mapped to a corresponding fluid element in the perturbed configuration by a Lagrangian displacement ξ^α. The fundamental variables describing the perturbation are then ξ^α and the perturbed metric $h_{\alpha\beta} := \delta g_{\alpha\beta}$.

Following our discussion of the action for the Einstein-Euler system in Section 2.2, we make this description precise by introducing a family of (time-dependent) solutions

$$\mathcal{Q}(\mathsf{s}) = \{g_{\alpha\beta}(\mathsf{s}), u^\alpha(\mathsf{s}), \rho(\mathsf{s}), s(\mathsf{s})\}, \tag{7.1}$$

indexed by a parameter s, and compare to first order in s the perturbed variables $Q(\mathsf{s})$ with their equilibrium values $Q(0)$.

Because oscillations of a neutron star proceed on a dynamical timescale, a timescale faster than that of heat flow, we now further require that the family of solutions $Q(\mathsf{s})$ be such that each member can be reached by an adiabatic deformation of the equilibrium $Q(0)$. That is, there is to be a family of diffeomorphisms χ_s mapping fluid trajectories of the equilibrium model $Q(0)$ to fluid trajectories of the solution $Q(\mathsf{s})$, with $s(\mathsf{s}) \circ \chi_\mathsf{s} = s(0)$.

The Eulerian and Lagrangian changes in the fluid variables are again expressed in terms of the Lagrangian displacement ξ^α associated with the family of diffeos χ_s by

$$\delta\mathcal{Q} := \frac{d}{d\mathsf{s}}\mathcal{Q}(\mathsf{s})\Big|_{\mathsf{s}=0}, \qquad \Delta\mathcal{Q} = \frac{d}{d\mathsf{s}}[\chi_{-\mathsf{s}}\mathcal{Q}(\mathsf{s})]\Big|_{\mathsf{s}=0} = (\delta + \mathcal{L}_\xi)\mathcal{Q}. \tag{7.2}$$

In particular, the Lagrangian change in s vanishes, and perturbations of u^α, ρ, and ϵ are expressed in terms of ξ^α, and $h_{\alpha\beta}$ by Eqs. (2.37), (2.35), and (2.64):

$$\Delta u^\alpha = \frac{1}{2}u^\alpha u^\beta u^\gamma \Delta g_{\beta\gamma}, \qquad \Delta\rho = -\frac{1}{2}\rho q^{\alpha\beta}\Delta g_{\alpha\beta}, \qquad \Delta\epsilon = -\frac{1}{2}(\epsilon+p)q^{\alpha\beta}\Delta g_{\alpha\beta}, \tag{7.3}$$

with $\Delta g_{\alpha\beta} = h_{\alpha\beta} + \nabla_\alpha\xi_\beta + \nabla_\beta\xi_\alpha$. Our restriction to adiabatic perturbations means that the Lagrangian perturbation in the pressure, Δp, is given by

$$\frac{\Delta p}{p} = \Gamma_1\frac{\Delta\rho}{\rho} = -\frac{1}{2}\Gamma_1\, q^{\alpha\beta}\Delta g_{\alpha\beta}, \tag{7.4}$$

where the adiabatic index Γ_1 is defined by Eq. (1.118).

The corresponding Eulerian changes in the fluid variables are given by

$$\delta u^\alpha = q^\alpha{}_\beta \mathcal{L}_u \xi^\beta + \frac{1}{2}u^\alpha u^\beta u^\gamma h_{\beta\gamma}, \tag{7.5a}$$

$$\delta\rho = -\frac{1}{2}\rho q^{\alpha\beta}\Delta g_{\alpha\beta} - \boldsymbol{\xi}\cdot\nabla\rho, \tag{7.5b}$$

$$\delta\epsilon = -\frac{1}{2}(\epsilon+p)q^{\alpha\beta}\Delta g_{\alpha\beta} - \boldsymbol{\xi}\cdot\nabla\epsilon, \tag{7.5c}$$

$$\delta p = -\frac{1}{2}\Gamma_1 p q^{\alpha\beta}\Delta g_{\alpha\beta} - \boldsymbol{\xi}\cdot\nabla p. \tag{7.5d}$$

The equations governing the perturbations of a perfect-fluid spacetime are the perturbed Einstein-Euler system,

$$\delta(G^{\alpha\beta} - 8\pi\, T^{\alpha\beta}) = 0, \tag{7.6}$$

$$q_\alpha{}^\gamma\,\delta(\nabla_\beta T^\beta{}_\gamma) = \delta\left[(\epsilon+p)u^\beta\nabla_\beta u_\alpha + q_\alpha{}^\beta\nabla_\beta p\right] = 0. \tag{7.7}$$

Only the projection orthogonal to u^α of the equation $\delta\nabla_\beta T^{\beta\gamma} = 0$ is included, because Eq. (7.3) for $\Delta\epsilon$ is a first integral of its projection along u^α: With $\Delta\epsilon$ given by Eq. (7.3), $u_\alpha\delta\nabla_\beta T^{\alpha\beta} = 0$ is an identity. As in the exact Einstein-Euler system, when all components of the perturbed Einstein equation are satisfied, the perturbed Bianchi identity implies $\delta\nabla_\beta T^{\beta\gamma} = 0$. In practice, however, one ordinarily chooses a gauge with six independent metric functions, solves six components of the perturbed Einstein equation for the metric, and then solves the perturbed Euler equation for the Lagrangian displacement ξ^α. The second kind of gauge freedom, allowing one to add to ξ^α a trivial displacement of the form gu^α, means that ξ^α has three independent components after one has used this gauge freedom to choose $\xi^\alpha u_\alpha$.

We first exhibit and then derive the perturbed equations in terms of the variables $h_{\alpha\beta}$ and ξ^α. As we will see in the next section, a self-adjoint symmetry of the system is associated with symmetric forms of the perturbed tensor densities $\delta(T^{\alpha\beta}\sqrt{|g|})$ and $\delta(G^{\alpha\beta}\sqrt{|g|})$. To avoid writing $\sqrt{|g|}$ in each expression below, we define operators $\boldsymbol{\Delta}$, $\boldsymbol{\delta}$, and $\boldsymbol{\mathcal{L}_\xi}$ by

$$\boldsymbol{\Delta} Q := \frac{1}{\sqrt{|g|}}\Delta(Q\sqrt{|g|}), \quad \boldsymbol{\delta} Q := \frac{1}{\sqrt{|g|}}\delta(Q\sqrt{|g|}),$$
$$\boldsymbol{\mathcal{L}_\xi} Q := \frac{1}{\sqrt{|g|}}\mathcal{L}_\xi(Q\sqrt{|g|}). \tag{7.8}$$

For a generic gauge, the equation

$$\boldsymbol{\delta} G^{\alpha\beta} = 8\pi\boldsymbol{\delta} T^{\alpha\beta} \tag{7.9}$$

has the form

$$-\frac{1}{2}\epsilon^{\alpha\gamma\epsilon\zeta}\epsilon^{\beta\delta\eta}{}_{\zeta}\nabla_\gamma\nabla_\delta h_{\epsilon\eta} + G^{\alpha\beta\gamma\delta}h_{\gamma\delta} = 8\pi\left(W^{\alpha\beta\gamma\delta}\Delta g_{\gamma\delta} - \boldsymbol{\mathcal{L}_\xi} T^{\alpha\beta}\right), \tag{7.10}$$

where the tensors $G^{\alpha\beta\gamma\delta}$ and $W^{\alpha\beta\gamma\delta}$ are given by

$$\begin{aligned} G^{\alpha\beta\gamma\delta} &= \frac{1}{2}R^{\alpha(\gamma\delta)\beta} + \frac{1}{4}(2R^{\alpha\beta}g^{\gamma\delta} + 2R^{\gamma\delta}g^{\alpha\beta} - 3R^{\alpha(\gamma}g^{\delta)\beta} - 3R^{\beta(\gamma}g^{\delta)\alpha}) \\ &\quad + \frac{1}{4}R(g^{\alpha\gamma}g^{\beta\delta} + g^{\alpha\delta}g^{\beta\gamma} - g^{\alpha\beta}g^{\gamma\delta}) && (7.11) \\ &= G^{\gamma\delta\alpha\beta} = G^{(\alpha\beta)(\gamma\delta)} && (7.12) \end{aligned}$$

and

$$\begin{aligned} W^{\alpha\beta\gamma\delta} &= \frac{1}{2}(\epsilon+p)u^\alpha u^\beta u^\gamma u^\delta + \frac{1}{2}p(g^{\alpha\beta}g^{\gamma\delta} - g^{\alpha\gamma}g^{\beta\delta} - g^{\alpha\delta}g^{\beta\gamma}) - \frac{1}{2}\Gamma_1 p q^{\alpha\beta}q^{\gamma\delta} \\ &= W^{\gamma\delta\alpha\beta} = W^{(\alpha\beta)(\gamma\delta)}, && (7.13) \end{aligned}$$

and where

$$\boldsymbol{\mathcal{L}_\xi} T^{\alpha\beta} = \nabla_\gamma(T^{\alpha\beta}\xi^\gamma) - T^{\gamma\beta}\nabla_\gamma\xi^\alpha - T^{\alpha\gamma}\nabla_\gamma\xi^\beta. \tag{7.14}$$

The left side of the field equation (7.10) – the perturbed Einstein tensor – involves $h_{\alpha\beta}$ and its first two derivatives; the perturbed stress tensor involves only $h_{\alpha\beta}$ and only in the combination $\Delta g_{\alpha\beta} = h_{\alpha\beta} + 2\nabla_{(\alpha}\xi_{\beta)}$; and only first derivatives of ξ^α appear. Note that, because the background spacetime satisfies the field equation, we have

$$\boldsymbol{\delta}(G^{\alpha\beta} - 8\pi T^{\alpha\beta}) = \left(\frac{\delta\sqrt{|g|}}{\sqrt{|g|}} + \delta\right)(G^{\alpha\beta} - 8\pi T^{\alpha\beta}) = \delta(G^{\alpha\beta} - 8\pi T^{\alpha\beta}). \tag{7.15}$$

The perturbed Euler equation (7.7) again involves only the combination $\Delta g_{\alpha\beta}$ of the perturbation variables:

$$\begin{aligned} \delta(q_\alpha{}^\beta\nabla_\gamma T_\beta{}^\gamma) &= (\epsilon+p)\mathcal{L}_{\mathbf{u}}(q_\alpha{}^\beta u^\gamma\Delta g_{\beta\gamma}) - \frac{1}{2}(\epsilon+p)q_\alpha{}^\beta\nabla_\beta(u^\gamma u^\delta\Delta g_{\gamma\delta}) \\ &\quad + \frac{1}{2}\nabla_\alpha p\left(1 + \frac{\Gamma_1 p}{\epsilon+p}\right)q^{\gamma\delta}\Delta g_{\gamma\delta} - \frac{1}{2}q_\alpha{}^\beta\nabla_\beta\left(\Gamma_1 p q^{\gamma\delta}\Delta g_{\gamma\delta}\right) \\ &= 0. && (7.16) \end{aligned}$$

For uniform rotation, we can replace $\mathcal{L}_{\mathbf{u}}$ in the first term by $u^t\mathcal{L}_{\mathbf{k}}$. Using the fact that $k^\alpha = t^\alpha + \Omega\phi^\alpha$ is a Killing vector of the background geometry, we have

$$\begin{aligned}(\epsilon+p)u^t q_\alpha{}^\beta u^\gamma \mathcal{L}_{\mathbf{k}}\Delta g_{\beta\gamma} &= \frac{1}{2}(\epsilon+p)q_\alpha{}^\beta\nabla_\beta(u^\gamma u^\delta\Delta g_{\gamma\delta})\\ &\quad -\frac{1}{2}\nabla_\alpha p\left(1+\frac{\Gamma_1 p}{\epsilon+p}\right)q^{\gamma\delta}\Delta g_{\gamma\delta}\\ &\quad +\frac{1}{2}q_\alpha{}^\beta\nabla_\beta\left(\Gamma_1 p q^{\gamma\delta}\Delta g_{\gamma\delta}\right). \end{aligned} \tag{7.17}$$

When the unperturbed model satisfies the same one-parameter EOS as the perturbation, the relation $\nabla p/p = \Gamma_1\nabla\epsilon/(\epsilon+p)$ gives a simpler form, related to the form (1.103), $\nabla_\alpha(h/u^t) = 0$, for the equilibrium equation of a uniformly rotating barotropic star:

$$\begin{aligned}u^t q_\alpha{}^\beta u^\gamma\mathcal{L}_{\mathbf{k}}\Delta g_{\beta\gamma} &= \frac{1}{2}q_\alpha{}^\beta\nabla_\beta\left(u^\gamma u^\delta\Delta g_{\gamma\delta}+\frac{\Gamma_1 p}{\epsilon+p}q^{\gamma\delta}\Delta g_{\gamma\delta}\right)\\ &= -q_\alpha{}^\beta\nabla_\beta\Delta\ln\frac{h}{u^t}. \end{aligned} \tag{7.18}$$

The derivations. We first derive the perturbed field equation (7.10), beginning with the perturbed stress-energy tensor. The definitions of $\boldsymbol{\delta}$ and $\boldsymbol{\Delta}$ give $\boldsymbol{\delta}T^{\alpha\beta} = \boldsymbol{\Delta}T^{\alpha\beta} - \boldsymbol{\mathcal{L}_\xi}T^{\alpha\beta}$. Using Eqs. (7.3) and (7.4), we have

$$\begin{aligned}\Delta T^{\alpha\beta} &= \Delta\epsilon u^\alpha u^\beta + 2\epsilon\Delta u^{(\alpha}u^{\beta)} + \Delta p q^{\alpha\beta} + p\Delta q^{\alpha\beta}\\ &= \left[-\frac{1}{2}(\epsilon+p)q^{\gamma\delta}u^\alpha u^\beta + \epsilon u^\alpha u^\beta u^\gamma u^\delta - \frac{1}{2}\Gamma_1 p q^{\alpha\beta}q^{\gamma\delta}\right]\Delta g_{\gamma\delta} + p\Delta q^{\alpha\beta}. \end{aligned} \tag{7.19}$$

Then, writing

$$\Delta\sqrt{|g|} = \frac{1}{2}g^{\alpha\beta}\Delta g_{\alpha\beta}, \qquad \Delta q^{\alpha\beta} = (-g^{\alpha\gamma}g^{\beta\delta} + u^\alpha u^\beta u^\gamma u^\delta)\Delta g_{\gamma\delta}, \tag{7.20}$$

and grouping coefficients of $\Delta g_{\alpha\beta}$, we obtain

$$\boldsymbol{\Delta}T^{\alpha\beta} = W^{\alpha\beta\gamma\delta}\Delta g_{\gamma\delta} \tag{7.21}$$

and recover the right side of Eq. (7.10).

For the left side, we need to show that the perturbed Einstein tensor has in a generic gauge the form

$$\boldsymbol{\delta}G^{\alpha\beta} := \frac{1}{\sqrt{|g|}}\delta\left(\sqrt{|g|}\,G^{\alpha\beta}\right) = -\frac{1}{2}\epsilon^{\alpha\gamma\epsilon\zeta}\epsilon^{\beta\delta\eta}{}_\zeta\nabla_{(\gamma}\nabla_{\delta)}h_{\epsilon\eta} + G^{\alpha\beta\gamma\delta}h_{\gamma\delta}. \tag{7.22}$$

To obtain Eq. (7.22), begin with

$$\boldsymbol{\delta}G^{\alpha\beta} = \frac{1}{\sqrt{|g|}}\delta\left[R_{\gamma\delta}\left(g^{\alpha\gamma}g^{\beta\delta} - \frac{1}{2}g^{\alpha\beta}g^{\gamma\delta}\right)\sqrt{|g|}\right] \tag{7.23}$$

and use Eqs. (2.86) and (2.91) to write

$$\delta R_{\alpha\beta} = \frac{1}{2}(\nabla_\gamma \nabla_\alpha h_\beta{}^\gamma + \nabla_\gamma \nabla_\beta h_\alpha{}^\gamma - \nabla_\alpha \nabla_\beta h - \nabla_\gamma \nabla^\gamma h_{\alpha\beta}). \tag{7.24}$$

Then

$$\begin{aligned}
&\delta R_{\gamma\delta}\left(g^{\alpha\gamma}g^{\beta\delta} - \frac{1}{2}g^{\alpha\beta}g^{\gamma\delta}\right) \\
&= \frac{1}{2}\left(g^{\alpha\gamma}g^{\beta\delta}g^{\epsilon\eta} - \frac{1}{2}g^{\alpha\beta}g^{\gamma\delta}g^{\epsilon\eta}\right) \\
&\qquad \times (\nabla_\epsilon \nabla_\gamma h_{\delta\eta} + \nabla_\epsilon \nabla_\delta h_{\gamma\eta} - \nabla_\gamma \nabla_\delta h_{\epsilon\eta} - \nabla_\epsilon \nabla_\eta h_{\gamma\delta}) \\
&= \frac{1}{2}\nabla_{(\gamma}\nabla_{\delta)} h_{\epsilon\eta}(g^{\alpha\beta}g^{\gamma\delta}g^{\epsilon\eta} + g^{\alpha\eta}g^{\gamma\beta}g^{\epsilon\delta} + g^{\alpha\delta}g^{\gamma\eta}g^{\epsilon\beta} \\
&\qquad - g^{\alpha\beta}g^{\gamma\eta}g^{\epsilon\delta} - g^{\alpha\delta}g^{\gamma\beta}g^{\epsilon\eta} - g^{\alpha\eta}g^{\gamma\delta}g^{\epsilon\beta}) \\
&\quad + \frac{1}{2}\nabla_{[\gamma}\nabla_{\delta]} h_{\epsilon\eta}(g^{\alpha\eta}g^{\beta\delta}g^{\gamma\epsilon} + g^{\alpha\delta}g^{\gamma\eta}g^{\beta\epsilon}) \\
&= -\frac{1}{2}\epsilon^{\alpha\gamma\epsilon\zeta}\epsilon^{\beta\delta\eta}{}_\zeta \nabla_{(\gamma}\nabla_{\delta)} h_{\epsilon\eta} - \frac{1}{2}R^{\alpha\gamma\beta\delta}h_{\gamma\delta} + \frac{1}{2}R^{\gamma(\alpha}h^{\beta)}{}_\gamma .
\end{aligned} \tag{7.25}$$

Adding the additional terms from the right side of Eq. (7.23), we obtain the claimed form,

$$\begin{aligned}
\delta G^{\alpha\beta} &= -\frac{1}{2}\epsilon^{\alpha\gamma\epsilon\zeta}\epsilon^{\beta\delta\eta}{}_\zeta \nabla_{(\gamma}\nabla_{\delta)} h_{\epsilon\eta} \\
&\quad - \frac{1}{2}R^{\alpha\gamma\beta\delta}h_{\gamma\delta} + \frac{1}{4}(2g^{\alpha\beta}R^{\gamma\delta} + 2g^{\gamma\delta}R^{\alpha\beta} - 3g^{\alpha\gamma}R^{\beta\delta} - 3g^{\beta\gamma}R^{\alpha\delta})h_{\gamma\delta} \\
&\quad + \frac{1}{4}R(g^{\alpha\gamma}g^{\beta\delta} + g^{\alpha\delta}g^{\beta\gamma} - g^{\alpha\beta}g^{\gamma\delta})h_{\gamma\delta} \\
&= -\frac{1}{2}\epsilon^{\alpha\gamma\epsilon\zeta}\epsilon^{\beta\delta\eta}{}_\zeta \nabla_{(\gamma}\nabla_{\delta)} h_{\epsilon\eta} + G^{\alpha\beta\gamma\delta}h_{\gamma\delta}.
\end{aligned} \tag{7.26}$$

To obtain the perturbed Euler equation (7.16), we first use the unperturbed Euler equation to write Eq. (7.7) as

$$0 = \delta\left[(\epsilon + p)u^\beta \nabla_\beta u_\alpha + q_\alpha{}^\beta \nabla_\beta p\right] = (\epsilon + p)q_\alpha{}^\gamma \Delta\left(u^\beta \nabla_\beta u_\gamma + q_\gamma{}^\beta \frac{\nabla_\beta p}{\epsilon + p}\right). \tag{7.27}$$

We next use the relation $u_\beta \nabla_\alpha u^\beta = 0$ to write $u^\beta \nabla_\beta u_\alpha = \mathcal{L}_{\mathbf{u}} u_\alpha$. It is because $\mathcal{L}_{\mathbf{u}}$ does not involve ∇_α, together with the fact that the perturbed fluid variables only involve $\Delta g_{\alpha\beta}$, that ξ^α and $h_{\alpha\beta}$ occur only in the form $\Delta g_{\alpha\beta}$. In particular, we have

$$\Delta(\mathcal{L}_{\mathbf{u}} u_\alpha) = \mathcal{L}_{\Delta \mathbf{u}} u_\alpha + \mathcal{L}_{\mathbf{u}} \Delta u_\alpha, \tag{7.28}$$

with Δu^α given by Eq. (7.3) and with

$$\Delta u_\alpha = \Delta(g_{\alpha\beta}u^\beta) = \Delta g_{\alpha\beta}u^\beta + g_{\alpha\beta}\Delta u^\beta = \Delta g_{\alpha\beta}u^\beta + \frac{1}{2}u_\alpha u^\beta u^\gamma \Delta g_{\beta\gamma}. \tag{7.29}$$

The projections orthogonal to u^α of the two terms on the right side of Eq. (7.28) are then

$$q_\alpha{}^\beta \mathcal{L}_{\mathbf{u}} \Delta u_\beta = q_\alpha^\beta \left[\mathcal{L}_{\mathbf{u}} \Delta g_{\beta\gamma} u^\gamma + \frac{1}{2} (\mathcal{L}_{\mathbf{u}} u_\beta) u^\gamma u^\delta \Delta g_{\gamma\delta} \right]$$
$$= q_\alpha^\beta u^\gamma \mathcal{L}_{\mathbf{u}} \Delta g_{\beta\gamma} + \frac{1}{2} \mathcal{L}_{\mathbf{u}} u_\alpha u^\gamma u^\delta \Delta g_{\gamma\delta} \tag{7.30}$$

and

$$q_\alpha{}^\beta \mathcal{L}_{\Delta \mathbf{u}} u_\beta = \frac{1}{2} \mathcal{L}_{\mathbf{u}} u_\alpha u^\gamma u^\delta \Delta g_{\gamma\delta} - \frac{1}{2} q_\alpha{}^\beta \nabla_\beta (u^\gamma u^\delta \Delta g_{\gamma\delta}); \tag{7.31}$$

adding Eqs. (7.30) and (7.31) and using $\mathcal{L}_{\mathbf{u}} q_\alpha{}^\beta = u^\beta \mathcal{L}_{\mathbf{u}} u_\alpha$, we have

$$q_\alpha{}^\beta \Delta (\mathcal{L}_{\mathbf{u}} u_\beta) = \mathcal{L}_{\mathbf{u}} \left(q_\alpha{}^\beta u^\gamma \Delta g_{\beta\gamma} \right) - \frac{1}{2} q_\alpha{}^\beta \nabla_\beta (u^\gamma u^\delta \Delta g_{\gamma\delta}). \tag{7.32}$$

The remaining term in the perturbed Euler equation is

$$(\epsilon + p) q_\alpha{}^\beta \Delta \frac{\nabla_\beta p}{\epsilon + p} = q_\alpha{}^\beta \left[\nabla_\beta \Delta p - \frac{\nabla_\beta p}{\epsilon + p} (\Delta \epsilon + \Delta p) \right]$$
$$= \frac{1}{2} q_\alpha{}^\beta \left[-\nabla_\beta (\Gamma_1 p q^{\gamma\delta} \Delta g_{\gamma\delta}) + \nabla_\beta p \left(1 + \frac{\Gamma_1 p}{\epsilon + p} \right) q^{\gamma\delta} \Delta g_{\gamma\delta} \right]. \tag{7.33}$$

Finally, adding Eqs. (7.32) and (7.33) gives the perturbed Euler equation in the form (7.16).

Two special cases. In two opposite situations, the perturbed field equation and the perturbed Euler equation, respectively, have simpler forms: Outside the star, the perturbed field equation is substantially simpler in a Lorenz gauge,[1] given by

$$\nabla_\beta \tilde{h}^{\alpha\beta} = 0, \text{ where } \tilde{h}_{\alpha\beta} := h_{\alpha\beta} - \frac{1}{2} g_{\alpha\beta} h. \tag{7.34}$$

The perturbed vacuum field equation has the form

$$\delta G^{\alpha\beta} = -\frac{1}{2} \nabla_\gamma \nabla^\gamma \tilde{h}^{\alpha\beta} - 2 R^{\alpha\gamma\beta\delta} \tilde{h}_{\gamma\delta} = 0, \tag{7.35}$$

with the more elaborate expression

$$\delta G^{\alpha\beta} = -\frac{1}{2} \nabla_\gamma \nabla^\gamma \tilde{h}^{\alpha\beta} - 2 R^{\alpha\gamma\beta\delta} \tilde{h}_{\gamma\delta}$$
$$+ \frac{1}{2} \Big[g^{\alpha\beta} R^{\gamma\delta} + g^{\gamma\delta} R^{\alpha\beta} - g^{\alpha\gamma} R^{\beta\delta} - g^{\alpha\delta} R^{\beta\gamma}$$
$$+ (g^{\alpha\gamma} g^{\beta\delta} - \frac{1}{2} g^{\alpha\beta} g^{\gamma\delta} R) \tilde{h}_{\gamma\delta} \Big] \tag{7.36}$$

in the star.

[1] The gauge is variously referred to as the deDonder, Einstein, Hilbert, Lorenz, or harmonic gauge. Its electromagnetic analog was introduced by Lorenz, and it and the gravitational gauge are often incorrectly ascribed to Lorentz. In gravity, the gauge was first used by Einstein and Hilbert. DeDonder and Fock introduced its nonlinear version, harmonic coordinates.

The second case is that of a perturbation of the matter for which the perturbed metric is negligible. This is the Cowling approximation, and it is ordinarily accurate for a perturbation of the matter that is confined to a small region or that has a short wavelength (see Section 8.4.2). With $h_{\alpha\beta} = 0$, then, the perturbed Euler equation for a homentropic, uniformly rotating model has the form

$$(u^t)^2(\mathcal{L}_{\mathbf{k}}^2\xi_\alpha + 2q_\alpha{}^\beta\mathcal{L}_{\mathbf{k}}\xi^\gamma\nabla_\gamma k_\beta) - q_\alpha{}^\beta\nabla_\beta\left[\frac{\xi^\gamma\nabla_\gamma p + \Gamma_1 p q^{\gamma\delta}\nabla_\gamma\xi_\delta}{\epsilon + p}\right] = 0. \tag{7.37}$$

This follows from Eq. (7.18) after some algebra using the relations $u^\alpha = u^t k^\alpha$,

$$\frac{1}{2}u^\alpha u^\beta \Delta g_{\alpha\beta} = u^\alpha u^\beta \nabla_\alpha \xi_\beta = -u^\alpha \xi^\beta \nabla_\alpha u_\beta = \xi^\beta \frac{\nabla_\beta p}{\epsilon + p}, \tag{7.38}$$

and

$$\Delta u_\alpha = u^t(\mathcal{L}_{\mathbf{k}}\xi_\alpha + 2\xi^\beta\nabla_\beta k_\alpha) + u_\alpha\xi^\beta\frac{\nabla_\beta p}{\epsilon + p}. \tag{7.39}$$

7.2 An action for the perturbation equations

The perturbed Einstein-Euler equations, (7.6) and (7.7), are self-adjoint in the weak sense that they are a symmetric system up to a total divergence: For any pairs $(\xi^\alpha, h_{\alpha\beta})$ and $(\widehat{\xi}^\alpha, \widehat{h}_{\alpha\beta})$, the symmetry relation has the form

$$\widehat{\xi}_\beta\delta(\nabla_\gamma T^{\beta\gamma}\sqrt{|g|}) + \frac{1}{16\pi}\widehat{h}_{\beta\gamma}\delta\left[(G^{\beta\gamma} - 8\pi T^{\beta\gamma})\sqrt{|g|}\right] = -\mathscr{L}(\widehat{\xi}, \widehat{h}; \xi, h) + \nabla_\beta\Theta^\beta, \tag{7.40}$$

where $\mathscr{L}$ is symmetric under interchange of (ξ, h) and $(\widehat{\xi}, \hat{h})$. A symmetry relation of the form (7.40) implies that $\mathscr{L}^{(2)}(\xi, h) := \frac{1}{2}\mathscr{L}(\xi, h; \xi, h)$ is a Lagrangian density and

$$I^{(2)} = \int d^4x \mathscr{L}^{(2)} \tag{7.41}$$

is an action for the perturbed system. That is, requiring that the variation of $I^{(2)}$ vanish along paths of the form $\xi^\alpha + \lambda\widehat{\xi}^\alpha, h_{\alpha\beta} + \lambda\widehat{h}_{\alpha\beta}$, for arbitrary $\widehat{\xi}^\alpha$ and $\widehat{h}_{\alpha\beta}$ that vanish outside a compact set, yields the perturbed Einstein-Euler equations:

$$\begin{aligned}\frac{d}{d\lambda}I^{(2)}|_{\lambda=0} &= \int d^4x \frac{1}{2}\left[\mathscr{L}(\widehat{\xi}, \widehat{h}; \xi, h) + \mathscr{L}(\xi, h; \widehat{\xi}, \widehat{h})\right] \\ &= \int d^4x \mathscr{L}(\widehat{\xi}, \widehat{h}; \xi, h) \\ &= \int d^4x \left\{\widehat{\xi}_\beta\delta(\nabla_\gamma T^{\beta\gamma}\sqrt{|g|}) + \frac{1}{16\pi}\widehat{h}_{\beta\gamma}\delta\left[(G^{\beta\gamma} - 8\pi T^{\beta\gamma})\sqrt{|g|}\right]\right\}.\end{aligned} \tag{7.42}$$

The symmetric second-order Lagrangian density $\mathscr{L}^{(2)}$ and the corresponding canonical energy and angular momentum of stellar perturbations underlie the study of stellar stability and play a significant role in stellar oscillations. As we show below following Friedman and Schutz [240], $\mathscr{L}$ and Θ^α are given by

$$\begin{aligned}\mathscr{L}(\widehat{\xi},\widehat{h};\xi,h)|g|^{-1/2} &= U^{\alpha\beta\gamma\delta}\nabla_\alpha\widehat{\xi}_\beta\nabla_\gamma\xi_\delta + V^{\alpha\beta\gamma\delta}(\widehat{h}_{\alpha\beta}\nabla_\gamma\xi_\delta + h_{\alpha\beta}\nabla_\gamma\widehat{\xi}_\delta)\\ &\quad - \frac{1}{32\pi}\epsilon^{\alpha\gamma\epsilon\eta}\epsilon^{\beta\delta\zeta}{}_\eta\nabla_\gamma\widehat{h}_{\alpha\beta}\nabla_\delta h_{\epsilon\zeta}\\ &\quad - T^{\alpha\beta}R_{\alpha\gamma\beta\delta}\widehat{\xi}^\gamma\xi^\delta + \left(\frac{1}{2}W^{\alpha\beta\gamma\delta} - \frac{1}{16\pi}G^{\alpha\beta\gamma\delta}\right)\widehat{h}_{\alpha\beta}h_{\gamma\delta}\\ &\quad - \frac{1}{2}\nabla_\gamma T^{\alpha\beta}(\widehat{h}_{\alpha\beta}\xi^\gamma + h_{\alpha\beta}\widehat{\xi}^\gamma) \qquad (7.43)\end{aligned}$$

and

$$\Theta^\alpha(\widehat{\xi},\widehat{h};\xi,h) = \Pi^{\alpha\beta}\widehat{\xi}_\beta + \pi^{\alpha\beta\gamma}\widehat{h}_{\beta\gamma}, \qquad (7.44)$$

where

$$\Pi^{\alpha\beta} = \frac{1}{2}\frac{\partial\mathscr{L}(\xi,h;\xi,h)}{\partial\nabla_\alpha\xi_\beta} = (U^{\alpha\beta\gamma\delta}\nabla_\gamma\xi_\delta + V^{\gamma\delta\alpha\beta}h_{\gamma\delta})\sqrt{|g|}, \qquad (7.45)$$

$$\pi^{\alpha\beta\gamma} = \frac{1}{2}\frac{\partial\mathscr{L}(\xi,h;\xi,h)}{\partial\nabla_\alpha h_{\beta\gamma}} = -\frac{1}{32\pi}\epsilon^{\alpha\epsilon(\gamma}{}_\eta\epsilon^{\beta)\delta\zeta\eta}\nabla_\delta h_{\epsilon\zeta}\sqrt{|g|}, \qquad (7.46)$$

$$\begin{aligned}U^{\alpha\beta\gamma\delta} &= (\epsilon+p)u^\alpha u^\gamma q^{\beta\delta} + p(g^{\alpha\beta}g^{\gamma\delta} - g^{\alpha\delta}g^{\beta\gamma}) - \Gamma_1 p\, q^{\alpha\beta}q^{\gamma\delta}\\ &= U^{\gamma\delta\alpha\beta}, \qquad (7.47)\\ 2V^{\alpha\beta\gamma\delta} &= (\epsilon+p)(u^\alpha u^\gamma q^{\beta\delta} + u^\beta u^\gamma q^{\alpha\delta} - u^\alpha u^\beta q^{\gamma\delta}) - \Gamma_1 p\, q^{\alpha\beta}q^{\gamma\delta}. \qquad (7.48)\end{aligned}$$

Before deriving these relations, it is helpful to note that the symmetry of the field equations follows quite generally from the fact that they arise from an unconstrained action. Let $\{\phi^I\}$ be a set of fields whose field equations $E_I(\phi) = 0$ are given by a variation of the action $I[\phi] = \int \mathscr{L}(\phi)d^4x$. That is, $E_I = \frac{\delta I}{\delta\phi^I}$, with $\delta I = \int \frac{\delta I}{\delta\phi^I(x)}\delta\phi^I(x)d^4x$, for any choice of perturbed fields $\delta\phi^I$ vanishing sufficiently rapidly at the boundary of the region of integration that no surface terms arise. Then the perturbed field equations, $\delta E_I = 0$, governing a perturbation $\delta\phi^I$, have the form

$$\delta E_I(x) = \int \frac{\delta^2 I}{\delta\phi^I(x)\delta\phi^J(x')}\delta\phi^J(x')d^4x',$$

and the symmetry of the system corresponds to an exchange of the order of the two derivatives of the action:

$$\begin{aligned}\int \hat{\delta}\phi^I(x)\delta E_I(x)d^4x &= \int \hat{\delta}\phi^I(x)\frac{\delta^2 I}{\delta\phi^I(x)\delta\phi^J(x')}\delta\phi^J(x')d^4x'd^4x\\ &= \int \delta\phi^J(x')\frac{\delta^2 I}{\delta\phi^J(x')\delta\phi^I(x)}\hat{\delta}\phi^I(x)d^4xd^4x'\\ &= \int \delta\phi^J(x')\hat{\delta}E_J(x')d^4x'. \qquad (7.49)\end{aligned}$$

We begin with the metric part of the second-order action. From Eq. (2.95), the first derivative of the gravitational action is given by

$$16\pi \frac{\delta}{\delta g_{\alpha\beta}(x)} I_G = \frac{\delta}{\delta g_{\alpha\beta}(x)} \int R\sqrt{|g|} d^4x = -G^{\alpha\beta}\sqrt{|g|}. \tag{7.50}$$

The symmetry relation (7.49) means in this case that $\widehat{h}_{\alpha\beta}\delta(G^{\alpha\beta}\sqrt{|g|})$ is, up to a divergence, symmetric in $h_{\alpha\beta}$ and $\widehat{h}_{\alpha\beta}$. This is immediate from the form of $\delta(G^{\alpha\beta}\sqrt{|g|})$ given in Eq. (7.26) and the index symmetry (7.12) $G^{\alpha\beta\gamma\delta} = G^{\gamma\delta\alpha\beta}$:

$$\begin{aligned} \widehat{h}_{\alpha\beta}\boldsymbol{\delta} G^{\alpha\beta} &= \frac{1}{2}\epsilon^{\alpha\gamma\epsilon\eta}\epsilon^{\beta\delta\zeta}{}_{\eta}\nabla_\alpha \widehat{h}_{\gamma\delta}\nabla_\beta h_{\epsilon\zeta} + G^{\alpha\beta\gamma\delta}\widehat{h}_{\alpha\beta}h_{\gamma\delta} \\ &\quad - \nabla_\alpha\left(\frac{1}{2}\epsilon^{\alpha\gamma\epsilon\eta}\epsilon^{\beta\delta\zeta}{}_{\eta}\widehat{h}_{\gamma\delta}\nabla_\beta h_{\epsilon\zeta}\right). \end{aligned} \tag{7.51}$$

What remains is to establish the part of the symmetry relation (7.40) associated with the matter, to show that the quantity

$$\widehat{\xi}_\beta \boldsymbol{\Delta}(\nabla_\alpha T^{\alpha\beta}) - \frac{1}{2}\widehat{h}_{\alpha\beta}\boldsymbol{\delta} T^{\alpha\beta} \tag{7.52}$$

is, up to a total divergence, symmetric in the hatted and unhatted quantities.

The index symmetry of $W^{\alpha\beta\gamma\delta}$ in Eq. (7.21) for $\boldsymbol{\Delta} T^{\alpha\beta}$ implies that the expression

$$\widehat{\Delta} g_{\alpha\beta}\boldsymbol{\Delta} T^{\alpha\beta} = W^{\alpha\beta\gamma\delta}\widehat{\Delta} g_{\alpha\beta}\Delta g_{\gamma\delta} \tag{7.53}$$

is symmetric. After writing the left side of this equation in the form $\widehat{h}_{\alpha\beta}(\boldsymbol{\delta} + \boldsymbol{\mathcal{L}}_{\boldsymbol{\xi}})T^{\alpha\beta} + 2\nabla_\alpha\widehat{\xi}_\beta\boldsymbol{\Delta} T^{\alpha\beta}$, we obtain

$$\begin{aligned} \widehat{\xi}_\beta\boldsymbol{\Delta}\nabla_\alpha T^{\alpha\beta} - \frac{1}{2}\widehat{h}_{\alpha\beta}\boldsymbol{\delta} T^{\alpha\beta} &= -\frac{1}{2}W^{\alpha\beta\gamma\delta}\widehat{\Delta} g_{\alpha\beta}\Delta g_{\gamma\delta} + \nabla_\alpha(\widehat{\xi}_\beta\boldsymbol{\Delta} T^{\alpha\beta}) \\ &\quad + \frac{1}{2}\widehat{h}_{\alpha\beta}\boldsymbol{\mathcal{L}}_{\boldsymbol{\xi}} T^{\alpha\beta} + \widehat{\xi}_\beta[\boldsymbol{\Delta}, \nabla_\alpha]T^{\alpha\beta}, \end{aligned} \tag{7.54}$$

giving the desired expression (7.52) in terms of an explicitly symmetric term, a divergence, and two remaining terms. To show that these last two terms are symmetric to within a divergence, we use the commutation relations

$$\begin{aligned} &[\mathcal{L}_{\boldsymbol{\xi}}, \nabla_\beta]\eta^\alpha = R^\alpha{}_{\beta\gamma\delta}\xi^\gamma\eta^\delta + \eta^\gamma\nabla_\gamma\nabla_\beta\xi^\alpha, \quad [\delta, \nabla_\beta]\eta^\alpha = \delta\Gamma^\alpha{}_{\gamma\beta}\,\eta^\gamma, \\ &[\Delta, \nabla_\beta]\sqrt{|g|} = -\nabla_\beta\Delta\sqrt{|g|} = -\nabla_\beta\left(\frac{1}{2}h + \nabla_\gamma\xi^\gamma\right)\sqrt{|g|} \end{aligned} \tag{7.55}$$

to obtain

$$\begin{aligned} \widehat{\xi}_\beta[\boldsymbol{\Delta}, \nabla_\alpha]T^{\alpha\beta} &= \widehat{\xi}_\beta\left[\frac{1}{2}(\nabla_\gamma h_\alpha{}^\beta + \nabla_\alpha h_\gamma{}^\beta - \nabla^\beta h_{\alpha\gamma}) + \nabla_\gamma\nabla_\alpha\xi^\beta - R^\beta{}_{\alpha\gamma\delta}\xi^\delta\right]T^{\alpha\gamma} \\ &= -T^{\alpha\beta}(\nabla_\beta\widehat{\xi}^\gamma\nabla_\alpha\xi_\gamma - R_{\alpha\gamma\beta\delta}\widehat{\xi}^\gamma\xi^\delta) + \frac{1}{2}h_{\alpha\beta}\boldsymbol{\mathcal{L}}_{\widehat{\boldsymbol{\xi}}}T^{\alpha\beta} + \nabla_\alpha A^\alpha, \end{aligned} \tag{7.56}$$

with $A^\alpha = T^{\alpha\beta}(\widehat{\xi}^\gamma h_{\beta\gamma} + \widehat{\xi}^\gamma \nabla_\beta \xi_\gamma) - \frac{1}{2} T^{\beta\gamma}\widehat{\xi}^\alpha h_{\beta\gamma}$. Finally, substituting the right side of Eq. (7.56) for the last term of Eq. (7.54), we have

$$\begin{aligned}\widehat{\xi}_\beta \boldsymbol{\Delta}\nabla_\alpha T^{\alpha\beta} &- \frac{1}{2}\widehat{h}_{\alpha\beta}\boldsymbol{\delta} T^{\alpha\beta} \\ &= -\frac{1}{2}W^{\alpha\beta\gamma\delta}\widehat{\Delta}g_{\alpha\beta}\Delta g_{\gamma\delta} - T^{\alpha\beta}\nabla_\beta\widehat{\xi}^\gamma\nabla_\alpha\xi_\gamma + T^{\alpha\beta}R_{\alpha\gamma\beta\delta}\widehat{\xi}^\gamma\xi^\delta \\ &\quad + \frac{1}{2}h_{\alpha\beta}\boldsymbol{\mathcal{L}}_{\widehat{\boldsymbol{\xi}}}T^{\alpha\beta} + \frac{1}{2}\widehat{h}_{\alpha\beta}\boldsymbol{\mathcal{L}}_{\boldsymbol{\xi}}T^{\alpha\beta} \\ &\quad + \nabla_\alpha(U^{\alpha\beta\gamma\delta}\widehat{\xi}_\beta\nabla_\gamma\xi_\delta + V^{\gamma\delta\alpha\beta}h_{\gamma\delta}\widehat{\xi}_\beta). \end{aligned} \tag{7.57}$$

This establishes the required symmetry relation. The symmetric density $\mathscr{L}(\widehat{\xi},\widehat{h};\xi,h)$, the associated Lagrangian $\mathscr{L}^{(2)}(\xi,h)$, and the divergence $\nabla_\alpha\Theta^\alpha$ are now obtained by adding Eq. (7.57) to $\widehat{h}_{\alpha\beta}\boldsymbol{\delta}G^{\alpha\beta}$, with $\boldsymbol{\delta}G^{\alpha\beta}$ given by Eq. (7.22).

To obtain the form of $\mathscr{L}$ given in Eq. (7.43), write $\Delta g_{\alpha\beta}$ as $h_{\alpha\beta} + \nabla_\alpha\xi_\beta + \nabla_\beta\xi_\alpha$, expand the Lie derivatives in Eq. (7.57), and then group each quadratic combination of ξ_α, $h_{\alpha\beta}$, $\widehat{\xi}_\alpha$, $\widehat{h}_{\alpha\beta}$, and their gradients.

7.3 Energy and angular momentum

Noether's theorem relates the stationarity and axisymmetry of the background (unperturbed) spacetime to conservation of a canonical energy and angular momentum. The stability of a rotating star is governed by the sign of the energy of its perturbations, and the evolution of a perturbation is governed by the rate at which its energy and angular momentum are radiated. We first present the conservation laws and then review their derivation.

The canonical energy and angular momentum on a hypersurface can each be written as a functional quadratic in the perturbation; the conservation laws express their change in terms of the flux of gravitational waves radiated to null infinity. On a spacelike hypersurface Σ with unit normal $n^\alpha = -\alpha\nabla^\alpha t$, the canonical momenta conjugate to ξ^α and $h_{\alpha\beta}$ are given in terms of the quantities $\Pi^{\alpha\beta}$ and $\pi^{\alpha\beta\gamma}$ of Eqs. (7.45) and (7.46) by

$$\Pi^\alpha = \Pi^{\gamma\alpha}\,\nabla_\gamma t, \qquad \pi^{\alpha\beta} = \pi^{\gamma\alpha\beta}\,\nabla_\gamma t, \tag{7.58}$$

where the minus sign is associated with the choice of a future pointing unit normal and a $-+++$ signature.

Associated with the Killing vector t^α is the conserved current,

$$j_t^\alpha = \Pi^{\alpha\beta}\mathcal{L}_{\boldsymbol{t}}\xi_\beta + \pi^{\alpha\beta\gamma}\mathcal{L}_{\boldsymbol{t}}h_{\beta\gamma} - t^\alpha\mathscr{L}^{(2)}. \tag{7.59}$$

Because j_t^α is a vector density (the factor $\sqrt{|g|}$ is already included in the densities $\Pi^{\alpha\beta}$ and $\pi^{\alpha\beta\gamma}$), the expression for the energy is written with the notation

$$d\sigma_\alpha := dS_\alpha/\sqrt{|g|}. \tag{7.60}$$

When Σ is a $t =$ constant surface, $d\sigma_\alpha = \nabla_\alpha t\, d^3x$. Then the canonical energy is given by

$$E_c = \int_S d\sigma_\alpha j_t^\alpha \tag{7.61}$$

$$= \int_S d^3x\,(\Pi^\alpha \pounds_t \xi_\alpha + \pi^{\alpha\beta}\pounds_t h_{\alpha\beta} - \mathscr{L}^{(2)}). \tag{7.62}$$

Explicitly,

$$E_c = \int_S dS_\epsilon \Big\{ U^{\epsilon\beta\gamma\delta}\dot{\xi}_\beta \nabla_\gamma \xi_\delta + V^{\gamma\delta\epsilon\beta} h_{\gamma\delta}\dot{\xi}_\beta - \frac{1}{32\pi}\epsilon^{\epsilon\gamma\zeta\iota}\epsilon^{\beta\delta\eta}{}_\iota \dot{h}_{\gamma\delta}\nabla_\beta h_{\zeta\eta}$$
$$-\frac{1}{2}t^\epsilon \Big[U^{\alpha\beta\gamma\delta}\nabla_\alpha\xi_\beta\nabla_\gamma\xi_\delta + 2V^{\gamma\delta\alpha\beta}h_{\gamma\delta}\nabla_\alpha\xi_\beta - \frac{1}{32\pi}\epsilon^{\alpha\gamma\zeta\iota}\epsilon^{\beta\delta\eta}{}_\iota \nabla_\alpha h_{\gamma\delta}\nabla_\beta h_{\zeta\eta}$$
$$-T^{\alpha\beta}R_{\alpha\gamma\beta\delta}\xi^\gamma\xi^\delta + \frac{1}{2}\Big(W^{\alpha\beta\gamma\delta} - \frac{1}{16\pi}G^{\alpha\beta\gamma\delta}\Big)h_{\alpha\beta}h_{\gamma\delta} - \nabla_\gamma T^{\alpha\beta}h_{\alpha\beta}\xi^\gamma\Big]\Big\}. \tag{7.63}$$

The conserved current associated with the Killing vector ϕ^α is similarly given by

$$j_\phi^\alpha = -[\Pi^{\alpha\beta}\pounds_\phi\xi_\beta + \pi^{\alpha\beta\gamma}\pounds_\phi h_{\beta\gamma} - \phi^\alpha\mathscr{L}^{(2)}]. \tag{7.64}$$

Because $\phi^\alpha n_\alpha = 0$, for an axisymmetric hypersurface, the corresponding canonical angular momentum has the form

$$J_c = \int_S d\sigma_\alpha j_\phi^\alpha \tag{7.65}$$

$$= -\int_S d^3x\,(\Pi^\alpha\pounds_\phi\xi_\alpha + \pi^{\alpha\beta}\pounds_\phi h_{\alpha\beta}). \tag{7.66}$$

We reviewed in Section 2.4 the derivation of a conserved current from a Lagrangian density that has a symmetry. In the present case, the symmetry arises from the invariance of the background spacetime under the symmetry vectors t^α and ϕ^α, and the form of the theorem is more familiar. We again use the short-hand ϕ^I for the dynamical variables, here the pair $\xi^\alpha, h_{\alpha\beta}$. The Lagrangian density $\mathscr{L}(\phi^I, \nabla_\alpha\phi^I)$ is constructed from ϕ^I and from a background that has symmetry vectors t^α and ϕ^α, and we denote the momenta conjugate to the fields ϕ^I by

$$\pi_I = \pi_I^\alpha\,\nabla_\alpha t \quad \text{where } \pi_I^\alpha = \frac{\partial\mathscr{L}}{\partial\nabla_\alpha\phi^I}. \tag{7.67}$$

The fact that $\mathscr{L}$ is constructed locally from a background symmetric under the action of a symmetry vector t^α implies

$$\pounds_t\mathscr{L}(\phi^I, \nabla_\alpha\phi^I) = \frac{\partial\mathscr{L}}{\partial\nabla_\alpha\phi^I}\pounds_t\nabla_\alpha\phi^I + \frac{\partial\mathscr{L}}{\partial\phi^I}\pounds_t\phi^I. \tag{7.68}$$

Using the fact that $\mathscr{L}$ is a scalar density to write $\pounds_t\mathscr{L} = \nabla_\alpha(t^\alpha\mathscr{L})$, we have

$$\nabla_\alpha(t^\alpha\mathscr{L}) = \Big(-\nabla_\alpha\Pi_I^\alpha + \frac{\partial\mathscr{L}}{\partial\phi^I}\Big)\pounds_t\phi^I + \nabla_\alpha(\Pi_I^\alpha\pounds_t\phi^I). \tag{7.69}$$

Here we have used $[\mathcal{L}_{\boldsymbol{t}}, \nabla_\alpha] = 0$, true because t^α is a Killing vector. Because the Euler-Lagrange equations (here the perturbed Einstein-Euler equations) are

$$E_I = -\nabla_\alpha \frac{\partial \mathscr{L}}{\partial \nabla_\alpha \phi^I} + \frac{\partial \mathscr{L}}{\partial \phi^I} = 0, \tag{7.70}$$

when they are satisfied, the current j_t^α is conserved:

$$\nabla_\alpha (\Pi_I^\alpha \mathcal{L}_{\boldsymbol{t}} \phi^I - t^\alpha \mathscr{L}) = 0. \tag{7.71}$$

(In the notation of Section 2.4, $\delta\Phi^I = \mathcal{L}_{\boldsymbol{t}}\phi^I$, $\Theta^\alpha = \Pi_I^\alpha \mathcal{L}_{\boldsymbol{t}}\phi^I$, and $\mathcal{A}^\alpha = t^\alpha \mathscr{L}$.) The derivation for ϕ^α is, of course, identical.

Along a family Σ_t of asymptotically spacelike hypersurfaces, energy and momentum are conserved. Along a family Σ_t of asymptotically null hypersurfaces, however, they change in time because of the radiation of energy and angular momentum to future null infinity. This radiated energy and angular momentum may again be expressed as surface integrals of j_t^α or j_ϕ^α, this time at null infinity. In particular, the canonical energy decreases monotonically from one asymptotically null hypersurface Σ_t to another $\Sigma_{t'}$ in its future.

To compute the radiated energy at future null infinity, we choose a surface Σ that is exactly null outside some radius R. In particular, we take Σ to be a surface of constant u where u is a coordinate having the asymptotic form of the outgoing null coordinate of the Schwarzschild geometry,

$$u = t - r^* + O(r^{-1}), \quad \text{with} \quad r^* := r + 2M \ln\left(\frac{r}{2M} - 1\right). \tag{7.72}$$

To order r^{-1} with respect to the basis

$$\boldsymbol{e}_{\bar{0}} = \boldsymbol{\partial}_u, \; \boldsymbol{e}_{\bar{1}} = \boldsymbol{\partial}_r, \; \boldsymbol{e}_{\bar{2}} = \frac{1}{r}\boldsymbol{\partial}_\theta, \; \boldsymbol{e}_{\bar{3}} = \frac{1}{r\sin\theta}\boldsymbol{\partial}_\phi, \tag{7.73}$$

the background metric agrees with the asymptotic form of the Schwarzschild geometry, having in coordinates u, r, θ, ϕ (from, for example, Eq. [3.27]), the Eddington-Finkelstein form

$$ds^2 = -(1 - 2M/r)du^2 - 2du dr + (1 + 2M/r)r^2(d\theta^2 + \sin^2\theta d\phi^2). \tag{7.74}$$

We will evaluate the surface term in a Lorenz gauge, for which the perturbed vacuum field equation has the form (7.35). Within the family of Lorenz gauges, one has in a vacuum the freedom to set $h_\alpha{}^\alpha = 0$, and we will do so.[2]

[2] Given $h_{\alpha\beta}$ satisfying the Lorenz gauge condition, $\nabla_\beta \tilde{h}_\alpha{}^\beta = 0$, one can choose a gauge vector ζ^α and its time derivative $\dot{\zeta}^\alpha = \mathcal{L}_{\boldsymbol{t}}\zeta^\alpha$ on an initial spacelike hypersurface $\widetilde{\Sigma}$ to enforce $h_\alpha^{\text{new }\alpha} = 0$ and $\dot{h}_\alpha^{\text{new }\alpha} = 0$ on $\widetilde{\Sigma}$, where $h_\alpha^{\text{new }\alpha} = h_\alpha{}^\alpha + 2\nabla_\alpha\zeta^\alpha$: For example, pick $\zeta^\alpha = 0, \dot{\zeta}^t = -h_\alpha{}^\alpha/2, \partial_\phi\dot{\zeta}^\phi = -\frac{1}{2}\dot{h}_\alpha{}^\alpha, \dot{\zeta}^r = \dot{\zeta}^\theta = 0$. Then $h_\alpha^{\text{new }\alpha} = 0$ and $\dot{h}_\alpha^{\text{new }\alpha} = 2\ddot{\zeta}^t$. The value of $\ddot{\zeta}^t$ on $\widetilde{\Sigma}$ is set by the Lorenz gauge condition, which gives the time evolution of ζ^α in the domain of dependence of $\widetilde{\Sigma}$:

$$0 = \nabla_\beta \tilde{h}^{\text{new }\alpha\beta} = \nabla_\beta(\nabla^\alpha\zeta^\beta + \nabla^\beta\zeta^\alpha - g^{\alpha\beta}\nabla_\gamma\zeta^\gamma) = \nabla_\beta\nabla^\beta\zeta^\alpha. \tag{7.75}$$

Here we have used the relation $[\nabla_\beta, \nabla^\alpha]\zeta^\beta = 0$, implied by the vacuum field equation $R_{\alpha\beta} = 0$. Finally, the trace of the field equation, $-\frac{1}{2}\nabla_\beta\nabla^\beta h = 0$, implies that h_α^α remains zero.

Because the Cartesian components of $h_{\alpha\beta}$ asymptotically satisfy the flat scalar wave equation, they are $O(r^{-1})$, implying that the components along the basis (7.73) are at most $O(r^{-1})$. For outgoing radiation, these components each have the form

$$h_{\bar{\mu}\bar{\nu}} = \frac{a_{\bar{\mu}\bar{\nu}}(u,\theta,\phi)}{r} + O(r^{-2}), \tag{7.76}$$

with

$$\nabla_{\bar{\lambda}} h_{\bar{\mu}\bar{\nu}} = \delta^{\bar{0}}_{\bar{\lambda}} \frac{\dot{a}_{\bar{\mu}\bar{\nu}}(u,\theta,\phi)}{r} + O(r^{-2}), \tag{7.77}$$

where $\dot{a}_{\mu\nu} = \mathcal{L}_{\mathbf{t}} a_{\mu\nu} = \partial_u a_{\mu\nu}$.

The Lorenz gauge condition sets to zero the longitudinal part of a plane wave, components of $h_{\alpha\beta}$ along the direction of propagation. Because outgoing radiation has the asymptotic behavior of a wave propagating in the radial direction, one expects the leading term in the radial components of $h_{\alpha\beta}$ to vanish, and we can easily check that this is true. Because a stationary perturbation – for example, an infinitesimal change in the mass of a star – is not transverse, the constraint is on $\dot{h}_{\alpha\beta}$, not on $h_{\alpha\beta}$, and it is the condition $\dot{a}_{\bar{1}\bar{\mu}} = \dot{a}_{r\bar{\mu}} = 0$. It follows from the relation $\nabla_i h_{\bar{\mu}}{}^i = O(r^{-2})$: We have $0 = \nabla_{\bar{\nu}} h_{\bar{\mu}}{}^{\bar{\nu}} = \partial_u h_{\bar{\mu}}{}^{\bar{0}} + O(r^{-2})$. Then $\dot{h}_{\bar{\mu}}{}^{\bar{0}} = O(r^{-2})$, and the asymptotic form (7.74) of the metric implies $\dot{h}_{\bar{1}\bar{\mu}} = O(r^{-2})$.

We can now evaluate the energy and angular momentum radiated to null infinity. The two computations are essentially identical, and we begin with the radiated energy. In Eq. (7.62), we choose a 3-surface $\mathcal{S}_r$ of constant r and consider the limit as $r \to \infty$ at fixed u of the flux $\dot{E}_c$ of energy. The radiated energy itself has the form

$$\Delta E_c = \int_{\mathcal{S}_r} j^\alpha_t \nabla_\alpha r d^3x = \int_{\mathcal{S}_r} \pi^{\alpha\beta\gamma} \dot{h}_{\beta\gamma} \nabla_\alpha r du d\theta d\phi, \tag{7.78}$$

with no contribution from the term $t^\alpha \mathscr{L}$, because $t^\alpha \nabla_\alpha r = 0$; and the corresponding flux at future null infinity is given by

$$-\dot{E}_c = \int_{S_\infty} \pi^{\alpha\beta\gamma} \mathcal{L}_{\mathbf{t}} h_{\beta\gamma} \nabla_\alpha r \, d\theta d\phi. \tag{7.79}$$

In the expression

$$\pi^{\alpha\beta\gamma} \dot{h}_{\beta\gamma} \nabla_\alpha r = -\frac{\sqrt{|g|}}{32\pi} \epsilon^{r\bar{\mu}\bar{\nu}\bar{\iota}} \epsilon^{\bar{\lambda}\bar{\sigma}\bar{\tau}}{}_{\bar{\iota}} \nabla_{\bar{\lambda}} h_{\bar{\nu}\bar{\tau}} \dot{h}_{\bar{\mu}\bar{\sigma}}, \tag{7.80}$$

because there can be no repeated index in a nonzero component of $\epsilon^{r\bar{\mu}\bar{\nu}\bar{\iota}}$, and $h_{r\bar{\mu}}$ is $O(r^{-2})$ and $\nabla_{\bar{\lambda}} h_{\bar{\mu}\bar{\nu}} = \delta^{\bar{0}}_{\bar{\lambda}} \dot{h}_{\bar{\mu}\bar{\nu}} + O(r^{-2})$, we have

$$\begin{aligned} \pi^{\alpha\beta\gamma} \dot{h}_{\beta\gamma} \nabla_\alpha r &= -\frac{r^2 \sin\theta}{32\pi} \epsilon^{r\bar{\mu}\bar{\nu}u} \epsilon^{u\bar{\sigma}\bar{\tau}}{}_u \, \dot{h}_{\bar{\mu}\bar{\sigma}} \dot{h}_{\bar{\nu}\bar{\tau}} + O(r^{-3}) \\ &= \frac{r^2 \sin\theta}{32\pi} [(\dot{h}_{\bar{2}\bar{2}})^2 + (\dot{h}_{\bar{3}\bar{3}})^2 + 2(\dot{h}_{\bar{2}\bar{3}})^2] + O(r^{-3}). \end{aligned} \tag{7.81}$$

Gravitational waves thus carry positive energy to null infinity, with the rate of energy loss given by

$$\dot{E}_c = -\frac{1}{32\pi}\int_{S_\infty}[(\dot{a}_{\bar{2}\bar{2}})^2 + (\dot{a}_{\bar{3}\bar{3}})^2 + 2(\dot{a}_{\bar{2}\bar{3}})^2]\sin\theta d\theta d\phi, \tag{7.82}$$

with a relation $a_{\bar{2}\bar{2}} + a_{\bar{3}\bar{3}} = a_\theta^\theta + a_\phi^\phi = 0$ implied by $h_\alpha{}^\alpha = 0$.

The flux of angular momentum is given by Eq. (7.79) with a sign change and with $\mathcal{L}_{\boldsymbol{\phi}} h_{\beta\gamma}$ replacing $\mathcal{L}_{\boldsymbol{t}} h_{\beta\gamma}$:

$$\begin{aligned}\dot{J}_c &= \int_{S_\infty}\pi^{\alpha\beta\gamma}\mathcal{L}_{\boldsymbol{\phi}} h_{\beta\gamma}\nabla_\alpha r d\theta d\phi\\ &= \frac{1}{32\pi}\int_{S_\infty}[\dot{a}_{\bar{2}\bar{2}}\mathcal{L}_{\boldsymbol{\phi}} a_{\bar{2}\bar{2}} + \dot{a}_{\bar{3}\bar{3}}\mathcal{L}_{\boldsymbol{\phi}} a_{\bar{2}\bar{3}} + 2\dot{a}_{\bar{2}\bar{3}}\mathcal{L}_{\boldsymbol{\phi}} a_{\bar{2}\bar{3}}]\sin\theta d\theta d\phi. \end{aligned}\tag{7.83}$$

It is then immediate that a real-frequency mode, with t- and ϕ-dependence of the form $e^{i(m\phi+\sigma t)}$, satisfies

$$\dot{J}_c = -\frac{m}{\sigma}\dot{E}_c. \tag{7.84}$$

Now the frequency σ is strictly real only for a mode with equal amounts of ingoing and outgoing radiation, in which case $\dot{J}_c$ means the flux of outgoing radiation at future null infinity. Equation (7.84), however, is an accurate approximation for outgoing modes of neutron stars, because they ordinarily have damping (or growth) times much longer than the period of the mode.

A radiating source at time t produces a radiation field that carries positive energy to null infinity along a $u =$ constant surface. The corresponding asymptotic $f(u,\theta,\phi)/r$ behavior of the transverse components of the perturbed metric is required of a source whose multipole moments with $l \geq 2$ that are changing at the present time $t \sim u$. In contrast, the behavior of the perturbed metric near *spatial* infinity reflects not the present behavior of the source but its behavior in the infinite past. Without restricting present source behavior, one is free to demand that the source be nonradiative in the infinite past, and standard definitions of asymptotic flatness, like that of Eq. (3.55), do so. Energy and angular momentum are then conserved from one asymptotically flat and spacelike hypersurface to another.

Gauge invariance. As we discussed in Section 2.3, the Lagrangian description of a perfect-fluid spacetime has two kinds of gauge freedom: the usual gauge freedom associated with an infinitesimal diffeo generated by a gauge vector ζ^α and the freedom to add to ξ^α a trivial displacement η^α, a displacement that leaves the Eulerian changes in all physical quantities unchanged. We show now that the integrals E_c and J_c are gauge invariant under the first kind of gauge transformations. They are *not*, however, invariant under addition of trivial displacements ([47, 242]); in the next section we relate the noninvariance to changes in circulation and define a class of canonical displacements for which the canonical energy is unique.

The proof of invariance of E_c and J_c under the usual gauge transformations is based on the fact that E_c and J_c are conserved. Let $\xi^\alpha, h_{\alpha\beta}$ be an asymptotically flat solution to the perturbed Einstein-Euler equations, and let E_c be its canonical energy on any spacelike (and asymptotically spacelike) hypersurface Σ_{t_1}. Let $\widehat{\xi}^\alpha, \widehat{h}_{\alpha\beta}$ be related to $\xi^\alpha, h_{\alpha\beta}$ by the gauge transformation

$$\widehat{\xi}^\alpha = \xi^\alpha - \zeta^\alpha, \quad \widehat{h}_{\alpha\beta} = h_{\alpha\beta} + \mathcal{L}_\zeta g_{\alpha\beta}. \tag{7.85}$$

To show that the associated canonical energy $\widehat{E}_c$ is equal to E_c on Σ, we consider a vector $\widetilde{\zeta}^\alpha$ that coincides with ζ^α in a neighborhood of Σ and that vanishes in the neighborhood of another spacelike hypersurface Σ' to the future of Σ.[3] Then the perturbation, $\widetilde{\xi}^\alpha = \xi^\alpha - \widetilde{\zeta}^\alpha, \quad \widetilde{h}_{\alpha\beta} = h_{\alpha\beta} + \mathcal{L}_{\widetilde{\zeta}} g_{\alpha\beta}$, coincides with $\xi^\alpha, h_{\alpha\beta}$ on Σ' and with $\widehat{\xi}^\alpha, \widehat{h}_{\alpha\beta}$ on Σ. Because the canonical energy $\widetilde{E}_c = E_c(\widetilde{\xi}^\alpha, \widetilde{h}_{\alpha\beta})$ is conserved, we have $E_c(\widehat{\xi}^\alpha, \widehat{h}_{\alpha\beta}) = E_c|_\Sigma = E_c|_{\Sigma'} = E_c(\xi^\alpha, h_{\alpha\beta})$. The proof for J_c is identical.

7.4 Canonical displacements and the symplectic form

We begin by introducing an antisymmetric product of two perturbations that is conserved by their evolution. This *symplectic* product is important both for our present discussion and for the behavior of quasinormal modes: Using the product to define orthogonality, we will find that displacements that conserve vorticity are orthogonal to the space of trivial displacements; that modes with different frequencies are orthogonal; and that any mode is orthogonal to the space of trivial displacements.

7.4.1 The symplectic form

Let Σ be an asymptotically flat spacelike hypersurface of the background spacetime. Given two solutions $\xi^\alpha, h_{\alpha\beta}$ and $\widehat{\xi}^\alpha, \widehat{h}_{\alpha\beta}$ to the perturbed Einstein-Euler system, their symplectic product, $W(\widehat{\xi}, \widehat{h}; \xi, h)$, is

$$W(\widehat{\xi}, \widehat{h}; \xi, h) := \int_\Sigma (\widehat{\Pi}_\alpha \xi^\alpha + \widehat{\pi}^{\alpha\beta} h_{\alpha\beta} - \Pi_\alpha \widehat{\xi}^\alpha - \pi^{\alpha\beta}\widehat{h}_{\alpha\beta}) d^3x. \tag{7.87}$$

From Eq. (7.44), W can be expressed in terms of the surface term Θ^α associated with the second-order Lagrangian as

$$W(\widehat{\xi}, \widehat{h}; \xi, h) = \int_\Sigma [\Theta^\alpha(\xi, h; \widehat{\xi}, \widehat{h}) - \Theta^\alpha(\widehat{\xi}, \widehat{h}; \xi, h)] d\sigma_\alpha. \tag{7.88}$$

[3] For example, choose a time function T for which $T = -2$ on Σ and $T = 2$ on Σ'. Let $F(T)$ be the smooth step function

$$F(T) = \begin{cases} 1, & T \leq -1, \\ \frac{1}{2}\left[1 - \tanh \frac{T}{1-T^2}\right], & -1 < T < 1, \\ 0, & T \geq 1. \end{cases} \tag{7.86}$$

Because $F(T)$ is smooth (C^∞), the gauge vector $\widetilde{\zeta}^\alpha = F\zeta^\alpha$ is smooth, equal to ζ^α in a neighborhood of Σ_0 and equal to zero in a neighborhood of Σ'. For $\zeta^\alpha = O(r^{-1})$, $\nabla_\beta \zeta^\alpha = O(r^{-2})$, and $\nabla_\beta \nabla_\gamma \zeta^\alpha = O(r^{-3})$, asymptotic regularity is preserved, and the canonical energy is again conserved.

To see that the product is conserved, we again use the shorthand $\phi^I = (\xi^\alpha, h_{\alpha\beta})$, writing the symplectic product in the form

$$W(\widehat{\phi}, \phi) = \int (\widehat{\pi}_I \phi^I - \pi_I \widehat{\phi}^I) d^3x = \int [\Theta^\alpha(\phi, \widehat{\phi}) - \Theta^\alpha(\widehat{\phi}, \phi)] d\sigma_\alpha, \tag{7.89}$$

with the fundamental symmetry relation (7.40) given by

$$\widehat{\phi}^I E_I = -\mathscr{L}(\widehat{\phi}, \phi) + \nabla_\alpha \Theta^\alpha(\widehat{\phi}, \phi). \tag{7.90}$$

When $\widehat{\phi}^I$ and ϕ^I satisfy the field equations (here the Einstein-Euler equations) $E_I(\phi) = 0 = E_I(\widehat{\phi})$, the symmetry of $\mathscr{L}$ immediately gives the conserved current

$$\nabla_\alpha [\Theta^\alpha(\phi, \widehat{\phi}) - \Theta^\alpha(\widehat{\phi}, \phi)] = 0, \tag{7.91}$$

and the corresponding conserved product is[4]

$$W(\widehat{\phi}, \phi) = \int_\Sigma [\Theta^\alpha(\phi, \widehat{\phi}) - \Theta^\alpha(\widehat{\phi}, \phi)] d\sigma_\alpha = \int_\Sigma (\pi_I \widehat{\phi}^I - \widehat{\pi}_I \phi^I)\, d^3x, \tag{7.92}$$

or

$$W(\widehat{\xi}^\alpha, \widehat{h}_{\alpha\beta}; \xi^\alpha, h_{\alpha\beta}) = \int_\Sigma (\Pi_\alpha \widehat{\xi}^\alpha + \pi^{\alpha\beta} \widehat{h}_{\alpha\beta} - \widehat{\Pi}_\alpha \xi^\alpha - \widehat{\pi}^{\alpha\beta} h_{\alpha\beta})\, d^3x. \tag{7.93}$$

If for the solution $\widehat{\xi}^\alpha, \widehat{h}_{\alpha\beta}$ one chooses $\mathcal{L}_{\boldsymbol{t}}\xi^\alpha, \mathcal{L}_{\boldsymbol{t}}\widehat{h}_{\alpha\beta}$ or $\mathcal{L}_{\boldsymbol{\phi}}\xi^\alpha, \mathcal{L}_{\boldsymbol{\phi}}h_{\alpha\beta}$, then the conserved product (7.93) provides an alternative form of the canonical energy or momentum of the perturbation:

$$\begin{aligned} E_c &= \frac{1}{2} W(\mathcal{L}_{\boldsymbol{t}}\xi^\alpha, \mathcal{L}_{\boldsymbol{t}}h_{\alpha\beta}, \xi^\alpha, h_{\alpha\beta}) \\ &= \frac{1}{2} \int_\Sigma (\Pi_\alpha \mathcal{L}_{\boldsymbol{t}}\xi^\alpha + \pi^{\alpha\beta} \mathcal{L}_{\boldsymbol{t}}h_{\alpha\beta} - \mathcal{L}_{\boldsymbol{t}}\Pi_\alpha \xi^\alpha - \mathcal{L}_{\boldsymbol{t}}\pi^{\alpha\beta} h_{\alpha\beta}) d^3x, \end{aligned} \tag{7.94}$$

$$\begin{aligned} J_c &= -\frac{1}{2} W(\mathcal{L}_{\boldsymbol{\phi}}\xi^\alpha, \mathcal{L}_{\boldsymbol{\phi}}h_{\alpha\beta}, \xi^\alpha, h_{\alpha\beta}) \\ &= -\frac{1}{2} \int_\Sigma (\Pi_\alpha \mathcal{L}_{\boldsymbol{\phi}}\xi^\alpha + \pi^{\alpha\beta} \mathcal{L}_{\boldsymbol{\phi}}h_{\alpha\beta} - \mathcal{L}_{\boldsymbol{\phi}}\Pi_\alpha \xi^\alpha - \mathcal{L}_{\boldsymbol{\phi}}\pi^{\alpha\beta} h_{\alpha\beta}) d^3x. \end{aligned} \tag{7.95}$$

For a normal mode, this is a useful form of E_c, because each Lie derivative gives simply a factor $i\sigma$: $\mathcal{L}_{\boldsymbol{t}}\xi^\alpha = i\sigma\xi^\alpha, \ldots, \mathcal{L}_{\boldsymbol{t}}\pi^{\alpha\beta} = i\sigma\pi^{\alpha\beta}$. Note that the defining equation (7.62) for the canonical energy expresses E_c in terms of initial data on a hypersurface, whereas the symplectic expression (7.94) involves second time derivatives of ξ^α and $h_{\alpha\beta}$ in $\mathcal{L}_{\boldsymbol{t}}\pi_\alpha$ and $\mathcal{L}_{\boldsymbol{t}}\pi^{\alpha\beta}$. The two expressions are thus equal only when the perturbed Einstein-Euler equations are satisfied: The symplectic expression is the canonical energy of a perturbation only when the second time derivatives are determined by the initial data.

[4] This is the Noether current associated with the symmetry $\phi^I \to \phi^I + \widehat{\phi}^I$, with $\delta\phi^I = \widehat{\phi}^I$ in the notation of Section 2.4. The symplectic product is the infinite-dimensional version of the symplectic form $dp_i \wedge dq^i$ of finite-dimensional Hamiltonian mechanics. Given two solutions δq^i and $\hat{\delta} q^i$ to the perturbed Euler-Lagrange equations, with conjugate momenta δp_i, $\hat{\delta} p_i$, the antisymmetric product $(dp_i \wedge dq^i)(\hat{\delta}p, \hat{\delta}q; \delta p, \delta q) = \hat{\delta}p_i \delta q^i - \delta p_i \hat{\delta} q^i$ is conserved.

The symplectic expression for J_c, on the other hand, involves only the initial data, and an integration by parts immediately converts it to the defining form (7.66). For a normal mode with behavior $e^{i(m\phi+\sigma t)}$, for real σ, these forms of E_c and J_c immediately imply the relation

$$J_c = -\frac{m}{\sigma}E_c, \tag{7.96}$$

whose time derivative is Eq. (7.84).

7.4.2 Perturbations that preserve circulation and canonical displacements

The first law of thermodynamics, in the form (3.127), implies that a perturbation joining two nearby equilibria generically changes the mass to first order if it changes the entropy, baryon number, or circulation of the fluid. Dynamical evolution conserves these quantities, and in considering stellar oscillations and stability, one obtains a perturbed model whose energy is second order in the perturbation by choosing as the corresponding unperturbed model an equilibrium with the same specific entropy, baryon number, and circulation [47, 242]. Writing the perturbed fluid variables in terms of ξ^α and $h_{\alpha\beta}$ both used and enforced conservation of entropy and baryons. What remains is to enforce conservation of circulation,

$$\Delta \int_c hu_\alpha dl^\alpha = 0, \tag{7.97}$$

where c is a closed curve in a surface of constant specific entropy. Equivalently, the requirement that the perturbation preserve the vorticity $\omega_{\alpha\beta}$ of each fluid element is

$$0 = \Delta\left[\omega_{[\alpha\beta}\nabla_{\gamma]}s\right] = \Delta\left[\nabla_{[\alpha}(hu_\beta)\nabla_{\gamma]}s\right] = \nabla_{[\alpha}\Delta(hu_\beta)\nabla_{\gamma]}s, \tag{7.98}$$

when $\nabla s \neq 0$, and

$$0 = \Delta\omega_{\alpha\beta} = \nabla_{[\alpha}\Delta(hu_{\beta]}) = 0 \tag{7.99}$$

for a homentropic star. Like the requirement that the Lagrangian perturbation conserve the baryon number of each fluid element, this equation is a first integral of the perturbed conservation of vorticity equation (1.51),

$$0 = \Delta\mathcal{L}_{\mathbf{u}}\omega_{\alpha\beta} = \mathcal{L}_{\mathbf{u}}\Delta\omega_{\alpha\beta}. \tag{7.100}$$

The last equality follows from the relation $[\Delta, \mathcal{L}_{\mathbf{u}}] = \mathcal{L}_{\Delta\mathbf{u}}$, together with the unperturbed Euler equation in the form (1.32): Writing $\Delta u^\alpha = Uu^\alpha$, we have

$$\mathcal{L}_{\Delta\mathbf{u}}\omega_{\alpha\beta} = U\mathcal{L}_{\mathbf{u}}\omega_{\alpha\beta} + \omega_{\gamma\beta}\nabla_\alpha \ln Uu^\gamma + \omega_{\alpha\gamma}\nabla_\beta \ln Uu^\gamma = 0, \tag{7.101}$$

with each term on the right separately vanishing.

Because baryon conservation implies $\Delta\left(\frac{1}{\rho}\epsilon^{\alpha\beta\gamma}\right)=0$, preservation of vorticity by the perturbation can be written in the form

$$\Delta\left[\frac{1}{\rho}\epsilon^{\alpha\beta\gamma}\nabla_\alpha(hu_\beta)\nabla_\gamma f\right]=0, \qquad (7.102)$$

where f is any scalar Lie-derived by u^α when $\nabla_\alpha s=0$, and $f=s$ when $\nabla_\alpha s\neq 0$. We have seen in Section 2.4 that conservation of vorticity is related by Noether's theorem to the invariance of the action under trivial displacements. For the second-order Lagrangian, the perturbed conserved quantity associated with a trivial displacement η^α is given by $W(\eta^\alpha,0;\xi^\alpha,h_{\alpha\beta})$, and conservation of vorticity in the form (7.102) is the requirement that this inner product vanish:

$$W(\eta^\alpha,0;\xi^\alpha,h_{\alpha\beta})=W(\eta^\alpha,0;\xi^\alpha,0)=0. \qquad (7.103)$$

That is, for a Lagrangian displacement ξ^α with $\xi^\alpha u_\alpha=0$, *the Lagrangian change in circulation (or vorticity) vanishes if and only if ξ^α is orthogonal to the space $\mathcal{T}$ of trivial displacements.*

Definition. A Lagrangian displacement ξ^α is *canonical* if $\xi^\alpha u_\alpha=0$ and it preserves circulation in surfaces of constant entropy.

Then ξ^α is canonical if and only if it is orthogonal to all trivial displacements. When ξ^α is canonical, the canonical energy E agrees with the second-order change in the energy of the spacetime. This implies that E must be invariant under the remaining gauge freedom – the freedom to add a trivial displacement that is itself orthogonal to the subspace of trivials (because W is antisymmetric, any real displacement is orthogonal to itself). This invariance is immediate from the expression (7.94) for E in terms of W and the fact that $\dot\eta^\alpha$ is trivial if η^α is trivial:

$$E_c(\xi+\eta,h)=E_c(\xi,h)+W(\dot\xi;\eta)+W(\dot\eta;\xi)+W(\dot\eta;\eta)=E_c(\xi,h), \qquad (7.104)$$

where we have suppressed the argument $h_{\alpha\beta}$ of W, because, for η^α trivial, $h_{\alpha\beta}$ and $\pi_{\alpha\beta}$ do not appear in the product.

For nonaxisymmetric perturbations (perturbations with no $m=0$ part), we show in the next section that there is no loss of generality in using canonical displacements. For axisymmetric perturbations, however, this is a physical constraint that eliminates perturbations that change the rotation law and/or the total angular momentum at linear order.

7.4.3 Existence of canonical displacements

We show here that any nonaxisymmetric perturbation of a uniformly rotating star that preserves baryon number and entropy can be described by a canonical displacement. That is, given a perturbation $(h_{\alpha\beta},\xi^\alpha)$, one can find a trivial displacement η^α for which $\xi^\alpha-\eta^\alpha$ is canonical.[5] The argument given here for a homentropic

[5] We thank Stephen Green for pointing out an error in and suggesting a simplification to an earlier version of this proof.

star generalizes an earlier Newtonian result [241]. The corresponding proof for a two-parameter equation of state is simpler and is given in [231].

We must show that there is a vector η^α with $\eta^\alpha u_\alpha = 0$ that satisfies the triviality conditions (2.147),

$$q^\alpha{}_\beta \nabla_\alpha(\rho\eta^\beta) = 0, \qquad q^\alpha{}_\beta \mathcal{L}_{\boldsymbol{u}}\eta^\beta = 0, \tag{7.105}$$

and for which the perturbation $(h_{\alpha\beta}, \xi^\alpha - \eta^\alpha)$ has vanishing perturbed vorticity,

$$0 = q_\alpha{}^\gamma q_\beta{}^\delta \nabla_{[\gamma}\Delta_{\boldsymbol{\xi}-\boldsymbol{\eta}}(hu_{\delta]}) = q_\alpha{}^\gamma q_\beta{}^\delta \nabla_{[\gamma}\Delta(hu_{\delta]}) - q_\alpha{}^\gamma q_\beta{}^\delta \nabla_{[\gamma}\mathcal{L}_{\boldsymbol{\eta}}(hu_{\delta]}), \tag{7.106}$$

where Δ is the Lagrangian change associated with the original perturbation $(h_{\alpha\beta}, \xi^\alpha)$ and $\Delta_{\boldsymbol{\xi}-\boldsymbol{\eta}}$ is the Lagrangian change associated with $(h_{\alpha\beta}, \xi^\alpha - \eta^\alpha)$.

Equation (7.106) will be satisfied if there is a scalar ψ satisfying

$$\mathcal{L}_{\boldsymbol{\eta}}(hu_\alpha) = \Delta(hu_\alpha) + \nabla_\alpha\psi. \tag{7.107}$$

The component of this equation along u^α determines $u^\alpha\nabla_\alpha\psi$ as follows. Using the relativistic Euler equation in the form (1.31) to write $\nabla_\alpha h = -h\mathcal{L}_{\boldsymbol{u}}u_\alpha$, a short algebraic manipulation gives

$$u^\alpha \mathcal{L}_{\boldsymbol{\eta}}(hu_\alpha) = 0. \tag{7.108}$$

and the component of Eq. (7.107) along u^α takes the form

$$u^\alpha\nabla_\alpha\psi = -u^\alpha\Delta(hu_\alpha). \tag{7.109}$$

The projection of Eq. (7.107) perpendicular to u^α is

$$q_\alpha{}^\beta \mathcal{L}_{\boldsymbol{\eta}}(hu_\beta) = q_\alpha{}^\beta[\Delta(hu_\beta) + \nabla_\beta\psi]. \tag{7.110}$$

Using the Cartan identity (A.38),

$$\mathcal{L}_{\boldsymbol{\eta}}\sigma = d(\sigma\cdot\eta) + \eta\cdot d\sigma, \tag{7.111}$$

for the covector $\sigma_\alpha = hu_\alpha$, we have

$$\omega_{\alpha\beta}\eta^\beta = -q_\alpha{}^\beta[\Delta(hu_\beta) + \nabla_\beta\psi], \tag{7.112}$$

with $\omega_{\alpha\beta} = 2\nabla_{[\alpha}(hu_{\beta]})$ the vorticity tensor.

To find a solution η^α to Eq. (7.112), note that the equation involves only the projection of η^α orthogonal to the vorticity vector ω^α, defined by $\omega^\alpha := \frac{1}{2}\epsilon^{\alpha\beta\gamma}\omega_{\beta\gamma}$. We therefore decompose η^α in the manner

$$\eta^\alpha = \zeta^\alpha + g\omega^\alpha, \tag{7.113}$$

where ζ^α is orthogonal to ω^α. With $\widehat{\omega}^\alpha$ the unit vector $\omega^\alpha/|\boldsymbol{\omega}|$, the projection operator onto the subspace orthogonal to ω^α and u^α is

$$q^\alpha{}_\beta - \widehat{\omega}^\alpha\widehat{\omega}_\beta = \omega^{\alpha\gamma}\omega_{\beta\gamma}|\boldsymbol{\omega}|^{-2}, \tag{7.114}$$

and the projection of Eq. (7.112) orthogonal to ω^α gives

$$\zeta^\alpha = |\boldsymbol{\omega}|^{-2}\omega^{\alpha\beta}[\Delta(hu_\beta) + \nabla_\beta\psi]. \tag{7.115}$$

The component of (7.112) along ω^α is

$$\omega^\alpha\nabla_\alpha\psi = -\omega^\alpha\Delta(hu_\alpha). \tag{7.116}$$

For any choice ψ_- for the value of ψ on the half of the star's surface that lies below the equatorial plane, we obtain ψ elsewhere by integrating Eq. (7.116) along each integral curve c of ω^α. The symmetry of the equilibrium star under reflection in the equatorial plane implies that an integral curve through a point of the star's interior intersects the star's surface in two symmetrically opposite points above and below the equator, and we integrate upward from the intersection point below the equator. We have

$$\psi = -\int_c \hat{\omega}^\alpha\Delta(hu_\alpha)dl + \psi_-. \tag{7.117}$$

Finally, we satisfy the first triviality condition in (2.154) with a choice of g for which

$$0 = \omega^\alpha\nabla_\alpha(\rho g) + q_\beta{}^\alpha\nabla_\alpha(\rho\zeta^\beta). \tag{7.118}$$

Beginning with $g = 0$ on the bottom half of the star's surface, we again integrate along the integral curves of ω^α to find g. Because ρ vanishes at the top surface of the star, g is finite there only if

$$\int_{\text{bottom}}^{\text{top}} |\boldsymbol{\omega}|^{-1}q_\beta{}^\alpha\nabla_\alpha(\rho\zeta^\beta)dl = 0; \tag{7.119}$$

and this requirement can be met by using our freedom in the choice of ψ_-. The second triviality condition gives a time-evolution of η,

$$\mathcal{L}_{\mathbf{k}}\eta^\alpha = 0, \tag{7.120}$$

with $k^\alpha = t^\alpha + \Omega\phi^\alpha$; because k^α is a Killing vector (for uniform rotation), the evolution preserves the conditions (7.106) and $q^\alpha{}_\beta\nabla_\alpha(\rho\eta^\beta) = 0$. We have in this way shown existence of a canonical displacement corresponding to a given perturbation.

The requirement that ψ vanish at the surface of the star uniquely fixes ψ_- by fixing the value of

$$\int_{\text{bottom}}^{\text{top}} |\boldsymbol{\omega}|^{-1}q_\beta{}^\alpha\nabla_\alpha(\rho|\boldsymbol{\omega}|^{-2}\omega^{\beta\gamma}\nabla_\gamma\psi_-)\,dl$$
$$= (\partial_\phi\psi_-)\int_{\text{bottom}}^{\text{top}} |\boldsymbol{\omega}|^{-1}q_\beta{}^\alpha\nabla_\alpha(\rho|\boldsymbol{\omega}|^{-2}\omega^{\beta\gamma})\nabla_\gamma\phi\,dl, \tag{7.121}$$

Here we have used the fact that ψ_- is constant along each curve c to bring it outside the integral. Only the ϕ derivative of ϕ_- appears because $\omega^{\alpha\beta}$ is nonzero only for indices in the subspace orthogonal to ω^α and u^α and the coefficient of $\nabla_\gamma\psi_-$ on the left side is axisymmetric. The fact that the construction uniquely specifies η^α and thus uniquely specifies a canonical displacement corrects a mistaken claim in [241] of ambiguity in a Newtonian analog of ψ_-.

7.4.4 Canonical energy in a rotating frame

Where gravitational radiation always decreases the energy in an inertial frame, viscosity dissipates the energy measured by a comoving observer. Then for a uniformly rotating star, viscosity decreases the energy measured in the rotating frame. When the rate of energy loss to gravitational waves is negligible compared to the energy loss to viscosity, secular stability is governed by the canonical energy in a rotating frame, $E_{c,r}$. This is the conserved quantity associated with the Killing vector $k^\alpha = t^\alpha + \Omega\phi^\alpha$ tangent to the trajectory of an comoving observer:

$$E_{c,r} = \int (j_t^\alpha - \Omega j_\phi^\alpha) d\sigma_\alpha = \frac{1}{2} W(\mathcal{L}_{\mathbf{k}}\xi^\alpha, \mathcal{L}_{\mathbf{k}} h_{\alpha\beta}, \xi^\alpha, h_{\alpha\beta}), \tag{7.122}$$

where the second equality follows from Eqs. (7.94) and (7.95).

Remarkably, although both the canonical energy and angular momenta each change when one changes ξ^α by a trivial displacement, we will see that the combination $E_{c,r} = E_c - \Omega J_c$ is invariant under this gauge freedom. The invariance implies that $E_{c,r}$ depends only on the Eulerian perturbations of the fluid variables and the metric, and we display its explicit form. In the Newtonian approximation, the general argument (due to Bardeen) and the explicit Eulerian form of $E_{c,r}$ are given in Friedman and Schutz [242]. In the full theory, the Eulerian form was obtained by Lindblom and Hiscock [423]. The Newtonian argument used an explicit computation of the second-order changes in the energy and angular momentum, showing that $E_{c,r}$ can be written as the explicitly gauge-invariant combination $\delta^{(2)} M - \Omega\delta^{(2)} J = E_c - \delta J_c \equiv E_{c,r}$. The symplectic expression (7.124) allows a simpler version.

Gauge invariance for trivial displacements η^α, given by Eq. (2.149), follows from that fact that $E_{c,r}$ involves derivatives of the Lagrangian displacement along the time-translation symmetry vector $k^\alpha = t^\alpha + \phi^\alpha$, and $\mathcal{L}_k \eta^\alpha = 0$. This last relation is immediate from Eq. (2.149), and it follows directly from the defining condition $q^\alpha_\beta \, \mathcal{L}_u \eta^\beta = 0$, using $u^\alpha = u^t k^\alpha$ and $\eta^\alpha u_\alpha = 0$. Then, writing $\xi^\alpha = \xi_c^\alpha + \eta^\alpha$, where ξ_c^α is canonical we have

$$\begin{aligned} E_{c,r}(\xi, h) &= \frac{1}{2} W(\mathcal{L}_{\boldsymbol{k}}\xi^\alpha, \mathcal{L}_{\boldsymbol{k}} h_{\alpha\beta}; \xi^\alpha + \eta^\alpha, h_{\alpha\beta}) \\ &= \frac{1}{2} W(\mathcal{L}_{\boldsymbol{k}}\xi_c^\alpha, \mathcal{L}_{\boldsymbol{k}} h_{\alpha\beta}; \xi_c^\alpha, h_{\alpha\beta} + \frac{1}{2} W(\mathcal{L}_{\boldsymbol{k}}\xi_c^\alpha, 0; \eta^\alpha, 0) \\ &= E_{c,r}(\xi_c, h), \end{aligned} \tag{7.123}$$

where, in the last equality, we have used the symplectic orthogonality of ξ_c^α and η^α that defines canonical displacements.

The explicitly Eulerian form of $E_{c,r}$ is [423]

$$E_{c,r} = \int_S \left\{ -\frac{1}{2}(\epsilon+p)u^\alpha g^{\beta\gamma}\delta u_\beta \delta u_\gamma \right.$$
$$+ \frac{1}{2}u^\alpha \frac{1}{\epsilon+p}\left[\left(\frac{\partial\epsilon}{\partial p}\right)_s (\delta p)^2 + \left(\frac{\partial\epsilon}{\partial s}\right)_p \frac{dp}{ds}(\delta s)^2\right]$$
$$+ \frac{1}{32\pi}u^\alpha [G^{\beta\gamma\delta\epsilon} - 4\pi(\epsilon+p)u^\beta(u^\gamma g^{\delta\epsilon} - 4u^\delta g^{\gamma\epsilon}) + 8\pi p g^{\beta\delta}g^{\gamma\epsilon}]h_{\beta\gamma}h_{\delta\epsilon}$$
$$\left. + \delta T^\alpha_\beta \delta u^\beta - \frac{1}{32\pi}(\delta^\alpha_\iota \mathcal{L}_{\mathbf{u}} h_{\gamma\delta} - \frac{1}{2}u^\alpha \nabla_\iota h_{\gamma\delta})\epsilon^{\iota\gamma\epsilon\zeta}\epsilon^{\beta\delta\eta}{}_\zeta \nabla_\beta h_{\epsilon\eta} \right\} \frac{1}{u^t} dS_\alpha. \tag{7.124}$$

From Eq. (3.140) for the locally measured viscous dissipation of energy $dE/d\tau$ in a comoving frame, and using $dE/d\tau = u^t dE/dt$, we immediately obtain the rate at which the perturbation's energy is damped by viscosity:

$$\dot{E}_{c,r} = -\int \left(2\eta\delta\sigma_{\alpha\beta}\delta\sigma^{\alpha\beta} + \zeta\delta\theta^2\right)\frac{1}{u^t}\sqrt{|g|}d^3x, \tag{7.125}$$

where the shear and divergence are given in Eqs. (1.57) and (3.130).

7.4.5 A stability criterion

We have shown that the canonical energy E_c is monotonically decreasing, and we now use that fact to state a stability criterion in terms of E_c:

Let Σ be a spacelike, asymptotically flat hypersurface with normal n_a. We will call an asymptotically flat initial data set $(\xi^\alpha, h_{\alpha\beta}, \dot{\xi}^\alpha, \dot{h}_{\alpha\beta})$ on Σ canonical data if it satisfies the perturbed constraint equations $\delta(G^\alpha{}_\beta - 8\pi T^\alpha{}_\beta)n_\alpha = 0$ and it is data for a canonical perturbation.

To characterize stability, we will consider a family Σ_u of spacelike hypersurfaces that asymptotically approach surfaces of constant outgoing null coordinate u. Equation (7.82) is the statement that gravitational radiation carries positive energy to null infinity, and it implies that E_c decreases from one asymptotically null hypersurface to the next:

$$\partial_u E_c \le 0. \tag{7.126}$$

Now there are canonical data with $E_c < 0$ on an asymptotically spacelike hypersurface Σ if and only if there are canonical data on an asymptotically null hypersurface Σ_u. To see this, note that asymptotic flatness (or the fact that E_c is finite) implies that the contribution to E_c from large values of r approaches zero as $r \to \infty$. In particular, there is a radius R large enough that the contribution to E_c from the part of Σ with $r < R$ is negative. Let S be the part of Σ with $r > R$. In the domain of dependence of S, deform Σ to an asymptotically null hypersurface Σ_u. Then $E_c|_{\Sigma_u} \le E_c|_\Sigma < 0$.

We then have the following characterization of stability:

1. If $E_c < 0$ for some canonical data on Σ, then the configuration is unstable or marginally stable: There exist perturbations on a family of asymptotically null hypersurfaces Σ_u that do not die away in time.
2. If $E_c > 0$ for all canonical data on Σ, the magnitude of E_c is bounded in time and only finite energy can be radiated.

For perturbations that are not spherical, stable perturbations have positive energy and die away in time; unstable perturbations have negative canonical energy and radiate negative energy to infinity, implying that E_c becomes increasingly negative. One would like to show that when $E_c < 0$, a perfect-fluid configuration is strictly unstable; that within the linearized theory, the time-evolved data radiate infinite energy; and that $|E_c|$ becomes infinite along a family Σ_u of asymptotically null hypersurfaces. There is no proof of this conjecture, but it is easy to see that if $E_c < 0$, the time derivatives $\dot{\xi}^\alpha$ and $\dot{h}_{\alpha\beta}$ must remain finitely large. Thus a configuration with $E_c < 0$ will be strictly unstable unless it admits nonaxisymmetric perturbations that are time dependent but *nonradiative*. The key fact here, that the perturbation's time derivatives are bounded away from zero in an integral norm when $E_c < 0$, follows immediately from the expression (7.94) for E_c in terms of the symplectic product. That is, because E_c is always less than its initial negative value, the value on successive hypersurfaces of $(\dot{\xi}^\alpha, \dot{h}_{\alpha\beta})$ is bounded away from zero by the relation

$$W(\xi, h; \dot{\xi}, \dot{h}) = 2E_c \leq 2E_c|_\Sigma < 0. \tag{7.127}$$

For spherical perturbations of spherical stars, there is no gravitational radiation, and the radiative argument is not relevant to a stability criterion. As we will see, however, the perturbation equations have the form of a Sturm-Liouville system, implying a complete set of normal modes and a precise statement of stability, namely that $E_c < 0$ if and only if there is an unstable mode.

7.5 Perturbations of spherical stars

Although our primary concern is with rotating stars, the description of slowly rotating stars and their perturbations involves the formalism for perturbations of a spherical star. We begin with radial perturbations of spherical stars and then develop a formalism for nonradial perturbations. The treatment includes a discussion of tensor spherical harmonics, a gauge-invariant description of the perturbations, and the equations governing nonradial oscillations.

7.5.1 Spherical (radial) perturbations

The study of oscillations of relativistic stars began with work by Chandrasekhar [126] that obtained the relativistic Euler equation for radial perturbations and used it to show that the stronger gravity of general relativity leads to an earlier onset of instability than is seen in a Newtonian context. Fowler [228, 227] independently suggested a destabilizing effect, using a rough argument based on the binding energy of hot equilibrium configurations that is critically discussed in Thorne's 1966 summary of relativistic astrophysics [687]. A more detailed review of radial oscillations from the same time is given by Bardeen, Thorne, and Melzer [49].

By 1965, Shklovsky [628] had suggested that Sco X-1 might be a neutron star (although it was two years before his more famous paper [629] linked the source of x-rays to accretion from a companion); Bardeen et al. wrote, "The discovery of galactic X-ray sources has given impetus to theoretic research on neutron stars, and supermassive stars have been studied in connection with the quasi-stellar radio sources." The latter connection had spurred the work of Chandrasekhar and Fowler, because a supermassive star is radiation dominated, and in the Newtonian limit, as we review below, a spherical star is unstable to radial oscillations if the pressure-weighted average of the adiabatic index Γ_1 governing its perturbations is less than $4/3$, its value for a gas of photons.

We begin with the familiar equations governing the equilibrium of a spherical star, using Schwarzschild coordinates, with metric

$$ds^2 = -e^{2\nu} dt^2 + e^{2\lambda} dr^2 + r^2 d\Omega^2, \tag{7.128}$$

where $e^{2\lambda} = (1 - 2m/r)^{-1}$ and where ν is given by the equation of hydrostatic equilibrium (the TOV equation),

$$\frac{p'}{\epsilon + p} = -\nu' = -\frac{m + 4\pi r^3 p}{r(r - 2m)}, \tag{7.129}$$

with $m = m(r)$.

For motion that maintains spherical symmetry, one can choose a gauge in which the metric retains the form (7.128). The perturbed configuration is described by the perturbed potentials $\delta\nu$ and $\delta\lambda$ and by a Lagrangian displacement with only an r-component, which we denote by ξ:

$$\xi := \xi^r. \tag{7.130}$$

We will show that the perturbed Euler equation (7.17) has the form

$$\begin{aligned}(\epsilon + p)e^{2\lambda - 2\nu}\ddot{\xi} = &-\left[\frac{4}{r}p' - \frac{p'^2}{\epsilon + p} + 8\pi\, e^{2\lambda} p(\epsilon + p)\right]\xi \\ &+ e^{-\lambda - 2\nu}\left[\Gamma_1 p \frac{e^{3\nu + \lambda}}{r^2}(e^{-\nu} r^2 \xi)'\right]',\end{aligned} \tag{7.131}$$

self-adjoint with respect to the volume element $\sqrt{|g|}\ d^3x = e^{\lambda+\nu} r^2 dr \sin\theta d\theta d\phi$ or, in 1 dimension, with respect to the measure $e^{\lambda+\nu} r^2 dr$.

We now derive the equation and obtain a variational principle for the frequency of normal modes. In Chapter 8, we use the results to discuss their stability.

Derivation. The perturbed Euler equation (7.17) has only a radial component, and it has the form

$$(\epsilon+p)e^{-2\nu}(\Delta g_{tr})^{\cdot} = \frac{1}{2}(\epsilon+p)\left(u^\alpha u^\beta \Delta g_{\alpha\beta}\right)' - \frac{1}{2}p'\left(1+\frac{\Gamma_1 p}{\epsilon+p}\right)q^{\alpha\beta}\Delta g_{\alpha\beta}$$
$$+\frac{1}{2}\left[\Gamma_1 p q^{\alpha\beta}\Delta g_{\alpha\beta}\right]', \tag{7.132}$$

where we denote partial derivatives along t and r by $(\dot{})$ and $(')$, respectively. To obtain Eq. (7.131), we need to express the terms involving $\Delta g_{\alpha\beta}$ in terms of ξ; to do so, we need to find the perturbed metric functions $\delta\lambda$ and $\delta\nu$ in terms of ξ. We first write the independent components of the exact Einstein equation and then linearize about a spherical equilibrium. With $\dot r = dr/dt$ the coordinate velocity of a fluid element moving along a path $t \mapsto (t, r(t), \theta, \phi)$, the fluid's 3-velocity measured by an observer at fixed r is $v = e^{\lambda-\nu}\dot r$, and its 4-velocity has components

$$(u^\mu) = \frac{e^{-\nu}}{\sqrt{1-v^2}}(1, \dot r, 0, 0) = \frac{1}{\sqrt{1-v^2}}(e^{-\nu}, e^{-\lambda}v, 0, 0). \tag{7.133}$$

The components of the field equation are then

$$G^t{}_t = 8\pi T^t{}_t: \quad e^{-2\lambda}\left(\frac{1}{r^2}-\frac{2}{r}\lambda'\right) - \frac{1}{r^2} = -8\pi\epsilon\frac{1}{1-v^2}, \tag{7.134a}$$

$$G^r{}_r = 8\pi T^r{}_r: \quad e^{-2\lambda}\left(\frac{1}{r^2}+\frac{2}{r}\nu'\right) - \frac{1}{r^2} = 8\pi\left(\epsilon\frac{v^2}{1-v^2}+p\right), \tag{7.134b}$$

$$G^t{}_r = 8\pi T^t{}_r: \quad -\frac{2}{r}e^{-2\nu}\dot\lambda = 8\pi e^{\lambda-\nu}(\epsilon+p)\frac{v}{1-v^2}. \tag{7.134c}$$

We will also need two equilibrium equations:

$\frac{r}{2}e^{2\lambda}(G^r{}_r - G^t{}_t) = 4\pi r e^{2\lambda}(T^r{}_r - T^t{}_t):$

$$\nu' + \lambda' = 4\pi(\epsilon+p)e^{2\lambda}r, \tag{7.135}$$

$e^{2\lambda}G^\theta{}_\theta = 8\pi e^{2\lambda}T^\theta{}_\theta:$

$$\nu'' + \nu'^2 - \nu'\lambda' + \frac{1}{r}(\nu' - \lambda') = 8\pi e^{2\lambda}p. \tag{7.136}$$

The perturbed 4-velocity δu^α is given by Eq. (7.5a),[6] but we can more quickly compute the perturbed stress-energy tensor by writing

$$\delta v = \delta(e^{\lambda-\nu}\dot r) = e^{\lambda-\nu}\dot\xi. \tag{7.137}$$

From Eq. (7.4), the perturbed pressure is given by

$$\frac{\Delta p}{\Gamma_1 p} = -\frac{1}{2}q^{\alpha\beta}\Delta g_{\alpha\beta} = -\frac{1}{2}(g^{rr}h_{rr} + 2\,{}^3g^{ij}\nabla_i\xi_j) = -\delta\lambda - \frac{e^{-\lambda}}{r^2}[e^\lambda r^2\xi]',$$
$$\delta p = \Delta p - \xi p'. \tag{7.138}$$

[6] For example, because $u^r = 0$ for the unperturbed configuration, we have $\Delta u^r = 0$, $\delta u^r = -(\mathcal{L}_{\boldsymbol\xi} u)^r = \mathcal{L}_{\mathbf u}\xi^r = e^{-\nu}\dot\xi$. Alternatively, write $\delta u^r = \delta(e^{-\nu}\dot r) = e^{-\nu}\dot\xi$.

The perturbed components of the field equation that we use are

$$\delta G^r{}_r = 8\pi\delta T^r{}_r : \quad 2e^{-2\lambda}\left[\frac{1}{r}\delta\nu' - \left(\frac{2}{r}\nu' + \frac{1}{r^2}\right)\delta\lambda\right] = 8\pi\delta p, \tag{7.139a}$$

$$\delta G^t{}_r = 8\pi\delta T^t{}_r : \quad -\frac{2}{r}e^{-2\nu}(\delta\lambda)^{\cdot} = 8\pi e^{2\lambda-2\nu}(\epsilon+p)\dot{\xi}, \tag{7.139b}$$

where terms involving v^2 do not appear because v^2 is quadratic in the perturbation. The second of these has the immediate solution

$$\delta\lambda = -4\pi(\epsilon+p)e^{2\lambda}r\xi = -(\nu'+\lambda')\xi, \tag{7.140}$$

where the second equality follows from Eq. (7.135). The first, (7.139a), together with Eq. (7.138) for δp (and again using [7.135]), then gives $\delta\nu'$:

$$\delta\nu' = \left[\frac{\Delta p}{\epsilon+p} - \left(\nu' + \frac{1}{r}\right)\xi\right](\nu'+\lambda'). \tag{7.141}$$

We can now write the terms involving $\Delta g_{\alpha\beta}$ in the perturbed Euler equation in terms of ξ, beginning with $q^{\alpha\beta}\Delta g_{\alpha\beta}$ (and thus Δp): From Eqs. (7.138) and (7.140), we have

$$q^{\alpha\beta}\Delta g_{\alpha\beta} = \frac{2}{r^2}e^{\nu}(e^{-\nu}r^2\xi)', \qquad \Delta p = -\Gamma_1 p\frac{1}{r^2}e^{\nu}(e^{-\nu}r^2\xi)'. \tag{7.142}$$

The remaining terms involve Δg_{tr} and $(u^\alpha u^\beta\Delta g_{\alpha\beta})'$:

$$\Delta g_{tr} = \mathcal{L}_{\boldsymbol{\xi}} g_{tr} = g_{rr}\partial_t\xi^r = e^{2\lambda}\dot{\xi}, \tag{7.143}$$

$$u^\alpha u^\beta\Delta g_{\alpha\beta} = e^{-2\nu}\Delta g_{tt} = -2(\delta\nu + \nu'\xi).$$

Using Eqs. (7.141) for $\delta\nu'$ and then (7.142) for Δp, we have

$$\begin{aligned}(u^\alpha u^\beta\Delta g_{\alpha\beta})' &= -2\frac{\Delta p}{\epsilon+p}(\nu'+\lambda') - 2\left[\nu'' - \left(\nu'+\frac{1}{r}\right)(\nu'+\lambda')\right]\xi - 2\nu'\xi' \\ &= 2\frac{\Gamma_1 p}{\epsilon+p}(\nu'+\lambda')\frac{e^\nu}{r^2}(e^{-\nu}r^2\xi)' - 2\nu'\xi' \\ &\quad - 4\left(4\pi e^{2\lambda}p - \frac{1}{r}\nu' - \nu'^2\right)\xi,\end{aligned} \tag{7.144}$$

where the equilibrium equation (7.136) was used to simplify the coefficient of ξ in the last expression.

Substituting expressions (7.142), (7.143), and (7.144) in the Euler equation (7.131), we obtain

$$\begin{aligned}(\epsilon+p)e^{2\lambda-2\nu}\ddot{\xi} &= \Gamma_1 p(\nu'+\lambda')\frac{e^\nu}{r^2}(e^{-\nu}r^2\xi)' - 2(\epsilon+p)\left(4\pi e^{2\lambda}p - \frac{1}{r}\nu' - \nu'^2\right)\xi \\ &\quad - (\epsilon+p)\nu'\xi' - p'\left(1+\frac{\Gamma_1 p}{\epsilon+p}\right)\frac{e^\nu}{r^2}(e^{-\nu}r^2\xi)' \\ &\quad + \left[\Gamma_1 p\frac{e^\nu}{r^2}(e^{-\nu}r^2\xi)'\right]'.\end{aligned} \tag{7.145}$$

Finally, combining terms with Γ_1 after replacing p' by $-(\epsilon+p)\nu'$, and combining terms without Γ_1 after replacing ν' by $-p'/(\epsilon+p)$, we recover the form (7.131) of the perturbed Euler equation.

Variational principle and dynamical stability criterion. In writing the linearized Euler equation in the form (7.131), we expressed the perturbed metric in terms of ξ by satisfying the linearized Einstein equation. The symmetry relation (7.40) then implies that $\widehat{\xi}^\alpha\,\delta(\nabla_\beta T_\alpha{}^\beta\sqrt{|g|})$ is symmetric in $\widehat{\xi}$ and ξ up to a divergence. It is for this reason that the right side of (7.131) is self-adjoint (ignoring domains) with respect to the measure $e^{\lambda+\nu}r^2dr$, the radial part of $\sqrt{|g|}drd\theta d\phi$.

The second-order Lagrangian is then

$$L/4\pi = A(\dot{\xi}) - C(\xi), \tag{7.146}$$

where

$$A(\dot{\xi}) = \int_0^R \frac{1}{2}e^{3\lambda-\nu}(\epsilon+p)\dot{\xi}^2r^2dr, \tag{7.147}$$

$$C(\xi) = \int_0^R \frac{1}{2}e^{\lambda+\nu}\left\{\left[\frac{4}{r}p' - \frac{p'^2}{\epsilon+p} + 8\pi p(\epsilon+p)\right]\xi^2 + \frac{e^{2\lambda}}{r^4}\Gamma_1 p\left[(e^{-\nu}r^2\xi)'\right]^2\right\}r^2dr, \tag{7.148}$$

and the factor 4π is present to include in the definition of the Lagrangian an integration over angles. The canonical energy $E(\dot{\xi},\xi)$ is then given by

$$E_c/4\pi = A(\dot{\xi}) + C(\xi). \tag{7.149}$$

The boundary conditions on ξ are fixed by requiring that ξ^α be differentiable at $r=0$ and that the pressure vanish at the perturbed surface: With ξ required to be smooth on [0,R] (assuming a smooth equation of state), the first condition is

$$\xi(r) = O(r), \quad \text{implying} \quad \xi(0) = 0; \tag{7.150}$$

the second

$$\Delta p|_{r=R} = -\Gamma_1 p\frac{e^\nu}{r^2}(e^{-\nu}r^2\xi)'|_{r=R} = 0, \tag{7.151}$$

where R is the radius of the unperturbed star. Because $p(R)=0$ (the unperturbed pressure vanishes at the surface), the second condition ordinarily requires only that ξ be finite at the surface. For uniform density models, the condition $\Delta p = 0$ restricts the value of ξ^r, but this is relevant only for nonradial oscillations: If an incompressible equation of state is used to model the perturbations as well as the equilibrium, there is no radial motion; if the perturbations are governed by a compressible equation of state, the boundary condition is again (7.151), and it again has the meaning that ξ is finite at R.

The perturbed Euler equation has a Sturm-Liouville form, implying a complete set of discrete normal modes. For $\xi = \zeta e^{i\sigma t}$, with ζ real, multiplying Eq. (7.131) by $e^{\lambda+\nu}r^2\zeta$ and integrating over r gives the equation

$$\sigma^2 A(\zeta) = C(\zeta), \tag{7.152}$$

It follows that a spherical perfect-fluid model is dynamically stable to radial adiabatic perturbations if and only $C(\zeta) > 0$ for all ζ that vanish at $r = 0$. Equivalently, the star is dynamically stable if and only if its canonical energy is positive for all nonzero data $\dot{\xi}, \xi$ vanishing at $r = 0$.

We use the criterion in Chapter 8 to discuss the stability of supermassive stars. As we will discuss in Chapter 9, secular instability along a sequence of cold spherical stars sets in where the mass is an extremum. The dynamical and secular stability criteria coincide in the usual models based on 1-parameter equations of state. Because of this, and because the secular timescale is shorter than times over which neutron stars are observed, the dynamical stability criterion is not generally used for spherical neutron stars.

7.5.2 Nonradial perturbations

Perturbations of spherical stars can be decomposed into parts associated with the angular harmonics Y_{lm}, each of which separately satisfies the perturbed Einstein equations. In particular, each mode belongs to an (l, m) representation of the rotation group. Its scalar, vector, and tensor parts involve scalar, vector, and tensor harmonics on the 2-sphere that are constructed from Y_{lm} and that transform under a rotation in the same way as Y_{lm}. Because of the spherical symmetry of the unperturbed star and metric, there is no coupling between perturbations belonging to different choices of (l, m). There is a further decomposition of each (l, m) subspace into perturbations with polar or axial parity. Again the decoupling of perturbations with opposite parity is implied by the symmetry of the unperturbed star and metric – in this case its symmetry under the parity map $\mathcal{P}$ that takes a point with coordinates (t, r, θ, ϕ) to the diametrically opposite point, with coordinates $(t, r, \pi - \theta, \phi + \pi)$.

The study of nonradial oscillations of static spherical stars was pioneered by Thorne and his students [692, 547, 688, 689, 112, 333], and references to the subsequent literature can be found in reviews by Kokkotas and Schmidt [364] and by Andersson and Kokkotas [25]. In our short review, we present the formalism of tensor spherical harmonics; give the equations governing polar and axial perturbations, including a form of the polar equations as a fourth-order system; and derive gauge-invariant quantities that characterize the perturbations. (Definitions of *polar* and *axial* are given after Eq. (7.159).) Quasinormal modes – outgoing modes – of spherical stars will be discussed briefly in the next chapter.

We will need indices for tensors on the 2-sphere S^2 and on the orthogonal t-r subspace. *In this section only:*

lowercase Latin indices a, b, c will be indices on S^2;
and uppercase Latin indices will be indices on the t-r subspace.

Thus concrete indices i, j, k take the values θ, ϕ, whereas I, J, K take the values t, r.

Tensor spherical harmonics

The tensor-harmonic description of gravitational perturbations was first presented by Regge and Wheeler (RW) [561] in their discussion of the stability of Schwarzschild black holes. A number of variants have been used (symmetric trace-free tensors, spin-weighted harmonics), and Thorne's careful review [691] translates among the alternatives.

We denote by $e_{ab}(r)$ the metric on a 2-sphere S^2 of radius r,

$$e_{ij}dx^i dx^j = r^2\left(d\theta^2 + \sin^2\theta d\phi^2\right). \tag{7.153}$$

We similarly denote by ϵ_{ab} the antisymmetric tensor associated with e_{ab} for which $\epsilon_{\theta\phi} = \sqrt{e} = r^2 \sin\theta$:

$$e_{ij} = g_{ij}, \quad e^{ij} = g^{ij}, \quad \epsilon_{ij} = \epsilon_{ij\hat{0}\hat{1}} \equiv \epsilon_{ij\mu\nu} e_{\hat{0}}{}^{\mu} e_{\hat{1}}{}^{\nu}, \tag{7.154}$$

where $\mathbf{e}_{\hat{0}}$ and $\mathbf{e}_{\hat{1}}$ are unit vectors along $\boldsymbol{\partial}_t$ and $\boldsymbol{\partial}_r$.

We could introduce a new symbol $\mathcal{D}_a$ for the covariant derivative operator associated with e_{ab}, but if we use the same symbol F for a scalar on spacetime that depends only on θ and ϕ and for its restriction to an $r =$ constant, $t =$ constant 2-sphere, then

$$\mathcal{D}_i F = \nabla_i F \quad \text{and} \quad \mathcal{D}_i \mathcal{D}_j F = \nabla_i \nabla_j F.$$

Because of this, we will simply use the symbol ∇_a for the covariant derivative operator of e_{ab}. The angular Laplacian is given by

$$e^{ab}\nabla_a\nabla_b F = \frac{1}{r^2}\left(\partial_\theta^2 + \cot\theta\partial_\theta + \frac{1}{\sin^2\theta}\partial_\phi^2\right)F, \tag{7.155}$$

with eigenvalues $-l(l+1)/r^2$. The corresponding eigenfunctions are the spherical harmonics Y_{lm}, which constitute an orthonormal basis for the space of square-integrable functions on S^2.

That is, a scalar F on S^2 can be decomposed into spherical harmonics,

$$F = \sum F_{lm} Y_{lm}(\theta, \phi), \tag{7.156}$$

where each F_{lm} is a constant and (in this section)

$$\sum := \sum_{l=0}^{\infty}\sum_{m=-l}^{l}. \tag{7.157}$$

A scalar f on spacetime then has the corresponding decomposition

$$f = \sum f_{lm}(t, r) Y_{lm}(\theta, \phi). \tag{7.158}$$

The parity of Y_{lm} is $(-1)^l$: $Y_{lm}(\theta,\phi) = (-1)^l Y_{lm}(\pi-\theta,\phi+\pi)$, or

$$Y_{lm}\circ\mathcal{P} = (-1)^l Y_{lm}, \tag{7.159}$$

where $\mathcal{P}(\theta,\phi) = (\pi-\theta,\phi+\pi)$. Quantities with the same parity as Y_{lm} are called *polar*, whereas quantities with opposite parity are called *axial.*[7] Because any scalar belonging to the (l,m) subspace is proportional to Y_{lm}, any scalar field is polar: Its angular harmonics have the same parity as Y_{lm}. Vector and tensor fields, however, can have polar or axial parity.

Any vector field v_a on S^2 can be written as a sum of the gradient of a scalar field and a vector field with vanishing divergence[8]:

$$v_a = \nabla_a f + \epsilon^b{}_a\nabla_b\tilde{f}. \tag{7.160}$$

The angular decompositions of f and $\tilde{f}$ then give a decomposition of v_a as a superposition of the *vector spherical harmonics*

$$\nabla_a Y_{lm} \text{ and } \epsilon^b{}_a\nabla_b Y_{lm}. \tag{7.161}$$

On an $r=$ constant sphere in flat space, these vectors are equivalent to the 3-vectors

$$\boldsymbol{\nabla} Y_{lm} \text{ and } \hat{\boldsymbol{r}}\times\boldsymbol{\nabla} Y_{lm} \tag{7.162}$$

(they are the pullbacks to S^2 of these 3-dimensional vectors). Their θ-ϕ components are

$$(\nabla_i Y_{lm}) = \begin{pmatrix}\partial_\theta Y_{lm}\\ \partial_\phi Y_{lm}\end{pmatrix}, \quad (\epsilon^j{}_i\nabla_j Y_{lm}) = \begin{pmatrix}-\dfrac{1}{\sin\theta}\partial_\phi Y_{lm}\\ \\ \sin\theta\,\partial_\theta Y_{lm}\end{pmatrix}. \tag{7.163}$$

The vector harmonic $\nabla_a Y_{lm}$ is polar: It has the same parity as Y_{lm}, because

$$\mathcal{P}^*\nabla_a Y_{lm} = \nabla_a(Y_{lm}\circ\mathcal{P}) = (-1)^l\nabla_a Y_{lm}. \tag{7.164}$$

The vector harmonic $\epsilon^b{}_a\nabla_b Y_{lm}$ is axial, because the antisymmetric tensor field ϵ_{ab} changes its sign under parity: That is, using $\epsilon_{\theta\phi}(\pi-\theta,\phi+\pi) = r^2\sin(\pi-\theta) = r^2\sin\theta = \epsilon_{\theta\phi}(\theta,\phi)$, we have

$$\mathcal{P}^*\epsilon_{\theta\phi} = \frac{\partial(\pi-\theta)}{\partial\theta}\frac{\partial(\phi+\pi)}{\partial\phi}\epsilon_{\theta\phi} = -\epsilon_{\theta\phi}. \tag{7.165}$$

A symmetric tensor field S_{ab} can be written as a sum of a symmetric trace-free tensor $\tilde{S}_{ab} := S_{ab} - \frac{1}{2}e_{ab}S_c{}^c$ and a tensor proportional to the metric e_{ab},

$$S_{ab} = \tilde{S}_{ab} + \frac{1}{2}e_{ab}S_c{}^c. \tag{7.166}$$

[7] In the Newtonian literature on stellar oscillations, polar perturbations are commonly called *spheroidal* and axial perturbations are called *toroidal.* In the Regge-Wheeler paper [561] and in many of the subsequent papers on perturbations of relativistic stars and black holes, polar perturbations are (confusingly) called *even parity* and axial perturbations are called *odd parity.*

[8] To see this, define f by $\nabla_a\nabla^a f = \nabla_a v^a$. Then $\hat{v}_a = v_a - \nabla_a f$ satisfies $d^*\hat{v} = 0$, implying $^*\hat{v} = -d\tilde{f}$ for some scalar $\tilde{f}$; in index notation, $\hat{v}_a = \epsilon^b{}_a\nabla_b\tilde{f}$.

The second term is a tensor field of the form $e_{ab}f$; because e_{ab} is invariant under rotations, $e_{ab}f$ transforms under rotations as a scalar, and it has the corresponding decomposition

$$e_{ab}f = \sum e_{ab} f_{lm} Y_{lm}. \tag{7.167}$$

Just as a vector field can be written in terms of first derivatives of Y_{lm}, a symmetric tensor field on S^2 can be written in terms of second derivatives of Y_{lm}. Because the covariant derivative operator is constructed from a spherically symmetric metric, $\nabla_a\nabla_b Y_{lm}$ transforms as Y_{lm}. That is, if $\mathcal{R}$ is the diffeo of S^2 associated with a rotation, the differential map $\mathcal{R}^*$ commutes with ∇_a. Again each (l,m) tensor harmonic can be further decomposed into polar and axial parts, proportional to the trace-free part of $\nabla_a\nabla_b Y_{lm}$,

$$\left(\nabla_a\nabla_b - \frac{1}{2}e_{ab}\nabla^2\right)Y_{lm} = \left[\nabla_a\nabla_b + \frac{l(l+1)}{2}e_{ab}\right]Y_{lm}, \tag{7.168}$$

and to

$$\epsilon_{(a}{}^c\nabla_{b)}\nabla_c Y_{lm}, \tag{7.169}$$

respectively. (The axial tensor $\epsilon_{(a}{}^c\nabla_{b)}\nabla_c Y_{lm}$ is already trace-free.) To summarize: Any symmetric tensor field S_{ab} on S^2 can be written in the form

$$S_{ab} = \sum \left\{ \hat{S}^{\text{scalar}}_{lm} Y_{lm} e_{ab} + S^{\text{polar}}_{lm}\left[\nabla_a\nabla_b + \frac{l(l+1)}{2}e_{ab}\right]Y_{lm} \right.$$
$$\left. + S^{\text{axial}}_{lm}\epsilon_{(a}{}^c\nabla_{b)}\nabla_c Y_{lm}\right\}. \tag{7.170}$$

Regge and Wheeler slightly modify this decomposition, grouping together the two terms proportional to e_{ab}:

$$S_{ab} = \sum\left[S^{\text{scalar}}_{lm} Y_{lm} e_{ab} + S^{\text{polar}}_{lm}\nabla_a\nabla_b Y_{lm} + S^{\text{axial}}_{lm}\epsilon_{(a}{}^c\nabla_{b)}\nabla_c Y_{lm}\right], \tag{7.171}$$

with $S^{\text{scalar}}_{lm} := \hat{S}^{\text{scalar}}_{lm} + \frac{1}{2}l(l+1)S^{\text{polar}}_{lm}$.

The θ, ϕ components of the second and third tensors in Eq. (7.171) are

$$[\nabla_i\nabla_j Y_{lm}] = \begin{bmatrix} \partial_\theta^2 Y_{lm} & (\partial_\theta - \cot\theta)\partial_\phi Y_{lm} \\ (\partial_\theta - \cot\theta)\partial_\phi Y_{lm} & (\partial_\phi^2 + \sin\theta\cos\theta\partial_\theta)Y_{lm} \end{bmatrix}, \tag{7.172a}$$

$$[\epsilon_{(i}{}^k\nabla_{j)}\nabla_k Y_{lm}] = \begin{bmatrix} \frac{1}{\sin\theta}(\partial_\theta - \cot\theta)\partial_\phi Y_{lm} & -\frac{1}{2}\sin\theta(\partial_\theta^2 - \cot\theta\partial_\theta - \frac{1}{\sin^2\theta}\partial_\phi^2)Y_{lm} \\ \text{Sym} & -\sin\theta(\partial_\theta - \cot\theta)\partial_\phi Y_{lm} \end{bmatrix}. \tag{7.172b}$$

For $l = 0$, only the scalar contribution S^{scalar}_{ab} is present because $Y_{00} = (4\pi)^{-1/2}$, a constant. For $l = 1$, the two polar tensors $\nabla_a\nabla_b Y_{1m}$ and $e_{ab}Y_{1m}$ are linearly dependent,

$$\nabla_a\nabla_b Y_{1m} = -\frac{1}{r^2}e_{ab}Y_{1m}, \tag{7.173}$$

so there is only a scalar contribution to the polar $l = 1$ part of S_{ab}.

Polar perturbations

We consider first the form of polar perturbations and then obtain the equations governing polar-parity oscillations of a spherical star. The perturbations of the metric, Einstein tensor, and stress-energy tensor are all symmetric two-index tensors and can therefore all be decomposed in the same way on each symmetric S^2. We will write the explicit decomposition for the perturbed metric $h_{\alpha\beta}$. Like much of the subsequent literature on stellar oscillations, we follow here the RW notation for the perturbed metric apart from an overall sign in the definition of their set of perturbed metric functions $\{H_0, H_1, H_2, h_0, h_1, G, K\}$ (defined below). The change of sign arose from a choice of $+---$ signature in the pioneering papers on stellar pulsations by Thorne and collaborators. Subsequent papers use signature $-+++$ and change the sign in the RW notation to make the absolute values $|g_{\mu\nu}+h_{\mu\nu}|$ of the metric components agree with their values for Thorne et al.[9]

Because the unit vector fields $\mathbf{e}_{\hat{0}}$ and $\mathbf{e}_{\hat{1}}$ are invariant under rotations, the components $h_{\hat{0}\hat{0}}$, $h_{\hat{0}\hat{1}}$, and $h_{\hat{1}\hat{1}}$ are rotational scalars:

$$\begin{bmatrix} h_{\hat{0}\hat{0}} & h_{\hat{0}\hat{1}} \\ \text{Sym} & h_{\hat{1}\hat{1}} \end{bmatrix} = -\begin{bmatrix} H_0(t,r)Y_{lm} & e^{-\lambda-\nu}H_1(t,r)Y_{lm} \\ \text{Sym} & H_2(t,r)Y_{lm} \end{bmatrix}. \tag{7.174}$$

Components with one index in the t-r subspace and one angular index comprise two vectors $h_{\hat{0}a}$ and $h_{\hat{1}a}$, the pullbacks to S^2 of $h_{\hat{0}\alpha}$ and $h_{\hat{1}\alpha}$:

$$e^{\nu}h_{\hat{0}a} = h_{ta} = -h_0^{\text{polar}}(t,r)\nabla_a Y_{lm}, \tag{7.175a}$$

$$e^{\lambda}h_{\hat{1}a} = h_{ra} = -h_1^{\text{polar}}(t,r)\nabla_a Y_{lm}. \tag{7.175b}$$

Finally, components with two angular indices comprise a symmetric two-index tensor, the pullback to S^2 of $h_{\alpha\beta}$:

$$h_{ab} = -K(t,r)e_{ab}Y_{lm} - r^2 G(t,r)\nabla_a\nabla_b Y_{lm}. \tag{7.176}$$

A perturbation belonging to the l, m angular harmonic and having polar parity then has a perturbed metric with coordinate components of the form

$$[h_{\mu\nu}] = \\ -\begin{bmatrix} e^{2\nu}H_0Y_{lm} & H_1Y_{lm} & h_0^{\text{polar}}\partial_\theta Y_{lm} & h_0^{\text{polar}}\partial_\phi Y_{lm} \\ \text{Sym} & e^{2\lambda}H_2Y_{lm} & h_1^{\text{polar}}\partial_\theta Y_{lm} & h_1^{\text{polar}}\partial_\phi Y_{lm} \\ \text{Sym} & \text{Sym} & r^2(K+G\partial_\theta^2)Y_{lm} & r^2G(\partial_\theta-\cot\theta)\partial_\phi Y_{lm} \\ \text{Sym} & \text{Sym} & \text{Sym} & r^2[K\sin^2\theta+G(\partial_\phi^2+\sin\theta\cos\theta\partial_\theta)]Y_{lm} \end{bmatrix} \tag{7.177}$$

The symbols h_0 and h_1 have the superscript *polar* here because the RW notation uses the same symbols for both polar and axial perturbations.

[9] Because the equations in the original RW paper are linear and source-free, the change of sign of the perturbed metric functions leaves the form of the equations unaltered.

Choice of gauge. The perturbed metric of a spherically symmetric system has a remarkable feature: It admits a gauge, the Regge-Wheeler gauge, in which each component of $h_{\alpha\beta}$ can be regarded as a local gauge-invariant: That is, each component can be written as an explicitly gauge-invariant linear combination of components of the metric perturbation and their derivatives in a generic gauge. We begin with a discussion of gauge transformations for polar perturbations and introduce the Regge-Wheeler gauge (henceforth, the *RW* gauge). Later, at the end of this section on polar perturbations, we construct the gauge-invariant versions of each of its metric potentials.

A gauge transformation of a polar (or axial) (l, m) perturbation maintains the parity and (l, m) angular behavior if the gauge vector ζ^α has polar (axial) parity and belongs to an (l, m) representation. A gauge transformation by a gauge vector $\tilde{\zeta}$ belonging to a different representation $(\tilde{l}, \tilde{m})$ (or different parity) simply adds a pure gauge perturbation for which the $(\tilde{l}, \tilde{m})$ part of the perturbed field equation is identically zero because the background spacetime satisfies the field equation:

$$\tilde{\delta}(G_{\alpha\beta} - 8\pi T_{\alpha\beta}) = \mathcal{L}_{\tilde{\zeta}}(G_{\alpha\beta} - 8\pi T_{\alpha\beta}) = 0. \tag{7.178}$$

A polar (l, m) gauge vector can be decomposed into two scalars ζ^t and ζ^r in the t-r subspace and a vector ζ^a on S^2:

$$\zeta^t = \hat{\zeta}^t(t, r) Y_{lm}, \quad \zeta^r = \hat{\zeta}^r(t, r) Y_{lm}, \quad \zeta^a = \hat{\zeta}_{\|}(t, r) \nabla^a Y_{lm}, \tag{7.179}$$

where the subscript $\|$ indicates a component parallel to the sphere. The three independent components of ζ^α can be chosen to make three metric functions vanish, and we adopt the RW gauge with

$$h_0^{\text{polar}} = h_1^{\text{polar}} = G = 0. \tag{7.180}$$

Note that the restriction to polar (l, m) perturbations changes the counting of degrees of freedom: The symmetric tensor $h_{\alpha\beta}$ has 7 instead of 10 independent components, of which 3 are eliminated by the 3 independent components of the gauge vector.

There are then four nonzero components of the metric, corresponding to the potentials H_0, H_1, H_2 and K. As we will see, the equation $\delta G^\theta_\theta - \delta G^\phi_\phi = 0$, Eq. (7.196), implies $H_0 = H_2$, and the independent metric variables are commonly taken to be

$$H_0, H_1, \text{ and } K.$$

Perturbed Euler equation. The perturbed fluid has two dynamical degrees of freedom, corresponding to two independent components of the Lagrangian displacement ξ^α, given by[10]

$$\xi^r = \hat{\xi}^r Y_{lm}, \qquad \xi^a = \hat{\xi} \nabla^a Y_{lm}. \tag{7.181}$$

[10] In the papers by Thorne et al. and by Detweiler, Ipser, and Price, whose treatments we are using, the components are written as $\hat{\xi}^r = e^{-\lambda} r^{-2} W$ and $\hat{\xi} = V$ (Detweiler, Ipser, Price) or $\hat{\xi} = -V$ (Thorne). Introducing W somewhat simplifies the perturbed Euler equation, replacing the expression $(\partial_r + 2/r + \lambda')\hat{\xi}^r$ by $e^{-\lambda} r^{-2} W'$.

We will write perturbed fluid variables in the form

$$\delta\epsilon = \hat{\delta}\epsilon Y_{lm}, \qquad \delta p = \hat{\delta}p Y_{lm} \tag{7.182}$$

(for example). They are related to ξ^α by Eqs. (7.5). Writing

$$q^{\alpha\beta}h_{\alpha\beta} = h_{\hat{1}\hat{1}} + h_{\hat{2}\hat{2}} + h_{\hat{3}\hat{3}} = -(H_0 + 2K)Y_{lm}, \tag{7.183}$$

$$q^{\alpha\beta}\nabla_\alpha\xi_\beta = \frac{1}{\sqrt{^3g}}\partial_i(\sqrt{^3g}\xi^i) = e^{-\lambda}\boldsymbol{\nabla}\cdot(e^\lambda\boldsymbol{\xi}), \tag{7.184}$$

with $\boldsymbol{\nabla}$ the flat 3-dimensional derivative operator, we have

$$\frac{\Delta\rho}{\rho} = \frac{\Delta p}{\Gamma_1 p} = \frac{\Delta\epsilon}{\epsilon + p} = -\frac{1}{2}q^{\alpha\beta}\Delta g_{\alpha\beta},$$

$$\frac{\widehat{\Delta}\rho}{\rho}(t,r) = \frac{1}{2}H_0 + K - (\partial_r + \frac{2}{r} + \lambda')\hat{\xi}^r + \frac{l(l+1)}{r^2}\hat{\xi}. \tag{7.185}$$

Then

$$\hat{\delta}p = \Gamma_1 p\left[\frac{1}{2}H_0 + K - \left(\partial_r + \frac{2}{r} + \lambda'\right)\hat{\xi}^r + \frac{l(l+1)}{r^2}\hat{\xi}\right] + (\epsilon + p)\nu'\hat{\xi}^r. \tag{7.186}$$

For a uniformly rotating barotropic star, we can use Eq. (7.18) to write the perturbed Euler equation in the form

$$e^{-2\nu}q_\mu{}^\nu\partial_t\Delta g_{t\nu} = -\partial_\mu\Delta\ln\frac{h}{u^t} = -\partial_\mu\delta\ln\frac{h}{u^t}, \tag{7.187}$$

where the unperturbed equilibrium equation, $\nabla_\alpha(h/u^t) = 0$, allowed us to replace Δ by δ in the second equality. Writing $\delta\ln h = \delta p/(\epsilon + p)$, and using

$$\Delta g_{tr} = h_{tr} + g_{rr}\partial_t\xi^r = [-H_1 + e^{2\lambda}\partial_t\hat{\xi}^r]Y_{lm}, \tag{7.188a}$$

$$\Delta g_{tt} = h_{tt} + \xi^r\partial_r g_{tt} = e^{2\nu}[-H_0 - 2\nu'\hat{\xi}^r]Y_{lm}, \tag{7.188b}$$

$$\Delta g_{ta} = g_{\theta\theta}\partial_t\xi^\theta = \partial_t\hat{\xi}D_a Y_{lm}, \tag{7.188c}$$

$$\delta\ln u^t = \frac{1}{2}u^\alpha u^\beta\Delta g_{\alpha\beta} + \nu'\xi^r = -\frac{1}{2}H_0 Y_{lm}, \tag{7.189}$$

we obtain for the radial and angular parts of the Euler equation

$$e^{-2\nu}(\partial_t^2\hat{\xi}_r - \partial_t H_1) = -\partial_r\left[\frac{\hat{\delta}p}{\epsilon + p} + \frac{1}{2}H_0\right],$$

$$e^{-2\nu}\partial_t^2\hat{\xi} = -\left[\frac{\hat{\delta}p}{\epsilon + p} + \frac{1}{2}H_0\right]. \tag{7.190}$$

Using the $\widehat{E}_{r\theta}$ equation given below to eliminate H_1 gives the radial component of the perturbed Euler equation in the form

$$e^{2\lambda-2\nu}\partial_t^2\hat{\xi}^r = -K' + \frac{1}{2}H_0' + \nu' H_0 - \partial_r\left[\frac{\hat{\delta}p}{\epsilon+p}\right]. \tag{7.191}$$

The equations are used in the form given above for the organization of the quasinormal-mode problem described below, due to Ipser and Price. For the more standard treatment, with the Euler equation used as a dynamical equation for ξ^α, $\hat{\delta}p$ is replaced by its expression (7.186) in terms of $\hat{\xi}^r$ and $\hat{\xi}$.

The dynamical system. The Einstein-Euler system (with $H_0 = H_2$) involves the five quantities ξ^r, ξ, H_0, H_1, and K. It can be reduced to two second-order differential equations that (together with boundary conditions) determine the modes of oscillation of a spherical star. The fact that polar oscillations can be described by a fourth-order system was pointed out by Ipser and Thorne [333], and Lindblom and Detweiler [421] first obtained an explicit fourth-order system. That system of equations was sometimes singular inside the star, and they obtained an improved, singularity-free, version in [184].

Using a gauge in which the metric is diagonal, Chandrasekhar and Ferrari [135] found a fifth-order system involving only metric perturbations; Ipser and Price [332] then obtained a system of two second-order equations for metric perturbations in a different (RW) gauge. Gundlach and Martín-García [281] use the Ipser-Price equations, generalized to a time-dependent background, as two of their dynamical equations in their gauge-invariant treatment of perturbations of spherical stellar collapse. We will use the Ipser-Price system for its relative simplicity and for the ease with which it can be obtained from the components of the perturbed field equation. The equations can be used to find the quasinormal modes of a rotating star, and they, together with the perturbed Euler equation, serve as dynamical equations for the initial value problem.

Subsequent work by Lindblom, Mendell, and Ipser [422] gives an alternative system of two second-order equations in the Ipser-Lindblom-Managan two-potential formalism that we review in the more general context of rotating stars in Section B.2. This latter system has much lengthier pulsation equations but a more direct relation to the Newtonian approximation (in the two-potential language).

Another more detailed review of Schwarzschild perturbation theory is given by Martel and Poisson [460]. In presenting the Ipser-Price version of the pulsation equations, we specialize to the case of a barotropic star whose perturbations satisfy the same one-parameter equation of state as the equilibrium star: $\Delta p = \frac{p'}{\epsilon'}\Delta\epsilon$. (Ipser and Price consider more general perturbations.) We first present the equations and the way one reconstructs the full solution from the two metric potentials, H_0 and K, that appear in the pulsation equations. We then give their derivation, after presenting the components of the perturbed field equation. In this reduction of the system to two second-order equations, the perturbed Euler equation is needed only to write the boundary condition at the surface of the star and to find the perturbed fluid velocity from the metric potentials H_0 and K.

The quasinormal modes (outgoing modes) can be found from the following two equations valid inside the star for angular harmonics $l \geq 2$:

$$\left[-e^{2\lambda-2\nu}\partial_t^2 + \partial_r^2 - \left(\frac{2}{r} - 5\nu' + \lambda'\right)\partial_r - (l-1)(l+2)\frac{e^{2\lambda}}{r^2}\right](K - H_0)$$
$$+ \left[\frac{6}{r}(\nu' - \lambda') - 4\nu'^2 + 8\pi(\epsilon - 3p)e^{2\lambda}\right] H_0 = 0, \quad (7.192)$$

$$\left\{-e^{2\lambda-2\nu}\partial_t^2 + v_s^2\partial_r^2 - \left[\frac{1}{r} - \nu' - v_s^2\left(\frac{3}{r} - \lambda'\right)\right]\partial_r\right.$$
$$\left. - \frac{1}{2}(l-1)(l+2)\frac{e^{2\lambda}}{r^2}\left(1 + v_s^2\right)\right\} K$$
$$+ \left\{\frac{1}{r}\left(1 - v_s^2\right)\partial_r - \frac{1}{r^2} + \frac{2}{r}\nu' - \left(\frac{1}{r^2} - \frac{2}{r}\lambda'\right)v_s^2\right.$$
$$\left. + \frac{1}{2}l(l+1)\frac{e^{2\lambda}}{r^2}\left(1 - v_s^2\right)\right\} H_0 = 0.$$
$$(7.193)$$

Here c_s is the sound speed:

$$c_s^2 := \frac{dp}{d\epsilon} = \frac{\Gamma_1 p}{\epsilon + p}. \quad (7.194)$$

Note that in each equation, the second derivatives act on only one of the two variables, and they give a hyperbolic equation with different characteristics: In the first equation, they are along the light cone. The second equation, however, has characteristics along the sound cone, despite the fact that K is a metric variable.

For harmonic time dependence, $H_0(t,r) = H_0(r)e^{i\sigma t}$, $K(t,r) = K(r)e^{i\sigma t}$, this is a system of two ODEs. Near $r = 0$, H_0 and K both behave as kr^l with the same coefficient k. Two additional pieces are required to complete the eigenvalue problem. First, the perturbed surface of the star is determined by the condition that the Lagrangian change Δp in the pressure vanish,

$$\Delta p|_S = 0, \quad (7.195)$$

and this condition must be written in terms of H_0 and K to complete the system (7.192), (7.193). How this is done is described after Eq. (7.205). Second, the gravitational potentials must be matched to an exterior outgoing wave solution to the perturbed vacuum field equation.

Derivation and components of field equation. The components of the perturbed field equation used to obtain Eqs. (7.192) and (7.193) are

$$0 = \delta G_\theta{}^\theta - \delta G_\phi{}^\phi = (H_2 - H_0)\frac{1}{2r^2}(\partial_\theta^2 - \cot\theta\,\partial_\theta - \frac{1}{\sin^2\theta}\partial_\phi^2)Y_{lm}, \quad (7.196)$$

or

$$H_0 = H_2; \tag{7.197}$$

$$\begin{aligned}
0 &= \widehat{E}_t{}^t Y_{lm} := \delta G_t{}^t - 8\pi\delta T_t{}^t, \\
-e^{2\lambda}\widehat{E}_t{}^t &= K'' + \left(\frac{3}{r} - \lambda'\right)K' - \frac{e^{2\lambda}}{2r^2}(l-1)(l+2)K \\
&\quad - \frac{1}{r}H_0' - \left[\frac{1}{r^2} - \frac{2}{r}\lambda' + \frac{e^{2\lambda}}{2r^2}l(l+1)\right]H_0 - 8\pi e^{2\lambda}\hat{\delta}\epsilon(t,r); \qquad (7.198)
\end{aligned}$$

$$\begin{aligned}
0 &= \widehat{E}_r{}^r Y_{lm} := \delta G_r{}^r - 8\pi\delta T_r{}^r, \\
\widehat{E}_r{}^r &= e^{-2\nu}\partial_t^2 K - 2\frac{e^{-2\nu-2\lambda}}{r}\partial_t H_1 - e^{-2\lambda}\left(\frac{1}{r} + \nu'\right)K' \\
&\quad + \frac{1}{2r^2}(l-1)(l+2)(K - H_0) \\
&\quad + \left[e^{-2\lambda}\left(\frac{1}{r}\partial_r + \frac{1}{r^2} + 2\frac{\nu'}{r}\right) - \frac{1}{r^2}\right]H_0 - 8\pi\hat{\delta}p, \qquad (7.199)
\end{aligned}$$

in which Eq. (7.197) has been used to replace H_2 by H_0;

$$\begin{aligned}
0 &= \widehat{E}_\theta{}^\theta Y_{lm} := \delta G_\theta{}^\theta - 8\pi\delta T_\theta{}^\theta, \\
-2e^{2\lambda}\widehat{E}_\theta{}^\theta &= \left[-e^{2\lambda-2\nu}\partial_t^2 + \partial_r^2 + \left(\frac{2}{r} + \nu' - \lambda'\right)\partial_r\right]K + 2e^{-2\nu}(\partial_r + \frac{1}{r} - \lambda')\partial_t H_1 \\
&\quad - \left[e^{2\lambda-2\nu}\partial_t^2 + \partial_r^2 + \left(\frac{2}{r} + 3\nu' - \lambda'\right)\partial_r + 16\pi e^{2\lambda}p\right]H_0 \\
&\quad + 16\pi e^{2\lambda}\hat{\delta}p; \qquad (7.200)
\end{aligned}$$

and

$$\begin{aligned}
0 &= \widehat{E}_{r\theta}\partial_\theta Y_{lm} := \delta G_{r\theta}, \\
2\widehat{E}_{r\theta} &= e^{-2\nu}\partial_t H_1 - H_0' - \nu'(H_0 + H_2) + \frac{1}{r}(H_0 - H_2) + K'. \qquad (7.201)
\end{aligned}$$

Or, using $H_0 = H_2$,

$$e^{-2\nu}\partial_t H_1 = H_0' + 2\nu' H_0 - K'. \tag{7.202}$$

After H_2 is replaced everywhere by H_0, the first of the two pulsation equations that comprise the fourth-order dynamical system, Eq. (7.192), is

$$-2e^{2\lambda}\left(\widehat{E}_r{}^r - \widehat{E}_\theta{}^\theta\right) + 4\left(\partial_r - \frac{1}{r} + 2\nu' - \lambda'\right)\widehat{E}_{r\theta} = 0. \tag{7.203}$$

Adding the term involving $\widehat{E}_{r\theta}$ is equivalent to using the $\widehat{E}_{r\theta}$ equation in the form (7.202) to eliminate H_1 in favor of H_0 and K in the equation $-2e^{2\lambda}\left(\widehat{E}_r{}^r - \widehat{E}_\theta{}^\theta\right) = 0$. To obtain Eq. (7.192) in the form given, one also needs to use the equilibrium equations (7.135) and (7.136) to replace terms involving ν'' and $\nu'+\lambda'$ in the coefficient of H_0.

Finally, the second pulsation equation, Eq. (7.193), is

$$-e^{2\lambda}\left(\widehat{E}_r{}^r + v_s^2\widehat{E}_t{}^t\right) - \frac{4}{r}\widehat{E}_{r\theta} = 0. \tag{7.204}$$

This equation can again be described as the result of using the $\widehat{E}_{r\theta}$ equation to eliminate H_1 from $-e^{2\lambda}\left(\widehat{E}_r{}^r + v_s^2\widehat{E}_t{}^t\right)$.

Boundary conditions and the eigenvalue problem. For perturbations of a spherical star of radius R, the boundary condition (7.195) on the fluid at the star's surface is

$$0 = \Delta p|_{r=R} = \delta p(t,R) + p'\xi^r(t,R). \tag{7.205}$$

To use it in conjunction with the fourth-order system for H_0 and K, one needs to write δp and ξ^r in terms of these variables. First use the $\widehat{E}_r{}^r$ equation, (7.199), to replace δp in this boundary condition and $\delta p/(\epsilon+p)$ in the radial component (7.191) of the Euler equation by expressions in H_0 and K. Then (7.191) gives ξ^r in terms of H_0 and K.

For harmonic time dependence, this is an efficient way to organize the equations. In more general contexts and for direct comparison to the literature and to the Newtonian equations written in terms of $\boldsymbol{\xi}$, it is helpful to write the radial and angular components of the perturbed Euler equation with δp replaced by its expression (7.186) in terms of $\hat{\xi}^r$ and $\hat{\xi}$.

There is an additional boundary condition, the requirement that the gravitational waves be outgoing. Outside the star, the metric perturbation (for $l \geq 2$) is determined by a single second-order differential equation, commonly chosen to be the Zerilli equation, Eq. (7.218). The outgoing-wave condition is then the condition that the Zerilli function Z, a linear combination of H_0, K and K', be asymptotically outgoing.

The eigenvalue problem can now be stated as follows: For a given complex value of σ, there are two linearly independent solutions to the pulsation equations, (7.192) and (7.193), that are regular at $r=0$. One linear combination of these (up to an overall scaling) is picked out by the boundary condition $\Delta p|_S = 0$. With data for Z determined by H_0, K, K' and their derivatives at the surface of the star, this solution then yields a sum of ingoing and outgoing waves at infinity. An outgoing mode has eigenvalue σ for which the solution is purely outgoing.

Gauge-invariant quantities and vacuum perturbations. The metric and fluid perturbations of spherically symmetric spacetimes can be written in terms of gauge-invariant quantities. In describing physical perturbations with polar parity, one considers only gauge transformations that have polar parity in order not to add to the perturbed metric a pure-gauge piece with parity opposite to that of the physical perturbation. Gauge invariance then means invariance under gauge transformations with the same parity as that of the physical perturbation. For polar perturbations, they have the form (7.179).

Beginning with work by Moncrief [482] and by Gerlach and Sengupta [255], a number of authors have discussed perturbations of spherically symmetric spacetimes in a gauge-invariant framework. In the RW gauge we have been using, each metric potential can be identified with an explicitly gauge-invariant combination of metric components (in a generic gauge). We adopt the notation introduced by Gerlach and Sengupta, because it is tailored to this identification. The same notation is used in subsequent work by, for example, Gundlach and Martín-García [281, 462] and in a review by Nagar and Rezzolla [489]. The Nagar-Rezzolla review has a comprehensive set of intervening references. We present gauge-invariant quantities with polar parity in this section and give their relation to the metric potentials in an RW gauge.

Perturbations of the vacuum Schwarzschild geometry with polar or axial parity can each be described by a single gauge-invariant quantity that satisfies a wave equation. For polar parity perturbations, the most commonly used variant is the Zerilli equation, and we conclude the section with a presentation of it. An alternate approach due to Press and Bardeen [543] and generalized to perturbations of rotating black holes by Teukolsky [681] involves an equation for a component of the Weyl tensor along a tetrad based on the principal null directions of the vacuum spacetime. The Weyl-tensor approach unifies axial and polar vacuum perturbations, but it has less interest for a treatment of stellar perturbations, and the formalism needed to derive it is outside the scope of this chapter. Finally, Ipser and Price [332] obtained a wave equation involving K alone.

The gauge-invariants identified with H_0, H_1, K and G in an RW gauge all involve adding parts with the same combination of potentials that vanish in the RW gauge, namely,

$$\mathrm{p}_I = h_I^{\text{polar}} - \frac{1}{2} r^2 \nabla_I G, \tag{7.206}$$

where $h_r = h_0$, $h_t = h_1$. The invariants are then

$$k_{AB} := -\hat{h}_{AB} + \nabla_A p_B + \nabla_B p_A, \tag{7.207a}$$

$$k := K - \frac{2}{r}\mathrm{p}^r = K^{\text{RW}}, \tag{7.207b}$$

where

$$(\hat{h}_{IJ}) = -\begin{pmatrix} e^{2\nu} H_0 & H_1 \\ H_1 & e^{2\lambda} H_2 \end{pmatrix}. \tag{7.208}$$

The RW potentials then have the explicitly gauge-invariant expressions

$$H_0^{\text{RW}} = H_0 + 2e^{-2\nu}\nabla_t p_t, \tag{7.209a}$$

$$H_1^{\text{RW}} = H_1 + \nabla_t p_r + \nabla_r p_t, \tag{7.209b}$$

$$H_2^{\text{RW}} = H_2 + 2e^{-2\lambda}\nabla_r p_r. \tag{7.209c}$$

Derivation. For a polar gauge vector (7.179), we first evaluate $\mathcal{L}_{\boldsymbol{\zeta}} g_{\mu\nu}$, with each index either along or orthogonal to the 2-sphere to find the change under a gauge transformation of the potentials H_0, H_1, h_0, h_1, K, and G. The change $\widetilde{h}_{IJ}$ in $\hat{h}_{IJ}$ follows from the form $\mathcal{L}_{\boldsymbol{\zeta}} g_{IJ} = 2\nabla_{(I}\zeta_{J)}$, implying

$$\begin{aligned} \mathcal{L}_{\boldsymbol{\zeta}} g_{IJ} &= 2\nabla_{(I}\hat{\zeta}_{J)} Y_{lm}, \\ \widetilde{h}_{IJ} &= 2\nabla_{(I}\hat{\zeta}_{J)}. \end{aligned} \tag{7.210}$$

Similarly, from

$$\mathcal{L}_{\boldsymbol{\zeta}} g_{Ij} = \nabla_I \hat{\zeta}_j + \nabla_j \hat{\zeta}_I = [\hat{\zeta}_I + r^2 \nabla_I (r^{-2} \hat{\zeta}_\parallel)] \nabla_j Y_{lm}, \tag{7.211}$$

we have

$$\widetilde{h}_I = -\hat{\zeta}_I - r^2 \nabla_I (r^{-2} \hat{\zeta}_\parallel). \tag{7.212}$$

For the change $\widetilde{h}_{ij}$ in $\hat{h}_{ij}$, it is quicker to use the form $\mathcal{L}_{\boldsymbol{\zeta}} g_{\mu\nu} = \zeta^\lambda \partial_\lambda g_{\mu\nu} + g_{\lambda\nu}\partial_\mu \zeta^\lambda + g_{\mu\lambda}\partial_\nu \zeta^\lambda$. Then

$$\mathcal{L}_{\boldsymbol{\zeta}} g_{ij} = \frac{2}{r}\zeta^r g_{ij} + 2\hat{\zeta}_\parallel \nabla_i \nabla_j Y_{lm}, \tag{7.213}$$

implying

$$\widetilde{K} = -\frac{2}{r}\hat{\zeta}^r, \quad \widetilde{G} = -\frac{2}{r^2}\hat{\zeta}_\parallel. \tag{7.214}$$

From Eqs. (7.212) and (7.214), it follows that the change in p_I, as defined in Eq. (7.206), is simply

$$\begin{aligned} \widetilde{p}_I &= \hat{\zeta}_I - r^2 \nabla_I (r^{-2} \hat{\zeta}_\parallel) - \frac{1}{2} r^2 \partial_I \left(\frac{2}{r^2} \hat{\zeta}_\parallel \right) \\ &= -\hat{\zeta}_I. \end{aligned} \tag{7.215}$$

From this equation and the changes in $\widetilde{h}_{IJ}$, G, and K given by Eqs. (7.210) and (7.214), gauge invariance of k_{AB} and k is immediate:

$$\begin{aligned} \widetilde{k}_{IJ} &= \widetilde{h}_{IJ} + 2\nabla_{(I}\widetilde{p}_{J)} = -2\nabla_{(I}\hat{\zeta}_{J)} + 2\nabla_{(I}\hat{\zeta}_{J)} = 0, \\ \widetilde{k} &= \widetilde{K} - \frac{2}{r}\widetilde{p}^r = 0. \end{aligned} \tag{7.216}$$

The Zerilli and Ipser-Price equations. The Zerilli equation has the form of the scalar wave equation on a Schwarzschild geometry with a potential V that falls off asymptotically as r^{-3}. The scalar wave equation, $\nabla_\alpha \nabla^\alpha \psi = 0$, has on each l subspace the form

$$\left[(1 - 2M/r)^{-1} (-\partial_t^2 + \partial_{r^*}^2) - l(l+1)/r^2 - \frac{2M}{r^3} \right] \Psi = 0,$$

where $\Psi = r\psi$ and, as in Eq. (7.72), r^* is the radial coordinate for which $u = t - r^* =$ constant is an outgoing null surface:

$$r^* := r + 2M \ln(r/2M - 1). \tag{7.217}$$

The Zerilli equation then has the form

$$[(1 - 2M/r)^{-1}(-\partial_t^2 + \partial_{r^*}^2) - l(l+1)/r^2 - V]Z = 0. \tag{7.218}$$

With $\mathfrak{l} := (l-1)(l+2)/2$, the potential is given by

$$V = -\frac{6M}{r^3}\,\frac{r^2\,\mathfrak{l}(\mathfrak{l}+2) + 3M(r-M)}{(\mathfrak{l}r + 3M)^2}. \tag{7.219}$$

The Zerilli function Z is the linear combination[11]

$$Z = rK + \frac{r - 2M}{\mathfrak{l} + 3M/r}\,(H_0 + rK'). \tag{7.220}$$

Outgoing solutions with harmonic time dependence have the asymptotic behavior of solutions to the scalar wave equation,

$$Z \sim e^{i\sigma u}/r. \tag{7.221}$$

A derivation by Martin-García and Gundlach [462] uses the two pulsation equations and the $\widehat{E}_t{}^t$ equation, each in its vacuum form, obtained by setting v_s, ϵ, and p to zero in Eqs. (7.192), (7.193), and (7.198). In a slightly streamlined version, one seeks a variable $Z = rK + F(r)H_0 + G(r)K'$ that satisfies an equation of the form (7.218) with F, G, and V to be determined. The steps are:

1. Substitute this form of Z in Eq. (7.218), with V kept as the symbol V. The result includes second derivatives of H_0 and both second and third derivatives of K.
2. Find the linear combination of Eqs. (7.192), (7.193), and ∂_r (7.193) that matches the terms with time derivatives in Eq. (7.218). The coefficients will be proportional to r, F, and G.
3. Add to this a linear combination of Eqs. (7.198) and ∂_r(7.198) to match the remaining second radial derivatives.
4. The terms that do not explicitly match the form (7.218) are now the terms involving K, K', H_0, and H_0'. Find F, G, and V for which these remaining terms have the form VZ.

The initial value equations, $\delta G_{tr} = 0$, $\delta G_{ta} = 0$, and (7.198), turn out to give H_0 and K in terms of Z and Z'. The expressions for H_0 and K are

$$K = \left[\frac{1}{r} - \frac{3}{l(l+1)}\frac{(e^{4\nu})'}{\mathfrak{l} + 3M/r}\right] Z + \frac{2e^{\nu}}{l(l+1)} Z', \tag{7.222a}$$

$$H_0 = K - \left(\frac{2}{r} - \frac{3(e^{2\nu})'/2}{\mathfrak{l} + 3M/r}\right) Z + e^{\nu} r \left[e^{-\nu}(K - Z/r)\right]'. \tag{7.222b}$$

[11] In Zerilli's paper, Z is written as a linear combination of K and H_0, with powers of σ in their coefficients, a form useful only for harmonic time dependence.

Ipser and Price found a single second-order wavelike equation for the metric potential K. Although the equation is somewhat more complicated than the Zerilli equation, the derivation is significantly easier; the equation may also be a more natural choice if one uses the Ipser-Price fourth-order system for computing quasi-normal modes. It is not hard to see how to get a single second-order equation for K outside the star. The second pulsation equation and the $\widehat{E}_t{}^t$ equation each contain H_0 and H_0', but not H_0''. If we use the $\widehat{E}_{r\theta}$ equation as before to eliminate H_1 in the $\widehat{E}_r{}^r$ equation, we obtain a third independent equation that contains H_0 and H_0' but not H_0''. There is therefore a linear combination of this last equation and the $\widehat{E}_t{}^t$ equation that has the same coefficients of H_0' and H_0 as those of the second pulsation equation. That allows one to eliminate H_0 and H_0' from the pulsation equation, turning it into a second-order equation for K.

Axial perturbations

Because any scalar has polar parity, axial-parity linear perturbations of spherical stars cannot involve changes in pressure or density:

$$\delta p = \delta\epsilon = \delta\rho = 0. \tag{7.223}$$

In an axial perturbation of a perfect fluid model, only the star's velocity field and metric can change. We will find that the only solutions to the perturbed Euler equation are stationary perturbations of the covariant fluid velocity. There are, however, additional axial-parity metric perturbations that describe gravitational waves decoupled from the fluid.

The fact that axial fluid perturbations are stationary is immediate in a Newtonian context. A fluid perturbation is described by the variables δp, $\delta\rho$, $\delta\Phi$, and $\delta\mathbf{v}$. Because the perturbed scalars vanish for an axial perturbation, only $\delta\mathbf{v}$ is nonzero. With $\mathbf{v} = 0$ for the unperturbed spherical star, the perturbed Euler equation is simply

$$0 = \delta\left[(\partial_t + \mathbf{v}\cdot\nabla)\mathbf{v} + \frac{1}{\rho}\nabla p + \nabla\Phi\right] = \partial_t\delta\mathbf{v}, \tag{7.224}$$

implying that any linear axial perturbation is a stationary flow.

The analogous result for the perturbed relativistic Euler equation follows from the form (7.18) and from the expression for δu_α. Because h and u^t are rotational scalars, the right side of Eq. (7.18) vanishes for axial perturbations. Similarly, because the t and r components of the left side of the equation are rotational scalars, they vanish identically, and only the angular part of the perturbed Euler equation survives. It has the form

$$0 = (u^t)^2\partial_t\Delta g_{ta} = u^t\partial_t\delta u_i, \tag{7.225}$$

where we have used Eq. (7.5a) to write

$$\delta u_i = \delta(g_{\mu i}u^\mu) = u^t(h_{ti} + \partial_t\xi_i) = u^t\Delta g_{ti}. \tag{7.226}$$

Thus the perturbed fluid flow is stationary,

$$\partial_t \delta u_\alpha = 0, \tag{7.227}$$

and it is tangent to the $r =$ constant surfaces. In particular, because any axial vector in the l, m subspace is proportional to $\epsilon^b{}_a \nabla_b Y_{lm}$, the perturbed 4-velocity can be written in the form

$$\delta u_a = e^{-\nu} U \epsilon^b{}_a \nabla_b Y_{lm}, \tag{7.228}$$

with the notation chosen to agree with Thorne and Campolattaro. Then the stationarity condition (7.227) is

$$\partial_t U = 0. \tag{7.229}$$

We turn next to the field equation for the perturbed axial-parity metric. This will lead to a gauge-invariant wave equation describing gravitational waves with axial parity; we will later use the system of equations to describe axial-parity perturbations of slowly rotating stars.

The quantities $h_{\alpha\beta}$, $\delta G_{\alpha\beta}$, and $\delta T_{\alpha\beta}$ are all symmetric tensors, and their projections along and orthogonal to symmetric 2-spheres have the same angular behavior. We choose $h_{\alpha\beta}$ to exhibit it. The components h_{tt}, h_{tr}, and h_{rr} vanish because they are rotational scalars. From Eqs. (7.161) and (7.171), the two rotational vectors, $h_{\hat{0}a}$ and $h_{\hat{1}a}$, are proportional to $\epsilon^b{}_a \nabla_b Y_{lm}$, and the rotational tensor h_{ab} is proportional to $\epsilon^c{}_{(a} \nabla_{b)} \nabla_c Y_{lm}$. The metric perturbation then has the following form, characteristic of any symmetric tensor belonging to an axial-parity (l, m) harmonic:

$$[h_{\mu\nu}] = \begin{bmatrix} 0 & 0 & h_0 \frac{1}{\sin\theta}\partial_\phi Y_{lm} & -h_0 \sin\theta \partial_\theta Y_{lm} \\ 0 & 0 & h_1 \frac{1}{\sin\theta}\partial_\phi Y_{lm} & -h_1 \sin\theta \partial_\theta Y_{lm} \\ \text{Sym} & \text{Sym} & -h_2 \frac{1}{\sin\theta}(\partial_\theta - \cot\theta)\partial_\phi Y_{lm} & h_2 \frac{1}{2}\sin\theta(\partial_\theta^2 - \cot\theta\partial_\theta - \frac{1}{\sin^2\theta}\partial_\phi^2) Y_{lm} \\ \text{Sym} & \text{Sym} & \text{Sym} & h_2 \sin\theta(\partial_\theta - \cot\theta)\partial_\phi Y_{lm} \end{bmatrix} \tag{7.230}$$

We again adopt the RW gauge for axial perturbations, in which the single free function $\zeta_\|(t, r)$ associated with an axial gauge vector,

$$(\xi^\mu) = (0, 0, \zeta_\| \epsilon^{ji} \nabla_j Y_{lm}), \tag{7.231}$$

is used to set $h_2 = 0$, leaving $h_{ta} = -h_0 \epsilon^b{}_a \nabla_b Y_{lm}$ and $h_{ra} = -h_1 \epsilon^b{}_a \nabla_b Y_{lm}$ as the only nonzero parts of the metric perturbation:

$$[h_{\mu\nu}] = \begin{bmatrix} 0 & 0 & h_0 \frac{1}{\sin\theta}\partial_\phi Y_{lm} & -h_0 \sin\theta \partial_\theta Y_{lm} \\ 0 & 0 & h_1 \frac{1}{\sin\theta}\partial_\phi Y_{lm} & -h_1 \sin\theta \partial_\theta Y_{lm} \\ \text{Sym} & \text{Sym} & 0 & 0 \\ \text{Sym} & \text{Sym} & 0 & 0 \end{bmatrix} \tag{7.232}$$

Because the rotational scalars $\delta T_{tt}, \delta T_{tr}$, and δT_{rr} vanish and the 3-velocity of the unperturbed star vanishes, the only nonzero parts of the perturbed stress-energy tensor are

$$\begin{aligned}\delta T_{ta} &= (\epsilon + p)u_t \delta u_a + p h_{ta}\\ &= -[(\epsilon + p)U + p h_0]\epsilon^b{}_a \nabla_b Y_{lm}\end{aligned} \tag{7.233}$$

and

$$\delta T_{ra} = p h_{ra} = -p h_1 \epsilon^b{}_a \nabla_b Y_{lm}. \tag{7.234}$$

As we have seen, an axial parity symmetric tensor has three independent components associated with the ta, ra, and ab projections. For time-dependent perturbations, the perturbed Bianchi identity, $\nabla_\beta \delta G^{a\beta} = 0$, relates δG^{ta} to δG^{ra} and δG^{ab}, and there are then only two independent field-equation components, namely,

$$\begin{aligned}0 &= \widehat{E}_{r\|}\epsilon^b{}_a \nabla_b Y_{lm} := \delta G_{ra} - 8\pi \delta T_{ra},\\ 2\widehat{E}_{r\|} &= \left[-e^{-2\nu}\partial_t^2 - \frac{(l-1)(l+2)}{r^2} + 16\pi p\right] h_1 + e^{-2\nu}\left(\partial_r - \frac{2}{r}\right)\partial_t h_0;\end{aligned} \tag{7.235}$$

and

$$\begin{aligned}0 &= \widehat{E}_\| \epsilon^c{}_{(a} \nabla_{b)} \nabla_c Y_{lm} := \delta G_{ab},\\ \widehat{E}_\| &= e^{-2\nu}\partial_t h_0 - e^{-2\lambda}(\partial_r + \nu' - \lambda')h_1.\end{aligned} \tag{7.236}$$

Using the last of these equations to write $\partial_t h_0$ in terms of h_1 in the second equation (7.235) immediately gives the axial-parity wave equation

$$-e^{-2\nu}\partial_t^2 h_1 + r^2 e^{-2\nu}\partial_r\left[\frac{1}{r^2}e^{\nu-\lambda}\partial_r(e^{\nu-\lambda}h_1)\right] - \left[\frac{(l-1)(l+2)}{r^2} - 16\pi p\right] h_1 = 0. \tag{7.237}$$

The equation describes gravitational waves that are not coupled to the perturbed matter and have characteristics along the light cone. Its vacuum form, known as the Regge-Wheeler equation, is that of the scalar wave equation (7.221) with a simple potential term, when written in terms of $Q := \dfrac{e^{2\nu}}{r} h_1$:

$$\left[(1 - 2M/r)^{-1}(-\partial_t^2 + \partial_{r^*}^2) - \frac{l(l+1)}{r^2} + \frac{6M}{r^3}\right] Q = 0. \tag{7.238}$$

For time-independent perturbations, the perturbed metric is related to the stationary velocity field by the remaining field equation,

$$\begin{aligned}0 &= \widehat{E}_{t\|}\epsilon^b{}_a \nabla_b Y_{lm} := \delta G_{ta} - 8\pi \delta T_{ta},\\ 2e^{2\lambda}\widehat{E}_{t\|} &= \left[\partial_r^2 - (\lambda' + \nu')\partial_r + \frac{4}{r}\nu' - \frac{l(l+1)}{r^2}e^{2\lambda} + 8\pi e^{2\lambda}(\epsilon + p)\right] h_0\\ &\quad - \left[\partial_r + \frac{2}{r} - (\lambda' + \nu')\right]\partial_t h_1 + 16\pi e^{2\lambda}(\epsilon + p)U.\end{aligned} \tag{7.239}$$

Because U is time independent, the time derivative of this equation is independent of U, and, as noted above, is already implied by the δG_{ra} and δG_{ab} equations.

Gauge-invariant quantities. Gauge-invariant combinations of the metric components are somewhat simpler to write and to derive for axial perturbations. In a general gauge, the perturbed metric has the form (7.230), with nonzero h_2. The gauge-invariant quantity that reduces to h_I in the RW gauge (for which h_2 vanishes) is

$$k_I := h_I + r^2 \partial_I \left(\frac{h_2}{r^2}\right). \tag{7.240}$$

Derivation: An axial parity gauge vector has the form

$$\zeta^a = \zeta_{\|} \epsilon^b{}_a \nabla_b Y_{lm}. \tag{7.241}$$

The changes in the perturbed metric are then

$$\pounds_\zeta g_{ta} = \partial_t \zeta_{\|} \epsilon^b{}_a \nabla_b Y_{lm}, \tag{7.242a}$$

$$\pounds_\zeta g_{ra} = \left(\partial_r - \frac{2}{r}\right) \zeta_{\|} \epsilon^b{}_a \nabla_b Y_{lm}, \tag{7.242b}$$

$$\pounds_\zeta g_{ab} = \zeta_{\|} 2\epsilon^c{}_{(a} \nabla_{b)} \nabla_c Y_{lm}, \tag{7.242c}$$

implying

$$\tilde{h}_0 = -\partial_t \zeta_{\|}, \qquad \tilde{h}_1 = -r^2 \partial_r \left(\frac{\zeta_{\|}}{r^2}\right), \qquad \tilde{h}_2 = \zeta_{\|}. \tag{7.243}$$

The gauge invariance of k_I immediately follows.

The Eulerian changes in mixed components of the stress-energy tensor given by Eqs. (7.233) and (7.234) are already gauge invariant:

$$\mathcal{L}_\zeta T^A_B = 0, \mathcal{L}_\zeta T^A_b = 0, \mathcal{L}_\zeta T^a_b = 0. \tag{7.244}$$

That time-dependent perturbations are decoupled from the perturbed matter then has the following gauge-invariant meaning: (1) $\delta T^A_b = 0$ for perturbations with no zero-frequency part; and (2) the perturbed metric is governed by the wave equation (7.237) for a gauge-invariant potential ($k_1 = h_1^{\rm RW}$) whose waves propagate only at speed c. Note the contrast here to the polar-parity case. There one can again describe perturbations by dynamical equations, (7.192) and (7.193), that involve only the perturbed metric, but they include waves that are coupled to the matter and that propagate at the sound speed; these are metric perturbations driven by matter oscillations.

A warning is needed here: One cannot simply say that the matter is unperturbed by an axial parity wave without clarifying the meaning of that claim, because whether the perturbation of the stress-energy tensor vanishes depends on whether one considers its covariant or contravariant form. The contravariant perturbations

δu^α and $\delta T^{\alpha\beta}$ differ from the corresponding covariant quantities by terms of the form $h^{\alpha\gamma}u_\gamma$ and $h^{\alpha\gamma}T_\gamma{}^\beta$; they are therefore gauge dependent and do not in general vanish for time-dependent perturbations.

7.6 Multipole decomposition of energy radiated in gravitational waves

We obtained in Section 7.3 equations for the energy and angular momentum radiated to null infinity by a perturbation of a rotating star, Eqs. (7.79) and (7.83). We have seen that the background spacetime is spherically symmetric to $O(1/r)$, and this allows one to decompose the leading order coefficients $a_{\theta\theta}$, $a_{\phi\phi}$, and $a_{\theta\phi}$ in the asymptotic metric into a sum of tensor spherical harmonics with even and odd parity. Orthogonality of the harmonics then gives the radiated energy (or angular momentum) as a sum of angular harmonics. As in the electromagnetic case, the lth axial and polar harmonics can be expressed in terms of $l+1$ time derivatives of mass and current multipole moments of a source [691],

$$\dot{E}_c = -\sum_{l\geq m} N_l \left(\left| \frac{d^{l+1}}{dt^{l+1}}\delta D_{lm} \right|^2 + \left| \frac{d^{l+1}}{dt^{l+1}}\delta J_{lm} \right|^2 \right). \tag{7.245}$$

To obtain a standard relation between the asymptotically defined mass multipole moments and their Newtonian form as integrals over the star, the constants N_l are defined by

$$N_l := \frac{4\pi(l+1)(l+2)}{l(l-1)[(2l+1)!!]^2}. \tag{7.246}$$

The relativistic multipoles δD_{lm} and δJ_{lm} can be written as integrals over a spacelike hypersurface, but, because the integrand involves the metric perturbation as well as the perturbation of the fluid, the region of integration extends beyond the star. As a result, it can be more efficient to compute the asymptotic multipole moments directly from the asymptotic form of the metric perturbation.

To lowest nonvanishing order in a post-Newtonian expansion (at, for example, 2.5 post-Newtonian order for mass-quadrupole radiation), however, the multipole moments entering the formula for $\dot{E}_c$ are mass and current multipoles of a Newtonian star. They have the form

$$\delta D_{lm} = \int r^l\, \delta\rho Y^*_{lm} d^3x, \tag{7.247a}$$

$$\begin{aligned}\delta J_{lm} &= 2\sqrt{\frac{l}{l+1}} \int r^l\, (\rho\delta v_a + \delta\rho v_a) Y^{Ba*}_{lm} dV \\ &= \frac{2}{l+1}\int r^{l+1}\, (\rho\delta v_a + \delta\rho v_a)\epsilon^{ba}\nabla_b Y^*_{lm} dV.\end{aligned} \tag{7.247b}$$

7.7 Eulerian perturbations

This final section supplements the Lagrangian perturbation theory developed in the previous sections with a brief description of an alternative Eulerian description that will be used in Section 8.7 to find the locations of neutral modes along sequences of models parametrized by angular velocity.

Generic perturbations and quasinormal modes of oscillation are described by the perturbed Einstein-Euler equations together with the energy conservation equations. Choose a gauge and search for eigensolutions that satisfy the outgoing-wave boundary conditions at infinity.

Defining the Eulerian perturbation in the metric tensor as $h_{\alpha\beta} := \delta g_{\alpha\beta}$, the perturbed field equation is

$$\delta R_{\alpha\beta} = 8\pi\left(\delta T_{\alpha\beta} - \frac{1}{2}g_{\alpha\beta}\delta T - \frac{1}{2}h_{\alpha\beta}T\right), \tag{7.248}$$

where $\delta R_{\alpha\beta}$ is given by Eq. (7.24),

$$\delta R_{\alpha\beta} = \nabla_\gamma\nabla_{(\alpha}h_{\beta)}{}^\gamma - \frac{1}{2}\Big(\nabla_\alpha\nabla_\beta h_\gamma{}^\gamma + \nabla^\gamma\nabla_\gamma h_{\alpha\beta}\Big), \tag{7.249}$$

and $h^\alpha{}_\beta = g^{\alpha\gamma}h_{\gamma\beta}$. Notice that $\delta g^{\alpha\beta} = -g^{\alpha\gamma}g^{\beta\delta}h_{\gamma\delta}$. The explicit form of the components of $\delta R_{\alpha\beta}$ prior to any choice of gauge were derived by Priou [550]. The perturbed metric $h_{\alpha\beta}$ can be expressed in terms of ten *perturbation functions*, $\bar{h}, \bar{p}, \bar{k}, \bar{w}, \bar{q}, \bar{a}, \bar{b}, \bar{L}, \bar{M}$, and $\bar{y}$, in the manner

$$h_{tt} = -2\bar{h}e^{2\nu} + (2\bar{y}\omega + 2\bar{w}\omega^2)e^{2\psi}, \tag{7.250}$$
$$h_{tr} = \bar{L} + \bar{a}\omega e^{2\psi}, \tag{7.251}$$
$$h_{t\theta} = \bar{M} + \bar{b}\omega e^{2\psi}, \tag{7.252}$$
$$h_{t\phi} = -e^{2\psi}(\bar{y} + 2\omega\bar{w}), \tag{7.253}$$
$$h_{rr} = 2\bar{k}e^{2\alpha}, \tag{7.254}$$
$$h_{r\theta} = \bar{q}, \tag{7.255}$$
$$h_{r\phi} = -\bar{a}e^{2\psi}, \tag{7.256}$$
$$h_{\theta\theta} = 2\bar{p}r^2e^{2\alpha}, \tag{7.257}$$
$$h_{\theta\phi} = -\bar{b}e^{2\psi}, \tag{7.258}$$
$$h_{\phi\phi} = 2\bar{w}e^{2\psi}, \tag{7.259}$$

or, equivalently

$$\begin{aligned} ds^2 &= -e^{2\nu}(1+2\bar{h})dt^2 + e^{2\psi}(1+2\bar{w})\Big[d\phi - (\omega+\bar{y})dt - \bar{a}dr - \bar{b}d\theta\Big]^2 \\ &\quad + e^{2\alpha}\Big[(1+2\bar{k})dr^2 + r^2(1+2\bar{p})d\theta^2\Big] \\ &\quad + 2\bar{q}drd\theta + 2\bar{L}dtdr + 2\bar{M}dtd\theta \end{aligned} \tag{7.260}$$

(where one retains only terms up to first order in the perturbation functions).

The perturbation of the stress-energy tensor is

$$\delta T_{\alpha\beta} = u_\alpha u_\beta(\delta\epsilon + \delta p) + (\epsilon + p)(u_\alpha \delta u_\beta + u_\beta \delta u_\alpha) + g_{\alpha\beta}\delta p + p h_{\alpha\beta}, \qquad (7.261)$$

whereas the perturbation of its contraction is

$$\delta T = 3\delta p - \delta\epsilon. \qquad (7.262)$$

Recall that for a stationary, axisymmetric background, (1.43) implies $u^\alpha \nabla_\alpha \epsilon = 0$ (and similarly for the pressure p and rest-mass density ρ). The energy conservation equation (1.27) then implies $\nabla_\beta u^\beta = 0$, whereas the Euler equation (1.30) becomes $(\epsilon + p)u^\beta \nabla_\beta u_\alpha + \nabla_\alpha p = 0$.

Perturbing the energy conservation equation (1.27), one obtains

$$u^\alpha \nabla_\alpha \delta\epsilon + \delta u^\alpha \nabla_\alpha \epsilon + (\epsilon + p)\nabla_\alpha \delta u^\alpha = -\frac{1}{2}(\epsilon + p) u^\alpha \nabla_\alpha h_\gamma{}^\gamma, \qquad (7.263)$$

whereas the perturbation of the Euler equation (1.30) becomes

$$(\epsilon + p)\left(u^\beta \nabla_\beta \delta u^\alpha + \delta u^\beta \nabla_\beta u^\alpha - u^\alpha \delta u^\beta u^\gamma \nabla_\gamma u_\beta\right) + (\delta\epsilon + \delta p) u^\beta \nabla_\beta u^\alpha + q^{\alpha\beta}\nabla_\beta \delta p$$
$$= -(\epsilon + p)\left[\frac{1}{2} u^\beta u^\gamma \left(\nabla_\gamma h^\alpha{}_\beta + \nabla_\beta h^\alpha{}_\gamma - \nabla_\alpha h^\gamma{}_\beta\right) + u^\gamma \nabla_\gamma u_\beta h^{\alpha\beta}\right]. \qquad (7.264)$$

If one defines $\delta\hat{u}^a := q^\alpha{}_\beta \delta u^\beta$, so that $u_\alpha \delta u^\alpha = -(1/2) u^\alpha u^\beta h_{\alpha\beta}$, then (7.263) and (7.264) become

$$u^\alpha \nabla_\alpha \delta\epsilon + \delta\hat{u}^\alpha \nabla_\alpha \epsilon + (\epsilon + p)\nabla_\alpha \delta\hat{u}^\alpha = -\frac{1}{2}(\epsilon + p) u^\alpha \nabla_\alpha (q^{\beta\gamma} h_{\beta\gamma}) \qquad (7.265)$$

and

$$(\epsilon + p)\left(q^\alpha{}_\gamma u^\beta \nabla_\beta \delta\hat{u}^\gamma + \delta\hat{u}^\beta \nabla_\beta u^\alpha\right) + (\delta\epsilon + \delta p) u^\beta \nabla_\beta u^\alpha + q^{\alpha\beta}\nabla_\beta \delta p = (\epsilon + p)\delta F^\alpha, \qquad (7.266)$$

where δF^α is linear in $h_{\alpha\beta}$ and was derived by Ipser and Lindblom [328].

Instead of the pressure perturbation δp, it is convenient to use another scalar function δU, defined by the relation [665]

$$\delta p := (\epsilon + p)\left(\delta U + \frac{1}{2} u^\alpha u^\beta h_{\alpha\beta}\right). \qquad (7.267)$$

For an adiabatic perturbation, we have

$$\delta\epsilon = \frac{(\epsilon + p)}{\Gamma_1 p}\delta p, \qquad (7.268)$$

and we will assume that the adiabatic index Γ_1 of the perturbation is equal to the adiabatic index of the equilibrium configuration (same EOS for equilibrium and perturbations).

For perturbations with harmonic t- and ϕ-dependence, Eqs. (7.264) and (7.265) are *algebraic* in $\delta\hat{u}^\alpha$, which can be eliminated in favor of δp and $h_{\alpha\beta}$ (see [328, 665] for details). Because of (7.268), one need not consider a perturbed equation of baryon number conservation, and a self-consistent perturbative solution $(h_{\alpha\beta}, \delta U)$

satisfies the system of the perturbed field equation (7.248) and perturbed energy-conservation equation (7.263), where δU enters through (7.267) and (7.268).

The complete system of equations (7.248) and (7.263), describing linear adiabatic oscillations of rapidly rotating stars (with outgoing-wave boundary conditions at infinity) has not yet been solved numerically, except for the case of neutral (zero-frequency) modes [665, 486] (see Section 8.7). However, oscillation modes have been computed under various approximations, such as the slow-rotation approximation, the Cowling approximation, and the IWM-CFC approximation (see Section 8.4).

8
Quasinormal modes

8.1 Quasinormal modes

Perturbations of rotating stars are commonly described in terms of *quasinormal modes*, modes of oscillation with complex eigenfrequencies. These are perturbations with harmonic time dependence and purely outgoing radiation, with Cartesian components of the perturbed metric in a Lorenz gauge having asymptotic behavior $\frac{f(\theta,\phi)}{r}e^{i\sigma u}$, with u an outgoing null coordinate. As we saw in Section 7.5.2, the spherical symmetry of a nonrotating star, together with its invariance under parity $(\theta \to \pi - \theta, \phi \to \phi + \pi)$, implied the decoupling of perturbations with different values of (l, m) and different parity. In the spacetime of a rotating star, spherical symmetry is weakened to axisymmetry, and the reflection symmetries of a spherical star are reduced to symmetry under reflection in the equatorial plane, $\theta \to \pi - \theta$. The axisymmetry and reflection symmetry of a rotating star and its geometry imply that the perturbed Einstein-Euler equations do not couple perturbations that are even and odd under reflection. The quasinormal modes of a rotating star are then partly characterized by the index m specifying their angular dependence $e^{im\phi}$ and by their even or odd parity under $\theta \to \pi - \theta$. Because modes are ordinarily continuous functions of Ω, one can also assign to a mode the value of l that it has for a nonrotating model. (As we discuss in Section 8.6.2, however, there is a class of modes of homentropic stars, the *inertial modes*, whose spherical limit is a superposition of zero-frequency perturbations belonging to different values of l.) For a given (l, m) assignment, modes can be classified by the number n of radial nodes of the radial component ξ^r of the Lagrangian displacement.

We have seen that a mode of a spherical star is described by a perturbation $(\delta\epsilon, \delta p, \delta u^\alpha, h_{\alpha\beta})$ for which all quantities can be written in terms of scalar, vector, and tensor harmonics associated with a single Y_{lm}: That is, the perturbation is associated with a single (l, m) representation of the rotation group and with a choice of polar or axial parity. In particular, the density has the form

$$\delta\epsilon = f_n(r)P_l^m(\cos\theta)e^{i(m\phi+\sigma t)}, \tag{8.1}$$

where σ is the frequency of the mode as measured by an inertial observer at infinity and $P_l^m(\cos\theta)$ is an associated Legendre polynomial. In spherical stars, the radial part of the eigenfunctions and the quasinormal mode frequencies are degenerate in m, and it suffices to study the axisymmetric ($m = 0$) case. The value of l coincides with the number of nodal lines (circles) that divide the surface of the star into regions of opposite phase. The azimuthal order m coincides with the number of roots of $\Re\{e^{im\phi}\}$ in the region $0 \le \phi < \pi$. *Axisymmetric* ($m = 0$) modes are also called *zonal* modes, as all the nodal lines are lines of constant latitude. *Nonaxisymmetric* modes with $l \neq m$ are also called *tesseral* modes, having both nodal lines of constant latitude and nodal lines of constant longitude, whereas those with $l = m$ are also called *sectoral* modes, having only nodal lines with constant longitude.

We denote by ω_i the real part of the frequency of a mode as measured by an *inertial observer* at infinity,

$$\omega_i := \Re\{\sigma\}, \tag{8.2}$$

and by ω_r the real part of the frequency measured by a *corotating observer*, an observer rotating with the angular velocity of the star. In their angular behavior, axisymmetric modes are *standing waves*, whereas nonaxisymmetric modes are *traveling waves* that move about the axis of symmetry, with phase velocity $(\partial\phi/\partial t)_{phase} = -\omega_i/m$, as seen by a distant inertial observer. Modes with $\omega_i > 0$ and $m > 0$ are thus *retrograde*, and those with $m < 0$ are *prograde*. To relate the two frequencies ω_i and ω_r, use the fact that a point at constant angular coordinate ϕ_r measured by a rotating observer has coordinate $\phi = \phi_r + \Omega t$ measured by an inertial observer. Then

$$e^{i(m\phi+\omega_i t)} = e^{i(m\phi_r+\omega_r t)}, \tag{8.3}$$

implying

$$\omega_r = \omega_i + m\Omega. \tag{8.4}$$

As we show in Section 9.5 later, a nonaxisymmetric mode of a perfect-fluid model is unstable when the mode is retrograde relative to the star but prograde relative to infinity – when ω_r and ω_i have opposite signs. For slow rotation, a stable retrograde mode has $m > 0$ if we choose $\omega_i > 0$. In a slowly rotating star, a mode that has the character of a wave in the fluid moves with nearly the same speed relative to the rotating observer that it had in the spherical star: If one neglects the dragging of inertial frames, the frequency seen by a corotating observer changes only at $O(\Omega^2)$. Because the mode is dragged along with the rotating fluid, its frequency seen by an inertial observer at infinity then changes at first order in Ω: As the star's rotation increases, Eq. (8.4) implies that ω_i decreases, and the onset of instability is at the angular velocity Ω for which ω_i vanishes. From Eq. (8.4), this neutral stability point is at $\Omega = \omega_r/|m| \sim \omega_{i0}/|m|$, with ω_{i0} the mode's frequency for the nonrotating star.

For a given triplet (l, m, n), a general solution of the system of perturbation equations can represent a mixture of ingoing and outgoing waves. *Outgoing modes* are defined as the discrete set of solutions for which there are no incoming waves at infinity. These are the modes that will be excited in various astrophysical situations. In the following, when discussing oscillation modes of relativistic stars we will always mean outgoing modes.

The main modes of pulsation that are known to exist in relativistic stars have been classified as follows [676, 364] (frequencies reported in hertz are the values of $f_i \equiv \omega_i/2\pi$):

1. *Polar modes*
   - f(undamental)-modes: *oscillations with no radial nodes* (Kelvin-modes). For incompressible spheres, the $l = 0$ mode does not exist because of mass conservation; the fundamental $l = 1$ modes (for both compressible and incompressible models) are just translations of the center of mass of the star. Fundamental modes of index l are called ${}^l f$-modes. Typical frequencies are of order kHz. Typical damping times of nonradial modes due to emission of gravitational radiation (imaginary part of complex quasinormal mode frequencies) are of order 0.1 s to 0.5 s. The fundamental mode can be viewed as the nodeless p- or g-mode.
   - p(ressure)-modes: restored by *pressure* and denoted by ${}^l p_n$ (where $n \geq 1$ is the radial order). Typical frequencies of the first overtone ${}^2 p_1$ are of order several KHz. Typical damping times of nonradial modes due to emission of gravitational radiation are of order seconds or longer. A subclass are the radial ($l = 0$) modes, for which the fundamental ($n = 0$) mode is called the F-mode, whereas overtones are denoted by H_1, H_2, ... Instability to gravitational collapse is the instability of the fundamental radial mode and is discussed in Chapter 9.
   - g(ravity modes): restored by gravity through *buoyancy*, when thermal or composition gradients exist in a star. In such cases, the adiabatic index Γ_1 (1.118) of the perturbed fluid elements differs from the corresponding expression of the equilibrium fluid, which we denote by Γ: That is, for a barotropic equilibrium, we write

$$\Gamma := \left.\frac{d \ln p}{d \ln \rho}\right|_\star = \frac{d \ln p/dr}{d \ln \rho/dr} = \left.\frac{dp}{d\epsilon}\right|_\star \frac{\epsilon + p}{p}. \tag{8.5}$$

     Particular modes are denoted as ${}^l g_n$. Typical frequencies are a few hundred Hz and lower (decreasing with increasing mode order), and gravitational-radiation damping times are substantially longer than for p-modes.

     Although the two branches of g- and p-modes join at the fundamental mode (which has no radial nodes in nonrotating stars), their disparate character is clear for wavelengths short compared to the radius of the star. We derive the frequencies of these two families of modes for spherical stars in the next section. The effect of rotation is discussed in subsequent sections of this chapter.
   - s(hear modes): restored by *elastic stresses* in the solid crust of the star.
   - a(lfvén modes): magnetohydrodynamic modes in a magnetized star, restored by *magnetic tension*. Frequencies increase almost linearly with increasing magnetic field strength.

2. *Axial and hybrid modes*
   - *Rotationally* restored modes, whose frequencies are proportional to the angular velocity of the star. One can regard the restoring force as the *Coriolis* force in a rotating frame. There are two subclasses.
     - r(Rossby)-modes: These are modes whose limit as $\Omega \to 0$ has axial parity. Generically unstable to the Chandrasekhar-Friedman-Schutz (CFS) instability with growth times as short as a few seconds at high rotation rates.
     - i(nertial) modes: A larger family of rotationally restored modes of barotropic models whose perturbations and equilibria obey the same one-parameter EOS: Both r-modes and g-modes of a spherical star then have zero frequency, and inertial modes are *hybrid*, having both axial and polar parts in the $\Omega \to 0$ limit. Frequencies range from zero (in nonrotating stars) to several kHz in rapidly rotating stars. Growth or damping times due to gravitational wave emission are inversely proportional to a high power of the star's angular velocity.
   - t(orsional) modes: axial modes restored by *elastic stresses* in the solid crust. The fundamental frequency of the $l = 2$ mode is around 30 Hz.
3. *Polar and axial spacetime modes*
   - w(ave)-modes: analogous to the quasinormal modes of a black hole (very weak coupling to the fluid). Three distinct families are known. *Curvature modes* (the standard w-modes) exist for all relativistic stars. Typical frequencies for the fundamental mode are 5 to 12 kHz, whereas higher-order modes have higher frequencies. Typical damping times are of order a few tenths of a millisecond (strongly damped modes). *Trapped modes* exist only for hypothetical supercompact stars ($R \leq 3M$) (when the surface of the star is inside the peak of the gravitational potential barrier). Only a finite number of trapped modes exist. Typical frequencies are of order a few hundred Hz to a few kHz; typical damping times are of order a few tenths of a second (slowly damped modes). *Interface modes* are similar to acoustic modes scattered off a hard sphere. Only a finite number (a few) interface modes exist. Typical frequencies are 2 to 15 kHz; typical damping times are of order less than a tenth of a millisecond.
4. *Superfluid modes*
   The macroscopic behavior of superfluid neutrons and protons in the core of a neutron star is described by a two-fluid model [349, 388], in which fluids associated with neutrons and with protons can have separate motions. (On timescales relevant for stellar oscillations, electrons and muons move with the protons.) Because of its two-fluid hydrodynamics, a superfluid has twice as many degrees of freedom as an ordinary fluid. The neutron and proton fluids are coupled or "entrained," and the coupling terms in the Lagrangian imply that the momentum associated with each fluid involves a linear combination of the two velocities: The neutron-dominated current includes a small proton-current part, and the proton-dominated current includes a small neutron-current part.

Lindblom and Mendell [424] and Epstein [206] first incorporated superfluid dynamics in the description of neutron star modes, and Epstein noticed that the doubling of degrees of freedom leads to a doubling of the modes of a star. (For a degenerate subspace of zero-frequency modes, the number of degrees of freedom of the subspace is doubled.) As these and subsequent authors have elucidated, there remains a set of hydrodynamic modes that mimic the modes of a one-fluid star, in which the two fluids move in tandem; however there is a second set of modes, the *superfluid modes*, in which the two fluids are countermoving and whose frequencies are sensitive to the strength of the entrainment. The modes might be observable in the frequency spectrum of oscillating neutron stars, but there is as yet no indication that they play an important role in stellar stability.

The formalism for describing the superfluid modes and the results that have been obtained are outside the scope of this book, and we simply give references for interested readers: Work has included modes of nonrotating Newtonian [401, 17, 552, 141] and relativistic [152, 20, 18, 415] stars; Newtonian studies of r-modes and inertial modes of slowly rotating stars [426, 307] and all modes of rapidly rotating Maclaurin spheroids [551]; and inertial modes of slowly rotating relativistic stars [402, 743].

For a more detailed description of various types of oscillation modes, see [676, 467, 114, 364].

Completeness and incompleteness of oscillation modes. It is a plausible conjecture that the modes are complete for fluid perturbations in the sense that any initial data for the perturbed fluid variables can be written as a sum of their data for outgoing modes, where convergence is in L_2(support of the star).[1] They are not complete for perturbations of the spacetime, because generic initial data describes a perturbation with both incoming and outgoing radiation. In the Newtonian approximation, there are proofs of completeness and discreteness of the normal modes of spherical polytropes [67], and some results for rotating stars by Dyson and Schutz [201]. Even in the Newtonian framework, criteria are not known for discreteness of the spectrum. For relativistic stars, little has been done beyond radial modes of spherical stars. One difficulty is that stable outgoing modes grow exponentially as $r \to \infty$ on a $t =$ constant surface. They are, however, asymptotically finite on asymptotically null hypersurfaces, and one can ask whether the outgoing modes are complete for all perturbations that have purely outgoing radiation. It is a problem that does not fall within standard spectral theory.

Some insight, however, has been gained by looking at toy models that mimic the features of the Einstein-Euler system, a system that couples gravitational waves and waves in the fluid. Kokkotas and Schutz [367] considered a model 1-dimensional system, replacing the spacetime by a semi-infinite string and the fluid by a finite spring coupled to the string, with the speed of waves in the string much faster

[1] The support of the star here means the region V of a $t =$ constant spatial slice on which the density is nonzero. A function f is in $L_2(V)$ if it is square-integrable, if $\int_V |f|^2 dV$ is finite.

than those in the spring. Like the fluid modes of the star, the normal modes of the uncoupled spring become quasinormal modes (outgoing modes) of the coupled string-spring system. With a very similar one-dimensional model, Price and Husain [548] found that the outgoing modes are, in fact, complete, forming a basis for the space of all purely outgoing solutions.

In addition to the modified spring modes, the initial model of Kokkotas and Schutz has a second set of modes: modes of the string modified by its coupling to the spring. It was the presence of these "spacetime" modes that led the authors to discover the analogous w-modes of perfect-fluid spacetimes. The image of outgoing modes of a stretched string (or a half-infinite stiff, light spring) attached to a more massive spring remains a useful way to think about the w-modes.

8.2 Frequency of short-wavelength p- and g-modes

The frequencies of g-modes and p-modes of spherical relativistic stars in a short-wavelength (Cowling) approximation were found by Finn [218]. As mentioned above, it is in the limit of short radial wavelength that the different character of these two classes of modes is most pronounced. Short-wavelength p-modes are sound waves whose frequencies as measured by a local inertial observer are given by $c_s k$, where k is the locally measured wavenumber. For short-wavelength g-modes, the change in the pressure of a fluid element is small, and the restoring force is buoyancy; the mode frequency is proportional to the Brunt-Väisälä frequency N, a frequency that is a measure of the difference between the EOS that describes the equilibrium star and the adiabatic EOS that governs oscillations too rapid for thermal and chemical equilibrium to be established.

We similarly denote by $c_\star$ the value for the star's EOS of the expression for the speed of sound: $c_\star := \sqrt{\frac{dp}{d\epsilon}|_\star} = \sqrt{\Gamma\frac{p}{\rho}}$. In particular, for a spherical star, $c_\star = \sqrt{\frac{dp/dr}{d\epsilon/dr}}$, whereas the actual speed of sound in the star is $c_s := \sqrt{\frac{\partial p}{\partial\epsilon}|_{s,Y_i}} = \sqrt{\Gamma_1\frac{p}{\rho}}$, with Γ_1 the adiabatic index. The Newtonian value of the Brunt-Väisälä frequency is

$$\mathrm{N} = \Phi'\left[\left(\frac{1}{\Gamma} - \frac{1}{\Gamma_1}\right)\frac{\rho}{p}\right]^{1/2} = \Phi'\left(c_\star^{-2} - c_s^{-2}\right)^{1/2},$$

where ($'$) denotes differentiation with respect to r. For a spherical relativistic star, we will see that the Newtonian relation for the locally measured g-mode frequency in terms of N is retained if Φ' is replaced by $\nabla_{\hat{1}}\nu = e^{-\lambda}\nu'$, the derivative of ν along the unit vector $\mathbf{e}_{\hat{1}}$ in the radial direction:

$$\mathrm{N} = e^{-\lambda}\nu'\left(c_\star^{-2} - c_s^{-2}\right)^{1/2}. \tag{8.6}$$

It follows that barotropic models whose perturbations obey the same one-parameter EOS as that of the equilibrium star are marginally unstable to convection, and in this case the g-modes are zero-frequency convection currents. When

$\Gamma > \Gamma_1$ ($c_\star > c_s$), the Brunt-Väisälä frequency is imaginary, and the star is unstable to convection. Rotation mixes modes with axial and polar parity – mixing r-modes and g-modes when the star is nearly homentropic. As we show in Section 9.3, however, the same criterion governs stability of rotating stars to convection. Note that N^2 is proportional to $A = (1/\Gamma - 1/\Gamma_1)p'/p$, the *Schwarzschild discriminant* for spherical stars, and the stability criterion can be stated in terms of the sign of A.

The formalism we have developed for perturbations of spherical stars lets us quickly derive the frequencies of the two families of modes in the short radial wavelength approximation. For short radial wavelengths, the Cowling approximation is valid: We can neglect the perturbation of the metric. We can also assume that the background coefficients appearing in the perturbed Euler equation are constant (independent of r) over the radial interval where the perturbation is nonzero. This implies that solutions have the form $e^{ik_r r}Y_{lm}(\theta,\phi)$, and that $\lambda \sim 1/k_r$ is small compared to characteristic lengths of the unperturbed fluid: $k_r \gg p'/p, \nu'/c_s^2$. An eigenvalue equation for the frequency σ then follows from the perturbed Euler equation, together with the expressions (7.5) for δp and $\delta\epsilon$ in terms of ξ^α.

In the Cowling approximation, the angular and radial parts of the perturbed Euler equation (7.190) have the form

$$- e^{-2\nu}\sigma^2\hat{\xi}_\parallel + \frac{\hat{\delta p}}{\epsilon + p} = 0, \tag{8.7}$$

$$-e^{-2\nu}\sigma^2\hat{\xi}_r + \frac{\nabla_r\hat{\delta p}}{\epsilon+p} - \frac{\nabla_r p}{\epsilon+p}\frac{\hat{\delta\epsilon}+\hat{\delta p}}{\epsilon+p} = 0, \tag{8.8}$$

with $\delta\epsilon$ and δp related to ξ^r and $\xi_\parallel$ by

$$\frac{\delta\epsilon}{\epsilon+p} = -q^{\alpha\beta}\nabla_\alpha\xi_\beta - \frac{\boldsymbol{\xi}\cdot\nabla\epsilon}{\epsilon+p}, \tag{8.9}$$

$$\frac{\delta p}{\Gamma p} = -q^{\alpha\beta}\nabla_\alpha\xi_\beta - \frac{\boldsymbol{\xi}\cdot\nabla p}{\Gamma p}. \tag{8.10}$$

It is helpful to use ξ^r and δp as independent variables instead of ξ^r and $\xi_\parallel$.[2] From the last two equations, we have

$$\frac{\delta\epsilon}{\epsilon+p} = \frac{\delta p}{\Gamma p} - \xi^r\nu'\left(c_\star^{-2} - c_s^{-2}\right), \tag{8.11}$$

and the radial component of the Euler equation takes the form

$$(\mathrm{N}^2 - e^{-2\nu}\sigma^2)e^{2\lambda}\xi^r + ik_r\frac{\delta p}{\epsilon+p} = 0. \tag{8.12}$$

The angular component of the Euler equation gives a second relation between ξ^r and δp, when Eq. (8.10) and the relation $q^{\alpha\beta}\nabla_\alpha\xi_\beta = \nabla_\alpha\xi^\alpha = ik_r\xi^r + \frac{l(l+1)}{r^2}\xi_\parallel$ are

[2] Using ξ^r and $\xi_\parallel$ as independent variables has the difficulty that the combination $\nabla_\alpha\xi^\alpha$ is small for g-modes by a factor of order $(kR)^{-1}$, compared to $k\sqrt{\xi_\alpha\xi^\alpha}$. As a result, one cannot always neglect, for example, $\nu'\xi^r$ in an expression that includes both $k\xi^r$ and $\xi_\parallel$. By using δp and ξ^r instead, one avoids subtleties in identifying terms of dominant order in kR.

used to replace $\xi_\parallel$ in favor of ξ^r and δp:

$$ik_r\xi^r + \left[\frac{1}{c_s^2} - \frac{l(l+1)}{\sigma^2 r^2}e^{2\nu}\right]\frac{\delta p}{\epsilon + p} = 0. \tag{8.13}$$

The two equations (8.12) and (8.13) are consistent only if the determinant of the coefficients of ξ^r and δp vanishes, giving the dispersion relation

$$k_{\hat{1}}^2 \equiv k_r^2 e^{-2\lambda} = (\mathrm{N}^2 - e^{-2\nu}\sigma^2)\left[\frac{l(l+1)}{\sigma^2 r^2}e^{2\nu} - \frac{1}{c_s^2}\right]. \tag{8.14}$$

The two families of solutions to Eq. (8.14) are now easily identified.

p-modes: When $\sigma \gg \mathrm{N}$, the term involving N can be neglected, and σ satisfies the dispersion relation for sound waves,

$$\hat{\sigma} := \sigma e^{-\nu} = c_s k, \tag{8.15}$$

where $\hat{\sigma}$ and $k = \sqrt{k_{\hat{1}}^2 + l(l+1)/r^2}$ are the frequency and wavenumber measured by a local inertial observer.

g-modes: When $\sigma r \ll c_s$, the term c_s^{-2} can be neglected and the frequency is proportional to N:

$$\hat{\sigma} = \sigma e^{-\nu} = \left[\frac{l(l+1)}{(k_{\hat{1}}r)^2 + l(l+1)}\right]^{\frac{1}{2}} \mathrm{N} = \frac{\sqrt{l(l+1)}}{kr}\mathrm{N}. \tag{8.16}$$

As noted above, with N defined by Eq. (8.6), this equation for $\hat{\sigma}$ has the form of the Newtonian dispersion relation.

8.3 Effect of rotation on quasinormal modes

Rotation has several dramatic effects on the structure and frequency of oscillation modes:

1. The degeneracy in the azimuthal order m is removed. The single frequency of modes with the same value of l and different values of m is split: Each of the $2l+1$ degenerate modes of a spherical star with $-l \le m \le l$ acquires a different frequency.

   In particular, prograde modes differ from retrograde modes. The frequency ω_i has the form $\omega_{i0} - Cm\Omega + O(\Omega^2)$, where ω_{i0} is the frequency of a mode in the nonrotating limit and $C > 0$ is a factor (different for each mode) that depends on the equilibrium structure of the star. Thus, the frequency of a prograde mode increases, whereas the frequency of a retrograde mode decreases with increasing rotation rate.
2. The eigenfunction of any mode in a rotating star is an infinite expansion of polar and axial spherical harmonics and is no longer described by a single term. For example, the eigenfunction of a nonrotating polar l-mode becomes an expansion

that includes polar $l \pm 2l'$ terms[3] beginning at $O(\Omega^2)$, and axial $l \pm 2l' + 1$ terms beginning at $O(\Omega)$. Similarly, the eigenfunction of a nonrotating axial l-mode becomes an expansion that includes axial $l \pm 2l'$ terms and polar $l \pm 2l' + 1$ terms. Here, $l' = 0, 1, ..$ with the constraints $l \pm 2l' \geq m$ for polar terms and $l \pm 2l' + 1 \geq m$ for axial terms.

3. Most oscillation modes reduce to a single spherical harmonic in the nonrotating limit (either spherical or polar) that labels the mode in a rotating star. As noted above, however, the nonrotating limit of a hybrid inertial mode of a barotropic star is a sum of polar and axial terms.
4. In rapidly rotating stars, *avoided crossings* between higher-order modes of different order can occur. This happens because no two modes with at least two coinciding l, m, n indices can have identical frequencies. Because higher-order modes can be affected by rotation much more drastically than low-order modes, a high-order mode can have a lower frequency compared to a lower-order mode when the star is rapidly rotating. However, the two mode sequences do not cross but display an avoided crossing, and modes are labeled according to the dominant terms in their eigenfunctions before and after the avoided crossing and not by continuity in frequency space. In the immediate region of the avoided crossing, the eigenfunctions of the two modes become similar and have a mixed character.
5. The damping times (due to gravitational-wave emission) of quasinormal modes depend moderately on rotation for polar modes. However, for axial-led inertial modes (where gravitational-wave emission is dominated by current rather than density variations), damping times are sensitive to rotation rate.

In the next section, we discuss various approximation methods that have been used for computing specific oscillation modes. Because of the complexity of the problem for rotating stars, certain approximations, such as the slow-rotation approximation, the post-Newtonian approximation, the Cowling approximation, and the spatial-conformal-flatness (IWM-CFC) condition, have been used, with varying degrees of accuracy.

8.4 Computational methods

8.4.1 Nonrotating stars

We reviewed in Section 7.5 the formalism that describes perturbations of spherical stars, a formalism initially developed to find stellar modes. Thorne, Campolattaro, Price, and subsequent collaborators pioneered the computation of quasinormal modes of spherical relativistic stars, reporting the work in a series of papers, beginning with [692, 547, 689]. To locate purely outgoing modes, one must distinguish the outgoing-wave part from the ingoing-wave part in a general solution. This

[3] These restrictions are implied by the odd or even character of a mode under reflection in the equatorial plane.

is typically done by matching an asymptotic expansion of the solution at spatial infinity to a numerical solution obtained by radial integration of the system on the interior. The w-modes pose a more challenging numerical problem because they are strongly damped and the techniques used for f- and p-modes fail to distinguish the outgoing-wave part. The first accurate numerical solutions were obtained by Kokkotas and Schutz [368] (see also [403]). In [26] a complex-coordinate integration method was used to obtain highly accurate complex frequencies for w-modes. In this method, the ingoing and outgoing solutions are separated more easily by calculating their analytic continuations to a place in the complex-coordinate plane where they have comparable amplitudes.

Because of the mixing of multipoles in the modes of rotating stars, the asymptotic behavior of a solution at infinity is not known – or, more accurately, it involves a set of free parameters associated with the amplitude and phase of each contributing multipole. This ambiguity in the asymptotic solution is one of the major difficulties in computing outgoing modes of rapidly rotating relativistic stars. It is thus important to devise computational methods that avoid using boundary conditions at infinity (at the cost of a reduced accuracy in the imaginary part of quasinormal mode frequencies). A method that uses near-zone boundary conditions for polar modes was presented in [422]. The pulsation equations were reformulated as a set of two second-order radial equations for two potentials – one corresponding to fluid perturbations and the other to the perturbations of the spacetime. The equation for the spacetime perturbation reduces to a scalar wave equation at infinity and to Laplace's equation for zero-frequency solutions. From these, an approximate boundary condition for outgoing modes was constructed and imposed in the near zone of the star (in fact, on its surface), yielding f-mode eigenfrequencies with errors of only 0.01% to 0.1% and imaginary parts with errors at the 10% to 20% level.

8.4.2 Rotating stars

The Newtonian approximation

In the Newtonian approximation, the fact that stable oscillation modes have real frequencies simplifies the boundary conditions of the eigenvalue problem. Despite this and the obvious simplification of a single gravitational potential, accurate numerical solutions of nonaxisymmetric modes were obtained only after a successful reformulation of the system of perturbation equations, by Ipser and Lindblom. The perturbation equations are written in terms of two perturbed scalar functions: $\delta\Phi$, the perturbation in the gravitational potential, and

$$\delta U = \delta h - \delta\Phi,$$

the difference between the perturbation of the enthalpy and $\delta\Phi$; they comprise a system of two second-order equations [325, 326, 327] (see also [331, 453]).[4] This

[4] Axisymmetric oscillation modes in the Newtonian limit were obtained previously by Clement [145].

reformulation was possible because for displacement vectors with harmonic t and ϕ dependence, the perturbation in the velocity, δv^a, can be written algebraically in terms of δU and $\delta\Phi$ (see [329] for a summary of the main equations in the two-potential formalism).[5] The system of equations can be further reduced to a single fourth-order equation for δU by eliminating $\delta\Phi$.

In searching for discrete modes, boundary conditions are imposed at the stellar surface (vanishing of the Lagrangian change in the enthalpy, $\Delta h = \Delta U + \Delta\Phi = 0$) and at infinity ($\delta\Phi$ must vanish sufficiently rapidly as $r \to \infty$). Notice that although δU has no physical meaning outside the star, the condition $\delta h = 0$ can be used to define a smooth extension in that region, which is useful for the numerical evaluation of δU. Discretizing the perturbation equations yields a linear system that can be solved using standard techniques, for discrete eigensolutions $(\omega_i, \delta U, \delta\Phi)$. Iterative techniques require a suitable initial guess, which can be obtained from the corresponding solution in the nonrotating limit. Another method for obtaining a suitable guess for the eigenfrequency is to use a variational principle, which can be constructed from the system of perturbation equations. For uniformly rotating stars, the expression for the variational principle reduces to a quartic polynomial in the frequency. For the $l = m = 2$ f-modes, it is easy to determine the physical root of this polynomial, which represents an estimate for the eigenfrequency of the mode. In [325], an iterative method is described that converges faster to the desired solution than standard methods. This method makes use, initially, of a trial eigenfunction and of an eigenfrequency estimate obtained via a variational principle, whereas in subsequent iterations, monitoring the change in a suitably defined spatial average of the eigenfunction, $\langle\delta U\rangle$, leads to faster convergence. We discuss the Newtonian two-potential formalism in more detail in Appendix B.

The post-Newtonian approximation

The effect of general relativity on oscillations can be estimated by solving the problem in the first post-Newtonian (1PN) approximation. The Newtonian two-potential formalism was extended to 1PN order by Cutler and Lindblom [168, 171, 418], who showed that the 1PN correction in the perturbed 4-velocity can be obtained analytically in terms of other variables. If a mode solution is known in the Newtonian limit, then the 1PN correction to its eigenfrequency, $\Delta\sigma$, can be obtained as an eigenvalue of a system of three second-order equations, with suitable boundary conditions. Cutler and Lindblom obtained an equation, which is based on the known 1PN equilibrium solution and the Newtonian mode solution, that yields $\Delta\sigma$ without actually solving the 1PN perturbation equations numerically. In the nonrotating limit, 1PN frequencies of $l = |m|$ f-modes for $N = 1$ polytropes with $M/R = 0.2$ differ by up to 8% from their general-relativistic counterparts.

Along a sequence of rotating stars, the frequency of a nonaxisymmetric mode is conveniently described by the ratio of the frequency of the mode in the comoving

[5] The potential δU was also introduced by Poincaré [537] in the study of oscillations of uniform density stars.

frame to the frequency of the mode in the nonrotating limit. For the fundamental $l = |m|$ modes of $N = 1$ polytropes with $M/R = 0.2$, this frequency ratio is reduced by as much as 12% in the 1PN approximation compared to its Newtonian counterpart, which indicates the order of magnitude of relativistic effects on the frequency of nonaxisymmetric modes in rotating stars.[6]

The slow-rotation approximation

We found in Section 6.1 that, although rotation changes the equilibrium pressure and density only at order Ω^2, the perturbed metric changes at order Ω, because of the dragging of inertial frames. This contrast with slowly rotating Newtonian equilibria has its counterpart in the perturbation spectra of slowly rotating stars: Because the frame-dragging frequency, ω, of an equilibrium spacetime is proportional to Ω, the frequency of oscillation modes in the full theory is already altered at $O(\Omega)$.

For slowly rotating stars, one computes oscillation modes by solving the system of perturbation equations to successive order in Ω. At $O(\Omega)$, the equilibrium fluid and all metric coefficients except $g_{t\phi}$ retain their spherical values; one can still formulate the problem as a system of ordinary differential equations in r and apply outgoing-wave boundary conditions. Because $g_{t\phi}$ at order Ω is the ϕ-component of an axial $l = 1$ harmonic, the perturbed Einstein-Euler system mixes the angular harmonic of a spherical perturbation with adjacent values of l. In particular, at $O(\Omega)$, the eigenfunction of a polar l-mode of an initially nonrotating star acquires axial $l \pm 1$ terms. Conversely, an axial l-mode acquires polar $l \pm 1$ terms [134]. The mode thereby retains its odd or even character under reflection in the equatorial plane: A polar (l, m) harmonic has the same behavior under reflection as an axial $(l \pm 1, m)$ harmonic.

The equations of nonaxisymmetric perturbations at $O(\Omega)$ were derived in a diagonal gauge by Chandrasekhar and Ferrari [134] and in the RW gauge by Kojima [356]. A different formulation, based on the ADM formalism, in which polar and axial equations are written in a similar way was presented in [573]. The resulting system is first-order hyperbolic and suitable for numerical simulations. Applications of the different versions of the slow-rotation formalism include the study of inertial and r-modes [16, 360, 361, 362, 442, 571, 572, 574, 444, 443, 538], f- and p-modes [358, 357, 742, 359, 654, 533], g-modes [215], w-modes [366], and crustal modes [709, 708].

The error in computed mode frequencies and eigenfunctions obtained in the $O(\Omega)$ slow-rotation approximation becomes appreciable for rapidly rotating stars, and it would be natural to extend the approximation to $O(\Omega^2)$. This, however, introduces a large number of couplings and a consistent description of oscillations to $O(\Omega^2)$ is likely to be at least as complex as solving the full problem without approximation.

6 Quasiradial modes in stars near the maximum mass model allowed by the high-density EOS are affected much more drastically by relativistic effects than are nonaxisymmetric modes.

The Cowling approximation

We have already noted for fluid perturbations whose wavelengths are short that the induced change in the metric is small, and we give a more formal argument below. Neglecting perturbations in the gravitational field – setting $h_{\alpha\beta} = 0$ in the perturbation equations – is a generalization of the corresponding *Cowling approximation* of the Newtonian theory, where perturbations in the gravitational potential are neglected [165, 468]. This standard Cowling approximation has been used for estimating frequencies of f-, p-, and r-modes (see, e.g., [431]). Finn argues that an extension of the approximation in which h_{tr} is kept nonzero is more accurate for g-modes [218], because these modes could have large fluid velocities, even though the variation in the gravitational field is weak.

The Cowling approximation introduces an error of roughly a factor of 2 in the frequency of the fundamental radial mode and an error of roughly 15% to 30% in the frequency of the fundamental quadrupole mode. For r-modes, the Cowling approximation gives more accurate frequencies, as these modes do not involve large density variations. The error made by the approximation decreases as the order of the mode increases (for high-order modes, the perturbations in the source average out, resulting in smaller metric perturbations). Yoshida and Kojima [742] examined the accuracy of the relativistic Cowling approximation in slowly rotating stars. For $N = 1$ polytropes, the relative error in the $l = 2$ f-mode is 30% for weakly relativistic stars, which is reduced to about 15% for more relativistic stars (higher compactness). Also, for p-modes, the Cowling approximation becomes more accurate for increasing radial mode number and less accurate for more relativistic stars.

Justification of the Cowling approximation. In the full theory, the perturbed metric is not gauge invariant, and it is helpful first to understand the justification in the Newtonian approximation where no issues of gauge arise. Let λ be a characteristic wavelength of the mode, and let $k = 2\pi/\lambda$ be the corresponding wavenumber. In particular, if k_r is a characteristic radial wavenumber, a mode belonging to the (l, m) angular harmonic has $k \sim \sqrt{k_r^2 + l(l+1)/R^2}$. Then $\nabla^2\Phi$ is of order $k^2\Phi$, and the solution to Laplace's equation, $\nabla^2\Phi = 4\pi\rho$, is of order $\Phi \sim \rho/k^2$, vanishing as $k \to \infty$. Formally, for a source ρ in $L^2(\mathbb{R}^3)$ with Fourier transform $\hat{\rho}(\mathbf{k})$, Φ satisfies the inequality

$$\int \Phi^2 dV \leq K \int \frac{|\hat{\rho}(\mathbf{k})|^2}{(1+k^2)^2} d^3k. \tag{8.17}$$

In the full theory, the corresponding justification for the approximation is readily seen in a Lorenz gauge (7.34), where $\widetilde{h}_{\alpha\beta} := h_{\alpha\beta} - \frac{1}{2}g_{\alpha\beta}h$ satisfies an equation of the form

$$\nabla_\lambda \nabla^\lambda \widetilde{h}_{\mu\nu} + (\text{terms involving } h_{\mu\nu} \text{ and its first derivatives}) = -16\pi\delta T_{\mu\nu}. \tag{8.18}$$

Again let $k = 2\pi/\lambda$ be a characteristic wavenumber of the source, and suppose that the characteristic speed of the wave is comparable to or smaller than the speed of sound. Then $\nabla_\lambda \nabla^\lambda \widetilde{h}_{\mu\nu}$ is of order $k^2 h_{\mu\nu}$, and we have

$$\widetilde{h}_{\mu\nu} \sim \frac{\delta T_{\mu\nu}}{k^2}. \tag{8.19}$$

The gauge ambiguity in stating a criterion for validity of the Cowling approximation can be avoided by requiring validity in a Lorenz gauge (or simply by requiring that there be some gauge in which the perturbed metric is negligibly small). The form of the Euler equation does not depend on the gauge used to state the criterion: It is simply the perturbed relativistic Euler equation with $h_{\alpha\beta}$ set to zero.

If the Cowling approximation is valid in a Lorenz gauge, it is ordinarily valid in other gauges used to describe stellar perturbations. There is, however, one gauge choice for which it is never valid: In a comoving gauge, the Lagrangian displacement vanishes, and all Eulerian changes in the fluid variables are expressed in terms of the perturbed metric. Comoving coordinates are often used to describe collapsing spherical stars, and in the corresponding comoving description of radial perturbations, one then cannot neglect h_{tr} for short-wavelength perturbations.

8.5 Axisymmetric modes

Axisymmetric ($m = 0$) pulsations in rotating relativistic stars are excited in a number of different astrophysical contexts, including immediately after core bounce in a core-collapse event; in star quakes induced by the secular spin-down of a pulsar or during a large phase transition in the high-density EOS; and in the merger of two relativistic stars in a binary system. In some of these scenarios, axisymmetric oscillations radiate substantial energy in gravitational waves.

The $l = 0$ modes (radial modes in the nonrotating limit) are called *quasi-radial* modes in rotating stars. Because the rotational correction to the eigenfunction of a radial mode is dominantly quadrupole, quasiradial modes emit primarily quadrupole gravitational waves. Quasiradial modes in rotating relativistic stars were initially studied in the slow-rotation approximation [300, 176], whereas a particular mode sequence for rapidly rotating polytropes in full general relativity was presented in [224]. More general, axisymmetric modes have been studied in the slow-rotation approximation [654], in the Cowling approximation [740, 225, 223, 661, 345], and in the IWM-CFC approximation [193] (which is very accurate for such modes), including differentially rotating polytropes.

The effect of rotation on the axisymmetric modes in rapidly rotating stars depends on several factors, such as the type of rotational sequence, the rotation law and the mass of the star. Figure 8.1 refers to a particular sequence of rotating $N = 1$ polytropes with a fixed rest mass of $M_0 = 1.506 M_\odot$. The dependence of the frequency of several axisymmetric modes on rotation is shown, choosing $T/|W|$ (which scales roughly as Ω^2) as the rotational parameter. For uniform rotation,

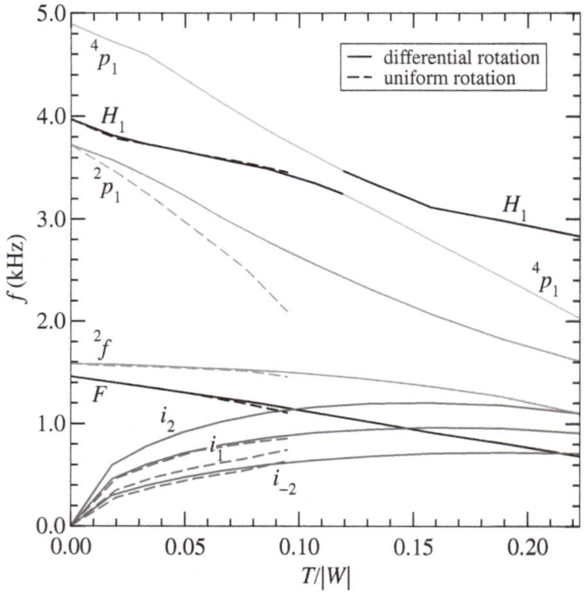

Figure 8.1. Frequencies for various axisymmetric modes along a sequence of uniformly rotating $N = 1$ polytropes with fixed rest mass (dashed lines) and along a corresponding sequence of differentially rotating models (solid lines), as a function of $T/|W|$. Note the avoided crossing between the H_1 and ${}^4p_1$ modes. (Figure from [193]. Reproduced by permission of John Wiley and Sons.)

the frequency of the fundamental quasiradial mode, F, decreases by 25% when the mass-shedding limit is reached. The frequency of its first overtone, H_1, decreases by 13%, whereas the decrease in the frequency of the fundamental quadrupole mode, ${}^2f$, and its first overtone, ${}^2p_1$, is 8% and 44%, correspondingly. If one follows a similar sequence with same rest mass, but rotating differentially with length scale $A/r_e = 1$ (A being the parameter in rotation law [1.125]), then stars with much higher $T/|W|$ exist (up to 0.22), and thus the effect of rotation on mode frequencies is stronger. The frequencies for the F, H_1, ${}^2f$, and ${}^2p_1$ modes decrease by 53%, 29%, 31%, and 57%, correspondingly.

The monotonic decrease in mode frequencies along sequences of fixed rest mass is explained by the fact that the central density of the star, along such a sequence, decreases as rotation increases, whereas frequencies of axisymmetric polar modes depend roughly on the compactness of the star. If, however, one chooses to study a different sequence of rotating stars, then the effect of rotation on mode frequencies can be drastically different than described above. For example, along a sequence of models with fixed central density, the frequencies of the ${}^2f$ and H_1 modes increase with rotation, instead of decreasing. If one constructs sequences of fixed M/R^3, then the effect of rotation on mode frequencies is sensitive to the mass of the star [745]. For example, in low-mass models the frequency of the fundamental quasiradial mode increases with rotation, whereas in high-mass models it decreases.

Figure 8.1 also includes an example of a strong avoided crossing between the H_1- and ${}^4p_1$-modes. The labeling of the two modes does not follow the continuity of frequency, but the structure of the eigenfunctions before and after the crossing is used to label the modes as shown in the figure. In the immediate region of the avoided crossing, the two eigenfunctions have very similar structure, so there is no single point along a sequence where the eigenfunction changes character, but rather a continuous transition. Going to higher-order modes, one finds a large number of avoided crossings as the mass-shedding limit ($\Omega = \Omega_K$) is approached.

The low-frequency modes, labeled i_1, i_2, and i_{-2} in Fig. 8.1, are axisymmetric inertial modes. In this frequency range there exist an infinite number of densely packed inertial modes. In a simulation of a perturbed star, a few of these modes stand out in the Fourier transform of the time evolution of velocity components, depending on the structure of the initial perturbation. In this particular example, a triplet of low-order inertial modes were excited, and the subscript in the labels simply indicates the number of nodes in the velocity component v_θ at slow rotation rates (with "−2" indicating a mode with two nodes but lower frequency than the i_1 mode). The frequency of inertial modes is linear in Ω in the limit of slow rotation, but along the differentially rotating sequence of fixed rest mass shown in Fig. 8.1, their frequency reaches a maximum value and then decreases with increasing $T/|W|$. This is because, along this sequence, the angular velocity at the center and at the surface also reaches a maximum value, even though $T/|W|$ continues to increase. In contrast, along a sequence of fixed central density there is a monotonic increase in inertial mode frequencies, as the angular velocity at the center and at the surface also increases monotonically.

When mode frequencies are expressed as a function of $T/|W|$ (as in Fig. 8.1), most low-order modes (such as the fundamental F and ${}^2f$ modes and the first quasi-radial overtone H_1) are insensitive to the degree of differential rotation, whereas other modes, such as the ${}^2p_1$ mode and inertial modes, do depend on the rotation law.

The frequencies of the two lowest-order quasiradial modes for a sequence of uniformly rotating, $N = 1$ polytropes, as a function of angular velocity, are shown in Fig. 8.2. The nonrotating member of the sequence is identical to the one in Fig. 8.1, but the sequence is formed by models having fixed central density, instead of fixed rest mass. The angular velocity is normalized by the angular velocity at the mass-shedding limit, Ω_K. Along such a sequence, the frequencies are affected less by rotation than along a sequence of fixed rest mass.

Frequencies in Fig. 8.2 are shown both in full general relativity (obtained with a nonlinear simulation code [224]) and in the Cowling approximation [223]. The error made by the Cowling approximation in the frequency of the fundamental quasiradial mode, F, is significant, roughly a factor of 2. For the first overtone, $H1$, the accuracy of the approximation improves.

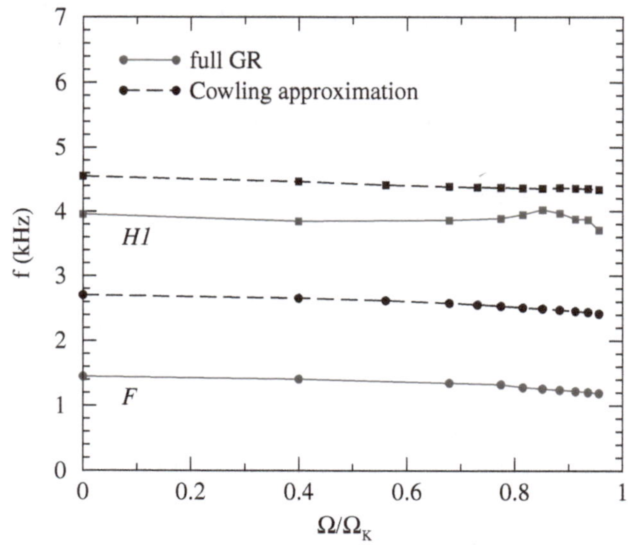

Figure 8.2. Frequencies of quasiradial modes for a sequence of uniformly rotating, $N = 1$ polytropes, as a function of angular velocity (see text for details). Filled circles correspond to the fundamental mode, F; filled squares, to the first overtone, H_1. The frequencies are shown both for a computation using full general relativity and in the Cowling approximation. The latter has a large error (roughly a factor of 2) for the fundamental mode. (Figure adapted from [224].)

8.6 Nonaxisymmetric modes

8.6.1 f- and p-modes

The nonaxisymmetric f- and p-mode oscillations are strong candidate sources of detectable gravitational waves, either from excitation of stable oscillations or through the onset of instabilities, such as the CFS gravitational-wave driven instability (see Chapter 9). In studying these oscillation modes, characteristics of key importance are the frequency range and the neutral points, where the frequencies of the $l = m$ counterrotating modes go through zero and the CFS instability sets in.

The properties of nonaxisymmetric f-modes in uniformly rotating compact stars have been studied extensively in various approximations: in the Newtonian limit [454, 325, 327, 329, 737, 531], in the post-Newtonian approximation [168, 171], in the slow-rotation approximation [358, 357, 742, 359, 654, 533], and in the Cowling approximation [739, 328, 251, 376]. In the Newtonian limit, the neutral point for the $l = m = 2$ instability in uniformly rotating stars is at $T/|W| = 0.14$, independently of the EOS. Only stiff EOSs, however, can reach such high values of $T/|W|$ before the mass-shedding limit. For Newtonian polytropes, $N < 0.808$ is required for the instability to set in. The Cowling approximation overestimates frequencies and

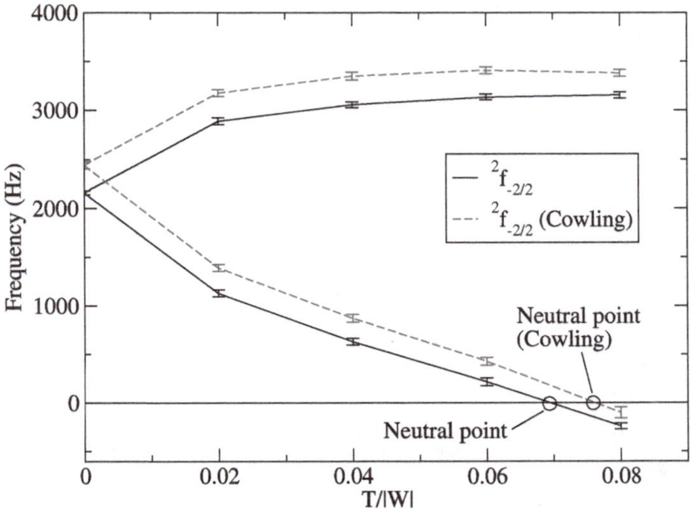

Figure 8.3. Inertial-frame frequencies of the fundamental $l = \pm m = 2$ f-modes along a sequence of the most relativistic, uniformly rotating $\Gamma = 2.0$ polytropes with fixed central density. The solid line is obtained in full general relativity, whereas the dashed line is obtained in the relativistic Cowling approximation. The rotational sequence terminates at the mass-shedding limit. (Figure from [752].)

the value of $T/|W|$ at neutral points. Typically, it yields neutral points very near the mass-shedding limit, leaving only a small range of rotation rates for which the CFS instability sets in. The slow-rotation approximation leads to relatively large errors in mode frequencies near the mass-shedding limit and thus cannot be used for accurately determining neutral modes.

Although neutral $l = m$ f-modes in full general relativity and rapid rotation were obtained in [665, 486] using a perturbative method, the *full range of frequencies* for fundamental modes was obtained by Zink et al. for relativistic polytropes [752], using a nonlinear time-evolution code. For the most relativistic sequence of $\Gamma = 2$, polytropes with fixed central density, the fundamental $l = 2$ f-mode frequency[7] starts at 2.16 kHz for the nonrotating member. As rotation increases, the $l = -m = 2$ corotating mode increases almost monotonically in frequency, reaching up to 3.2 kHz, with a slight decrease as the mass-shedding limit is reached. In contrast, the counterrotating $l = m = 2$ f-mode decreases monotonically in frequency, becomes neutral well before the mass-shedding limit, and reaches -0.24 kHz at the mass-shedding limit (i.e., it becomes corotating in the inertial frame) (Fig. 8.3). The neutral point is at $T/|W| = 0.07$, which is half of the Newtonian classical result. The Cowling approximation overestimates this by about 10%, whereas the perturbative results in [665] somewhat underestimate the rotation at the neutral point.[8]

[7] Here, we refer to frequencies in the inertial frame.
[8] In [665] the accuracy was affected by the limited set of available trial eigenfunctions.

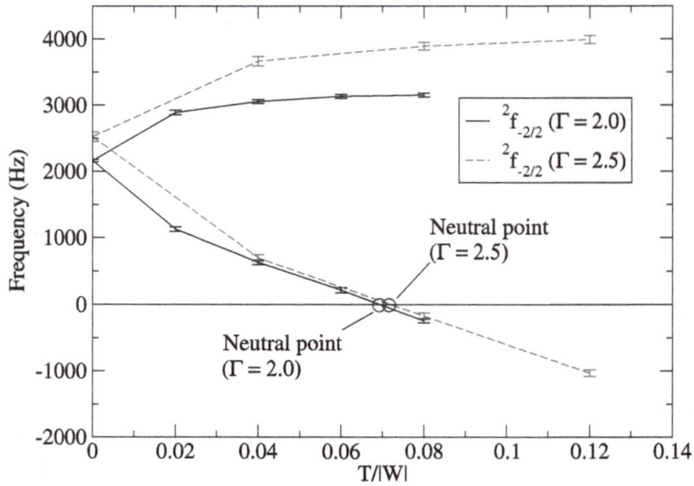

Figure 8.4. Comparison of inertial-frame frequencies of the fundamental $l = \pm m = 2$ f-modes along fixed-central-density sequences of the most relativistic, uniformly rotating $\Gamma = 2.0$ and $\Gamma = 2.5$ polytropes. The rotational sequences terminate at the mass-shedding limit. (Figure from [752].)

For a stiffer polytrope of $\Gamma = 2.5$, which more accurately produces relativistic models of rotating neutron stars than the usual choice of $\Gamma = 2.0$, the $l = -m = 2$ corotating mode increases monotonically in frequency along a similar sequence of rotating models, reaching 4.0 kHz at the mass-shedding limit. The counterrotating $l = m = 2$ f-mode decreases monotonically in frequency and reaches -1.02 kHz at the mass-shedding limit (Fig. 8.4), whereas the Cowling approximation underestimates this by nearly 30%. The neutral point along this sequence is at only slightly higher $T/|W|$ than for the corresponding sequence of $\Gamma = 2.0$ polytropes; that is, in terms of $T/|W|$, the neutral point along the most relativistic rotational sequence of fixed central density is somewhat insensitive to the EOS for polytropes in this range of stiffness.[9]

The $l = m = 3$ counterrotating f-mode along the same rotational sequence for the $\Gamma = 2.0$ polytropes as in Fig. 8.3 decreases monotonically in frequency, reaching -1.3 kHz at the mass-shedding limit, whereas the neutral point is at the significantly smaller $T/|W|$ of 0.0467, compared to the $l = m = 2$ f-mode (Fig. 8.5). This is in good agreement with the perturbative result of [665].

Frequencies of $l = m$ f-modes in differentially rotating compact stars[10] have been obtained in the Newtonian limit [384, 344, 613], in the slow-rotation approximation [533], and in the Cowling approximation [744, 376]. In general, the actual mode frequencies are somewhat higher than in comparable uniformly rotating models, and the neutral points for the CFS instability occur at somewhat larger $T/|W|$.

[9] This changes when a larger sample of tabulated EOs is considered; see [486].
[10] The differential rotation law (1.125) is usually assumed.

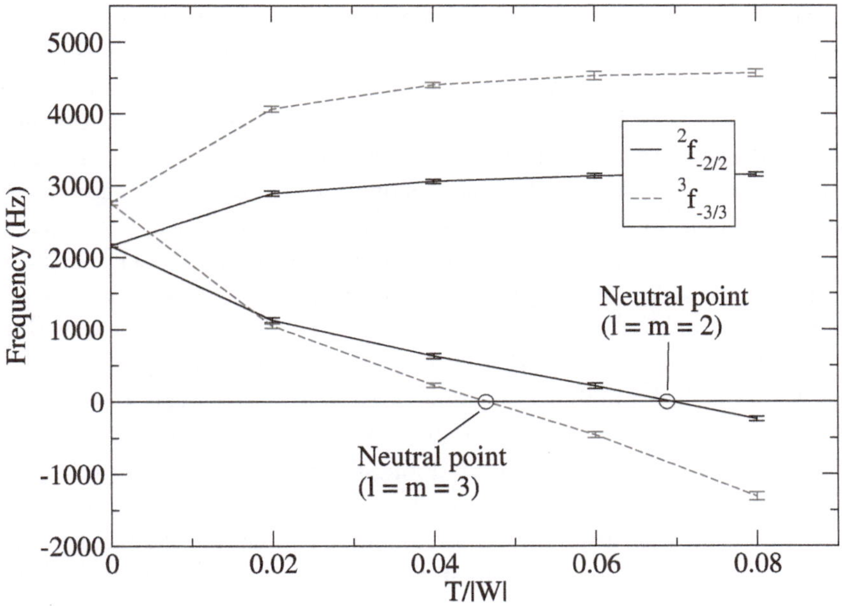

Figure 8.5. Comparison of inertial-frame frequencies of the fundamental $l = \pm m = 2$ and $l = \pm m = 3$ f-modes along a fixed-central-density sequence of the most relativistic, uniformly rotating $\Gamma = 2.0$ polytropes. The rotational sequence terminates at the mass-shedding limit. (Figure from [752].)

At the same time, however, the maximal $T/|W|$ along a rotational sequence can increase significantly because of differential rotation, as the mass-shedding limit is less restrictive when the outer layers are rotating at less than the mean rotation rate. This allows for a larger parameter space in which the CFS instability could operate than in the case of uniform rotation.

8.6.2 Rotationally restored modes: r-modes and inertial modes

We have seen for nonrotating stars that linear perturbations with axial parity are stationary flows. Nonrotating homentropic stars (more generally, barotropic models whose perturbations obey the same one-parameter EOS as the equilibrium star) have a still larger zero-frequency subspace that also includes stationary convection currents with polar parity. In each case, there is no restoring force, because the linear perturbation involves no change in a fluid element's density or pressure; there is also no radiation reaction, because the flow is stationary. In the corresponding rotating models, these zero-frequency perturbations become modes with frequencies proportional to the angular velocity Ω, with a restoring force associated with the Coriolis force in a rotating frame.

A subclass of these modes with purely axial parity for slowly rotating stars was first discussed by Papaloizou and Pringle [525], working in a Newtonian framework.

They termed them *r-modes* or Rossby modes, because they generalize to stellar models the Rossby waves of oceans and planetary atmospheres. As we will see in the next chapter, these modes are particularly interesting because they can be unstable for angular velocities well below the maximum angular velocity of a rotating neutron star. This is related to the fact that a nonaxisymmetric mode becomes unstable along a sequence of rotating stellar models only at a value of the angular velocity for which the mode's frequency vanishes (see Section 9.5). With a frequency that already vanishes for a spherical star, r-modes of perfect-fluid models are unstable for arbitrarily slow rotation. In neutron stars, instability requires a radiation-reaction growth time shorter than the viscous damping time, and the r-mode growth times may be short enough to be unstable in old neutron stars spun up by accretion, in neutron stars that are rapidly rotating at birth, and in the hypermassive neutron stars formed in the coalescence of a double–neutron-star system (this last possible only if the EOS is stiff enough to allow a long-lived postmerger remnant).

In this section, we briefly review the rotationally restored modes in the Newtonian approximation and then turn to the corresponding modes in the full theory. The Newtonian description is simpler, but it misses features present in the full theory that may alter the nature of rotationally restored oscillations.

The difference between the zero-frequency subspaces of nonrotating models for barotropic and for nonbarotropic models leads to a corresponding difference in the number and character of rotationally restored modes. For the nonbarotropic case, because all zero-frequency modes (time-independent perturbations) of a nonrotating star have axial parity, the rotationally restored modes have velocity fields that have purely axial parity in the $\Omega = 0$ limit. To order Ω^2, their form is that of Eq. (7.228), the axial vector field of a single angular harmonic:

$$\delta u_a = e^{-\nu} U(r) \epsilon^b{}_a \nabla_b Y_{lm}\, e^{i\omega_i t}[1 + O(\Omega^2)]. \tag{8.20}$$

In the Newtonian limit, the frequency of the mode has an simple analytic expression that we derive below,

$$\omega_i = -\frac{(l-1)(l+2)}{l(l+1)} m\Omega. \tag{8.21}$$

In barotropic models, because the zero-frequency subspace is larger, including velocity fields with polar as well as axial parity, there is a correspondingly larger set of rotationally restored modes. Rotation mixes the zero-frequency axial- and polar-parity perturbations, and the $\Omega = 0$ limit of the velocity field is in general a sum of contributions with axial and polar parity. In the Newtonian approximation, there is a residual class of r-modes with $l = m$, a remaining set of modes whose velocity field satisfies Eq. (8.20) and whose frequency is again given by Eq. (8.21). In the full theory, however, even these modes acquire a polar parity piece: All rotationally restored modes of barotropic stars have parts with both polar and axial parity in the $\Omega = 0$ limit.

There is one additional difference between the relativistic perturbations and their Newtonian versions, this time for nonbarotropic models. The eigenvalue equation for

Table 8.1. *Inertial modes in barotropic and nonbarotropic models*

	Nonbarotropic stars	Barotropic stars
Newtonian	inf. set of r-modes for each (l, m) inf. set of g-modes	r-modes only for $l = m$ inf. set of inertial modes
General relativity	inf. set of r-modes for each (l, m) inf. set of g-modes continuous spectrum?	no pure r-modes inf. set of inertial modes

the frequency, at lowest nonvanishing order in Ω, has a singular term that results in a continuous spectrum. It remains an open question whether the continuous spectrum is present in nonbarotropic models when one works to all orders in Ω.

Table 8.1, taken from Lockitch, Andersson, and Watts [443], summarizes these properties of rotationally restored modes.

The (l, m) assignment in this table refers to the angular behavior of a mode to lowest order in Ω, or, equivalently, to its $\Omega \to 0$ limit. In a rapidly rotating star, modes are superpositions of angular harmonics with $l \geq |m|$, but with definite parity under reflection in the equatorial plane. Because

$$Y_{lm}(\theta, \phi) = (-1)^{l+m} Y_{lm}(\pi - \theta, \phi), \tag{8.22}$$

modes odd under reflection are a superposition of axial contributions with $l = |m| + 2k$ and of polar contributions with $l = |m| + 2k + 1$, $k \geq 0$; modes even under reflection are a superposition of polar contributions with $l = |m| + 2k$ and axial contributions with $l = |m| + 2k + 1$.

In the remainder of this section, we derive key features of the rotationally restored modes that have been summarized above. We obtain the frequency of Newtonian r-modes (and, for barotropic stars, find their radial dependence). We show that nearly all the rotational modes of Newtonian barotropic models are inertial modes. We obtain the Kojima equation governing the r-modes of relativistic nonbarotropic models, and we review work on whether it implies the existence of a continuous spectrum.

Rotationally restored modes of Newtonian stars

Nonbarotropic stars. We begin with nonbarotropic Newtonian stars. All rotationally restored modes are then r-modes, modes whose $\Omega = 0$ limit is an axial perturbation of the velocity field. We saw that an axial (toroidal) perturbation of a spherical star is a stationary perturbation of its velocity field, and a Newtonian perturbation belonging to the (l, m) angular harmonic is then given by $\delta \mathbf{v} = U(r)\, \mathbf{r} \times \boldsymbol{\nabla} Y_{lm} e^{i\omega_i t}$ or

$$\delta v^a = U(r) \epsilon^{ba} \nabla_b Y_{lm} e^{i\omega_i t}, \tag{8.23}$$

where $\epsilon^{ab} = \epsilon^{abc} \nabla_c r$ is the Newtonian limit of Eq. (7.154).

The equilibrium pressure and density of a slowly rotating star differ from their values in the corresponding spherical configuration by terms of order Ω^2; there can be no term of order Ω, because changing the sign of Ω gives an equivalent description of the same star. For a nonbarotropic star, this implies that, to order Ω, the perturbed velocity field of a slowly rotating star again belongs to a single angular harmonic, again having angular dependence given by Eq. (8.23). All that changes is its time dependence: The mode acquires a nonzero frequency.

From the Euler equation perturbed about a uniformly rotating configuration,

$$\delta E_a := \delta\left[(\partial_t + \pounds_{\mathbf{v}})v_a + \frac{\nabla_a p}{\rho} + \nabla_a\left(-\frac{1}{2}v^2 + \Phi\right)\right] = 0, \tag{8.24}$$

we can quickly obtain an equation for the frequency of an axial mode. As we have seen, surfaces of constant p and ρ (isopycnic and isobaric surfaces) coincide in the equilibrium star, and this implies that the component of $\boldsymbol{\nabla}\times\delta\mathbf{E}$ along $\boldsymbol{\nabla}p$ involves only $\delta\mathbf{v}$. Because $p = p(r) + O(\Omega^2)$, the equation has the form

$$\begin{aligned} 0 &= \epsilon^{ab}[\nabla_a(\partial_t + \pounds_{\mathbf{v}})\delta v_b + \nabla_a \pounds_{\delta\mathbf{v}} v_b], \\ &= i\omega_r \epsilon^{ab}\nabla_a \delta v_b + \pounds_{\delta\mathbf{v}}\epsilon^{ab}\nabla_a v_b + O(\Omega^2), \end{aligned} \tag{8.25}$$

where ω_r is the Newtonian frequency in a corotating frame, and we are taking $\delta\mathbf{v}$ to be $O(\Omega^0)$. Using Eq. (8.23), we have

$$e^{-iwt}\epsilon^{ab}\nabla_a\delta v_b = U\epsilon^{ab}\epsilon^c{}_b\nabla_a\nabla_c Y_{lm} = U\nabla^2 Y_{lm} = -l(l+1)r^{-2}UY_{lm}, \tag{8.26}$$

$$e^{-iwt}\pounds_{\delta\mathbf{v}}\epsilon^{ab}\nabla_a v_b = e^{-iwt}\delta v^\theta\partial_\theta(2\Omega\cos\theta) = 2im\Omega r^{-2}UY_{lm}, \tag{8.27}$$

and Eq. (8.25) then becomes

$$i[\omega_r l(l+1) - 2m\Omega]UY_{lm} = 0, \tag{8.28}$$

implying for ω_r the value

$$\omega_r = \frac{2m}{l(l+1)}\Omega. \tag{8.29}$$

The frequency in an inertial frame is then

$$\omega_i = \frac{2m}{l(l+1)}\Omega - m\Omega = -\frac{(l-1)(l+2)}{l(l+1)}m\Omega. \tag{8.30}$$

Because the signs of ω_r and ω_i are opposite when $l \geq 2$ and $m \neq 0$, these modes are all unstable. The imaginary part of the frequency is higher order in Ω and is evaluated in Section 9.6.2.

Solution to next order in Ω. When δp, $\delta\rho$, and $\delta\Phi$ are expressed in terms of a Lagrangian displacement $\xi^a = (i\omega_r)^{-1}\delta v^a$, either of the two remaining components of the Euler equation can be solved to obtain the radial dependence $U(r)$ of $\delta\mathbf{v}$. The components are not independent, and the r-component of the curl, which we used to find ω_i, is the integrability condition, guaranteeing the consistency of these two remaining equations. The equations are second order in Ω, and their solution was first found by Provost, Berthomieu, and Rocca [553], with the restriction to the Cowling approximation and to $m \neq 0$.

In slowly rotating, Newtonian uniform-density stars in the Cowling approximation, one can describe the main features of the instability analytically to $O(\Omega^2)$ [369]. Here we briefly outline the derivation of the main results.[11]

The $O(\Omega)$ rotational corrections to the mode depend on the distortion of the equilibrium configuration, written in terms of a distorted radial function $a(r)$ with

$$r = a(1+\epsilon), \tag{8.31}$$

where ϵ is of order is $\mathrm{O}(\Omega^2)$ and represents the deformation of the equilibrium structure when compared to a corresponding nonrotating configuration. Surfaces of constant a are equipotential surfaces of the rotating equilibrium. At order Ω^2, the distortion of the star has only monopole and quadrupole parts:

$$\epsilon = \left[\tilde{D}_1(a) + \tilde{D}_2(a)P_2(\cos\theta)\right]\bar{\Omega}^2, \tag{8.32}$$

where $\bar{\Omega}$ is a dimensionless angular velocity, as defined below.

A normal-mode solution to the perturbation equations satisfies the perturbed Euler equation, the perturbed continuity equation, and an equation of state $p = p(\rho)$, here replaced by an assumption of uniform density. The perturbed velocity field of Eq. (8.23) acquires at $O(\Omega)$ a polar (spheroidal) correction:

$$\delta v^a = \left(U\epsilon^{ba}\nabla_b Y_{lm} + \frac{1}{r}WY_{l+1\,m}\nabla^a r + V\nabla^a Y_{l+1\,m}\right)e^{i\omega_i t}. \tag{8.33}$$

Along the basis of unit vectors $(\hat{\mathbf{r}}, \hat{\boldsymbol{\theta}}, \hat{\boldsymbol{\phi}})$, δv^a has components

$$e^{-iwt}r(\delta v^{\hat{r}}, \delta v^{\hat{\theta}}, \delta v^{\hat{\phi}}) = \left(0, U\frac{1}{\sin\theta}\partial_\phi, -U\partial_\theta\right)Y_{lm} + \left(W, V\partial_\theta, V\frac{1}{\sin\theta}\partial_\phi\right)Y_{l+1\,m}. \tag{8.34}$$

The (scalar) perturbation in the pressure is expressed in terms of a dimensionless function $\zeta(a)$ as

$$\delta p = \rho g a \zeta Y_{l+1}^m e^{i\omega_i t}, \tag{8.35}$$

where $g = -\rho^{-1}dp/dr$ is the gravitational acceleration.

We define a dimensionless frequency and angular velocity by

$$\bar{\omega}_r = \omega_r\left(\frac{R^3}{GM}\right)^{1/2}, \quad \bar{\Omega} = \Omega\left(\frac{R^3}{GM}\right)^{1/2} \tag{8.36}$$

and expand the mode frequency in the rotating frame as

$$\bar{\omega}_r = \sigma_0\bar{\Omega} + \sigma_2\bar{\Omega}^3. \tag{8.37}$$

[11] Note that, in our presentation, the perturbed velocity field is $O(\Omega^0)$, nonzero for the spherical star, and the scalar δp is $O(\Omega)$. Much of the Newtonian literature adopts a convention in which $\delta v^a = O(\Omega)$ and $\delta p = O(\Omega^2)$. To translate from one convention to another, one simply multiplies or divides all perturbed quantities by Ω.

The corresponding dimensionless versions of U, V, and W are

$$\bar{U} = \frac{U}{\sqrt{GMR}}, \quad \bar{V} = \frac{V}{\sqrt{GMR}}, \quad \bar{W} = \frac{W}{\sqrt{GMR}}. \tag{8.38}$$

The perturbed Euler and continuity equations for the variables $\bar{U}$, $\bar{V}$, $\bar{W}$, and ζ have for $l = m$ the form [582]

$$r\frac{d\zeta}{dr} = (l-1)\,\zeta, \tag{8.39}$$

$$r\frac{d\bar{W}}{dr} + (l+2)\,\bar{W} = ik\left(\frac{r}{R}\right)^2 \bar{\Omega}\,\zeta, \tag{8.40}$$

$$\bar{V} = \bar{W} - i\frac{(2l+1)(l+1)}{8l}\left[(l+1)\sigma_2 + 6\tilde{D}_2\right]\left(\frac{r}{R}\right)^2 \bar{\Omega}\,\zeta, \tag{8.41}$$

and

$$\zeta = \frac{2l}{(l+1)\sqrt{2l+3}}\left(\frac{R}{r}\right)^2 \bar{\Omega}\,\bar{U}, \tag{8.42}$$

where

$$k = \frac{1}{\sigma_0}\left\{\frac{(l+1)}{l}\left[(2l+3)\frac{\sigma_2}{\sigma_0} + 6(l-1)\tilde{D}_2\right] + 3\left(3\tilde{D}_2 + a\frac{d\tilde{D}_2}{da}\right)\right\}. \tag{8.43}$$

Equation (8.39) implies that ζ has the form

$$\zeta = Cr^{l-1}. \tag{8.44}$$

Then, Eq. (8.42) gives $\bar{U} \propto r^{l+1}$ and allows us to express the order-Ω change in the pressure in terms of the zeroth-order r-mode velocity field. Choosing the normalization

$$\bar{U} = \left(\frac{r}{R}\right)^{l+1}, \tag{8.45}$$

we have

$$\zeta = \frac{2l}{(l+1)\sqrt{2l+3}}\left(\frac{R}{r}\right)^{l-1} \bar{\Omega}. \tag{8.46}$$

These are the only two eigenfunctions needed in our later discussion of r-mode stability.

The $O(\Omega^2)$ frequency correction σ_2 can be found from an approximate integral relation that is constructed from the perturbed Euler equations (see the appendix of [582]). Integrating by parts and using the fact that $\tilde{D}_2 = -(5/6)\bar{\Omega}^2$ for uniform-density stars (Appendix II of [553]), one obtains the $O(\Omega^3)$ correction to the normalized mode frequency in the rotating frame:

$$\bar{\omega}_r = \frac{2}{l+1}\bar{\Omega} + \frac{5l}{(l+1)^2}\bar{\Omega}^3. \tag{8.47}$$

Barotropic stars. As we have noted, because the degenerate subspace of zero-frequency modes of barotropic stars includes the polar-parity g-modes, there is a larger set of rotationally restored modes. The Einstein-Euler system of barotropic stars mixes modes with axial and polar parity, and the $\Omega \to 0$ limit of nearly all modes with frequency proportional to Ω is a time-independent perturbation that mixes axial and polar parity. This larger class of rotationally restored perturbations are called *inertial modes.*

That slowly rotating barotropic stars have r-modes only for angular harmonics with $l = m$ can be seen as follows. In finding the r-mode frequency, we used the fact that the radial component (8.25) of $\delta \nabla \times \mathbf{E}$ involved only $\delta \mathbf{v}$. For barotropic stars, $\nabla \times \mathbf{E} = 0$ is the vorticity-conservation equation $(\partial_t + \mathcal{L}_{\mathbf{v}})\nabla \times v = 0$, and all components of its perturbed form,

$$q^a := \epsilon^{abc}(i\omega_r \nabla_b \delta v_c + \Omega \mathcal{L}_{\delta \mathbf{v}} \phi_c) = 0, \tag{8.48}$$

are satisfied. For nonaxisymmetric perturbations, any two components imply the third, and we can take $q^\theta = 0$ as the additional constraint on the perturbation. The perturbation is a superposition of different values of l, and a purely axial perturbation has the form

$$\delta v^a = \sum_{l=\ell}^{\infty} U_l(r)\epsilon^{ba}\nabla_b Y_{lm}, \tag{8.49}$$

where ℓ, the minimum value of l for which U_l is nonzero, satisfies $\ell \geq m$. We first show that $\ell = m$. To do this, we look at the projection of q^θ along $Y_{l-1,m}$, $\int q^\theta Y^*_{l-1,m} d\Omega$. Using the relations

$$\sin\theta \partial_\theta Y_{lm} = l Q_{l+1\,m} Y_{l+1\,m} - (l+1) Q_{l\,m} Y_{l-1\,m}, \tag{8.50a}$$

$$\cos\theta Y_{lm} = Q_{l+1\,m} Y_{l+1\,m} + Q_{l\,m} Y_{l-1\,m}, \tag{8.50b}$$

where

$$Q_{lm} := \left[\frac{(l-m)(l+m)}{(2l-1)(2l+1)}\right]^{1/2}, \tag{8.51}$$

we obtain

$$Q_{lm}\left\{[2m\Omega + (l+1)\omega_r]U_l' + 2(l+1)m\Omega\frac{U_l}{r}\right\} = 0. \tag{8.52}$$

The relation gives no new constraint for $l = m$ because $Q_{mm} = 0$. For $l \neq m$, however, the relation implies the vanishing of the expression in curly brackets. Using Eq. (8.29) for ω_r, that constraint becomes $U_l' + \dfrac{l}{r}U_l = 0$, implying

$$U_l = K r^{-l}. \tag{8.53}$$

Thus the velocity field is singular unless U_m is nonzero.

We next show that there is a solution with only the single angular harmonic $l = m$; that is, $U_l = 0$ for $l > m$. For $l > m$, the angular components of the vorticity conservation equation (8.48) each give the same recursion relation for U_{l+2} in terms of U_l, and the relation has a solution with only U_m nonzero: The harmonics of

$q^\theta = 0$ with $l > m+1$ vanish identically when $U_l = 0$ for $l > m$. We are left with the harmonic

$$\int q^\theta Y^*_{m+1,m} d\Omega = -i\frac{m^2}{m+1} Q_{m+1\,m}\left[U'_m - (m+1)\frac{U_m}{r}\right] = 0, \tag{8.54}$$

with immediate solution

$$U_m = u_0 \frac{r^{m+1}}{R^m}, \qquad \delta\mathbf{v} = u_0\left(\frac{r}{R}\right)^m \boldsymbol{r}\times\boldsymbol{\nabla}Y_{mm}, \tag{8.55}$$

with the constant u_0 an average velocity measuring the amplitude of the mode.

The remaining rotationally restored modes are hybrids, whose $\Omega = 0$ limit is a mixture of polar and axial contributions. The velocity field now has the form

$$\delta v^a = \sum_{l=m}^{\infty}\left\{\frac{1}{r}W_l Y_{lm}\nabla^a r + V_l\nabla^a Y_{lm} + U_l\epsilon^{ba}\nabla_b Y_{lm}\right\} e^{i\omega_i t}, \tag{8.56}$$

with the polar contributions W_l and V_l and axial contribution U_l nonzero for alternating values of l. For uniform-density models, there is a maximum nonzero value l_0 of l; for the more realistic models that have been considered [444], l_0 serves as an approximate cutoff, with larger angular harmonics having much smaller coefficients.

Rotationally restored modes of relativistic stars

In discussing the rotationally restored modes of slowly rotating relativistic stars, we use the formalism developed by Kojima and Hosonuma [360, 361, 362] and adopted by Lockitch et al. [442, 444, 443]. We refer briefly to related work [16, 571, 572, 574, 365, 743] mentioned previously in Section 8.4.2.

Rotational modes of nonbarotropic stars. As in the Newtonian case, we use conservation of vorticity in surfaces of constant s, Eq. (7.98). Again, to $O(\Omega^2)$, $\nabla_\alpha s$ is along $\nabla_\alpha r$, and conservation of circulation has the single component

$$0 = \frac{1}{h}\Delta\omega_{\theta\phi} = \partial_\theta\Delta u_\phi - \partial_\phi\Delta u_\theta, \tag{8.57}$$

where δu_a is given in terms of the single radial function $U(r)$ by Eq. (8.20). The θ and ϕ components of Δu_α involve the perturbed metric, which has the form of an axial perturbation of the Schwarzschild background. Recall from Section 7.5.2 that the axial perturbation with fluid source has h_{ta} nonzero in a Regge-Wheeler gauge. For harmonic time dependence, it has the form

$$h_{ta} = -h_0(r)\epsilon^b{}_a\nabla_b Y_{lm} e^{i\omega_i t}[1 + O(\Omega^2)]. \tag{8.58}$$

To find the components of Δu_α, write

$$\Delta u_\alpha = h_{\alpha\beta}u^\beta + u^\beta\nabla_\beta\xi_\alpha + u^\beta\nabla_\alpha\xi_\beta + u_\alpha u^\beta u^\gamma\nabla_\beta\xi_\gamma, \tag{8.59}$$

take $\xi_t = 0$, and use Eq. (7.5a),

$$\delta u^\alpha = q^\alpha{}_\beta\mathcal{L}_{\mathbf{u}}\xi^\beta = i\omega_r e^{-\nu} q^\alpha{}_\beta\xi^\beta, \tag{8.60}$$

to obtain

$$\Delta u_\theta = e^{-\nu}\left[h_{t\theta} + i\omega_r\xi_\theta + \Omega\partial_\theta\xi_\varphi - 2\bar\omega\cot\theta\xi_\varphi\right], \tag{8.61}$$

$$\Delta u_\varphi = e^{-\nu}\left[h_{t\varphi} + i\omega_r\xi_\varphi + \Omega\partial_\varphi\xi_\varphi + 2\bar\omega\sin\theta\cos\theta\xi_\theta\right]. \tag{8.62}$$

Finally, using Eqs. (8.20), (8.60), and (8.58) to express ξ_a and h_{ta} in terms of U and h_0, we can write the vorticity conservation relation (8.57) in the form

$$i\omega_r l(l+1)(h_0+U) - 2im\bar\omega U = 0. \tag{8.63}$$

In the Newtonian limit, we have $h_0 + U \to U$, $\bar\omega \to \Omega$, and Eq. (8.63) becomes Eq. (8.28). Here, however, the equation involves both h_0 and U, and the field equation $\delta G_t{}^\phi = 8\pi\delta T_t{}^\phi$ again relates h_0 to U, changing the character of the system and giving an eigenvalue equation for ω_i. This field equation has the form

$$h_0'' - (\nu'+\lambda')h_0' - \left[\frac{(l-1)(l+2)}{r^2}e^{2\lambda} + \frac{2}{r}(\nu'+\lambda') + \frac{2}{r^2}\right]h_0 = \frac{4}{r}(\nu'+\lambda')U. \tag{8.64}$$

The Kojima equation is obtained by using the first equation (8.63) to express U in terms of h_0,

$$U = -\frac{\omega_r}{(\omega_r-\alpha)}h_0, \quad \text{where} \quad \alpha = \frac{2m}{l(l+1)}\bar\omega, \tag{8.65}$$

and then replacing U in Eq. (8.64) by this expression. The resulting relation is slightly simpler if we add $\frac{4}{r}(\nu'+\lambda')h_0 = 16\pi(\epsilon+p)e^{2\lambda}h_0$ to both sides of Eq. (8.64); after multiplying by $(\omega_r - \alpha)$, we obtain

$$(\omega_r-\alpha)\left\{h_0'' - (\nu'+\lambda')h_0' - \left[\frac{(l-1)(l+2)}{r^2}e^{2\lambda} - \frac{2}{r}(\nu'+\lambda') + \frac{2}{r^2}\right]h_0\right\}$$
$$= -16\pi e^{2\lambda}(\epsilon+p)\omega_r h_0. \tag{8.66}$$

As long as $\omega_r \neq \alpha(r)$ inside the star, this equation admits a discrete set of modes for each l, m pair with $l \geq 2$, and the frequency of each mode is proportional to Ω. This is always the case for models with uniform density [442]. For relativistic polytropes with stiffness comparable to that of neutron stars, however, the eigenvalue is such that $\omega_r - \alpha(r_0) = 0$ at some radius r_0 inside the star [571, 572, 736, 741]. At this order in the slow-rotation approximation, one then faces a singular eigenvalue problem: The eigenfunctions on either side of the singular point r_0 are logarithmically singular at r_0, and there is in general a finite jump in their first derivative. Allowing the class of solutions with discontinuous derivative gives a continuous spectrum [360, 66] and raises the question of whether a discrete r-mode will be seen in the evolution of a perturbed nonbarotropic model for which the Kojima equation is singular inside the star.

Lockitch, Andersson, and Watts [443] consider the question of the continuous spectrum and the existence of r-modes in some detail, following earlier work on differentially rotating fluid shells by Watts et al. [717, 719]. They argue that the singularity in the Kojima equation is an artifact of the slow-rotation approximation and that it is not present if one includes terms of order Ω^2. Away from r_0, the order-Ω^2 terms are negligible for slow rotation, but near r_0 they dominate. The higher-order terms mix polar and axial contributions to the velocity field and yield a system of differential equations – equivalently, an ODE of higher-derivative order. In a toy model in which higher-order terms are added, the result is to smooth singular eigenfunctions near $r = r_0$. This fact, together with the nonsingular behavior of the system of equations governing rotational modes of barotropic stars, is a strong argument for the existence of r-modes in nonbarotropic models.

The existence of discrete r-modes, however, does not preclude a continuous part of the spectrum. Whether a continuous spectrum is present outside of the slow-rotation approximation is an open question.

Rotational modes of barotropic stars. In contrast to the nonbarotropic case, the barotropic mode problem is not singular at lowest order in Ω. All modes are now inertial, mixing axial and polar parity at zeroth order in Ω, with a velocity field of the form

$$\delta u^\alpha = \sum_{l=m}^{\infty} \left[\frac{1}{r} W_l(r) Y_{lm} r^\alpha + V_l(r) \nabla^\alpha Y_{lm} + U_l(r) \epsilon^{\alpha\beta\gamma\delta} \nabla_\beta Y_{lm} \nabla_\gamma t \nabla_\delta r \right] e^{i\sigma t}. \tag{8.67}$$

The perturbed metric similarly has polar and axial parts that are a sum of tensor harmonics of the form given in Eqs. (7.177) and (7.230), with h_0, H_0, H_1, H_2, and K nonzero in a Regge-Wheeler gauge and all but H_1 and h_0 of order Ω.

The residual $l = m$ r-modes of Newtonian barotropes acquire a polar part: Lockitch [441] shows that with h_0 nonzero and related to U by the $\delta G_t{}^\phi$ equation, the two independent components of the vorticity equation cannot be satisfied unless the polar contributions V and W to the velocity field are nonzero. The additional terms, however, are surprisingly small; as Fig. 8.6 shows, in stars for which the pressure distribution differs by 20% from the Newtonian configuration, the relativistic inertial mode differs by only 1% from its Newtonian form for the model, a uniform-density star with $(2M_0/R) = 0.2$.

The work of Lockitch, Andersson, and Watts [443] finds only discrete modes for relativistic (and Newtonian) rotational modes of barotropic stars. The earlier work by Ruoff et al., however, had suggested a continuous spectrum for this case as well as for nonbarotropic models. This appears to be a result of their use of the Cowling approximation together with the truncation of equations for spherical harmonics at a maximum value $l_{\max}$ of l. For each value of $l_{\max}$ there are continuous frequency bands in the spectrum; but as $l_{\max}$ is increased, the continuous parts of the spectrum are replaced by discrete modes. Finally, Ruoff, Stavridis, and Kokkotas [574] show that the inclusion of both polar and axial parts at order $O(\Omega)$, allows for discrete modes to be computed, in agreement with post-Newtonian [442] and nonlinear, rapid-rotation [663] calculations.

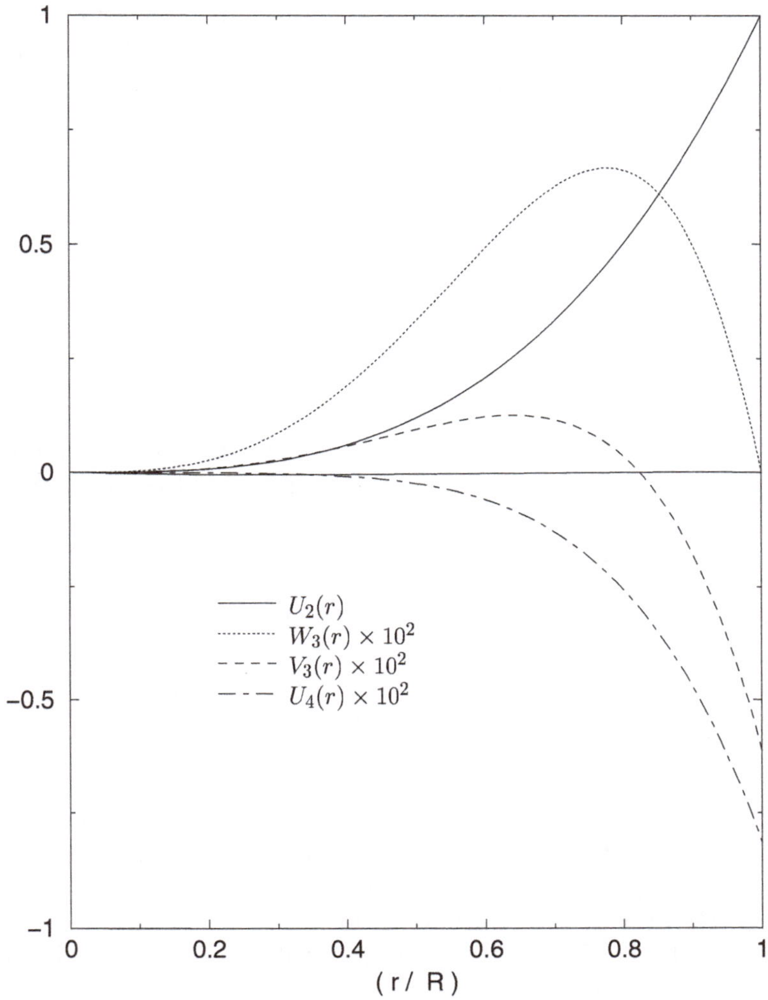

Figure 8.6. The dominant metric potential $U_2(r)$ of the relativistic $l = m = 2$ r-mode is plotted, together with the leading corrections W_2, V_2, and U_4, each of which is multiplied by 100 to be visible on the same scale. The vertical scale is set by the surface normalization $U_2(R) = 1$. (Figure adapted from [442].)

8.6.3 *g*-modes

In Section 8.2, we found the frequency of short-wavelength g-modes in terms of the Brunt-Väisälä frequency (8.6), a measure of the buoyancy of a displaced fluid element. In hot stars, the pressure depends on more parameters than just the rest mass density, and in the simplest model, one can consider the pressure to depend on both rest mass density and specific entropy, $p = p(\rho, s)$. Because the cooling timescale is much longer than the dynamical timescale, perturbed fluid elements oscillate adiabatically with frozen specific entropy, and the Lagrangian perturbations in pressure and energy density are related by

$$\Delta p = \frac{\Gamma_1 p}{\epsilon + p} \Delta \epsilon, \tag{8.68}$$

where

$$\Gamma_1 \equiv \left. \frac{\partial \ln p}{\partial \ln \rho} \right|_s \tag{8.69}$$

is the adiabatic index for the oscillating fluid.

The equilibrium star, though evolving constantly due to cooling, can be considered to be in an approximate equilibrium on timescales short compared to the cooling timescale and hence can be considered approximately as barotropic, with $s = s(\rho)$. When the adiabatic index Γ_1 differs from Γ of the barotropic equilibrium model, nontrivial g-mode oscillations arise. Then, the Eulerian perturbation in pressure, δp, is given by

$$\frac{\delta p}{\Gamma_1 p} = \frac{\delta \epsilon}{(\epsilon + p)} + \xi^\alpha A_\alpha, \tag{8.70}$$

where

$$A_\alpha \equiv \left(\frac{1}{\Gamma} - \frac{1}{\Gamma_1} \right) \nabla_\alpha \ln p, \tag{8.71}$$

is the Schwarzschild discriminant (see [442]), proportional, as we saw for spherical stars, to the squared frequency of short-wavelength g-modes. More generally, we show in Section 9.3 that when A_α points *inward* ($\Gamma_1 > \Gamma$), g-mode oscillations are *convectively stable* (see also [468]), whereas for $\Gamma_1 < \Gamma$, g-modes are unstable (see also [475, 477, 214]). For $\Gamma_1 = \Gamma$, the barotropic limit (marginal convective stability) is recovered (see [702] and Chapter 9).

In cold compact stars, nontrivial g-modes can also arise because of density discontinuities [466, 217, 671, 474]. In stars with a continuous composition gradient, the pressure can be considered (in a simple n, p, e model) to depend on the rest mass density and the proton fraction x_p. Equilibrium is governed by β-equilibrium of the weak interaction processes, but in an oscillating star, the timescale for these processes is much too long for matter to equilibrate within one oscillation period. One can then assume that the composition of a perturbed fluid element is frozen and define the adiabatic index Γ_1 as

$$\Gamma_1 \equiv \left. \frac{\partial \ln p}{\partial \ln \rho} \right|_{s, x_\mathrm{p}} . \tag{8.72}$$

When $\Gamma_1 \neq \Gamma$, one obtains core g-modes in cold, stratified, compact stars [532].

The frequency of g-modes in compact stars is sensitive to the difference of $\Gamma_1 - \Gamma$, which, in general, will depend on position. In the simplest models, one can assume that both Γ_1 and Γ are constant throughout the star, in order to obtain rough estimates. The frequency can be estimated from an averaging of the Brunt-Väisälä frequency [532]

$$< \mathrm{N} > \simeq \sqrt{\frac{4\pi}{3}} \left(1 - \frac{\Gamma}{\Gamma_1}\right)^{1/2} \sqrt{G\rho_c}, \tag{8.73}$$

where ρ_c is the central density, which agrees with numerical results to better than 5%. For example, for a typical nonrotating, relativistic polytrope with $\Gamma = 2.0$ and Γ_1 between 2.05 and 2.4, the frequency of the first $l = m = 2$ g-mode (denoted as ${}^2g_1$) ranges between roughly 200 and 600 Hz [252]. The next overtone, ${}^2g_2$, has a smaller frequency that ranges between roughly 150 and 400 Hz; higher overtones have decreasing frequencies.[12] In very hot proto–neutron stars, g-mode frequencies can be close to 1 kHz in the nonrotating limit, but within roughly 10 s they decrease to the values mentioned above [215].

In rapidly rotating stars, rotation increases the frequencies of corotating g-modes up to a few kHz, while the frequencies of counterrotating modes go through zero at a rotation rate of roughly $\Omega \sim 0.12\sqrt{G\rho_c}$ [382, 532, 252]. Thus, in rapidly rotating stars, g-modes can satisfy the CFS-instability condition, but it is estimated that the instability will be damped by viscous effects, because g-modes are weak gravitational wave emitters [215]. Another distinct feature of g-modes in rapidly rotating stars is that when the Coriolis force dominates over buoyancy, their character becomes similar to specific barotropic inertial modes [532, 252].

8.6.4 Crustal modes

Crustal shear modes for Newtonian models have been investigated in detail in [467]. In the crust, both polar and axial (torsional) oscillation modes can exist, and there is also a core/crust interfacial mode. Torsional modes in relativistic stars were first presented in [593] and have been considered as possibly detectable via electromagnetic emission (see [586] and references therein). The crust-core interfacial modes have been shown to be relevant for the shattering of the crust just prior to merger in binary neutron star systems, when tidally excited during the inspiral phase [701].

8.6.5 Alfvén modes

In a magnetized star, polar oscillations comprise a discrete spectrum, with frequencies nearly proportional to the strength of the magnetic field. In nonrotating magnetized stars, frequencies have been computed for relativistic models in [645]. For strongly magnetized models, rotation is usually ignored because of the rapid spin-down due to electromagnetic emission. Purely axial magnetohydrodynamic

[12] Recall that p-mode frequencies increase as the order of the mode increases.

oscillations in a star with a purely poloidal field form a continuous spectrum [406]. If a crust is present, its torsional shear modes will be effectively absorbed by the continuous spectrum in the core (see [406, 249] and references therein). Nevertheless, at the edges or turning points of continuous parts of the spectrum, one can have localized long-lived oscillations (Alfvén oscillations) [407]. The latter also appear as harmonics at integer multiples of a fundamental frequency [646, 123, 147], and for strong magnetic fields, the surface of the star (and its magnetosphere) can be modulated by interior Alfvén oscillations [249, 248]. Because of these properties, these oscillations (as well as torsional modes of the crust, modified by the presence of the magnetic field) have been investigated as possible models for the explanation of observed quasi-periodic oscillation in the aftermath of giant bursts in soft gamma repeaters (SGRs) [335, 721, 716]. If it is confirmed that the mechanism for the modulation of the X-ray spectra originates from the interior magnetars, this would be the first identification of neutron star oscillations.

8.7 Neutral $l = m$ f-modes

Counterrotating nonaxisymmetric modes can be dragged forward by rotation so that their inertial-frame frequency goes through zero, triggering the gravitational-wave–driven CFS instability that grows on secular timescales (see Chapter 9). Both the real and imaginary parts of the frequency of quasinormal modes go through zero at the same point along a sequence, and the zero-frequency mode is a *neutral mode* for the instability. For the CFS instability, the fastest-growing mode for each value of l is the $l = m$ mode, because its frequency goes through zero at a lower rotation rate than that of the $l \neq m$ modes. Note that when the real part of the frequency vanishes, the imaginary part vanishes as well: A neutral mode is a time-independent solution to the linearized Einstein-Euler system.

Although the problem of computing quasinormal modes via perturbative methods in full general relativity is understood only within approximation methods mentioned in Section 8.4, the restricted problem of finding the *neutral* f-modes was solved in [659, 665, 486]. Here we give a summary of the method, based on the formalism introduced in Section 7.7. The numerical method completes the Eulerian formalism developed by Ipser and Lindblom in the Cowling approximation [328], by finding an appropriate gauge in which the time independent perturbation equations can be solved numerically for $h_{\alpha\beta}$.

8.7.1 Numerical method

Definition of real perturbation functions. The perturbation in the metric, $h_{\alpha\beta}$, was written in Section 7.7 in terms of ten *complex* functions $\bar{h}, \bar{p}, \bar{k}, \bar{w}, \bar{q}, \bar{a}, \bar{b}, \bar{L}, \bar{M}$, and $\bar{y}$. However, for neutral modes, the system of perturbation equations (7.248) and

(7.263) becomes time independent. If one defines the following ten *real* and ϕ-*independent* functions

$$\hat{h} = \bar{h}e^{-im\phi}, \tag{8.74}$$

$$\hat{k} = \bar{k}e^{-im\phi}, \tag{8.75}$$

$$\hat{L} = -i\ \bar{L}\ e^{-im\phi}, \tag{8.76}$$

$$\hat{M} = -i\ \bar{M}\sin\theta\ e^{-im\phi}, \tag{8.77}$$

$$\hat{q} = \bar{q}\sin\theta\ e^{-im\phi}, \tag{8.78}$$

$$\hat{y} = \bar{y}\sin^2\theta\ e^{-im\phi}, \tag{8.79}$$

$$\hat{a} = -i\ \bar{a}\sin^2\theta\ e^{-im\phi}, \tag{8.80}$$

$$\hat{b} = -i\ \bar{b}\sin^3\theta\ e^{-im\phi}, \tag{8.81}$$

$$\hat{p} = \bar{p}\sin^2\theta\ e^{-im\phi}, \tag{8.82}$$

$$\hat{w} = \bar{w}\sin^2\theta\ e^{-im\phi}, \tag{8.83}$$

then the system of perturbation equations becomes a system of ten real equations for ten real functions of (r, θ).

The ten real perturbation functions obey specific symmetries with respect to the equatorial plane, which follow from the following schematic expansions in terms of Legendre polynomials:

$$\hat{h},\ \hat{k},\ \hat{L},\ \hat{y},\ \hat{a},\ \hat{p},\ \hat{w} \sim \sin^m\theta \sum_{l'=0}^{\infty} a_{2l'}(r)P_{2l'}(\cos\theta), \tag{8.84}$$

$$\hat{M},\ \hat{q},\ \hat{b} \sim \sin^m\theta \sum_{l'=0}^{\infty} a_{2l'+1}(r)P_{2l'+1}(\cos\theta), \tag{8.85}$$

that is, $\hat{M}, \hat{q}$ and $\hat{b}$ have opposite reflection symmetry with respect to the other perturbation functions. These specific symmetries allow for boundary conditions to be placed in the equatorial plane, and the system of equations needs to be solved only in the upper plane.

Gauge choice. A crucial step in finding a complete neutral-mode solution is the choice of gauge.[13] A useful guide in choosing the gauge is to require that the gauge reduce to the well-known RW gauge for polar modes [561] in the nonrotating limit.

[13] Recall that in general relativity, linear perturbations of the metric tensor are subject to a gauge freedom (four gauge conditions), which reduces the maximum number of independent functions to only six, instead of ten.

This is achieved by choosing a gauge defined by the following four conditions:

$$h_{r\theta} = 0 \quad \Longrightarrow \bar{q} = 0, \tag{8.86}$$

$$h_{\theta\phi} = 0 \quad \Longrightarrow \bar{b} = 0, \tag{8.87}$$

$$h_{t\phi} = -\omega h_{\phi\phi} \quad \Longrightarrow \bar{y} = 0, \tag{8.88}$$

$$h_{\phi\phi} = \frac{h_{\theta\theta}}{r^2} e^{2(\psi-\alpha)} \quad \Longrightarrow \bar{w} = \bar{p}. \tag{8.89}$$

Only six perturbation functions, $\hat{h}$, $\hat{p}$, $\hat{k}$, $\hat{L}$, $\hat{M}$, and $\hat{a}$, are then nonzero and can be found by solving six components of the perturbed field equation.[14] The six equations in which the above perturbation functions are dominant and can be used for finding a solution are the (tt), (rr), $(\theta\theta)$ (tr), $(t\theta)$, and $(r\theta)$ components of $\delta R_{\alpha\beta} = 8\pi\delta(T_{\alpha\beta} - \frac{1}{2}g_{\alpha\beta}T)$.

An approximate gauge. Although one can use all six components of the perturbed field equation to obtain an accurate neutral-mode solution, there turns out to be only moderate or little loss of accuracy in adopting an approximate gauge with fewer metric potentials and a correspondingly smaller set of equations. In the weak-field limit, the dominant component of the perturbed metric $h_{\alpha\beta}$ is h_{tt}, which reduces to $-2\bar{h}$. In addition, the metric functions $\bar{h}$, $\bar{k}$, $\bar{p}$ and $\bar{w}$ satisfy the relations

$$\bar{h} = -\bar{k} = -\bar{p} = -\bar{w}. \tag{8.90}$$

In even the most relativistic stars, the h_{tt} component continues to dominate all other components of $h_{\alpha\beta}$, and (8.90) still holds approximately in the gauge defined by (8.86)–(8.89).[15] One can thus use the relation

$$\bar{h} \simeq -\bar{k} \simeq -\bar{p} \tag{8.91}$$

to define an approximate gauge in which only four perturbation functions appear. Furthermore, because $h_{t\theta}$ and $h_{r\phi}$ are zero in the standard RW gauge for polar modes in nonrotating stars, these perturbation functions are less important than those that survive in the nonrotating limit. This allows the further simplification:

$$h_{t\theta} \simeq h_{r\phi} \simeq 0 \quad \Longrightarrow \quad \hat{M} \simeq \hat{a} \simeq 0. \tag{8.92}$$

With Eqs. (8.91) and (8.92) restricting the gauge defined by (8.86)–(8.89), one obtains an *approximate gauge* in which the perturbed metric has the form

$$h_{\alpha\beta} = \begin{pmatrix} -2\bar{h}(e^{2\nu} - \omega^2 e^{2\psi}) & \bar{L} & 0 & -2\omega\bar{h}e^{2\psi} \\ & -2\bar{h}e^{2\alpha} & 0 & 0 \\ & \text{sym.} & -2\bar{h}r^2e^{2\alpha} & 0 \\ & & & -2\bar{h}e^{2\psi} \end{pmatrix}. \tag{8.93}$$

[14] Because of the implicit dependence of $e^{2\psi}$ on $\sin^2\theta$, the term $\sin^2\theta$ is dropped, in this gauge, from the definition of $\hat{p}$.

[15] The last relation in (8.90) holds exactly in the gauge defined by Eqs. (8.86)–(8.89).

Only two perturbation functions, $\bar{h}$ and $\bar{L}$, now need to be determined, and these are found from the (tt) and (tr) components of the perturbed field equation. This approximate gauge, essentially a weak-field, slow-rotation limit of the full gauge (8.86)–(8.89), yields neutral-mode frequencies of acceptable accuracy.

Finite-differencing. The task of solving the coupled system of six partial differential equations is not straightforward. In the gauge specified above, the (tt) and $(\theta\theta)$ equations are of elliptic type for $\hat{h}$ and $\hat{p}$, respectively. The (rr) equation is of parabolic type for $\hat{k}$ (it lacks a $\partial^2\hat{k}/\partial r^2$ derivative). The (tr) and $(r\phi)$ equations are second-order ODEs for $\hat{L}$ and $\hat{a}$ in the angular direction, whereas the $(t\theta)$ equation is a second-order ODE for $\hat{M}$ in the radial direction. Each type of equation requires its own finite-differencing scheme and its own type of boundary conditions. Still, all six equations can be solved simultaneously on a 2-dimensional grid. The finite-differencing of the system of perturbed field-equation components yields a large linear system in which the unknowns can be ordered to make the matrix band diagonal.

Trial solutions. Assuming that the perturbation in all metric components vanishes at infinity, one can construct numerical trial solutions of the perturbed field equation (7.248) given a suitably chosen *trial function* δU. A trial solution $(h_{\alpha\beta}, \delta U)$ obtained in this way does not correspond to a correct neutral mode solution, since it does not satisfy the remaining equation, the perturbed energy conservation equation (7.263). One must therefore devise a method to determine δU that satisfies the complete system of both the perturbed field equations (7.248) and the perturbed energy-conservation equation (7.263).

Criterion for neutral modes. For a given oscillation mode, the complete system of equations (7.248) and (7.263) can have a zero-frequency solution (if any) only for a particular star along a rotational sequence. For other stars along this sequence, one can still solve only the perturbed field equation (7.248) for a set of chosen trial functions δU and use the perturbed energy conservation equation (7.263) to construct a criterion for locating the marginally stable star for which the neutral mode exists.

The perturbed energy conservation equation (7.263) is implicitly linear in δU. Hence, (7.263) can be represented as a linear operator L acting on δU

$$L(\delta U) = 0. \tag{8.94}$$

Equation (8.94) represents an eigenvalue problem for the linear operator L with zero eigenvalue and eigenfunction δU.

In the Newtonian limit, it was shown that the eigenfunction δU of an $l = m$ neutral mode can be approximated by expanding it in terms of a set of basis functions δU_i,

$$\delta U = \sum_i a_i \delta U_i, \tag{8.95}$$

where a_i are expansion coefficients. The basis functions δU_i were chosen to have the form [325, 635]

$$\delta U_i = \delta U_i^{(jk)} = r^{l+2(j+k)} Y_{l+2k}^{m}(\cos\theta), \tag{8.96}$$

with $j, k = 0, 1, \ldots$ and with each set of indices (j, k) yielding a particular δU_i. In practice, the eigenfunction δU is represented with reasonable accuracy by only a few terms, and the $(0, 0)$ term $r^l Y_{lm}$ dominates the expansion (8.96).

Substituting the expansion (8.95) in (8.94) yields

$$\sum_i a_i L(\delta U_i) = 0. \tag{8.97}$$

If we define the inner product[16]

$$\langle \delta U_j | L | \delta U_i \rangle = \int i \frac{\delta U_j}{\sigma u^t} L(\delta U_i) \sqrt{|g|} d^3x, \tag{8.98}$$

then taking the inner product of (8.97) with respect to δU_j yields

$$\sum_i a_i \langle \delta U_j | L | \delta U_i \rangle = 0. \tag{8.99}$$

Even though no assumption on orthogonality of the basis functions δU_i was made, nonzero $\langle \delta U_j | \delta U_i \rangle$ terms do not appear on the right side of (8.99), because of its zero eigenvalue. Equation (8.99) is thus a linear, homogeneous system for the coefficients a_i, and a nontrivial solution exists only if its determinant vanishes. This yields a criterion for locating the zero-frequency modes:

Criterion: *A stationary, axisymmetric star has a nonaxisymmetric neutral mode, if*

$$\det \langle \delta U_j | L | \delta U_i \rangle = 0. \tag{8.100}$$

In practice, one starts with a slowly rotating star and computes the matrix elements $\langle \delta U_j | L | \delta U_i \rangle$. For slowly rotating stars, the determinant in (8.100) always has a large value. Moving along a sequence of rotating stars of increasing angular velocity, one can locate the model for which the determinant goes through zero, when a neutral mode exists. In this numerical method, the accuracy in locating the neutral modes thus depends on how well the chosen set of basis functions δU_i in the expansion (8.95) can approximate the true eigenfunction δU.

[16] The motivation for defining the inner product as in (8.98) will become apparent in Section 8.7.2.

An Eulerian expression for the inner product. The perturbed energy conservation equation (7.263) is equivalent to

$$\begin{aligned}
L(\delta U) &\equiv \delta\Big(u_\beta \nabla_\alpha T^{\alpha\beta}\Big) \\
&= \Delta\Big(u_\beta \nabla_\alpha T^{\alpha\beta}\Big) \\
&= -\Delta\Big[(\epsilon + p)\nabla_\beta u^\beta + u^\beta \nabla_\beta \epsilon\Big] \\
&= -u^\gamma \nabla_\gamma \Big[\Delta\epsilon + \frac{1}{2}(\epsilon + p) q^{\alpha\beta} \Delta g_{\alpha\beta}\Big] \\
&= -i\sigma u^t \Big[\Delta\epsilon + \frac{1}{2}(\epsilon + p) q^{\alpha\beta} \Delta g_{\alpha\beta}\Big],
\end{aligned} \tag{8.101}$$

(cf. [235]). Then,

$$L(\delta U) = -i\sigma u^t \left\{\delta\epsilon + \xi_\perp^\alpha \nabla_\alpha \epsilon + \frac{1}{2}(\epsilon + p)\Big[h_\gamma{}^\gamma + u^\alpha u^\beta h_{\alpha\beta} + 2\nabla_\alpha \xi^\alpha + 2u^\alpha u^\beta \nabla_\alpha \xi_\beta\Big]\right\}, \tag{8.102}$$

where the Lagrangian displacement vector ξ^α was decomposed into components normal and parallel to the 4-velocity

$$\xi^\alpha = \xi_\perp^\alpha - (\xi_\gamma u^\gamma) u^\alpha, \tag{8.103}$$

With this decomposition, one obtains

$$\nabla_\alpha \xi^\alpha = \nabla_\alpha \xi_\perp^\alpha - \delta U \tag{8.104}$$

and

$$u^\alpha u^\beta \nabla_\alpha \xi_\beta = \delta U - \xi_\perp^\beta u^\alpha \nabla_\alpha u_\beta. \tag{8.105}$$

Using (1.30), (7.267), and (7.268), the perturbed energy conservation equation becomes

$$\begin{aligned}
L(\delta U) = -i\sigma u^t \Bigg\{ &(\epsilon + P)\left[\frac{(\epsilon + p)}{p\Gamma}\delta U + \frac{1}{2}\left(1 + \frac{\epsilon + p}{p\Gamma}\right) u^\alpha u^\beta h_{\alpha\beta} + \frac{h_\gamma{}^\gamma}{2}\right] \\
&+ \nabla_\alpha \Big[\xi_\perp^\alpha (\epsilon + p)\Big]\Bigg\}.
\end{aligned} \tag{8.106}$$

The inner product (8.98) now takes the form

$$\langle \delta U_j|L|\delta U_i\rangle = \int \frac{i\delta U_j}{\sigma u^t} L(\delta U_i)\sqrt{|g|}d^3x$$
$$= \int (\epsilon+p)\left\{ \delta U_j \left[\frac{(\epsilon+p)}{p\Gamma}\delta U_i + \frac{1}{2}\left(1+\frac{\epsilon+p}{p\Gamma}\right) u^\alpha u^\beta h_{\alpha\beta} + \frac{h_\gamma{}^\gamma}{2}\right] - \xi_\perp^\alpha \nabla_\alpha \delta U_j \right\} \sqrt{|g|}d^3x, \tag{8.107}$$

where the time independence of ξ^α was used to eliminate the term $\int \nabla_\alpha[\xi_\perp^\alpha(\epsilon + p)\delta U]\sqrt{|g|}d^3x$ as an integral of a spatial divergence.

Finally, the component of ξ^α normal to the 4-velocity u^α is related to the component of δu^α normal to u^α by

$$\xi_\perp^\alpha = \frac{\delta u_\perp^\alpha}{i\sigma u^t} = \frac{\delta u^\alpha - \frac{1}{2}u^\beta u^\gamma h_{\beta\gamma} u^\alpha}{i\sigma u^t}, \tag{8.108}$$

(cf. [235]), and the inner product can thus be written entirely in terms of Eulerian perturbation variables (which is the form required in numerical computations):

$$\langle \delta U_j|L|\delta U_i\rangle = \int (\epsilon+P)\left\{ \delta U_j \left[\frac{(\epsilon+p)}{p\Gamma}\delta U_i + \frac{1}{2}\left(1+\frac{\epsilon+p}{p\Gamma}\right) u^\alpha u^\beta h_{\alpha\beta} + \frac{h_\gamma{}^\gamma}{2}\right] - \frac{(\delta u^\alpha - \frac{1}{2}u^\beta u^\gamma h_{\beta\gamma} u^\alpha)}{i\sigma u^t} \nabla_a \delta U_j \right\} \sqrt{|g|}d^3x, \tag{8.109}$$

where $h_{\alpha\beta}$ and δu^α are computed with δU_i.

8.7.2 A variational principle

The criterion presented in Section 8.7.1 for locating a neutral mode solution is not a variational principle, because the matrix of the inner product $\langle \delta U_j|L|\delta U_i\rangle$ is not symmetric. However, a variational principle for locating neutral modes also exists, and its derivation is outlined here.

We found in Section 7.2 that for two pairs of trial solutions $(\xi^\alpha, h_{\alpha\beta})$ and $(\hat\xi^\alpha, \hat h_{\alpha\beta})$ that satisfy the perturbed energy conservation equation $\delta(u_\beta \nabla_\alpha T^{\alpha\beta}) = 0$, one can construct the symmetry relation (7.40)

$$\hat\xi_\beta\ \delta\Big(\nabla_\gamma T^{\beta\gamma}\sqrt{|g|}\Big) + \frac{1}{16\pi}\hat h_{\beta\gamma}\ \delta\left[\Big(G^{\beta\gamma} - 8\pi T^{\beta\gamma}\Big)\sqrt{|g|}\right] = -\mathcal{L}\Big(\hat\xi, \hat h; \xi, h\Big) + \nabla_\beta \Theta^\beta, \tag{8.110}$$

where $G^{\alpha\beta} \equiv R^{\alpha\beta} - \frac{1}{2}g^{\alpha\beta}R_\gamma{}^\gamma$, and $\nabla_\beta\Theta^\beta$ is a divergence constructed from ξ^α, $\hat\xi^\alpha$, $h_{\alpha\beta}$, and $\hat h_{\alpha\beta}$ (see Section 7.2). The term $\mathcal{L}(\hat\xi^\alpha, \hat h_{\alpha\beta}; \xi^\alpha, h_{\alpha\beta})$ is a function symmetric under the interchange of $(\xi^\alpha, h_{\alpha\beta})$ and $(\hat\xi^\alpha, \hat h_{\alpha\beta})$; hence the right side of Eq. in (8.110) is symmetric up to a divergence. If the perturbed energy conservation equation is *not satisfied*, then Eq. (8.110) still holds, but with a redefined divergence term $\nabla_\beta \Theta^\beta$, if the component of ξ^α along u^α is set to[17]

$$\xi_\alpha u^\alpha = (u\nabla)^{-1}\delta U$$
$$= \frac{1}{i\sigma u^t}\delta U; \tag{8.111}$$

see [235].

In the Eulerian approach to solving the perturbation equations, the perturbed Euler equation

$$\delta\Big(q^\alpha{}_\beta \nabla_\alpha T^{\beta\gamma}\Big) = 0, \tag{8.112}$$

is solved algebraically. For trial functions δU, one obtains trial solutions $h_{\alpha\beta}$ of six components of the perturbed field equation and is left with only the perturbed energy conservation equation unsolved. The symmetry in (8.110) can be exploited in order to construct a variational principle for this remaining unsolved equation.

The algebraically solved perturbed Euler equation can be eliminated from (8.110) by decomposing the Lagrangian displacement vector ξ^α into vectors normal and parallel to the 4-velocity, so that the first term in (8.110) becomes

$$\hat{\xi}_\beta\ \delta\Big(\nabla_\alpha T^{\alpha\beta}\Big) = -(\hat{\xi}_\gamma u^\gamma)\ \delta\Big(u_\beta \nabla_\alpha T^{\alpha\beta}\Big),$$
$$\equiv \frac{i\delta\hat{U}}{\sigma u^t} L(\delta U). \tag{8.113}$$

Next, we define

$$\mathcal{F}(\delta\hat{U};\delta U) := \frac{1}{16\pi}\hat{h}_{\beta\gamma}\ \delta\Big(G^{\beta\gamma} - 8\pi T^{\beta\gamma}\Big)$$
$$= \frac{1}{16\pi}\left(\hat{h}^{\beta\gamma} - \frac{1}{2}g^{\beta\gamma}\hat{h}_\delta{}^\delta\right)\ \delta\left[R_{\beta\gamma} - 8\pi\left(T_{\beta\gamma} - \frac{1}{2}g_{\beta\gamma}T_\delta{}^\delta\right)\right], \tag{8.114}$$

which is implicitly bilinear in $\delta\hat{U}$ and δU. The gauge freedom in $h_{\alpha\beta}$ leaves only six components of the perturbed field equation independent. Thus, if a trial solution satisfies six components of the perturbed field equation in (8.114), the other four components will be an implicit functional of the perturbed energy conservation equation, the only equation that remains to be satisfied. Schematically, one can thus write

$$\mathcal{F}(\delta\hat{U};\delta U) = \delta\hat{U}\mathbf{F}\Big[L(\delta U)\Big], \tag{8.115}$$

[17] The Lagrangian displacement ξ^α has a gauge freedom in its component along u^α: Adding a vector field fu^α (where f is some arbitrary scalar function) to ξ^α leaves the Eulerian perturbations unchanged.

where $\mathbf{F}$ is a functional of $L(\delta U)$, and the bilinearity of $\mathcal{F}(\delta\hat{U};\delta U)$ was used to factor out $\delta\hat{U}$. The symmetry relation (8.110) then becomes

$$\begin{aligned}
-\mathscr{L}\left(\hat{\xi}^\alpha,\hat{h}_{\alpha\beta};\xi^\alpha,h_{\alpha\beta}\right) &= \frac{i\delta\hat{U}}{\sigma u^t}L(\delta U)\sqrt{|g|}+\delta\hat{U}\mathbf{F}\left[L(\delta U)\right]\sqrt{|g|}-\nabla_\beta\Theta^\beta \\
&= \delta\hat{U}\left[\frac{i}{\sigma u^t}+\mathbf{F}\right]L(\delta U)\sqrt{|g|}-\nabla_\beta\Theta^\beta \\
&=: -\mathscr{L}(\delta\hat{U};\delta U), \qquad (8.116)
\end{aligned}$$

where $\mathscr{L}(\delta\hat{U};\delta U)$ is symmetric, up to a divergence,[18] in the interchange of $\delta\hat{U}$, δU.

Using (8.116), a variational principle for the remaining perturbed energy conservation equation is constructed, by requiring that the following integral (which is implicitly quadratic in δU) vanish:

$$I := \int -\mathscr{L}(\delta U;\delta U)d^3x, \qquad (8.117)$$

$$= \int \delta U\left[\frac{i}{\sigma u^t}+\mathbf{F}\right]L(\delta U)\sqrt{|g|}d^3x, \qquad (8.118)$$

$$= 0. \qquad (8.119)$$

The integral in (8.118) is stationary with respect to first-order variations in δU, when the perturbed energy conservation equation is satisfied, that is, when $L(\delta U)=0$

$$\frac{\delta I}{\delta(\delta U)} = \left[\frac{i}{\sigma u^t}+\mathbf{F}\right]L(\delta U)=0, \qquad (8.120)$$

Thus, $I=0$ is a variational principle for the perturbed energy conservation equation $L(\delta U)=0$.

Expanding the trial function δU in terms of a set of basis functions, as in (8.95), the variational principle becomes

$$\sum_i\sum_j a_i a_j A_{(ij)} = 0, \qquad (8.121)$$

where the symmetric matrix A has elements

$$A_{ij} := \int \delta U_j\left[\frac{i}{\sigma u^t}+\mathbf{F}\right]L(\delta U_i)\sqrt{|g|}d^3x, \qquad (8.122)$$

and (8.120) implies

$$\frac{\delta I}{\delta a_i}=0 \implies \sum_i a_j A_{(ij)} = 0. \qquad (8.123)$$

[18] The divergence $\nabla_\beta\tilde{R}^\beta$ is implicitly dependent only on $\delta\hat{U}, \delta U$.

The last equation is a homogeneous linear system for the coefficients a_j, which can have a nontrivial solution only when

$$\det A_{(ij)} = 0. \tag{8.124}$$

The last equation can be used as an alternative criterion to locate neutral modes along a sequence of rotating stars. However, the term $\mathbf{F}[L(\delta U_i)]$ involves a large number of second-order angular and radial derivatives, and thus the criterion (8.100) is simpler. It is interesting to note that in numerical applications, the inner product $\langle \delta U_j | L | \delta U_i \rangle$ used in (8.100) was shown to be *nearly* symmetric under the interchange of δU_i and δU_j [659, 665]. Thus, the simpler criterion (8.100) nearly coincides with a variational principle.

8.7.3 Neutral-mode configurations

Neutral-mode (critical) configurations for $l = m = 2$, 3, 4, and 5 nonaxisymmetric modes and for $N = 1.0$, 1.5 and 2.0 polytropes as well as for tabulated EOs were presented in [659, 665, 486]. Here we discuss a representative subset of these results. The neutral mode configurations were computed in the *approximate gauge* introduced in Section 8.7.1. with a numerical grid of 801 × 12 (radial × angular) grid points. The trial function δU was expanded as in (8.96) using eight basis functions, corresponding to the indices $j = 0, ..., 3$ and $k = 0, 1$.

For $N = 1.0$ polytropes, Fig. 8.7 shows the ratio of the critical angular velocity Ω_c to the *angular velocity* Ω_K *at mass-shedding at same central energy density* as a function of a dimensionless central energy density[19] for the four modes examined. The lowest central energy density in the figure corresponds to a mildly relativistic star. The highest central energy density shown is the central energy density of the most massive (and thus most relativistic) star allowed by this equation of state. The filled circles on the left vertical axis represent the values of Ω_c/Ω_K in the Newtonian limit. As the central energy density increases and the star becomes more relativistic, Ω_c/Ω_K decreases, and it decreases at a faster rate as it approaches the most relativistic configuration. In contrast to the Newtonian limit, where $N = 1.0$ polytropes do not have an unstable $m = 2$ mode, in relativistic $N = 1.0$ polytropes the $m = 2$ mode becomes unstable when the central energy density exceeds roughly one-tenth the central energy density of the most massive star. For the most relativistic configuration, the $m = 2$ mode becomes neutral for $T/|W|_c = 0.065$ or $\Omega_c/\Omega_K = 0.91$.

The value of Ω_c/Ω_K decreases by up to 15% for the $m = 3$, 4, and 5 modes, compared to the corresponding results in the Newtonian limit. The critical ratio $T/|W|_c$ is about 40% smaller, for the $m = 3$, 4, and 5 modes, than the corresponding ratio in the Newtonian limit. When the decrease in the Keplerian value $T/|W|_K$

[19] The dimensionless central energy density $\bar{\epsilon}_c$ is defined by setting $c = G = 1$ and taking $\tilde{K}^{N/2}$ as the fundamental length scale, where $\tilde{K} = K/m_B^{1+1/N}$; see [659, 665].

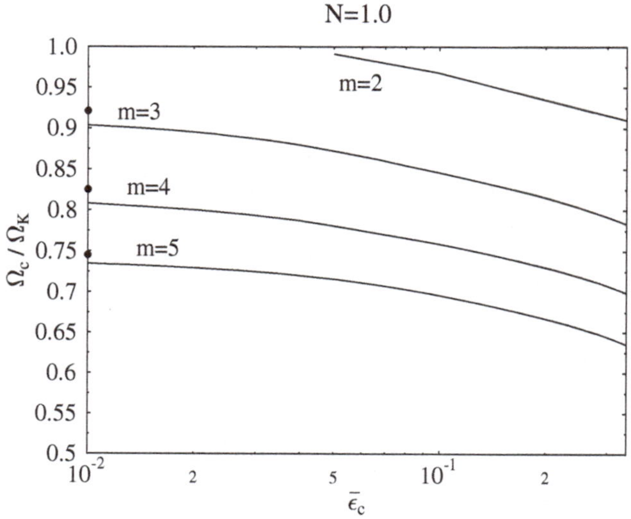

Figure 8.7. Critical angular velocity Ω (in terms of the angular velocity Ω_K at the mass-shedding limit) versus the dimensionless central energy density $\bar{\epsilon}_c$ for the $m = 2$, 3, 4, and 5 neutral modes of $N = 1.0$ polytropes. The largest value of $\bar{\epsilon}_c$ shown corresponds to the most relativistic stable configurations, whereas the lowest $\bar{\epsilon}_c$ corresponds to less relativistic configurations. The filled circles on the vertical axis represent the Newtonian limit. (Figure from [665]. Reproduced by permission of the AAS.)

is taken into account and one looks at the ratio $(T/|W|_c)/(T/|W|_K)$, the most relativistic values are still 25% to 30% lower than the Newtonian values.

The maximum-mass model of polytropes with $N > 1.0$ is less relativistic compared to the corresponding $N = 1$ model. Thus, the effect of general relativity in decreasing the rotation rates at which nonaxisymmetric modes become neutral is less pronounced. For example, the $m = 2$ mode does not become neutral even for the most relativistic $N = 1.5$ polytropes. For the $m = 3$, 4, and 5 modes, the value of Ω_c/Ω_K decreases by 7% to 10% for the most relativistic models, compared to the Newtonian limit. The corresponding decrease for $T/|W|_c$ is 30% to 35%, and for $(T/|W|_c)/(T/|W|_K)$ it is 13% to 19%.

For tabulated EOs and for $1.4M_\odot$ stars, the $l = m = 2$ mode becomes neutral for $\Omega_c/\Omega_K = 0.83$ for the softest EOS A and for $\Omega_c/\Omega_K = 0.93$ for EOS C. For maximum-mass stars, the $l = m = 2$ mode becomes neutral for $\Omega_c/\Omega_K = 0.69$ at $M = 2.9M_\odot$ for the stiffest EOS L and $\Omega_c/\Omega_K = 0.77$ at $M = 2.0M_\odot$ for EOS C.

In terms of $T/|W|$, the $l = m = 2$ neutral mode sequence can be approximated by the following linear empirical formula (independent of the details of the EOS):

$$(T/|W|)_2 = 0.115 - 0.048\frac{M}{M_{\rm max}^{\rm sph}}, \qquad (8.125)$$

where $M_{\rm max}^{\rm sph}$ is the maximum mass for a spherical star allowed by a given EOS.[20] The empirical formula has an uncertainty of order 5% for all values of M, except for stars near the maximum mass where the uncertainty is somewhat larger. These results show that in relativistic stars, the $l = m = 2$ mode becomes neutral at significantly smaller $T/|W|$ than the Newtonian value of $T/|W| \sim 0.14$.

[20] Similar empirical formulae can be written for $m > 2$ modes.

9
Stellar stability

9.1 Introduction

Neutron stars more than a few minutes old are uniformly rotating and satisfy to high accuracy the equation of state that describes cold neutron-star matter. As described in Chapter 5, their 2-dimensional family of equilibria is bounded by four curves: at low central density by a sequence of marginally stable configurations having minimum mass at constant angular momentum; at high density by a sequence of marginally stable configurations having maximum mass at constant angular momentum; by the sequence of nonrotating stars; and by the sequence of stars rotating at the Kepler (mass-shedding) limit on angular velocity. A nonaxisymmetric instability driven by gravitational radiation (the CFS instability) is likely to set an upper limit on rotation more stringent than the Kepler limit, drawing a more restrictive boundary at large rotation on the surface of stable equilibria.

Finding these bounding lines of marginal stability is a primary focus of work on the stability theory of relativistic stars. This chapter is devoted to a detailed presentation of this theory and several of its principal results. The presentation here is restricted to linear stability theory, to finding criteria for stability of first-order perturbations of an equilibrium. Numerical treatments of the nonlinear evolution of stable and unstable modes are discussed in Chapter 10.

Stellar instabilities fall into two classes: *Dynamical instabilities* are unstable solutions to the linearized Einstein-Euler system without dissipation; these are solutions whose timescale is set by the dynamical equations and is ordinarily comparable to the period of oscillation of a mode. *Secular instabilities* are instabilities that arise from dissipation – from gravitational radiation or viscosity – and that grow on a timescale set by the dissipation, the time in which the energy dissipated is comparable to the energy of the mode. This timescale is ordinarily long compared to dynamical times, and the unstable model approximately moves through a family of equilibrium configurations. Although the terms *dynamical* and *secular* refer to timescales, the physical distinction is between *conservative* and *dissipative* instabilities.

Dissipation ordinarily damps stellar oscillations, and it may seem paradoxical that an unstable mode can be driven rather than damped by viscosity or gravitational radiation. What happens is this: A mode is secularly unstable if there are nearby equilibria that have lower energy but that are not accessible by an evolution that obeys the dynamical equations without dissipation. By violating conservation of circulation, viscosity allows a transition to lower-energy equilibria with circulation different from that of the initial configuration; and by violating conservation of angular momentum, gravitational radiation allows a transition to configurations with lower energy and different angular momentum.

Gravitational waves carry energy from any system whose asymptotic quadrupole moments, or any higher multipole moments, change with time. Because dissipation due to gravitational radiation is thus present in essentially any oscillation of a rotating star, the distinction between dynamical and secular instability is not precise, and it becomes increasingly blurred as the radiation-reaction timescale approaches the period of oscillation of a mode.[1]

We begin with a discussion of instability to collapse that sets the upper limit on the mass of relativistic stars and the related instability that sets the lower mass limit. For barotropic stars, a *turning-point* method accurately locates the instability points (it is exact for spherical stars). Then, before discussing the nonaxisymmetric instability that may limit neutron-star rotation, we complete our discussion of axisymmetric instabilities with a brief treatment of local instability – of instability to convection – arising from a superadiabatic entropy gradient or from extreme differential rotation.

9.2 Axisymmetric stability and turning points

The best-known instability result in general relativity is the statement that instability to collapse sets in at a point of maximum mass, along a sequence of spherical barotropic models. The configuration with maximum mass is called a turning point along the sequence, and it is also the configuration with maximum baryon mass. A similar result holds for uniformly rotating stars [238]: Instability to collapse is implied by a point of maximum mass and maximum baryon mass, along a sequence of uniformly rotating barotropic models with fixed angular momentum. As in the spherical case, stars with higher central density than that of the maximum-mass configuration are unstable. For rotating stars, however, the turning point is a sufficient but not a necessary condition for instability: The onset of instability is at a configuration with slightly lower central density (for fixed angular momentum) than that of the maximum-mass star. A formal symmetry in the way baryon mass and angular momentum occur in the first law implies that the line of turning points is also the line of extrema of angular momentum along sequences of fixed baryon mass.

For dynamical oscillations of neutron stars the adiabatic index does not coincide with the polytropic index, $\Gamma_1 \neq \Gamma := \dfrac{d\log p(r)/dr}{d\log \rho/dr}$, and the turning point criterion

[1] In a post-Newtonian context, one can separate out the radiation-reaction terms. An equilibrium is then dynamically unstable if there is an unstable solution to the linearized post-Newtonian equations without radiation reaction – a solution to the equations in which only terms of only integer post-Newtonian order are kept and terms of post-Newtonian order $2.5, 3.5, 4, 5, \ldots$ are discarded. In the full theory, it may make sense to define a dynamically unstable mode as an unstable mode that has equal amounts of ingoing and outgoing radiation, so that there is no net dissipation and no net energy pumped in to the system.

implies *secular* instability – an instability whose growth time is long compared to the typical dynamical time of stellar oscillations. For spherical stars, the turning-point instability proceeds on a timescale slow enough to accommodate the nuclear reactions and energy transfer that accompany the change to a nearby equilibrium. For rotating stars, the timescale must also be long enough to accommodate a transfer of angular momentum from one fluid ring to another. That is, the growth rate of the instability is limited by the time required for viscosity to redistribute the star's angular momentum. For neutron stars, this is expected to be short, probably comparable to the spin-up time following a glitch, and certainly short compared to the lifetime of a pulsar or an accreting neutron star. For this reason, it is the secular instability that sets the upper and lower limits on the mass of spherical and uniformly rotating neutron stars.

For spherical stars in Newtonian gravity, instability sets in when the matter becomes relativistic, when the adiabatic index Γ_1 (more precisely, its pressure-weighted average $\bar{\Gamma}_1$) reaches the value 4/3 characteristic of zero rest–mass particles. This can be seen from the Newtonian limit of Eq. (7.149) for the canonical energy,

$$E_c = \int \left\{ \frac{1}{2}\rho\dot{\xi}^2 + \frac{2}{r}p'\xi^2 + \frac{\Gamma_1 p}{2r^4}\left[(r^2\xi)'\right]^2 \right\} dV. \tag{9.1}$$

Choosing as initial data $\xi = r$, $\dot{\xi} = 0$ gives

$$E_c = \int \left(2rp' + \frac{9}{2}\Gamma_1 p\right) dV = \frac{9}{2}\int \left(\Gamma_1 - \frac{4}{3}\right) p\, dV, \tag{9.2}$$

implying instability for $\bar{\Gamma}_1 < 4/3$. This shows only that $\Gamma < 4/3$ is a sufficient condition for instability, but spherical Newtonian polytropes with $\Gamma_1 > 4/3$ are stable.

In the stronger gravity of general relativity, even models with the stiffest equation of state must be unstable to collapse for some value of $R/M > 9/8$, the ratio for the most relativistic model of uniform density. The more stringent relativistic constraint on Γ_1 for a star to be stable against radial perturbations has the form

$$\Gamma_1 < \frac{4}{3} + K\frac{M}{R}, \tag{9.3}$$

where K is positive and of order unity [126]. Because a gas of photons has $\Gamma_1 = 4/3$ and massive stars are radiation dominated, the instability can be important for stars with $M/R \ll 1$ [126, 229].

If one considers perturbations conforming to the effective equation of state satisfied by the equilibrium star, then Chandrasekhar's canonical energy criterion coincides with the turning-point criterion for spherical stars. The turning-point criterion, however, has a longer history. In their 1939 paper [503], Oppenheimer and Volkoff had already used a turning point method to locate the stable part of a sequence of model neutron stars; and Misner and Zapolsky [478] noticed that, along a sequence of neutron star models, the configuration at which the functional E_c first becomes negative appeared to be the model with maximum mass. In each case, they used models in which the equilibrium configuration and its perturbations are governed by the same one-parameter equation of state. Then the turning-point method, due initially to Poincaré [537], implies that points at which the stability of a mode changes are extrema of the mass [296]. See Thorne [687] for a review of the turning point criterion applied to spherical neutron stars and [690] for later references; a somewhat different treatment is given by Zel'dovich and Novikov [750].

Although the secular instability diagnosed by the turning-point criterion sets the upper limit on the mass of currently observed neutron stars, there are several contexts in which the distinction between dynamical and secular instability to collapse can be important and may be observable: In nascent neutron stars, in the hypermassive neutron stars formed in the merger of binary black holes, and (with direct observation seeming less likely) in neutron stars driven by accretion to their upper mass limit. We begin with the turning-point criterion and then present criteria and mention results on dynamical instability.

9.2.1 Turning-point theorem

We turn now to a derivation of the theorem that, along a sequence of stellar models with constant angular momentum, the sequence is secularly unstable on one side of an extremum of the mass. The proof also shows that, along sequences of constant baryon number, the sequence is secularly unstable on one side of an extremum of angular momentum. These criteria hold for small perturbations that obey the zero-temperature EOS $p = p(\rho), \epsilon = \epsilon(\rho)$ that governs the equilibrium stars. Because the star's composition depends on density, this one-parameter equilibrium equation of state can describe only stellar perturbations whose timescale is long enough to accommodate the nuclear reactions and energy loss, via neutrino emission and heat flow, needed to reach the cold, equilibrium EOS.

We can easily understand why the instability sets in at an extremum of the mass for spherical stars by looking at a radial mode of oscillation of a nonrotating star with an equation of state $p = p(\rho), \epsilon = \epsilon(\rho)$. Along the sequence of spherical equilibria, a radial mode changes from stability to instability when its frequency σ changes from real to imaginary, with $\sigma = 0$ at the point of marginal stability. Now a zero-frequency mode is just a time-independent solution to the linearized Einstein-Euler equations – a perturbation from one equilibrium configuration to a nearby equilibrium with the same baryon number. From the first law of thermodynamics, a perturbation that keeps the star in equilibrium satisfies

$$\delta M = \mathcal{E} \delta M_0, \tag{9.4}$$

implying that, for a zero frequency perturbation that involves no change in baryon number ($\delta M_0 = 0$), the change δM in mass must vanish. This is exactly the requirement that the mass be an extremum along the sequence of equilibria.

Which side of an extremum of mass is stable depends on whether the extremum is a maximum or a minimum and on whether the injection energy is an increasing or decreasing function of central density. Consider first the maximum-mass, nonrotating configuration. The injection energy $\mathcal{E}$ can be computed at the surface of the star,

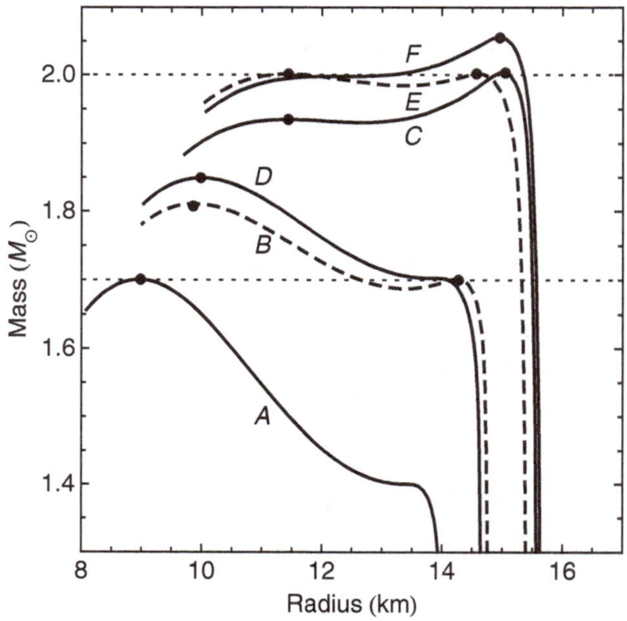

Figure 9.1. Mass-radius curves are plotted for several EOSs. Black dots on each curve mark maxima of mass. At these points, the curve turns clockwise as one moves to smaller radii and stability is lost. At mass minima (on curves B, C, and E), the curves turn counterclockwise, and stability is regained. (Figure from [558].)

where $h = \lim_{p\to 0} \dfrac{\epsilon + p}{\rho} = 1$ and $u^t = (1 - 2M/R)^{-1/2}$: From Eq. (1.110) we obtain $\mathcal{E} = \sqrt{1 - 2M/R}$. Because R is a decreasing function of ρ_c at the turning point, $\mathcal{E}$ is also a decreasing function of ρ_c. Models on the *high-density* side of the maximum-mass instability point are then unstable: Because the turning point is a star with maximum baryon number as well as maximum mass, there are models on opposite sides of the turning point with the same baryon number. Because $\mathcal{E}$ is a decreasing function of central density, the model on the high-density side of the turning point has greater mass than the corresponding model with smaller central density.

At the minimum-mass nonrotating star, R and $\mathcal{E}$ are again decreasing functions of central density. In this case, however, it is the *low-density* side that is unstable: Because the mass is a minimum, the model on the low-density side of the turning point has greater mass than the corresponding model with the same baryon mass on the high-density side.

On a mass-radius curve, stability is lost in the direction for which the curve turns counterclockwise about the maximum mass, regained when it turns clockwise. In Fig. 9.1, from [558], mass-radius curves are plotted for six EOSs, labeled A–F, chosen to illustrate possible islands of stability associated with a temporary softening of the EOS as the density increases (as might result from a phase transition in the core). The sequences associated with EOSs B, C, and E have two maximum masses (marked by black dots) separated by a minimum mass. As one moves along the sequence from larger to smaller radius – from lower to higher density – stability is temporarily lost at the first maximum mass, regained at the minimum mass, and permanently lost at the second maximum mass. In candidates for the neutron-star EOS, the stable configuration with minimum radius is the maximum-mass model. As the mass-radius curve C shows, however, the stable configuration with minimum radius (and largest central density) can in principle be only a local, not a global, maximum of mass.

For rotating stars, the turning-point criterion is a sufficient condition for secular instability to collapse. In general, however, collapse can be expected to involve differential rotation, and the turning point identifies only nearby uniformly rotating configurations with lower energy. Rotating stars are therefore likely to be secularly unstable to collapse at densities slightly lower than the turning-point density. The onset of secular instability to collapse is at or before the onset of dynamical instability along a sequence of uniformly rotating stars of fixed angular momentum, and recent work by Takami, Rezzolla, and Yoshida [675] appears to show that rapidly rotating stars can also be dynamically unstable to collapse just prior to the turning point.

The theorem itself holds for a two-parameter family of uniformly rotating barotropic stars, and it uses the variation of the mass given by the first law of thermodynamics. For perturbations $(\delta M_0, \delta J, h_{\alpha\beta}(x), \xi^\alpha(x))$ joining two zero-temperature equilibrium configurations of a zero-temperature, uniformly rotating star, the first law, Eq. (3.125), has the form

$$\delta M = \mathcal{E}\delta M_0 + \Omega\delta J. \tag{9.5}$$

The theorem is an immediate corollary of the turning point criterion in the form given by Sorkin [642, 643]. Its proof requires no assumptions about existence or completeness of normal modes.

Theorem 9.1 (Friedman, Ipser, Sorkin [238]). *Consider a continuous sequence of uniformly rotating stellar models based on an equation of state of the form* $p = p(\epsilon)$. *Let* λ *be the sequence parameter and denote the derivative* $d/d\lambda$ *along the sequence by* $(\dot{\ })$.

(i) Suppose that the total angular momentum is constant along the sequence and that there is a point λ_0 *where* $\dot{M} = 0$ *and where* $\mathcal{E} > 0$, $(\dot{\mathcal{E}}\dot{M})\dot{} \neq 0$. *Then the part of the sequence for which* $\dot{\mathcal{E}}\dot{M} > 0$ *is unstable for* λ *near* λ_0.

(ii) Suppose that the total baryon mass M_0 *is constant along the sequence and that there is a point* λ_0 *where* $\dot{M} = 0$ *and where* $\Omega > 0$, $(\dot{\Omega}\dot{M})\dot{} \neq 0$. *Then the part of the sequence for which* $\dot{\Omega}\dot{M} > 0$ *is unstable for* λ *near* λ_0.

Reference [238] points out the symmetry between M_0 and J that implies the maximum-J form of the theorem, and Cook, Shapiro, and Teukolsky [157] first use the theorem in this form.

Proof. From Eq. (3.123), with $T = 0$, every equilibrium configuration is an extremum of the mass among perturbations that preserve total baryon mass and total angular momentum and that satisfy the perturbed field equations. A configuration is secularly unstable if there is a linear perturbation that preserves total baryon mass and total angular momentum, satisfies the field equations, and lowers the mass M. Because every equilibrium is an extremum of the mass under such perturbations, the change in mass due to a perturbation $(h_{\alpha\beta}(x), \xi^\alpha(x))$ of $(g_{\alpha\beta}(x), \chi(x))$ is then of second order in the perturbation. To make the notation more concise and readable, we will denote the perturbation by $\delta\phi^I$,

$$(\delta\phi^I) := (h_{\alpha\beta}(x), \xi^\alpha(x)), \tag{9.6}$$

and will denote functional derivatives of M by $\dfrac{\partial M}{\partial \phi^I}$,

$$\left(\frac{\partial M}{\partial \phi^I}\right) := \left(\frac{\delta M}{\delta g_{\alpha\beta}(x)}, \frac{\delta M}{\delta \chi^\alpha(x)}\right). \tag{9.7}$$

Then $M = M(M_0, J, \phi^I)$, and, from Eq. (3.125) when the constraint (9.6) is satisfied, the partial derivatives of M have the values

$$\frac{\partial M}{\partial M_0} = \mathcal{E}, \qquad \frac{\partial M}{\partial J} = \Omega, \tag{9.8}$$

$$\frac{\delta M}{\delta g_{\alpha\beta}(x)} = \frac{1}{16\pi}(G^{\alpha\beta} - 8\pi T^{\alpha\beta})\sqrt{|g|}, \qquad \frac{\delta M}{\delta \chi^\alpha(x)} = \nabla_\beta T_\alpha{}^\beta \sqrt{|g|}. \tag{9.9}$$

In the remainder of the proof, we restrict consideration to perturbations $\delta\phi^I$ that satisfy the constraint (9.6). Then Eq. (3.125) can be written compactly as

$$\delta M = \mathcal{E}\delta M_0 + \Omega\delta J + \frac{\partial M}{\partial \phi^I}\delta\phi^I, \tag{9.10}$$

where the sum over I includes the integral over x (thought of as a sum over the continuous index x).

To prove the theorem in the form (i), consider a sequence of equilibria with fixed J. Because the Einstein-Euler equations, $\partial M/\partial\phi^I = 0$, are satisfied at each point along a sequence of equilibrium solutions, the change in M from a perturbation $\delta\phi^I$ is second-order in $\delta\phi^I$, and a configuration will be stable only if, for all $\delta\phi^I$,

$$\delta^2 M \equiv \frac{\partial^2 M}{\partial\phi^I \partial\phi^J}\delta\phi^I\delta\phi^J > 0. \tag{9.11}$$

For perturbations $\delta\phi^I$ that join equilibria separated by parameter distance $\delta\lambda$ along the sequence, the first-order changes in M_0 and ϕ^I along the sequence are $\delta M_0 = \dot{M}_0\delta\lambda$, $\delta\phi^I = \dot{\phi}^I\delta\lambda$. We will show that at a turning point, $\delta^2 M = 0$ for this choice of $\delta\phi^I$ and that it changes sign from one side of the turning point to the other.

By the first law (9.5), at each point of the sequence,

$$\dot{M} = \mathcal{E}\dot{M}_0, \tag{9.12}$$

and at the turning point, $\mathcal{E} \neq 0$ implies

$$\dot{M}\ |_{\lambda=\lambda_0} = 0, \quad \dot{M}_0\ |_{\lambda=\lambda_0} = 0. \tag{9.13}$$

Because $\partial M/\partial\phi^I = 0$ for each equilibrium solution, we have

$$0 = \frac{d}{d\lambda}\left(\frac{\partial M}{\partial \phi^I}\right) = \frac{\partial^2 M}{\partial M_0 \partial \phi^I}\dot{M}_0 + \frac{\partial^2 M}{\partial \phi^I \partial \phi^J}\dot{\phi}^J. \tag{9.14}$$

Contracting with $\dot{\phi}^I$ and using Eq. (9.8) yields

$$\frac{\partial^2 M}{\partial \phi^I \partial \phi^J}\dot{\phi}^I\dot{\phi}^J = -\frac{\partial^2 M}{\partial M_0 \partial \phi^I}\dot{\phi}^I \dot{M}_0 = -\frac{\partial \mathcal{E}}{\partial \phi^I}\dot{\phi}^I \dot{M}_0. \tag{9.15}$$

Because $\dot{M}_0 = 0$ at λ_0, we have $\left.\left(\frac{\partial^2 M}{\partial \phi^I \partial \phi^J}\dot{\phi}^I\dot{\phi}^J\right)\right|_{\lambda=\lambda_0} = 0$ and

$$\left.\frac{d}{d\lambda}\left(\frac{\partial^2 M}{\partial \phi^I \partial \phi^J}\dot{\phi}^I\dot{\phi}^J\right)\right|_{\lambda=\lambda_0} = -\left.\frac{\partial \mathcal{E}}{\partial \phi^I}\dot{\phi}^I \ddot{M}_0\right|_{\lambda=\lambda_0}. \tag{9.16}$$

Finally, using the relations

$$\dot{\mathcal{E}} = \frac{\partial \mathcal{E}}{\partial M_0}\dot{M}_0 + \frac{\partial \mathcal{E}}{\partial \phi^I}\dot{\phi}^I, \qquad \dot{\mathcal{E}}|_{\lambda=\lambda_0} = \left.\frac{\partial \mathcal{E}}{\partial \phi^I}\dot{\phi}^I\right|_{\lambda=\lambda_0}, \tag{9.17}$$

we obtain

$$\left.\frac{d}{d\lambda}\left(\frac{\partial^2 M}{\partial \phi^I \partial \phi^J}\dot{\phi}^I\dot{\phi}^J\right)\right|_{\lambda=\lambda_0} = -\dot{\mathcal{E}}\ddot{M}_0\ |_{\lambda=\lambda_0} = -(\dot{\mathcal{E}}\dot{M}_0)^{\cdot}\ |_{\lambda=\lambda_0}. \tag{9.18}$$

Then,

$$\left.\frac{\partial^2 M}{\partial \phi^I \partial \phi^J}\dot{\phi}^I\dot{\phi}^J\right|_{\lambda=\lambda_0} > 0, \tag{9.19}$$

positive for λ near λ_0 on the side of λ_0 for which $\dot{\mathcal{E}}\dot{M}_0 < 0$, and the sequence is unstable, as claimed in (i): For each equilibrium solution near λ_0 on the side of λ_0 for which $\dot{\mathcal{E}}\dot{M}_0 < 0$, there are nearby configurations with the same M_0 and J whose mass is smaller than the mass of the equilibrium solution.

From the symmetry between M_0 and J in the first law, the proof of (ii) is identical to the proof of (i) with the replacements $M_0 \leftrightarrow J$, $\mathcal{E} \leftrightarrow \Omega$. □

Because the perturbation that lowered the mass of the unstable configuration was a perturbation along the sequence, the proof also confirms the statement that corresponding equilibria that have the same baryon mass M_0 and angular momentum J and that lie on opposite sides of a turning point have smaller mass on the stable side.

An important feature of the turning-point criterion is its independence of the parameters chosen to label the equilibrium models, and one can state the result in a way that does not restrict the parameterization. This is the form in which the multi-parameter version of the turning-point theorem is originally given by Sorkin [643].

Corollary 9.1. *Consider a two-parameter family of uniformly rotating stellar models having an equation of state of the form* $p = p(\epsilon)$. *Suppose that along a continuous sequence of models labeled by a parameter* λ, *there is a point* λ_0 *at which both* $\dot{M}_0$ *and* $\dot{J}$ *vanish and where* $(\dot{\Omega}\dot{J} + \dot{\mathcal{E}}\dot{M}_0)\dot{} \neq 0$. *Then the part of the sequence for which* $\dot{\Omega}\dot{J} + \dot{\mathcal{E}}\dot{M}_0 > 0$ *is unstable for* λ *near* λ_0.

Proof. For a general path, Eq. (9.18) becomes

$$\frac{d}{d\lambda}\left(\frac{\partial^2 M}{\partial\phi^I\partial\phi^J}\dot{\phi}^I\dot{\phi}^J\right)\bigg|_{\lambda=\lambda_0} = -(\dot{\mathcal{E}}\dot{M}_0 + \dot{\Omega}\dot{J})\dot{}\,|_{\lambda=\lambda_0}, \tag{9.20}$$

and the result is immediate. □

We could have proved the theorem in this form first, deducing forms (i) and (ii) as special cases. But it is forms (i) and (ii) that have been used in the literature, and the equations are shorter in the proof as given.

Searches to determine the line of turning points have covered the set of models with sequences of constant rest mass M_0, extremizing J on each one, or vice versa. This is a computationally expensive procedure, and a more efficient way is summarized in the following corollary due to Jocelyn Read [559]:

Corollary 9.2. *Regard* M_0 *and* J *as functions on the two-dimensional space of equilibria. Turning points are the points at which* ∇M_0 *and* ∇J *are parallel.*

Equivalent statements of this characterization of turning points are:

The wedge product of the gradients vanishes: $dM_0 \wedge dJ = 0$.
With the space of equilibria embedded in a 3-dimensional space as in Fig. 5.7, $\nabla M_0 \times \nabla J = 0$.

In particular, with the space of equilibria parametrized by the central energy density ϵ_c and axis ratio $\mathfrak{r} = r_p/r_e$, the turning points satisfy

$$\frac{\partial(M_0, J)}{\partial(\epsilon_c, \mathfrak{r})} \equiv \frac{\partial M_0}{\partial\epsilon_c}\frac{\partial J}{\partial\mathfrak{r}} - \frac{\partial J}{\partial\epsilon_c}\frac{\partial M_0}{\partial\mathfrak{r}} = 0. \tag{9.21}$$

Proof of Corollary 9.2. Consider a sequence of configurations of constant J. The tangent to the sequence is orthogonal to ∇J at all points of the sequence; it is orthogonal to ∇M_0 at a turning point P because a turning point is an extremum of M_0. Because the space of equilibria is 2-dimensional, ∇M_0 and ∇J are parallel. □

For a given EOS, the model with maximum angular velocity on the low-density side of the turning point may be determined, without finding sequences of constant J and M_0, by considering a sequence of fixed central energy density ϵ_c. First, decrease the axis ratio $\mathfrak{r}$ until the Kepler limit is found. Second, vary ϵ_c and $\mathfrak{r}$ around this point to estimate the partial derivatives of Eq. (9.21). The sign of $\partial(M_0, J)/\partial(\epsilon_c, \mathfrak{r})$ will change as the Kepler limit sequence crosses the line of turning points.

9.2.2 Extensions and applications

Goussard, Haensel, and Zdunik [276] extend the stability criterion to hot proto–neutron stars with nonzero total entropy. In this case, the loss of stability is marked by the configuration with minimum angular momentum along a sequence of both constant rest mass and total entropy. In the nonrotating limit, Gondek, Haensel, and Zdunik [267] compute frequencies and eigenfunctions of radial pulsations of hot proto–neutron stars and verify that the secular instability sets in at the maximum-mass turning point, as in the case of cold neutron stars.

Although no neutron stars are known to have mass below $1.24M_\odot$, several authors have considered the possibility that a neutron star would be tidally stripped as it spirals in to a more compact neutron star (a neutron star with higher mass) or to a low-mass black hole. It would then lose mass to its companion and reach its lower mass limit [73, 150, 144, 292]. Below and slightly above the lower mass limit, the binding energy of the star is negative (compared to dispersed ^{56}Fe) [519]. As a result, the references listed above (and others mentioned therein) described an instability not to collapse but, as the point of dynamical instability (just below the minimum mass) is reached, to an explosion, with gamma-ray and high antineutrino luminosity from beta decay of neutrons as the star expands.

Recent general relativistic simulations of quasicircular binary inspiral of neutron stars with differing masses [623, 621] and of inspiral of neutron stars to low-mass black holes [624, 622, 436, 212] do show tidal stripping and high antineutrino luminosity from protonization of the unbound neutron matter [601]. In contrast to the possibility of slow stripping suggested by the low-mass instability papers, however, stripping occurs on a timescale of about one orbit, and it is accompanied by tidal disruption that pulls what remains of the star into a spiral and then a disk. Because the dispersal is already rapid, there is no evidence that one can distinguish an explosion due to the instability from the rapid tidal disruption of the star. For NS-NS inspiral with significant differences in neutron-star mass and, more dramatically, for NS-BH inspiral with black-hole spin aligned with the orbit of the small black hole, a substantial disk is formed. This, together with high enough neutrino luminosity to produce a fireball (see [601] and references therein), supports the identification of NS-NS and/or NS-BH mergers as sources of short hard–gamma-ray bursts.

Recent work by Stephens, East, and Pretorius [657] suggests a remaining possibility for a minimum-mass explosion with an observable signature: In dense stellar systems (globular clusters or galactic nuclei), BH-NS binaries form by capture of an unbound neutron star, and they typically have eccentric orbits. For some range of parameters, tidal stripping at successive periastrons may be able to remove mass over several orbits, leading to a final distinguishable orbit in which the neutron star falls below its minimum mass and suddenly explodes.

9.2.3 Dynamical instability to collapse

A secular instability to collapse, mediated by viscosity and nuclear reactions, can proceed when there is a lower energy configuration with the same total baryon number and angular momentum. This is a weaker condition than the condition for dynamical instability, which requires a lower energy configuration that can be reached without a change in the composition or angular momentum of a ring of fluid. As a result, dynamical instability to collapse sets in at a slightly higher density than secular instability, and the low-mass dynamical instability (to explosion) sets in a lower density than that of the corresponding marginally secularly unstable configuration.

From the canonical energy stability criterion of Section 7.4.5, it follows that the onset of dynamical instability along a sequence occurs when E_c vanishes for canonical initial data. The existence of accurate 2+1-evolution codes, however, makes it possible to find instability points by evolving initial data to see whether or not the star collapses, and this is the method that has so far been used to diagnose dynamical instability of rotating stars. The simulations, however, maintain the one-parameter equation of state of the background star, formally finding the dynamical instability points of a barotrope. That is, they in effect allow the nuclear reactions required to maintain the equilibrium relation between density and composition. For a spherical star, this is equivalent to finding points of secular instability. For a rotating star, it finds the stability point associated with an evolution that allows nuclear reactions but forbids transfer of angular momentum between fluid rings.

Numerical simulations by Shibata et al. and Baiotti et al. considered models with slightly higher central densities than secularly unstable models, finding that they do in fact collapse on dynamical timescales, whereas those at slightly smaller central densities are stable; see, for example, [41, 612]. These simulations found that the line of turning points was too close to the dynamical instability boundary to distinguish distinct lines. The subsequent work by Takami, Rezzolla, and Yoshida [675] looks more carefully at the difference between the line of turning points (the maximum-mass ridge) and the boundary between dynamically stable and dynamically unstable barotropic models. They find a dynamical instability line that coincides with the turning-point line for spherical stars and that, for rapid uniform rotation, has a central density about 5% below that of the turning point, see Fig. 9.2. This result somewhat overstates the difference between the two lines, because it ignores the difference $\Gamma - \Gamma_1$ between the indices governing dynamical oscillations and the equilibrium equation of state. The actual dynamical instability line begins at a spherical star with higher density than the marginally unstable turning-point star and probably crosses the turning-point line to lower density at some angular velocity less than Ω_K.

The greater significance of the Takami et al. result, however, is that stars along the line determined by using the equilibrium equation of state are guaranteed to be *secularly* unstable, because the diagnosed instability guarantees that the configurations have lower energy than equilibria with the same baryon mass and angular momentum. This means that the line of secular instability runs through rapidly rotating configurations with central densities more than 5% below those along the line of turning points.

The likelihood that the turning-point criterion does not locate the exact instability line for uniformly rotating stars prompts the following problem.

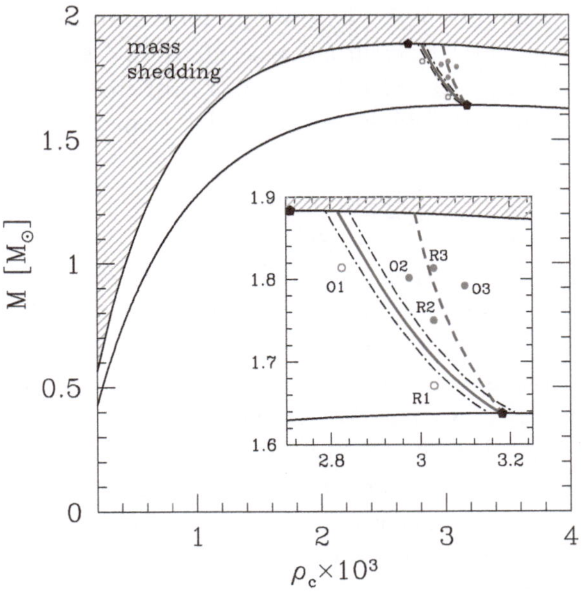

Figure 9.2. Stability lines in a (ρ_c, M) diagram. The two solid black lines mark sequences with either zero (lower line) or mass-shedding angular momentum (upper line), with the filled symbols marking the corresponding maximum masses. The solid gray line is the neutral-stability line, "thickened" by the error bar (dot-dashed lines). The gray dashed line is instead the turning-point criterion for secular stability. Marked with empty or filled circles are representative models with constant angular velocity O1, O2, O3, or constant initial central rest-mass density R1, R2, R3. (Figure from Takami, Rezzolla, and Yoshida [675]. Reproduced by permission of John Wiley and Sons.)

Research Problem. Find the point of secular instability for uniformly rotating stars with a range of candidate EOSs by finding the line along which the canonical energy can first become negative. This is also the line marking dynamical instability of models whose perturbations satisfy the same EOS as that of the equilibrium star.

9.3 Stability to convection (local stability)

The criterion for the stability of a spherical star against convection is easy to understand. When a fluid element is displaced upward, if its density decreases more rapidly than the density of the surrounding fluid, then the element will be buoyed upward and the star will be unstable. If, on the other hand, the fluid element expands less than its surroundings, it will fall back, and the star will be stable to convection.

As this argument suggests, criteria for convective stability are *local*, involving perturbations restricted to an arbitrarily small region of the star or, for axisymmetric perturbations, to an arbitrarily thin ring. For local perturbations, as noted in Section 8.4.2, the Cowling approximation is valid: The change in the gravitational field can be ignored. The argument in the earlier section involved short wavelengths, and we give a version here for perturbations confined to a small region of the star. A perturbation in density of order $\delta\epsilon/\epsilon$ that is restricted to a region of volume $V \ll R^3$ can be regarded as adding or subtracting from the source a mass δm of order $\delta\epsilon V$. Then

$$\frac{\delta m}{M} \sim \frac{V}{R^3}\frac{\delta\epsilon}{\epsilon} \ll \frac{\delta\epsilon}{\epsilon}. \tag{9.22}$$

The change in the metric is then also smaller than $\delta\epsilon/\epsilon$ by a factor V/R^3, arbitrarily small when the support of the matter perturbation is arbitrarily small. Note that, because the metric perturbation is gauge dependent, this statement about the smallness of the metric is also gauge dependent. A more precise way of stating this property of a local perturbation is that a gauge can be chosen in which the metric perturbation is smaller than the density perturbation by a factor of order V/R^3.

For a uniformly rotating star, we have seen that the surfaces of constant p and ϵ coincide, and the same argument can be made for a displacement perpendicular to the surfaces of constant pressure. We begin by making this heuristic argument more carefully, to obtain a criterion for dynamical instability to convection. We then formally obtain the criterion from the canonical energy of Section 7.3.

Convective instability of spherical relativistic stars was discussed by Thorne [686] and subsequently, with greater rigor, by Kovetz [375] and Schutz [597]. An initial heuristic treatment by Bardeen [45] of convective instability of differentially rotating stars was made more precise and extended to models with heat flow and viscosity by Seguin [599].

Consider a fluid element displaced radially outward from an initial position with radial coordinate r to $r+\xi$. The displacement vector then has components $\xi^\mu = \delta^\mu_r\xi$. The fluid element expands (or, if displaced inward, contracts) with its pressure adjusting immediately – in sound travel time across the fluid element – to the pressure outside:

$$\Delta p = \boldsymbol{\xi}\cdot\nabla p = \frac{dp}{dr}\xi. \tag{9.23}$$

Heat diffuses more slowly, and the analysis assumes that the motion is faster than the time for heat to flow into or out of the fluid element. The perturbation is thus *adiabatic*:

$$\begin{aligned}\Delta\epsilon &= \left(\frac{\partial\epsilon}{\partial p}\right)_s \Delta p\\ &= \left(\frac{\partial\epsilon}{\partial p}\right)_s \frac{dp}{dr}\xi = \frac{\epsilon+p}{\Gamma_1 p}\frac{dp}{dr}\xi,\end{aligned} \tag{9.24}$$

where we have used the adiabatic conditions (7.3) and (7.4).

The difference $\Delta_\star\epsilon$ in the density of the surrounding star between r and $r+\xi$ is given by

$$\Delta_\star\epsilon = \xi\frac{d\epsilon}{dr}. \tag{9.25}$$

The displaced fluid element falls back if $|\Delta\epsilon| < |\Delta_\star\epsilon|$ – if, that is, the fluid element's density decreases more slowly than the star's density:

$$\left(\frac{\partial\epsilon}{\partial p}\right)_s \left|\xi\frac{dp}{dr}\right| < \left|\xi\frac{d\epsilon}{dr}\right|. \tag{9.26}$$

The star is then stable against convection if the inequality

$$\left(\frac{dp}{d\epsilon}\right)_\star := \frac{dp/dr}{d\epsilon/dr} < \left(\frac{\partial p}{\partial\epsilon}\right)_s \tag{9.27}$$

(equivalently, $\Gamma < \Gamma_1$), is satisfied, and unstable if the inequality is in the opposite direction.

In particular, in a homentropic star with no composition gradient, the adiabatic value of $dp/d\epsilon$ coincides with its value in the equilibrium star,

$$\left(\frac{\partial\epsilon}{\partial p}\right)_s = \left(\frac{dp}{d\epsilon}\right)_\star, \tag{9.28}$$

implying that the star is marginally unstable. Because studies of neutron-star oscillations and stability often consider barotropic stars whose oscillations satisfy the same one-parameter equation of state as the equilibrium star, the space of linearized solutions includes a degenerate subspace of zero-frequency modes – time-independent solutions to the Einstein-Euler system linearized about the equilibrium star [301].

The convective stability criterion (Schwarzschild criterion) can also be stated in terms of the temperature gradient: If the temperature gradient is superadiabatic – if T decreases faster in the star than in an adiabatically displaced fluid element – then the star is unstable against convection. This second version of the condition may be intuitively obvious. To obtain it formally, write $\epsilon = \epsilon(p,T)$ or $\epsilon(p,s) = \epsilon(p,T(p,s))$. Then $\Delta\epsilon$ for a displaced fluid element is given by

$$\Delta\epsilon = \left.\frac{\partial\epsilon}{\partial p}\right|_T \Delta p + \left.\frac{\partial\epsilon}{\partial T}\right|_p \Delta T \tag{9.29}$$

$$= \left.\frac{\partial\epsilon}{\partial p}\right|_T \Delta p + \left.\frac{\partial\epsilon}{\partial T}\right|_p \left.\frac{\partial T}{\partial p}\right|_s \Delta p; \tag{9.30}$$

and $\Delta_\star\epsilon$ is given by

$$\Delta_\star\epsilon = \left.\frac{\partial\epsilon}{\partial p}\right|_T \Delta p + \left.\frac{\partial\epsilon}{\partial T}\right|_p \Delta_\star T \tag{9.31}$$

$$= \left.\frac{\partial\epsilon}{\partial p}\right|_T \Delta p + \left.\frac{\partial\epsilon}{\partial T}\right|_p \frac{dT_\star/dr}{dp_\star/dr}\Delta p. \tag{9.32}$$

Thus, as claimed, the star is stable if $|\Delta T|_{\mathrm{adiabatic}} > |\Delta_\star T|$.

For spherical stars, any perturbation can be written as a superposition of spherical harmonics that are axisymmetric about some axis, and one therefore need only consider stability of axisymmetric perturbations. In fact, Detweiler and Ipser [182] (generalizing a Newtonian result due to Lebovitz [398]), show that, apart from local instability to convection, one need only consider radial perturbations: *If a nonrotating star is stable to radial oscillations and stable against convection, the star is stable.* The Detweiler-Ipser argument shows that Schwarzschild criterion (9.27) for stability against convection implies that there are no zero-frequency nonradial modes with polar parity, and no time-independent polar-parity solutions to the perturbed Einstein-Euler system. The argument is then by continuity of the frequency of outgoing modes: For each relativistic model, there is a smooth sequence of nonrotating stars, starting from a nearly Newtonian star and ending with the relativistic model. For the nearly Newtonian star, the Lebovitz result should imply that all modes are stable. Then if a mode is unstable for the relativistic model, the imaginary part of its frequency must change sign along the sequence, becoming real for some nonrotating configuration. But if the mode is nonradial, this can happen only if the frequency vanishes (assuming that any nonradial, real-frequency mode is radiative), and that cannot happen if the Schwarzschild criterion is satisfied.

The argument is compelling but not rigorous. It could be made more cleanly and without assumptions about normal modes if one could show directly that the canonical energy was always positive. This may follow from an integral inequality (associated with Eq. (42) of [182]) that is central to the Detweiler-Ipser argument.

Research Problem. Consider a nonrotating equilibrium solution to the Einstein-Euler system. With reasonable assumptions about the EOS, show explicitly that the canonical energy of a nonradial perturbation is negative only if the Schwarzschild criterion is violated.

9.3.1 Canonical energy of local perturbations

For uniformly rotating stars, because surfaces of constant p and ϵ coincide, the derivative $(dp/d\epsilon)_\star$ is again well defined, given by

$$\left(\frac{dp}{d\epsilon}\right)_\star \nabla\epsilon := \nabla p. \tag{9.33}$$

In this more general case, as we now show, stability against convection is again governed by the Schwarzschild condition

$$\left(\frac{dp}{d\epsilon}\right)_\star < \left(\frac{\partial p}{\partial\epsilon}\right)_s. \tag{9.34}$$

To obtain this more general result, we replace the heuristic derivation of the last section by a proof that the sign of the canonical energy E_c is negative for initial data violating the Schwarzschild condition. This follows from a particularly simple form for E_c for local perturbations.

Proposition. *Consider an initial perturbation of a uniformly rotating star, with ξ^α time-independent and axisymmetric and $h_{\alpha\beta} = 0$. Then the canonical energy has the form*

$$E_c = \int \frac{1}{(\epsilon + p)} \left\{ \left[\left(\frac{\partial p}{\partial \epsilon}\right)_s - \left(\frac{dp}{d\epsilon}\right)_\star \right] \Delta\epsilon^2 + \left(\frac{dp}{d\epsilon}\right)_\star \delta\epsilon^2 \right\} \alpha dV. \tag{9.35}$$

Thus, for an initial perturbation with $\delta\epsilon = 0, \Delta\epsilon \neq 0$, we have

$$E_c = \int \frac{1}{\epsilon + p} \left[\left(\frac{\partial p}{\partial \epsilon}\right)_s - \left(\frac{dp}{d\epsilon}\right)_\star \right] \Delta\epsilon^2 N dV, \tag{9.36}$$

and there are time-independent, axisymmetric initial displacements ξ^α for which the canonical energy E_c of a rotating barotropic star is negative if and only if the generalized Schwarzschild criterion is violated.

We obtain the form (9.35) of E_c as follows. The canonical energy in a rotating frame, $E_{c,r} = E_c - \Omega J_c$, is invariant under the gauge transformations associated with trivial displacements; for axisymmetric perturbations, $E_{c,r} = E_c$, because the canonical angular momentum vanishes. For perturbations with $h_{\alpha\beta} = 0$ and ξ^α time independent on a hypersurface S, the canonical energy has the form

$$E_c = \int_S \xi_\alpha \Delta \nabla_\beta T^{\alpha\beta} N dV - \{\text{terms with second time derivatives of } \xi^\alpha\}. \tag{9.37}$$

Recall that Eq. (2.64) for $\Delta\epsilon$ in terms of a Lagrangian displacement,

$$\Delta\epsilon = (\epsilon + p) q^{\alpha\beta} \nabla_\alpha \xi_\beta, \tag{9.38}$$

is a first integral of the projection $u_\alpha \Delta \nabla_\beta T^{\alpha\beta} = 0$. We can therefore replace $\Delta \nabla_\beta T^{\alpha\beta}$ in E_c by its projection orthogonal to u_α to write E_c in the form

$$E_c = \int \xi^\alpha (\epsilon + p) \Delta \left[u^\beta \nabla_\beta u_\alpha + q_\alpha{}^\beta \frac{\nabla_\beta p}{\epsilon + p} \right] N dV$$
$$- \{\text{terms with second time derivatives of } \xi^\alpha\}. \tag{9.39}$$

Because E_c is unchanged by a gauge transformation of the form $\xi^\alpha \to \xi^\alpha + f u^\alpha$, we can set $\xi^\alpha u_\alpha = 0$, with no loss of generality. We first show the relation

$$\Delta(u^\beta \nabla_\beta u_\alpha) = -\nabla_\alpha \left(\frac{\boldsymbol{\xi} \cdot \nabla p}{\epsilon + p} \right). \tag{9.40}$$

We begin by using the relation $u^\beta \nabla_\beta u_\alpha = \mathcal{L}_{\mathbf{u}} u_\alpha$ to write

$$\Delta(u^\beta \nabla_\beta u_\alpha) = \Delta(\mathcal{L}_{\mathbf{u}} u_\alpha) = \mathcal{L}_{\Delta \mathbf{u}} u_\alpha + \mathcal{L}_{\mathbf{u}} \Delta u_\alpha. \tag{9.41}$$

Next, from Eq. (2.37) and the fact that ξ^α is orthogonal to u^α, we have

$$\Delta u^\alpha = u^\alpha u^\beta u^\gamma \nabla_\beta \xi_\gamma = -u^\alpha u^\beta \xi^\gamma \nabla_\beta u_\gamma = u^\alpha \frac{\boldsymbol{\xi} \cdot \nabla p}{\epsilon + p} \tag{9.42}$$

and

$$\Delta u_\alpha = \Delta(g_{\alpha\beta} u^\beta) = \Delta g_{\alpha\beta} u^\beta + g_{\alpha\beta} \Delta u^\beta = (\nabla_\alpha \xi_\beta + \nabla_\beta \xi_\alpha) u^\beta + u_\alpha \frac{\boldsymbol{\xi} \cdot \nabla p}{\epsilon + p}. \tag{9.43}$$

The two terms on the right side of Eq. (9.41) are then given by

$$\mathcal{L}_{\Delta \mathbf{u}} u_\alpha = \frac{\boldsymbol{\xi}\cdot\nabla p}{\epsilon+p}\mathcal{L}_{\mathbf{u}} u_\alpha + u_\beta \nabla_\alpha\left(\frac{\boldsymbol{\xi}\cdot\nabla p}{\epsilon+p}\right)u^\beta = -\frac{\boldsymbol{\xi}\cdot\nabla p}{(\epsilon+p)^2}\nabla_\alpha p - \nabla_\alpha\left(\frac{\boldsymbol{\xi}\cdot\nabla p}{\epsilon+p}\right) \tag{9.44}$$

and (writing $u^\alpha = u^t k^\alpha$)

$$\begin{aligned}\mathcal{L}_{\mathbf{u}}\Delta u_\alpha &= u^t \mathcal{L}_k \Delta u_\alpha + \Delta u_\beta \nabla_\alpha (u^t) k^\beta = \nabla_\alpha (\ln u^t) u^\beta \Delta u_\beta \\ &= \frac{\nabla_\alpha p}{\epsilon+p}\left[2u^\alpha u^\beta \nabla_\alpha \xi_\beta - \frac{\boldsymbol{\xi}\cdot\nabla p}{\epsilon+p}\right] \\ &= \frac{\boldsymbol{\xi}\cdot\nabla p}{(\epsilon+p)^2}\nabla_\alpha p;\end{aligned} \tag{9.45}$$

their sum has the form of Eq. (9.40), as claimed.

The term in the integrand of E_c involving $\Delta(u^\beta \nabla_\beta u_\alpha)$ can now be written as

$$(\epsilon+p)\xi^\alpha \Delta(u^\beta \nabla_\beta u_\alpha) = \nabla_\alpha[(\epsilon+p)\xi^\alpha]\frac{\boldsymbol{\xi}\cdot\nabla p}{\epsilon+p} - \nabla_\alpha(\boldsymbol{\xi}\cdot\nabla p\;\xi^\alpha); \tag{9.46}$$

using Eq. (7.5c) in the form

$$\delta\epsilon = -\nabla_\alpha[(\epsilon+p)\xi^\alpha], \tag{9.47}$$

we have

$$(\epsilon+p)\xi^\alpha \Delta(u^\beta \nabla_\beta u_\alpha) = -\delta\epsilon\frac{\boldsymbol{\xi}\cdot\nabla p}{\epsilon+p} - \nabla_\alpha(\xi\cdot\nabla p\;\xi^\alpha). \tag{9.48}$$

We write the second term in the integrand of E_c as

$$\begin{aligned}(\epsilon+p)\xi^\alpha \Delta\frac{\nabla_\alpha p}{\epsilon+p} &= \xi^\alpha \nabla_\alpha \Delta p - \frac{\boldsymbol{\xi}\cdot\nabla p}{\epsilon+p}(\Delta\epsilon + \Delta p) \\ &= -\nabla\cdot\boldsymbol{\xi}\Delta p - \frac{\boldsymbol{\xi}\cdot\nabla p}{\epsilon+p}(\Delta\epsilon+\Delta p) + \nabla_\alpha(\Delta p\;\xi^\alpha).\end{aligned} \tag{9.49}$$

Combining Eqs. (9.48) and (9.49) and noting that, for a time-independent perturbation, the spatial integral of a divergence vanishes, we have

$$E_c = \int\left[-\frac{\boldsymbol{\xi}\cdot\nabla p}{\epsilon+p}(\Delta\epsilon+\delta\epsilon) - (\nabla\cdot\boldsymbol{\xi} + \frac{\boldsymbol{\xi}\cdot\nabla p}{\epsilon+p})\Delta p\right]\alpha dV. \tag{9.50}$$

Finally, writing

$$\Delta p = \left(\frac{\partial p}{\partial\epsilon}\right)_s \Delta\epsilon, \qquad \frac{\Delta\epsilon}{\epsilon+p} = -\left(\nabla\cdot\boldsymbol{\xi} + \frac{\boldsymbol{\xi}\cdot\nabla p}{\epsilon+p}\right), \tag{9.51}$$

and using Eq. (9.33), we obtain for E_c the form

$$E_c = \int\left[-\frac{1}{\epsilon+p}\left(\frac{dp}{d\epsilon}\right)_\star(\Delta\epsilon^2 - \delta\epsilon^2) + \frac{1}{\epsilon+p}\left(\frac{\partial p}{\partial\epsilon}\right)_s \Delta\epsilon^2\right], \tag{9.52}$$

and the proposition follows.

9.3.2 Convective instability due to differential rotation: The Solberg criterion

Differentially rotating stars have one additional kind of convective (local) instability. If the angular momentum per unit rest mass, $j = hu_\alpha \phi^\alpha$, decreases outward from the axis of symmetry, the star is unstable to perturbations that change the differential rotation law.[2] The simplest example of the instability, considered by Rayleigh [557], is for an incompressible fluid in a cylindrical container. Its specific angular momentum (angular momentum per unit mass) is $j = \varpi^2\Omega(\varpi)$, and the flow is unstable when $dj/d\varpi < 0$. For differentially rotating stars with uniform entropy per baryon, it is not difficult to obtain the same criterion in a Newtonian context. This extension of the Rayleigh criterion is due to Solberg [638] and is commonly called the Solberg or Solberg-Høiland criterion in the astrophysics literature; Høiland [317] showed that a nonbarotropic star is stable to axisymmetric perturbations if and only if s decreases outward and j increases from pole to equator on each $s =$ constant surface.

For relativistic stars, the same criterion ordinarily holds, where the specific angular momentum $j = hu_\phi$ is the angular momentum per unit rest mass introduced in Eq. (1.47). Bardeen [45] gives a heuristic argument for a restricted version of this criterion, and a subsequent comprehensive and more precise treatment, including heat flow and viscosity, is due to Seguin [599]. Abramowicz [3] provides a much quicker and more intuitive derivation for a homentropic star with no dissipation. This is the appropriate context for dynamical instability of neutron stars, and we roughly follow his discussion, making it more precise by again showing that the canonical energy can be made negative when the generalized Solberg criterion is violated.

We first give Bardeen's two-line argument, using the first law (3.124) to argue that, if, in a homentropic, differentially rotating star, j has an extremum in going outward in the equatorial plane, then there are perturbations that conserve baryon number and that lower the energy of the system – for which $\delta M < 0$. The argument, for a homentropic star, is this: On opposite sides of the extremum, there are two rings, 1 and 2, with the same value of j and with $\Omega_2 > \Omega_1$. A perturbation that transfers matter with baryon mass δM_0 from ring 2 to ring 1 then gives $\delta M = (\Omega_1 - \Omega_2) j\, \delta M_0 < 0$.

It makes sense to speak of the angular momentum increasing or decreasing *outward*, because, as we have seen, the angular velocity of barotropic stars is stratified on cylinders: It is a function of a single parameter, $u^t u_\phi$, implying in the Newtonian limit the relation $\Omega = \Omega(\varpi)$. In particular, j increases outward if it is an increasing function of r in the equatorial plane. In the stability criterion, dj/dr appears with a coefficient that is positive definite for relativistic stars. As Abramowicz and Prasanna [4] point out, however, the coefficient changes sign near black holes, and we comment on that at the end of this subsection, after Eq. (9.65).

[2] We adopt in this section the convention that Ω is positive. This involves no loss of generality, because the stability criterion implies for a stable star that Ω cannot change sign. One can, if desired, avoid the need for a convention by working with j^2 instead of j.

We begin with an intuitive version of the argument for a Newtonian star. A ring of fluid in the equatorial plane at radius r_0 has angular momentum per unit mass $j_0 = v(r_0)r_0$. Whether or not differential rotation implies instability depends on whether or not a ring that is displaced outward, conserving angular momentum and mass, will fall back. In the background star, the centripetal acceleration $-v^2/r$ of a fluid ring at radius r is equal to the force per unit mass $\Phi' + p'/\rho$. The ring of fluid displaced from r_0 to $r_0 + \xi^r$ will fall back if its centripetal acceleration v^2/r is smaller than that of the ring of fluid at $r_0 + \xi^r$ in the background star; that is, it will fall back if the change $\Delta(v^2/r)$ for the displaced ring is smaller than the corresponding difference in the acceleration of fluid rings at $r_0 + \xi^r$ and at r_0 in the background star. The difference in acceleration for the background star is $\Delta_\star(v^2/r) = \xi^r \dfrac{d}{dr}(v^2/r)$, and stability then requires

$$\Delta_\star v^2/r - \Delta\frac{v^2}{r} > 0, \tag{9.53}$$

for $\xi^r > 0$.

Because $\Delta j = 0$ and $v(j, r) = j(r)/r$, we have

$$\Delta\frac{v^2}{r} = \Delta\frac{j^2}{r^3} = j^2\Delta\frac{1}{r^3} = j^2\xi^r\frac{d}{dr}\frac{1}{r^3}, \tag{9.54}$$

while

$$\Delta_\star\frac{v^2}{r} = \xi^r\frac{d}{dr}\frac{j^2}{r^3} = \xi^r\frac{1}{r^3}\frac{dj^2}{dr} + j^2\xi^r\frac{d}{dr}\frac{1}{r^3}. \tag{9.55}$$

Then

$$\Delta_\star\frac{v^2}{r} - \Delta\frac{v^2}{r} = \xi^r\frac{1}{r^2}\frac{dj^2}{dr}; \tag{9.56}$$

and the star is stable only if $\dfrac{dj}{dr} > 0$ in the equatorial plane, or, equivalently, only if $\partial_\varpi(\varpi^2\Omega) > 0$.

Before extending the argument to the full theory in a way that allows us to compute the canonical energy, it is helpful to note that, for an axisymmetric displacement of an axisymmetric fluid ring that preserves angular momentum per unit mass, the same Newtonian computation gives the form

$$\Delta_\star\left(\frac{v^2}{\varpi}\right) - \Delta\left(\frac{v^2}{\varpi}\right) = \frac{\partial}{\partial j}\left(\frac{v^2(j,\varpi)}{\varpi}\right)\xi^a\nabla_a j. \tag{9.57}$$

In the full theory, conservation of angular momentum and baryon number of a displaced ring implies that the angular momentum per unit baryon mass, $j = hu_\alpha\phi^\alpha = hu_\phi$, is conserved. From Eq. (1.84) for u_ϕ,

$$j = h\frac{e^\psi v}{\sqrt{1-v^2}}, \tag{9.58}$$

with inverse

$$v(j,h,\psi) = \frac{j}{\sqrt{j^2 + h^2 e^{2\psi}}}. \tag{9.59}$$

We will use the fluid equilibrium equation in essentially the form given in Eq. (1.99),

$$\nabla \ln h + \nabla \nu - \gamma^2 v^2 \nabla(\psi - \nu) + \gamma^2 v e^{\psi-\nu} \nabla\omega = 0, \tag{9.60}$$

where $\gamma = \dfrac{1}{\sqrt{1-v^2}}$. Because e^ψ is the radial coordinate for which $2\pi e^\psi$ is the proper circumference of circles about the axis of symmetry, the first three terms have as their Newtonian limit $\nabla p/\rho + \nabla\Phi - \dfrac{v^2}{\varpi}\nabla\varpi$, whereas the Newtonian limit of the final term is zero.

The heuristic discussion parallels the Newtonian version, with the last two terms of Eq. (9.60) regarded as the centripetal acceleration. We again consider a radial displacement ξ^α of a thin ring of fluid in the equatorial plane. Because a displacement of a thin ring of fluid is a local perturbation, in the sense that the volume of the displaced fluid is small, we have $\delta\psi = \delta\nu = 0$. The pressure of the displaced ring is that of the background star at the new location, implying $\delta h = 0$, and we have

$$\Delta\nu = \Delta_\star \nu = \xi^\alpha \nabla_\alpha \nu, \quad \Delta\psi = \Delta_\star \psi = \xi^\alpha \nabla_\alpha \psi, \quad \Delta h = \Delta_\star h = \xi^\alpha \nabla_\alpha h. \tag{9.61}$$

The displaced ring then falls back if, for an outward displacement,

$$(\Delta_\star - \Delta)[\gamma^2 v^2 \partial_r(\psi - \nu) - \gamma^2 v e^{\psi-\nu} \partial_r \omega] > 0. \tag{9.62}$$

We have

$$(\Delta_\star - \Delta)(\gamma^2 v^2) = \frac{\partial}{\partial v}(\gamma^2 v^2)\frac{\partial v}{\partial j}\xi^\alpha \nabla_\alpha j \tag{9.63a}$$

$$(\Delta_\star - \Delta)(\gamma^2 v) = \frac{\partial}{\partial v}(\gamma^2 v)\frac{\partial v}{\partial j}\xi^\alpha \nabla_\alpha j, \tag{9.63b}$$

and we need compute only

$$\frac{\partial}{\partial v}(\gamma^2 v) = (1+v^2)\gamma^4, \qquad \frac{\partial}{\partial v}(\gamma^2 v^2) = 2v\gamma^4. \tag{9.64}$$

The generalized Solberg criterion (9.62) then has, for $\xi^r > 0$, the form

$$\left[2v\partial_r(\psi - \nu) - (1+v^2)e^{\psi-\nu}\partial_r\omega\right]\gamma^4 \frac{\partial v}{\partial j}\xi^r \partial_r j > 0. \tag{9.65}$$

From Eq. (9.59), $\partial v/\partial j > 0$ for $j > 0$; for models of relativistic stars, the bracketed coefficient is positive. Thus a necessary condition for local stability is that, in the equatorial plane,

$$\partial_r j > 0. \tag{9.66}$$

A special case of the relativistic Solberg criterion (9.65) that we have just derived is the condition for stability of circular orbits. When $\nabla h = 0$, Eq. (9.60) determines the angular velocity distribution of a rotating pressureless disk; because each particle (each fluid element) moves in a circular geodesic, the equation gives the relation between Ω and r for circular orbits, and their stability is again governed by the sign of the left-hand side of (9.65), with orbits ordinarily stable for $dj/dr > 0$.

As noted earlier, near rotating black holes, the bracketed coefficient, $\mathcal{C} := 2v\partial_r(\psi - \nu) - (1+v^2)e^{\psi-\nu}\partial_r\omega$, becomes negative. This would seem to imply that, within the critical radius where the coefficient vanishes, an orbit is *unstable* if $\partial_r j > 0$, contradicting Bardeen's heuristic argument that simply gave $\partial_r j > 0$ as the criterion. There is, however, no contradiction if a fluid in equilibrium cannot include a region where $\mathcal{C}$ is negative.[3] This is likely, because the condition for equilibrium Eq. (9.60) can be written in the equatorial plane in the form

$$\partial_r \ln h + \partial_r \nu + \frac{1}{2} v e^{\psi-\nu}\partial_r\omega - v\gamma^2\mathcal{C} = 0. \tag{9.67}$$

A fluid in equilibrium outside a black hole cannot intersect the horizon, because its 4-velocity at the horizon would point inward, implying a nonzero flux. Where it terminates in the equatorial plane, Eq. (9.60) is satisfied with $\partial_r h \geq 0$. Now $\partial_r\nu + \frac{1}{2}ve^{\psi-\nu}\partial_r\omega > \partial_r\nu + \frac{1}{2}e^{\psi-\nu}\partial_r\omega > 0$ outside the horizon of a Kerr black hole, implying that no equilibrium is possible with $\mathcal{C} < 0$. The geometry of a black hole with matter outside is, of course, not exactly Kerr; to turn the argument into a proof that no equilibrium can extend in the equatorial plane to a region with $\mathcal{C} < 0$, one would need to show that the inequality $\partial_r\nu + \frac{1}{2}e^{\psi-\nu}\partial_r\omega > 0$ holds outside the horizon of a distorted black hole.

Canonical energy. Using Eqs. (9.63), we can quickly turn the heuristic discussion given above into a precise statement involving the sign of the canonical energy: For a differentially rotating homentropic star, the canonical energy of a local axisymmetric perturbation with $\delta p = 0$ is given by

$$E_c = \int (\epsilon + p)\left[2v\xi^\alpha\nabla_\alpha(\psi-\nu) - (1+v^2)e^{\psi-\nu}\xi^\alpha\nabla_\alpha\omega\right]\gamma^4\frac{\partial v}{\partial j}\xi^\alpha\nabla_\alpha j\ \alpha dV, \tag{9.68}$$

implying that there are perturbations for which $E_c < 0$ unless

$$\xi^\alpha\nabla_\alpha j > 0, \quad \text{for } \xi^\alpha \text{ outward-directed}, \tag{9.69}$$

where *outward-directed* is defined by

$$\xi^\alpha\left[\nabla_\alpha(\psi-\nu) - \frac{(1+v^2)}{2v}e^{\psi-\nu}\nabla_\alpha\omega\right] > 0. \tag{9.70}$$

We obtain the result as follows. For a uniformly rotating homentropic star, we have seen that the canonical energy of perturbations with ξ^α time independent is positive, vanishing only when $\delta p = 0$. To avoid that term, we set $\delta p = 0$; as in

[3] We are indebted to J. Bardeen for supplying the argument given here.

the preceding discussion (and to avoid a first-order change in the energy), we also require $\Delta j = 0$. This condition implies that the angular velocity of the perturbed fluid element changes and that $\dot{\xi}^\alpha \neq 0$. Once ξ^α is chosen on the hypersurface where E_c is evaluated, one satisfies the condition by writing $\dot{\xi}^\alpha = \zeta u^\alpha$ and determining ζ algebraically from $\Delta j = 0$. Then the kinetic term $\frac{1}{2}(\epsilon + p)q_{\alpha\beta}\dot{\xi}^\alpha \dot{\xi}^\beta$ in the integrand of E_c vanishes, and we can again write, as in Eq. (9.39)

$$E_c = \int (\epsilon + p)\xi^\alpha \delta \left[u^\beta \nabla_\beta u_\alpha + q_\alpha{}^\beta \frac{\nabla_\beta p}{\epsilon + p} \right] \alpha dV. \tag{9.71}$$

(Recall that the equilibrium Euler equation implies $\Delta \nabla_\beta T^{\alpha\beta} = \delta \nabla_\beta T^{\alpha\beta}$.) Now, however, with $\delta p = 0 = \left(\frac{\partial p}{\partial \epsilon}\right)_s - \left(\frac{dp}{d\epsilon}\right)_\star$, and with $\mathcal{L}_{\mathbf{u}}\xi^\alpha$ nonzero, we obtain a set of terms that are complementary to the terms in Eq. (9.35):

$$\begin{aligned} E_c &= \int (\epsilon + p)\xi^\alpha \delta \left[\nabla_\alpha h + \nabla_\alpha \nu - \gamma^2 v^2 \nabla_\alpha (\psi - \nu) + \gamma^2 v e^{\psi - \nu} \nabla_\alpha \omega \right] \alpha dV \\ &= \int (\epsilon + p)\xi^\alpha \left[-\delta(\gamma^2 v^2) \nabla_\alpha (\psi - \nu) + \delta(\gamma^2 v) e^{\psi - \nu} \nabla_\alpha \omega \right] \alpha dV \\ &= \int (\epsilon + p)\xi^\beta \left[2v \nabla_\beta (\psi - \nu) - (1 + v^2) e^{\psi - \nu} \nabla_\beta \omega \right] \gamma^4 \frac{\partial v}{\partial j} \xi^\alpha \nabla_\alpha j \, \alpha dV, \end{aligned} \tag{9.72}$$

where, in the last equality, we have used Eqs. (9.63) and the fact that the action of δ on a scalar Q is given by $\delta Q = (\Delta - \xi^\alpha \nabla_\alpha) Q = (\Delta - \Delta_\star) Q$. Note that, in choosing data on the hypersurface where E_c is evaluated, one can simultaneously demand that $\delta p = 0$ and that ξ^α be outward directed in the sense of Eq. (9.70). One first arbitrarily chooses a positive component of ξ^α along the vector $\nabla_\alpha (\psi - \nu) - \frac{(1 + v^2)}{2v} e^{\psi - \nu} \nabla_\alpha \omega$ of (9.70); the condition $\delta p = 0$ then determines (up to the ambiguity of the integration) the component of ξ^α orthogonal to that vector.

9.4 Instability sets in through zero-frequency modes

We can use a form of the symmetry relation (7.40) to obtain an equation for the frequency of a mode with time dependence $e^{i\sigma t}$ of the form

$$A\sigma^2 - (B + iD)\sigma - C = 0, \tag{9.73}$$

where A, B, C, and D are real, with $A > 0$ and $D \geq 0$; the dissipative coefficient D, is proportional to the flux of energy radiated to future null infinity. We derive the equation and then use it to describe the onset of instability of an outgoing mode.

Derivation. To derive the equation, we choose as the arguments ξ^α and $h_{\alpha\beta}$ of $\mathscr{L}(\widehat{\xi}, \widehat{h}; \xi, h)$ an outgoing mode, writing $\xi^\alpha = \tilde{\xi}^\alpha e^{i\sigma t}$, $h_{\alpha\beta} = \tilde{h}_{\alpha\beta} e^{i\sigma t}$, where $\tilde{\xi}^a$ and $\tilde{h}_{ab}$ are independent of t (Lie-derived by t^α). For the hatted quantities, we choose a time dependence that turns $\nabla_\alpha \Theta^\alpha$ into a spatial divergence. That is, with

$(\widehat{\xi}^\alpha, \widehat{h}_{\alpha\beta}) = (\xi^{\alpha *}e^{-i\sigma t}, \tilde{h}^*_{\alpha\beta}e^{-i\sigma t})$, all terms in the symmetry relation are time independent, implying

$$\nabla_\beta \Theta^\beta = \partial_t \Theta^t + \partial_j \Theta^j = \partial_j \Theta^j = D_b \Theta^b,$$

where we have used the form (A.47) of the divergence of a vector density. (We use an asterisk to denote complex conjugation in this section, to avoid writing "$\bar{\tilde{\xi}}^\alpha$".) Note that, because σ is in general complex, the hatted pair does not satisfy the Einstein-Euler equations.

With this choice of time dependence, the derivatives in $\mathscr{L}$ and Θ^a are given by

$$\nabla_\alpha \xi^\beta = (\nabla_\alpha \tilde{\xi}^\beta + i\sigma \nabla_\alpha t\, \tilde{\xi}^\beta)e^{i\sigma t}, \quad \nabla_\alpha \widehat{\xi}^\beta = (\nabla_\alpha \tilde{\xi}^{\beta *} - i\sigma \nabla_\alpha t\, \tilde{\xi}^{\beta *})e^{-i\sigma t},$$

with the corresponding replacements for $\nabla_\gamma h_{ab}$ and $\nabla_\gamma \widehat{h}_{ab}$.

The density $\mathscr{L}$ then has the form

$$2\mathscr{L}(\tilde{\xi}^* e^{-i\sigma t}, \tilde{h}^* e^{-i\sigma t}; \tilde{\xi} e^{i\sigma t}, \tilde{h} e^{i\sigma t}) = \sigma^2 \mathcal{A} - \sigma \mathcal{B} - \mathcal{C}, \tag{9.74}$$

where

$$|g|^{-1/2}\mathcal{A} = U^{\alpha\beta\gamma\delta} \nabla_\alpha t\, \nabla_\gamma t \tilde{\xi}^*_\beta \tilde{\xi}_\delta - \frac{1}{2}\epsilon^{\alpha\gamma\epsilon\eta}\epsilon^{\beta\delta\zeta}{}_\eta \nabla_\alpha t \nabla_\beta t\, \tilde{h}^*_{\gamma\delta}\tilde{h}_{\epsilon\zeta} \tag{9.75a}$$

$$\begin{aligned}|g|^{-1/2}\mathcal{B} = {} & iU^{\alpha\beta\gamma\delta}\nabla_\alpha t(\tilde{\xi}^*_\beta \nabla_\gamma \tilde{\xi}_\delta - \tilde{\xi}_\beta \nabla_\gamma \tilde{\xi}^*_\delta) + iV^{\alpha\beta\gamma\delta}\nabla_\gamma t(\tilde{h}_{\alpha\beta}\tilde{\xi}^*_\delta - \tilde{h}^*_{\alpha\beta}\tilde{\xi}_\delta) \\ & - \frac{i}{32\pi}\epsilon^{\alpha\gamma\epsilon\eta}\epsilon^{\beta\delta\zeta}{}_\eta \nabla_\alpha t(\tilde{h}^*_{\gamma\delta}\nabla_\beta \tilde{h}_{\epsilon\zeta} - \tilde{h}_{\gamma\delta}\nabla_\beta \tilde{h}^*_{\epsilon\zeta}),\end{aligned} \tag{9.75b}$$

$$\begin{aligned}|g|^{-1/2}\mathcal{C} = {} & -U^{\alpha\beta\gamma\delta}\nabla_\alpha \tilde{\xi}^*_\beta \nabla_\gamma \tilde{\xi}_\delta - V^{\alpha\beta\gamma\delta}(\tilde{h}^*_{\alpha\beta}\nabla_\gamma \tilde{\xi}_\delta + \tilde{h}_{\alpha\beta}\nabla_\gamma \tilde{\xi}^*_\delta) \\ & + \frac{1}{32\pi}\epsilon^{\alpha\gamma\epsilon\eta}\epsilon^{\beta\delta\zeta}{}_\eta \nabla_\gamma \tilde{h}^*_{\alpha\beta}\nabla_\delta \tilde{h}_{\epsilon\zeta} + \frac{1}{2}\nabla^\gamma T^{\alpha\beta}(\tilde{h}^*_{\alpha\beta}\tilde{\xi}_\gamma + \tilde{h}_{\alpha\beta}\tilde{\xi}^*_\gamma) \\ & + T^{\alpha\beta}R_{\alpha\gamma\beta\delta}\tilde{\xi}^{\gamma *}\tilde{\xi}^\delta + \frac{1}{16\pi}\left(G^{\alpha\beta\gamma\delta} - \frac{1}{2}W^{\alpha\beta\gamma\delta}\right)\tilde{h}^*_{\alpha\beta}\tilde{h}_{\gamma\delta}.\end{aligned} \tag{9.75c}$$

The symmetry of $\mathcal{A}$ and $\mathcal{C}$ and the antisymmetry of $\mathcal{B}$ under the interchange $(\tilde{\xi}^\alpha, \tilde{h}_{\alpha\beta}) \leftrightarrow (\tilde{\xi}^{\alpha *}, \tilde{h}^*_{\alpha\beta})$ imply that $\mathcal{A}$, $\mathcal{B}$, and $\mathcal{C}$ are all real. For a mode $(h_{\alpha\beta}, \xi^\alpha)$ that satisfies the perturbed Einstein-Euler equations, Eq. (7.40) implies $2\mathscr{L} = \nabla_\alpha \Theta^\alpha$, where

$$\begin{aligned}|g|^{-1/2}\Theta^\alpha = {} & U^{\alpha\beta\gamma\delta}\tilde{\xi}^*_\beta(\nabla_\gamma + i\sigma\nabla_\gamma t)\tilde{\xi}_\delta + V^{\gamma\delta\alpha\beta}\tilde{h}_{\gamma\delta}\tilde{\xi}^*_\beta \\ & - \frac{1}{32\pi}\epsilon^{\alpha\gamma\epsilon\eta}\epsilon^{\beta\delta\zeta}{}_\eta \tilde{h}^*_{\gamma\delta}(\nabla_\beta + i\sigma\nabla_\beta t)\tilde{h}_{\epsilon\zeta}.\end{aligned} \tag{9.76}$$

Equations (7.40) and (9.74) then give the relation

$$\sigma^2\mathcal{A} - \sigma\mathcal{B} - \mathcal{C} - D_a\Theta^a = 0;$$

and integrating this equation over a hypersurface that is asymptotically null, we have

$$\sigma^2 A - \sigma B - C - \int_{S_\infty} \Theta^a dS_a = 0, \tag{9.77}$$

where $A = \int_V \mathcal{A} d^3x$, $B = \int_V \mathcal{B} d^3x$, $C = \int_V \mathcal{C} d^3x$. The surface integral here is essentially identical to that in Eq. (7.79) for $\dot{E}_c$ if one omits the time derivative $\mathcal{L}_t$. The integrand is then the right side of Eq. (7.81), with $\dot{h}_{\bar{\mu}\bar{\sigma}}\dot{h}_{\bar{\nu}\bar{\tau}}$ replaced by $\dot{\tilde{h}}_{\bar{\mu}\bar{\sigma}}\tilde{h}^*_{\bar{\nu}\bar{\tau}}$, giving

$$\int_{S_\infty} \Theta^a dS_a = i\sigma \int_{S_\infty} \frac{1}{32\pi}\left[|\tilde{h}_{\bar{2}\bar{2}}|^2 + |\tilde{h}_{\bar{3}\bar{3}}|^2 + 2|\tilde{h}_{\bar{2}\bar{3}}|^2\right] r^2 d\Omega =: i\sigma D,$$

with $D \geq 0$ and with $D > 0$ for a radiative mode. The frequency thus satisfies Eq. (9.73).

What remains is to show that $A > 0$ or, more precisely, that $A > 0$ for an appropriate choice of hypersurface and gauge. We are free to make these choices because the frequency of a mode is gauge invariant. In particular, we will assume the existence of a null hypersurface $u =$ constant (to conform to the usual convention we change our notation from t to u) for which u agrees asymptotically with the Schwarzschild null coordinate and will specify the gauge by setting $h_{\alpha\beta}\nabla^\beta u = 0$.[4] In Eq. (9.75a), the term in $\mathcal{A}$ that involves $h_{\alpha\beta}$ then vanishes because $\nabla_\alpha u \nabla^\alpha u = 0$:

$$\epsilon^{\alpha\gamma\epsilon\eta}\epsilon^{\beta\delta\zeta}{}_\eta \nabla_\alpha u \nabla_\beta u \tilde{h}^*_{\gamma\delta}\tilde{h}_{\epsilon\zeta} = \nabla_\alpha u \nabla^\alpha u(\tilde{h}^*\tilde{h} - \tilde{h}^*_{\beta\gamma}\tilde{h}^{\beta\gamma}) = 0.$$

The remaining term,

$$U^{\alpha\beta\gamma\delta}\nabla_\alpha u \nabla_\gamma u \tilde{\xi}^*_\beta \tilde{\xi}_\delta = (\epsilon + p)(u^\alpha \nabla_\alpha u)^2 q^{\beta\delta}\tilde{\xi}^*_\beta\tilde{\xi}_\delta - \Gamma_1 p q^{\alpha\beta}q^{\gamma\delta}\nabla_\alpha u \nabla_\gamma u \tilde{\xi}^*_\beta\tilde{\xi}_\delta,$$

is positive definite by causality, $c_s = \Gamma_1 p/(\epsilon + p) < 1$, and by the inequality

$$|q^{\alpha\beta}\nabla_\alpha u \tilde{\xi}_\beta|^2 \leq q^{\alpha\beta}\nabla_\alpha u \nabla_\gamma u q^{\gamma\delta}\tilde{\xi}^*_\beta\tilde{\xi}_\delta = u^\alpha u^\beta \nabla_\alpha u \nabla_\beta u\, q^{\gamma\delta}\tilde{\xi}^*_\gamma\tilde{\xi}_\delta.$$

Implications. Along a sequence of relativistic stars, a mode becomes unstable when the imaginary part of its frequency, $\mathrm{Im}(\sigma)$, changes sign – when, that is, the frequency is purely real. For any radiative mode, the dissipative term D is nonzero when the frequency is nonzero. Then, because the only imaginary term in the frequency equation (9.73) is $-i\sigma D$, we have the following fundamental result:

Theorem. *If the frequency of a radiative mode varies continuously along a sequence of equilibrium configurations, then instability to that mode can set in only at a point along the sequence where its frequency vanishes.*

This does *not* describe the onset of instability of rotating stars in the Newtonian approximation. There $D = 0$, the frequency equation has roots

$$\sigma_\pm = [B \pm \sqrt{B^2 + 4AC}]/2A,$$

and, at the onset of instability to a nonaxisymmetric mode of a rotating star, σ_- has the nonzero real value $B/2A$ (overstability). In the exact theory, it is likely that every outgoing mode of a rotating star is radiative – that every mode has a

[4] The Lorenz gauge was used to evaluate D, but all that is needed to obtain the expression is the vanishing to leading order of $h_{r\bar{\mu}}$, which follows from $h_\alpha{}^\beta\nabla_\beta u = 0$, together with the fact that the only $O(1)$ component $g_{r\mu}$ is g_{ur}.

nonzero mass or current multipole with $l \geq 2$. For spherical stars, modes with $l = 0$ and $l = 1$ are nonradiative; however, because $B = 0$ for spherical stars, σ again vanishes at an instability point. The implication is that instability of any mode of a relativistic star sets in when the frequency of the mode vanishes.

When the frequency of a mode vanishes, $h_{\alpha\beta}$ and ξ^α are real, and Eq. (7.63) for the canonical energy takes the form $E_c = C/2$. Because the frequency equation implies $C = 0$, we have $E_c = 0$ as well. Then the $E_c < 0$ instability criterion implies that *instability sets in along a sequence when the minimum value of C for time-independent canonical initial data goes through zero.* That $C < 0$ for an unstable mode can be seen directly from the form of the roots of the relativistic frequency equation (9.73),

$$\sigma_\pm = \left[B + iD \pm (X + 2BiD)^{1/2}\right]/2A, \tag{9.78}$$

where $X = B^2 + 4AC - D^2$. To show this, note for an unstable mode that $\mathrm{Im}(\sigma) < 0$ and that the imaginary parts of the roots are given by

$$2A\,\mathrm{Im}(\sigma_\pm) = D \pm \frac{1}{\sqrt{2}}\left[-X + (X^2 + 4B^2D^2)^{1/2}\right]^{1/2}.$$

Then $\mathrm{Im}(\sigma_+) \geq 0$, and a mode can be unstable only if it corresponds to σ_-, with $\mathrm{Im}(\sigma_-) < 0$, implying $\mathrm{Im}(\sigma_-)\mathrm{Im}(\sigma_+) < 0$. Finally, with $Y := B^2 + 4AC + D^2$, we have

$$8A^2\,\mathrm{Im}(\sigma_-)\mathrm{Im}(\sigma_+) = Y - (Y^2 - 16ACD^2)^{1/2},$$

negative only if $C < 0$.

9.5 Nonaxisymmetric stability

Rapidly rotating stars and drops of water are unstable to a bar mode that leads to fission in the water droplets; a similar nonaxisymmetric instability is likely to be the reason most stars in the universe are in close binary systems. Galactic disks are unstable to nonaxisymmetric perturbations that lead to bars and to spiral structure. And a related instability of a variety of nonaxisymmetric modes, driven by gravitational waves, the Chandrasekhar-Friedman-Schutz (CFS) instability [130, 231]), may limit the rotation of neutron stars. The existence of the CFS instability in rotating stars was first found by Chandrasekhar [130] in the case of the $l = 2$ mode in uniformly rotating, uniform-density Maclaurin spheroids. Subsequently, Friedman and Schutz [242, 231] showed that this instability also appears in compressible stars and that all rotating self-gravitating perfect fluid configurations are generically unstable to the emission of gravitational waves. We have seen that, along a sequence of stellar models, a mode changes from stable to unstable when its frequency vanishes. The generic-instability result means that zero-frequency nonaxisymmetric modes of rotating perfect-fluid stellar models are marginally stable.

Whereas axisymmetric instability to collapse sets in at points that are nearly independent of the magnitude of viscosity or the strength of gravitational waves, the opposite is true for the nonaxisymmetric case. Gravitational radiation drives a nonaxisymmetric instability that, if no other dissipation is present, makes *every* rotating star unstable. Viscosity can drive a nonaxisymmetric instability in rapidly rotating stars for which gravitational radiation is negligible. For slowly rotating stars, however (and nearly all neutron stars rotate slowly compared to the Kepler limit), viscosity simply damps out the gravitational-wave–driven instability. That is, for slow rotation, we will see that the timescale of the CFS instability is longer than the timescale for viscous damping. On the other hand, for rapidly rotating neutron stars, the instability's timescale may be short enough that it limits the rotation of young neutron stars and of old neutron stars spun up by accretion.

We begin with a heuristic description of the way gravitational waves drive nonaxisymmetric modes unstable in perfect-fluid stars. Our starting point is the intuitive expectation that a local wave on a slowly rotating star (like an ocean wave on the rotating Earth) moves relative to the fluid with the speed of the corresponding wave on the corresponding nonrotating star. From the fact that a mode changes from stable to unstable when its frequency vanishes, we will argue that instability sets in when a mode that travels backward relative to the rotating fluid is dragged forward relative to an inertial observer at infinity.

Consider first a stable spherical star. All its modes have positive energy, and the sign of a mode's angular momentum J_c about an axis depends on whether the mode moves clockwise or counterclockwise around the star. That is, a mode with angular and time dependence of the form $\cos(m\phi + \omega_0 t)e^{-\alpha_0 t}$, has positive angular momentum J_c about the z-axis if and only if the mode moves in a positive direction: The pattern speed, $-\dfrac{\omega_0}{m}$, is positive. Because the wave moves in a positive direction relative to an observer at infinity, the star radiates positive angular momentum to infinity, and the mode is damped. Similarly, a mode with negative angular momentum has negative pattern speed, $-\dfrac{\omega_0}{m} < 0$, and radiates negative angular momentum to infinity; the mode is again damped.

Now consider a slowly rotating star with a backward-moving mode, a mode that moves in a direction opposite to the star's rotation. Because a short-wavelength fluid mode (a mode with a Newtonian counterpart, not a w-mode) is essentially a wave in the fluid, the wave moves with nearly the same speed relative to the rotating observer that it had in the spherical star, $\omega_r \approx \omega_0$. As we noted in Section 8.1, when the star rotates with an angular velocity $\Omega \gtrsim |\omega_r/m|$, the backward-going mode is dragged *forward* relative to an observer at infinity: ω_r and $\omega_i = \omega_r + m\Omega$ have opposite signs.

Because the pattern speed is now positive, the mode radiates positive angular momentum to infinity. However, the canonical angular momentum is still negative, because the mode is moving backward relative to the fluid: The angular momentum of the perturbed star is smaller than the angular momentum of the star without the backward-going mode. As the star radiates positive angular momentum to infinity, J_c becomes increasingly negative, implying that the amplitude of the mode grows in time: Gravitational radiation now drives the mode instead of damping it.

The fact that ω_r and ω_i have opposite signs when $\Omega \approx |\omega_0/m|$ suggests two classes of modes that are unstable for arbitrarily slow rotation: backward-moving modes with large values of m, and modes with any m whose frequency is zero in a spherical star. Both classes of perturbations exist. The usual p-modes and g-modes have finite frequencies for a spherical star and are unstable for $\Omega \gtrsim \omega_0/m$; r-modes, which have zero frequency for a nonrotating barotropic star, are unstable for all values of m and Ω (that is, those r-modes are unstable that are backward-moving in the rotating frame of a slowly rotating star).

We have so far not mentioned the canonical energy, but our key criterion for the onset of instability is a negative E_c. If we ignore the imaginary part of the frequency, the change in the sign of E_c follows immediately from the relation $E_c = -\frac{\omega_i}{m} J_c$ of Eq. (7.96). To take the imaginary part $\operatorname{Im}\sigma = \alpha \neq 0$ of the frequency into account, we need to use the fact that energy is lost at a rate $\dot{E}_c \propto \dddot{Q}^2 \propto \omega_i^6$ for quadrupole radiation, with $\dot{E}_c$ proportional to higher powers of ω_i for radiation in higher multipoles. Because E_c is quadratic in the perturbation, it is proportional to $e^{-2\alpha t}$, implying $\alpha \propto \omega_i^6$. Thus $\alpha/\omega_i \to 0$ as $\omega_i \to 0$, implying that for a mode E_c changes sign when ω_i changes sign.

Although the argument we have given so far is heuristic, there is a precise form of the statement that a stable, backward-moving mode becomes unstable when it is dragged forward relative to an inertial observer:

Proposition. *Consider an outgoing mode $(h_{\alpha\beta}(\lambda), \xi^\alpha(\lambda))$, that varies smoothly along a family of uniformly rotating perfect-fluid equilibria, labeled by λ. Assume that it has t and ϕ dependence of the form $e^{i(m\phi+\sigma t)}$, that $\omega_i = \operatorname{Re}\{\sigma\}$ satisfies $\omega_i/m + \Omega > 0$ for all λ, and that the sign of ω_i/m is positive for $\lambda < \lambda_0$ and negative for $\lambda > \lambda_0$. Then in a neighborhood of λ_0, $\alpha := \operatorname{Im}\{\sigma\} \leq 0$, and if the mode has at least one nonzero asymptotic multipole moment with $l \geq 2$ at future null infinity, the mode is unstable ($\alpha < 0$) for $\lambda > \lambda_0$.*

Proof. It follows from Eq. (7.245) that F is $O(\omega_i^6)$ for $l \geq 2$. C must change sign from $C > 0$ for $\lambda < \lambda_0$ to $C < 0$ for $\lambda > \lambda_0$, in some neighborhood of λ_0. Then, for λ near λ_0, α is positive when $\lambda < \lambda_0$, negative or zero for $\lambda > \lambda_0$. If there is a nonzero asymptotic multipole, then F is strictly positive, implying $\alpha < 0$. □

A stronger version of the result that does not restrict the instability to a neighborhood of λ_0 is likely to be true: that a mode is unstable if it is backward-moving relative to the fluid and forward-moving relative to an inertial observer.

Research Problem. Prove that a mode of a uniformly rotating perfect-fluid equilibrium with t and ϕ dependence of the form $e^{i(m\phi+\sigma t)}$ has negative canonical energy if $\omega_i/m + \Omega > 0$ and $\omega_i/m < 0$, where $\omega_i = \operatorname{Re}\sigma$.

The discussion of this section has been in part heuristic and is confined to stability of normal modes. Unstable perturbations, however, may not always have harmonic time dependence. In the next section we address these deficiencies by showing that the canonical energy can be made negative for arbitrarily small angular velocity Ω, by looking at initial data with large values of m.

9.5.1 Generic instability

Gravitational radiation lowers the energy in the inertial frame of an observer at infinity, while conserving circulation. The aim of this section is to show that every rotating equilibrium configuration is unstable (or marginally unstable) by showing that there is a nearby configuration with the same circulation and with lower energy in the inertial frame.

The nonaxisymmetric modes whose growth times are short enough to be of interest to astrophysics (and to be shorter than the viscous damping time) are global modes with small values of m. But the argument for a generic instability involves large values of m and is essentially local: It holds for initial perturbations that are confined to an arbitrarily small ring of fluid and for which the perturbed metric is negligible. For p-modes, the theorem matches the fact that, along a sequence of models with increasing rotation, only modes with the highest values of m are initially unstable, with modes having successively smaller values of m becoming unstable at successively larger values of Ω. The stronger result for r-modes, that there are unstable perturbations with all values of m and for arbitrarily small angular velocity Ω is given in a post-Newtonian context by Andersson [16] and in a relativistic context, using the canonical energy, by [239].

The sign of the canonical energy governs stability only if it is evaluated for perturbations that are canonical – that preserve vorticity in surfaces of constant entropy. We have already seen in Section 7.4.3 that, by adding a trivial displacement, we can describe any nonaxisymmetric physical perturbation by a canonical displacement ξ^α. We will ultimately need to pay attention to the restriction the vorticity condition imposes on ξ^α. Before doing so, it is helpful (and easy) to see why the canonical energy is negative for large values of m, by evaluating E_c for a displacement ξ^α that is not canonical. We will then show that E_c remains negative for the corresponding canonical displacement.

Consider a perturbation for which $h_{\alpha\beta}$ is negligible, and choose $\xi^\alpha u_\alpha = 0$ on our initial hypersurface Σ. The canonical energy is then given by the local form, Eq. (9.39),

$$E_c = \int_\Sigma \Big\{ (\epsilon+p)[\dot{\xi}^\alpha \dot{\xi}_\alpha - \Omega^2 \mathcal{L}_\phi \xi^\alpha \mathcal{L}_\phi \xi_\alpha] - \frac{\boldsymbol{\xi}\cdot\nabla p}{\epsilon+p}(\Delta\epsilon + \delta\epsilon) - (\nabla\cdot\boldsymbol{\xi} + \frac{\boldsymbol{\xi}\cdot\nabla p}{\epsilon+p})\Delta p \Big\} \alpha dV. \qquad (9.79)$$

Now for initial data of the form $\xi^\alpha = \mathrm{Re}\{\zeta^\alpha(\varpi, z)e^{im\phi}\}$, with $\dot{\xi}^\alpha = 0$, we have

$$E_c = \int_\Sigma \Big\{ -m^2\Omega^2(\epsilon+p)\xi^\alpha\xi_\alpha - \frac{\boldsymbol{\xi}\cdot\nabla p}{\epsilon+p}(\Delta\epsilon+\delta\epsilon) - (\nabla\cdot\boldsymbol{\xi} + \frac{\boldsymbol{\xi}\cdot\nabla p}{\epsilon+p})\Delta p \Big\} \alpha dV. \qquad (9.80)$$

Because the first term is negative definite and proportional to m^2, whereas the remaining terms are $O(m)$ or smaller, there is an m_0 for which E_c is negative for all $m > m_0$.

We must now show that E_c remains negative for a canonical displacement $\tilde{\xi}^\alpha = \xi^\alpha + \eta^\alpha$ that describes the same physical perturbation as ξ^α. We present a detailed outline of the proof for models whose entropy per baryon s is not constant. The proof is simpler in that case than for homentropic models, because the perturbed vorticity is required to vanish only in surfaces of constant s. We then discuss the differences that arise in the homentropic case.

We state the result as a proposition and include a proof that the contribution to E_c from a perturbed metric satisfying the initial value equations is $O(m)$. The proposition is stated for uniform rotation and constant entropy per baryon, because that is the case of greatest interest. The initial proof in [231] gives the result for models based on a two-parameter EOS, with nonzero gradient ∇s.

Proposition. *Let Σ be an asymptotically flat spacelike hypersurface of a uniformly rotating perfect-fluid equilibrium for which ∇s vanishes only at isolated points. For each integer m greater than some $m_0 > 0$, there is a canonical perturbation $h_{\alpha\beta}, \xi^\alpha, \dot{h}_{\alpha\beta}, \dot{\xi}^\alpha$ with angular dependence $e^{im\phi}$, for which the canonical energy E_c is negative.*

Proof. The proof here is a somewhat simpler version of that given in [231]. We evaluate E_c on a $t =$ constant surface Σ, using when needed the spherical coordinates of our standard form (1.59) of the metric and orthonormal frame (1.87). We have seen from Eq. (9.80) that E_c is negative for all sufficiently large m if one can find on a hypersurface Σ canonical data having the following properties:

$$(\xi^\alpha, h_{\alpha\beta}; \mathcal{L}_{\mathbf{n}}\xi^\alpha, \mathcal{L}_{\mathbf{n}}h_{\alpha\beta}) = (\widehat{\xi}^\alpha e^{im\phi}, \widehat{h}_{\alpha\beta}e^{im\phi}; \mathcal{L}_{\mathbf{n}}\widehat{\xi}^\alpha e^{im\phi}, \mathcal{L}_{\mathbf{n}}\widehat{h}_{\alpha\beta}e^{im\phi}), \qquad (9.81)$$

where n^α is the unit normal to Σ and the hatted quantities are independent of ϕ;

$$\xi^\alpha \phi_\alpha = 0 = \xi^\alpha t_\alpha; \qquad (9.82)$$

and

$$\|\xi^\alpha\| < k, \quad \|\nabla_\alpha \widehat{\xi}^\beta\| < k, \quad \|[1 + (r/a)^2]^{-1/2} h_{\alpha\beta}\| < k, \quad \|\nabla_\alpha h_{\beta\gamma}\| < k, \qquad (9.83)$$

where k is a constant independent of the integer m and $\|T^{\alpha\cdots\beta}{}_{\gamma\cdots\delta}\|$ is the L^2 norm on Σ of the components of the tensor $T^{\alpha\cdots\beta}{}_{\gamma\cdots\delta}$ with respect to the frame $\widehat{\mathbf{e}}_\mu$:

$$\|T^{\alpha\cdots\beta}{}_{\gamma\cdots\delta}\|^2 := \sum_{\hat{\mu},\ldots\hat{\nu}\,\hat{\sigma},\ldots\hat{\tau}} \int_\Sigma |T^{\hat{\mu}\cdots\hat{\nu}}{}_{\hat{\sigma}\cdots\hat{\tau}}|^2 dV. \qquad (9.84)$$

In the function $1 + (r/a)^2$, a is an arbitrary length introduced to make r/a dimensionless.

Note that the requirement that the metric perturbation be negligible for large m is enforced in Eq. (9.83) by an m-independent integral bound on components of $\nabla_\alpha h_{\beta\gamma}$, implying an m-independent bound on the norm of $m\, h_{\beta\gamma}$. In contrast, the norm of $\nabla_a \widehat{\xi^b}$ is bounded, but $\nabla_a \xi^b$ is $O(m)$. This condition will be enforced by writing the initial-value equations as elliptic equations for quantities corresponding roughly to the perturbed lapse and shift. Solutions to second-order elliptic equations have the property that the norm of the second derivatives of their solutions is bounded by a constant proportional to the norm of their source. The source, constructed from $\delta T^{\alpha\beta}$, will be $O(m)$, implying that the perturbed metric will be $O(m^{-1})$.

We must show that there are tensor fields $(\xi^\alpha, h_{\alpha\beta}; \mathcal{L}_{\mathbf{n}}\xi^\alpha, \mathcal{L}_{\mathbf{n}}h_{\alpha\beta})$ on Σ satisfying Eqs. (9.81)–(9.83) and satisfying the perturbed initial-value equations, $\delta(G^{\alpha\beta} - 8\pi T^{\alpha\beta})n_\beta = 0$, in a gauge for which the perturbed Einstein-Euler equations are hyperbolic. We choose a Lorenz gauge,[5]

$$\nabla_\beta \widetilde{h}^{\alpha\beta} = 0, \text{ where } \widetilde{h}_{\alpha\beta} := h_{\alpha\beta} - \frac{1}{2} g_{\alpha\beta} h, \tag{9.85}$$

and will specify initial values for a perturbed metric $h_{\alpha\beta}$ of the form

$$\widetilde{h}_{\alpha\beta} := h_{\alpha\beta} - \frac{1}{2} g_{\alpha\beta} h = \mathfrak{a} n_\alpha n_\beta + \mathfrak{b}_\alpha n_\beta + \mathfrak{b}_\beta n_\alpha, \tag{9.86}$$

with

$$\mathfrak{b}^\alpha n_\alpha = 0. \tag{9.87}$$

(We use the symbol $\widetilde{h}_{\alpha\beta}$ instead of the usual $\overline{h}_{\alpha\beta}$ to avoid confusion with complex conjugation.) In a chart T, X^i for which Σ is a surface of constant T and the unperturbed lapse and shift are $\alpha = 1, \beta^i = 0$, the perturbed lapse and shift have components $\delta\alpha = -\mathfrak{a}$, $\delta\beta^i = -\mathfrak{b}^i$.

Because we are dealing with an initial-value problem on the hypersurface Σ, we will use the 3+1 notation of Section 2.1, writing D_a for the covariant derivative operator associated with the 3-metric γ_{ab} and writing the initial data in terms of the 3-vectors ξ_a and $\mathfrak{b}_a$ on Σ associated with ξ_α and $\mathfrak{b}_\alpha$,

$$\mathfrak{b}_a = \gamma_a^\alpha \mathfrak{b}_\alpha|_\Sigma, \quad \xi_a = \gamma_a^\alpha \xi_\alpha|_\Sigma. \tag{9.88}$$

(ξ_a and $\mathfrak{b}_a$ are the pullbacks to Σ of ξ_α and $\mathfrak{b}_\alpha$). Then, with the notation

$$\dot{\mathfrak{a}} := n^\beta \nabla_\beta \mathfrak{a}|_\Sigma, \quad \dot{\mathfrak{b}}_a := \gamma_a^\alpha n^\beta \nabla_\beta \mathfrak{b}_\alpha|_\Sigma, \quad \dot{\xi}_a := \gamma_a^\alpha n^\beta \nabla_\beta \xi_\alpha|_\Sigma, \tag{9.89}$$

an initial data set specifies $(\xi^a, \dot{\xi}^a;\ \mathfrak{a}, \dot{\mathfrak{a}}, \mathfrak{b}_a, \dot{\mathfrak{b}}_a)$ on Σ.

The Lorenz gauge condition gives the time derivatives of $\mathfrak{a}$ and $\mathfrak{b}_a$ in terms of their spatial derivatives on Σ. It looks simplest for time derivatives of quantities on Σ defined in terms of derivatives along the unit normal $n^\alpha = -e^\nu \nabla^\alpha t$ to Σ. The

[5] Lorenz introduced the analogous gauge for electromagnetism. For gravity it was first used by Einstein and its existence first proved by Hilbert in 1916. deDonder and Fock introduced and used its nonlinear version, the harmonic gauge. The linearized gauge is variously referred to in the literature as the deDonder, harmonic, transverse, Fock, Lorenz, Lorentz (mistakenly), Einstein, or Hilbert gauge.

projections along n^α and orthogonal to n^α of the Lorenz condition (7.34) have the form

$$\dot{\mathfrak{a}} = K\mathfrak{a} - D_b\mathfrak{b}^b - 2D_a\nu\,\mathfrak{b}^b, \tag{9.90a}$$

$$\dot{\mathfrak{b}}_a = -\mathfrak{a}D_a\nu + K_{ab}\mathfrak{b}^b + K\mathfrak{b}_a, \tag{9.90b}$$

where we have used the definition (2.15) of extrinsic curvature to write the derivatives of n_α in terms of K_{ab} and $\nabla_a\nu$.

We already know that no second derivatives normal to Σ (no second derivatives along n^α of metric components) occur in the initial value equations (2.21) and (2.22). It follows that no second derivatives along n^α of $\mathfrak{a}$ and $\mathfrak{b}_a$ occur in their linearized form, $(\delta G_{\alpha\beta} = 8\pi\delta T_{\alpha\beta})n^\beta$. Using Eq. (7.24), we can write the perturbed Einstein tensor, with no gauge condition imposed, as

$$\begin{aligned}\delta G_{\alpha\beta} &= \frac{1}{2}(-\nabla_\gamma\nabla^\gamma\widetilde{h}_{\alpha\beta} + \nabla_\alpha\nabla^\gamma\widetilde{h}_{\beta\gamma} + \nabla_\beta\nabla^\gamma\widetilde{h}_{\alpha\gamma} - g_{\alpha\beta}\nabla^\gamma\nabla^\delta\widetilde{h}_{\gamma\delta})\\ &\quad + \left(-R_{\alpha\gamma\beta\delta} + g_{\gamma(\beta}R_{\alpha)\delta} + \frac{1}{2}g_{\alpha\beta}R_{\gamma\delta} - \frac{1}{2}Rg_{\alpha\gamma}g_{\beta\delta}\right)\widetilde{h}^{\gamma\delta}.\end{aligned} \tag{9.91}$$

When the Lorenz gauge conditions (9.90) are used to replace first time derivatives $n^\beta\nabla_\beta\mathfrak{a}$ and $n^\beta\nabla_\beta\mathfrak{b}_a$ by expressions linear in $\mathfrak{a}$, $\mathfrak{b}_a$ and $D_b\mathfrak{b}_a$, the initial-value equations take the form

$$D_bD^b\mathfrak{a} + A^bD_b\mathfrak{a} + B^{bc}D_b\mathfrak{b}_c + E\mathfrak{a} + F^b\mathfrak{b}_b = -16\pi n^\alpha n^\beta\delta T_{\alpha\beta}, \tag{9.92a}$$

$$D_bD^b\mathfrak{b}_a + \hat{A}_a^bD_b\mathfrak{a} + \hat{B}_a^{bc}D_b\mathfrak{b}_c + \hat{E}_a\mathfrak{a} + \hat{F}_a^b\mathfrak{b}_b = -16\pi\gamma_a^\alpha n^\beta\delta T_{\alpha\beta}, \tag{9.92b}$$

where the quantities $A^b, \cdots, \hat{F}_a^b$ are tensors on Σ constructed from the background geometry – from n_α and $g_{\alpha\beta}$. It is easy to see from Eq. (9.91) that the coefficients A^a, B^a, $\hat{A}_a^b$, and $\hat{B}_a^{bc}$ involve first derivatives of the background metric and are therefore $O(r^{-2})$, whereas the coefficients $E, F^a, \hat{E}_a$, and $\hat{F}_a^b$ are constructed from second derivatives of the background metric and terms quadratic in first derivatives and are therefore $O(r^{-3})$. To prove existence of solutions to the initial-value equations, we will need the elliptic character of the system and this asymptotic behavior of the coefficients, but not their explicit form.

A Lagrangian displacement is canonical if it preserves vorticity in surfaces of constant specific entropy s, satisfying Eq. (7.102). Using $\Delta s = \Delta t = 0$, we can write the condition in the form

$$W := |g|^{1/2}\epsilon^{\alpha\beta\gamma\delta}\nabla_\alpha\Delta(hu_\beta)\nabla_\gamma s\nabla_\delta t = 0. \tag{9.93}$$

We must show that we can find initial data on Σ satisfying the initial-value equations (9.92) and the vorticity condition (9.93) and obeying the bounds (9.83). We begin by choosing as the Lagrangian displacement a smooth vector ξ^a of the form

$$\xi^a|_\Sigma = \hat{\xi}^a(\varpi, z)e^{im\phi}, \tag{9.94}$$

with the support of $\widehat{\xi^a}$ confined to the fluid. Because ξ^a is smooth with compact support, there is a constant k for which

$$\|\xi^a\| < k \text{ and } \|D_b\widehat{\xi^a}\| < k. \tag{9.95}$$

We next solve the vorticity condition for components of $u \cdot \nabla\xi_\alpha$ in terms of the remaining initial data. That is, taking as the only nonvanishing component

$$u^\nu \nabla_\nu \xi_\varpi|_\Sigma = f\nabla_z s,\ u^\nu \nabla_\nu \xi_z = -f\nabla_\varpi s, \tag{9.96}$$

we will solve the vorticity equation for f.

Now f and ξ^a determine $\dot{\xi}_a$ through the 3+1 decomposition of $u^\beta\nabla_\beta\xi_\alpha$: From

$$\gamma_a^\alpha u^\beta \nabla_\beta \xi_\alpha = \gamma_a^\alpha u^\beta(-n_\beta n^\delta + \gamma_\beta^\delta)\nabla_\delta\xi_\alpha \tag{9.97}$$

and the decomposition $u^\alpha = u^{\hat{0}} n^\alpha + \gamma_a^\alpha u^a$, we have

$$\dot{\xi}_a = \frac{1}{u^{\hat{0}}}\left(\gamma_a^\alpha u^\beta \nabla_\beta \xi_\alpha - u^b D_b \xi_a\right). \tag{9.98}$$

Specifying ξ^a and $u^\beta\nabla_\beta\xi^\alpha$ is then equivalent to specifying ξ^a and $\dot{\xi}^a$. This is useful in writing the vorticity equation as an equation for f, because it is much simpler to decompose the expressions for Δh and Δu_α in terms of derivatives along and orthogonal to u^α than into derivatives along and orthogonal to n^α. In particular, from Eq. (7.4) and the relation $\Delta h = \Delta p/\rho$ we have

$$\Delta h = \frac{\Gamma_1 p}{\rho}\gamma^{\alpha\beta}\left(\nabla_\alpha\xi_\beta + \frac{1}{2}h_{\alpha\beta}\right), \tag{9.99}$$

an expression involving only derivatives orthogonal to u^α and thus independent of f. Next, because $u^\beta\nabla_\beta\xi^\alpha$ is orthogonal to u^α, Eq. (7.3) for Δu^α implies

$$\Delta u^\alpha = u^\alpha u^\beta u^\gamma\left(\nabla_\beta\xi_\gamma + \frac{1}{2}h_{\beta\gamma}\right) = \frac{1}{2}u^\alpha u^\beta u^\gamma h_{\beta\gamma}, \tag{9.100}$$

again independent of f. Then

$$\begin{aligned}\Delta(hu_\alpha) &= hu^\beta \Delta g_{\alpha\beta} + \{\text{ terms independent of } f\}\\ &= hu^\beta\nabla_\beta\xi_\alpha + \{\text{ terms independent of } f\}.\end{aligned} \tag{9.101}$$

Finally, denoting by $\widehat{W}$ an expression independent of f, we have

$$\begin{aligned}W &= |g|^{1/2}\epsilon^{\alpha\beta\gamma\delta}\nabla_\alpha(hu\cdot\nabla\xi_\beta)\nabla_\gamma s\nabla_\delta t + \widehat{W}\\ &= \partial_\phi(hu\cdot\nabla\xi_\varpi)\partial_z s - \partial_\phi(hu\cdot\nabla\xi_z)\partial_\varpi s + \widehat{W} \qquad (9.102)\\ &= imh(\nabla s)^2 f + \widehat{W}. \qquad (9.103)\end{aligned}$$

We thus make ξ^α canonical by choosing

$$f = -\frac{\widehat{W}}{imh(\nabla s)^2}. \tag{9.104}$$

Equations (9.95) and (9.104) express $u\cdot\nabla\xi^a$ in terms of $\mathfrak{a}$, and $\mathfrak{b}_a$.

To complete the proof, we need to show that we can satisfy the initial-value equations with $\delta T^{\alpha\beta}$ given by Eqs. (7.5) and that the resulting initial data obey the bounds (9.83). With f given by (9.104), Eqs. (7.5) give $\delta T^{\alpha\beta}$ in terms of ξ^a, $\mathfrak{a}$, and $\mathfrak{b}_a$. Because $\delta T^{\alpha\beta}$ involves only these variables, the initial-value equations retain the form (9.92) of an elliptic system for $\mathfrak{a}$ and $\mathfrak{b}^a$. With the terms in $\delta T^{\alpha\beta}$ involving $\mathfrak{a}$ and $\mathfrak{b}^a$ moved to the left side of the equations, they have the form

$$D_b D^b \mathfrak{a} + A^b D_b \mathfrak{a} + B^{bc} D_b \mathfrak{b}_c + \mathcal{E}\mathfrak{a} + \mathcal{F}^b \mathfrak{b}_b = c_a \xi^a + d_a^b D_b \xi^a, \tag{9.105a}$$

$$D_b D^b \mathfrak{b}_a + \hat{A}_a^b D_b \mathfrak{a} + \hat{B}_a^{bc} D_b \mathfrak{b}_c + \hat{\mathcal{E}}_a \mathfrak{a} + \hat{\mathcal{F}}_a^b \mathfrak{b}_b = \hat{c}_{ab} \xi^b + \hat{d}_{ac}^b D_b \xi^c. \tag{9.105b}$$

Outside the star, $\mathcal{E} = E, \mathcal{F}^a = F^a, \hat{\mathcal{E}}_a = E_a$, and $\hat{\mathcal{F}}_a^b = F_a^b$.

We now have an elliptic system for $\mathfrak{a}$ and $\mathfrak{b}_a$, whose source is $O(m)$ because it involves only ξ^a and its first derivatives. A standard theorem on elliptic operators (see [231] with a cited theorem by Cantor subsequently published in [113]) implies that the system has a finite-dimensional kernel – only a finite number of linearly independent solutions $\mathfrak{a}, \mathfrak{b}_a$ on Σ satisfy Eqs. (9.105) with zero source; and that there is an asymptotically flat solution for which the norms of $\nabla_a \nabla_b \mathfrak{a}$ and $\nabla_a \nabla_b \mathfrak{b}_c$ have bounds proportional to the norm of the source. In particular,

$$\|m^2 \mathfrak{a}\| < km, \quad \|m^2 \mathfrak{b}_a\| < km. \tag{9.106}$$

Because the kernel of the system is finite dimensional, if we restrict the system to the subspace of functions with angular dependence $e^{im\phi}$, the kernel vanishes for sufficiently large m, implying that the the elliptic operator is an isomorphism and thus that there is a solution satisfying the bounds (9.84). □

9.6 Implications of the CFS instability

Evidence for an upper limit on neutron-star spin smaller than the Keplerian frequency Ω_K comes from nearly 30 years of observations of neutron stars with millisecond periods, seen as pulsars and as X-ray binaries. At the time of writing, the observations reveal rotational frequencies ranging upward to 716 Hz and densely populating a range of frequencies below that, suggesting the possible existence of a sharp cutoff. Selection biases against detection of the fastest millisecond radio pulsars have made conclusions about an upper limit on spin uncertain, but Chakrabarty argues that the class of sources whose pulses are seen in nuclear bursts (nuclear powered accreting millisecond X-ray pulsars) constitute a sample without significant bias [124].

A magnetic field of order 10^8 G can limit the spin of an accreting millisecond pulsar. Because matter within the magnetosphere corotates with the star, only matter that accretes from outside the magnetosphere can spin up the star, leading to an equilibrium period given approximately by Ghosh and Lamb [259][6]:

$$P_{\rm eq} \sim 2 \times 10^{-3}{\rm s} \left(\frac{B}{10^8 {\rm G}}\right)^{6/7} \left(\frac{\dot{M}}{10^{-10} M_\odot {\rm yr}^{-1}}\right)^{-3/7}. \tag{9.107}$$

[6] Shapiro and Teukolsky [603] give a clear, simplified version, and Eq. (9.107) is their Eq. (15.2.22), with $M = 1.4 M_\odot$, $R = 10$ km, and a ratio ω_s of the angular velocity to Ω_K at the inner edge of the disk set to 1.

Because this period depends on the magnetic field, a sharp cutoff in the frequency of accreting stars is not an obvious prediction of magnetically limited spins. For a magnetically set maximum rotation rate of order 700–800 Hz, the range of magnetic fields would need to have a corresponding minimum cutoff value of about 10^8 G, and the highest observed spin rates should correspond to the lowest magnetic fields.

The required cutoff and a fairly narrow range of observed frequencies has made gravitational-wave–limited spin a competitive possibility for accreting neutron stars. Arguments for and against this based on available observations are given by White and Zhang [726] and by Patruno, Haskell, and D'Angelo [535], respectively. Although our concern here is with a limit set by instability, as mentioned in Section 5.1.3, the asymmetry of accretion can also lead to gravitational waves from nonaxisymmetric crust deformations associated with temperature and composition gradients or magnetic confinement of accreted matter.

A fundamental question is whether the instability survives the complex physics of a real neutron star. Although this has been the focus of intensive study by many researchers, there is no definitive answer. Studies have focused on:

dissipation from bulk and shear viscosity and from mutual friction;
magnetic field wind-up;
nonlinear evolution and the saturation amplitude; and
the possibility that a continuous spectrum replaces r-modes in relativistic stars.

Viscous dissipation eliminates the instability for slowly rotating stars [423]. For rapid rotation, as we discuss in subsequent sections, the competition between the radiation-reaction growth times of polar and axial modes and the damping times set by viscosity appears to allow an instability of polar modes in nascent neutron stars and of axial modes in both nascent stars and in stars spun up by accretion. In the following sections, we discuss the competing growth and damping times and the corresponding instability windows. At the end of this section, we briefly discuss work on magnetic field wind-up; in Section 10.3, we review work on nonlinear evolution, mode couplings, and the saturation amplitude. The question of a continuous spectrum was discussed in Section 8.6.2.

9.6.1 Instability of polar modes

Summary of results. We begin with a brief summary of the expected behavior of the CFS instability of polar modes. The remainder of this section is a more detailed review of the computations that lead to this scenario and their uncertainties. The $l = m$ f-modes with $m = 2$–5 have growth times shorter than the viscous damping during the first months after the formation of a neutron star, if the star's initial rotation is near the maximum angular velocity Ω_K. At temperatures somewhat above 10^{10} K, bulk viscosity (energy dissipation from beta decay, inverse beta decay, and related weak interactions) suppresses the instability of all modes. For T less than

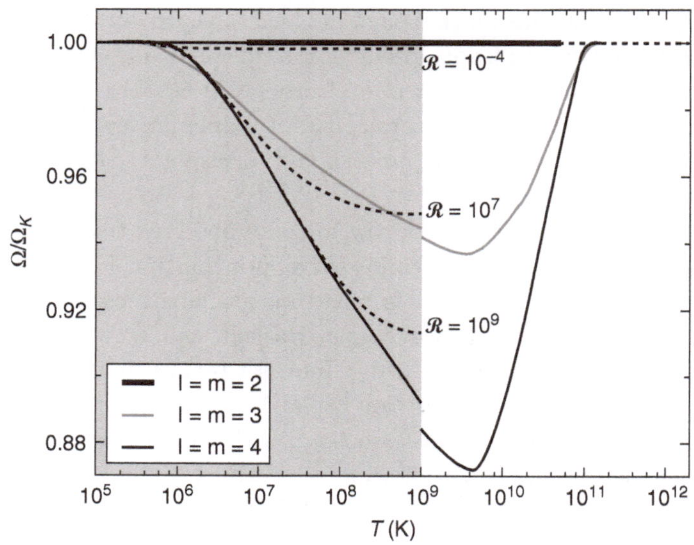

Figure 9.3. The relativistic f-mode instability window. The curves are the solutions of $-\tau_{\rm GR} = \tau_{\rm diss}$ for an $N = 0.73$ polytropic model with $M = 1.48\,M_\odot$, $R = 10.47\,$km (in the $\Omega = 0$ limit). The shaded area indicates the presence of superfluidity (assuming a critical temperature of 10^9 K). The dashed curves represent different choices for the mutual friction (shown only for the $m = 4$ multipole). (Reprinted figure with permission from Gaertig et al. [250]. Copyright 2011 by the American Physical Society.)

the transition temperature $T_s \lesssim 10^9\,$K at which neutrons become superfluid, mutual friction (associated with the scattering of electrons by neutron vortices) suppresses the instability of all f-modes [425, 418, 29, 21]. Between these two temperatures is an instability window. In Fig. 9.3, from Gaertig et al. [250], the curve on the right shows the critical angular velocities Ω_c above which the $m = 3$ and $m = 4$ f-modes are unstable. In the shaded region to the left of $T = T_s$, no modes are unstable. (There is also a curve on the left showing the critical angular velocity set by shear viscosity in the absence of mutual friction.) A nascent neutron star would rotate differentially, and for a differentially rotating star the Kepler limit is substantially higher, leaving more room for the mode to be unstable. More realistic candidates for the equation state, with npeμ cores, are stiffer at high density than the polytropic model of the figure, and they would typically also allow the $m = 2$ mode to be unstable.

In a star that forms with angular velocity near Ω_K, the dominant f-modes will be unstable by the time the temperature has dropped below 10^{10} K. They then grow, with growth times of minutes to days, until they reach their saturation amplitude. Over the next several months, as the star cools, gravitational waves emitted by the unstable modes (r-modes as well as f-modes) radiate angular momentum, and the star spins down from a combination of gravitational radiation, magnetic dipole radiation, and the spin-down mechanisms associated with the star's magnetic field. Finally, the instability is shut off, either when $\Omega < \Omega_c$ or when $T < T_s$.

In addition to the uncertainties listed above for the growth of r-modes in accreting neutron stars, there is a possibility that nascent neutron stars with rapid initial rotation are the progenitors of magnetars – that they generically develop large magnetic fields. A magnetic field of order 10^{15} G would spin down the star in a time comparable to the f-mode growth time, effectively eliminating the instability.

Instability points. In the Newtonian approximation, explicit points of instability for the $l = m$ modes along the Maclaurin sequence of uniformly rotating, uniform-density models were found by Comins [153, 154] and for $l \neq m$ by Baumgart and Friedman [53]. For polytropic Newtonian models with stiffness comparable to that of neutron stars, neutral modes have been determined for several polytropic EOSs in [322, 331, 453, 325, 327, 326, 737]. The instability first sets in through $l = m$ modes. Modes with larger l become unstable at lower rotation rates, but viscosity limits the interesting ones to $l \leq 5$. For an $N = 1$ polytrope, the critical values of $T/|W|$ for the $l = 3$, 4, and 5 modes are respectively 0.079, 0.058, and 0.045, and these values become smaller for softer polytropes. The $l = m = 2$ mode, the *bar mode*, has a critical $T/|W|$ ratio of 0.14 that is almost independent of the polytropic index. Because soft EOSs cannot produce models with high values of $T/|W|$, the bar mode is unstable only for stiff Newtonian polytropes, only for polytropic index $N \leq 0.808$ [340, 635].

The neutral modes of $N = 1$ polytropes in the exact theory [659, 665] are shown in Fig. 8.7 of Chapter 8. The computation of neutral modes for polytropes of indices $N = 1.0$, 1.5, and 2.0 shows that relativity significantly strengthens the instability. For the $N = 1.0$ polytropes, the critical angular velocity ratio Ω_c/Ω_K, where Ω_K is the angular velocity at the mass-shedding limit at the same central energy density, is reduced by as much as 15% for the most relativistic configuration. A surprising result (not found in computations that used the post-Newtonian approximation) is that the $l = m = 2$ bar mode is unstable even for relativistic polytropes of index $N = 1.0$. The classical Newtonian result for the onset of the bar mode instability ($N_{crit} < 0.808$) is replaced by

$$N_{crit} < 1.3 \tag{9.108}$$

in general relativity. These results are in qualitative agreement with the instability points found in the relativistic Cowling approximation by Yoshida and Eriguchi [738]; they are also consistent with a previous post-Newtonian computation of neutral modes by Cutler and Lindblom [171, 418] showing that general relativity strengthens the CFS instability. Compared to their Newtonian counterparts, critical angular velocity ratios Ω_c/Ω_K of post-Newtonian $N = 1$ polytropes are lowered by as much as 10%.

Morsink, Stergioulas, and Blattnig [486] extend the method presented in [665] to a wide range of candidates for the neutron-star equation of state (which usually have a stiff high-density region, corresponding to polytropes of index $N =$ 0.5–0.7). For such equations of state, the $l = m = 2$ mode becomes unstable for stars with gravitational mass as low as 1.0–1.2$M_\odot$. For 1.4$M_\odot$ neutron stars, the mode becomes unstable at 80% to 95% of the maximum allowed rotation rate. The $m = 2$ instability sets in at a ratio of rotational to gravitational energies $T/|W| \sim 0.08$ for $1.4M_\odot$ stars and $T/|W| \sim 0.06$ for maximum-mass stars, substantially lower than the Newtonian value $T/|W| \sim 0.14$. The empirical formula Eq. (8.125) gives the critical value of $T/|W|$ for the $m = 2$ mode with an accuracy of 4% to 6% for the widely varying EOS candidates.

In newly born neutron stars, the CFS instability can develop while the background equilibrium star is still differentially rotating, unless uniform rotation is enforced on a short timescale by, for example, magnetic braking [438]. For a differentially rotating background, the critical value of $T/|W|$ is only somewhat larger than the corresponding value for uniform rotation [744], but the mass-shedding limit of $T/|W|$ is dramatically relaxed, because Ω in the interior is significantly larger than the limiting frequency Ω_K at the surface. The f-mode instability window is then significantly larger than for uniform rotation.

Radiation-reaction growth times. When gravitational radiation is the only dissipative mechanism, the damping time $\tau_{\rm GR} = 1/\alpha$ of a stable mode (where $\alpha = \operatorname{Im}\sigma$) and the growth time $(-\tau_{\rm GR})$ of a radiation-driven unstable mode are given by

$$\frac{1}{\tau_{\rm GR}} = -\frac{\dot{E}_c}{2E_c}, \tag{9.109}$$

where $\dot{E}_c \equiv dE_c/dt$ is the rate of energy loss to gravitational waves. With the exception of w-modes, all quasinormal modes of neutron stars have damping times long compared to the period of the mode. From the expression (7.245) for $\dot{E}_c$ in terms of multipoles, we then have

$$\dot{E}_c = -\sum_{l\geq m} \omega_i^{2l+2} N_l \left(|\delta D_{lm}|^2 + |\delta J_{lm}|^2\right), \tag{9.110}$$

where terms smaller by a factor α/ω_i are neglected.

In the Newtonian approximation, δD_{lm} and δJ_{lm} are given by Eqs. (7.247). The additional factor of the velocity in the current multipoles compared to the corresponding mass multipoles implies in Eq. (9.110) a relative factor of order $(v/c)^2$ in the rate of radiated energy for a given current multipole compared to that of the corresponding mass multipole with the same value of l. Then at leading post-Newtonian order, only the mass multipole moments contribute; and, for a given value of m, the leading contribution is from δD_{lm} with the lowest allowable value l. This would ordinarily be $l = |m|$ or $l = |m| + 1$ for a mode even or odd, respectively, under reflection in the equatorial plane. In particular, for the dominant unstable $l = m$ f-modes,

$$\dot{E}_c = \omega_i^{2l+2} N_l \, |\delta D_{ll}|^2, \text{ with } \delta D_{ll} = \int r^l \, \delta\rho Y_{ll}^* d^3x. \tag{9.111}$$

From Eq. (7.124), the canonical energy in a rotating frame has the form [241]

$$E_c = \frac{1}{2}\int \left[\rho\delta v^a \delta v_a + \left(\frac{\delta p}{\rho} + \delta\Phi\right)\delta\rho\right] d^3x, \tag{9.112}$$

with $\delta\Phi$ the Eulerian change in the gravitational potential.

For nonrotating Newtonian polytropes, the damping times of stable f-modes due to gravitational-wave emission scale with the inverse compactness R/M and the mean density $\bar{\rho}$ as [327]

$$\tau_{GR} \propto (R/M)^{(2l+1)/2}(\bar{\rho})^{-1/2}. \tag{9.113}$$

For nonrotating relativistic models constructed with various tabulated EOSs, an empirical fit for the damping time of the $l = 2$ f-mode is [24]

$$\tau_{\rm GR}^{-1} \simeq M_{1.4}^3 R_{10}^{-4}(22.9 - 14.7 M_{1.4} R_{10}^{-1}) \ \ [\text{s}^{-1}], \tag{9.114}$$

where $M_{1.4} = M/(1.4 M_\odot)$ and $R_{10} = R/(10 \text{ km})$. Thus, a characteristic damping time is 0.1s or roughly 200 oscillation periods (taking 2 kHz as a characteristic oscillation period).

As we have previously noted, for the $l = m = 2$ mode, Eq. (9.111) implies $dE/dt|_{\rm GR} \propto \omega_i^6$; also, because the energy E_c scales as ω_i^2, the growth time scales as $\tau_{\rm GR}^{-1} \propto \omega_i^4$. In a study of the $l = m = 2$ f-modes in rapidly rotating relativistic stars (in the Cowling approximation), Gaertig and Kokkotas [253] find a numerical fit for the growth timescale of unstable modes that can be simplified to

$$\tau_{\rm GR} \sim -4(\omega_i/\omega_0)^{-4}\tau_0, \tag{9.115}$$

where τ_0 and ω_0 are the damping timescale and mode frequency in the nonrotating model. With the exponent replaced by $-2l$ for the higher modes, the growth time of the instability for all f-modes is highly sensitive to the ratio ω_i/ω_0 at birth. Because ω_i vanishes at the critical angular velocity Ω_c that marks the instability point, it is proportional to $\Omega - \Omega_c$, for Ω near Ω_c. The maximum growth rate is therefore highly sensitive to the maximum value Ω_K of the angular velocity.

The fact that $\Omega_K - \Omega_c$ is somewhat larger for relativistic than for Newtonian models then leads to dramatically smaller growth times, smaller by a factor of 10 to 100 in the full theory than in the Newtonian approximation [250]. For models compact enough to allow the $m = 2$ mode to be unstable for Ω significantly below Ω_K, the ratio ω_i/ω_0 can be of order 0.1, corresponding to a characteristic growth time $\tau_{\rm GR} \sim 4000$ s. For equations of state yielding more compact models and for differentially rotating stars, this ratio can increase to ~0.5 (or even somewhat higher), corresponding to much shorter growth times of ~6 s. In a model with compactness typical of candidate EOSs, the $m = 2$ mode of a $1.4 M_\odot$ star is stable or only marginally unstable, but f-modes with $l = 3$–5 are unstable (in the absence of viscosity) for all EOS candidates. In particular, for an $N = 0.73$ polytropic model with $M = 1.48 M_\odot$, Gaertig et al. [250] find a growth time $\tau_{\rm GR} \approx 2000$ s at $\Omega = \Omega_K$ for both the $m = 3$ and $m = 4$ modes.

For highly differentially rotating models, with $T/|W| > 0.25$, the growth times could be significantly shorter [384], but such models will likely be unstable to dynamical or low-$T/|W|$ instabilities, which will compete against the secular f-mode instability and will quickly spin down the star by gravitational-wave emission before the latter has time to grow.

Whereas gravitational radiation decreases the canonical energy in an inertial frame, we have seen in Eq. (7.125) that viscosity decreases the canonical energy $E_{c,r}$ in a *rotating* frame. To consistently combine the effects of radiation reaction and viscosity, we use the relations $\dot{E}_{c,r} = \dot{E}_c - \Omega \dot{J}_c$ and Eq. (7.84), $\dot{J}_c = -\frac{m}{\omega_i} \dot{E}_c$, to obtain an expression for $\dot{E}_{c,r}$ due to gravitational radiation. Recalling that $\omega_r = \omega_i + m\Omega$ is the frequency in a rotating frame and using Eq. (9.110), we have

$$\dot{E}_{c,r} = \frac{\omega_r}{\omega_i} \dot{E}_c = -\omega_r \omega_i^{2l+1} \sum_{l \geq 2} N_l \left(|\delta D_{lm}|^2 + |\delta J_{lm}|^2 \right) \tag{9.116}$$

(again neglecting terms of order α/ω_i). Equation (7.96) implies the same ratio for $E_{c,r}/E_c$,

$$E_{c,r} = \frac{\omega_r}{\omega_i} E_c, \tag{9.117}$$

and one can therefore replace E_c by $E_{c,r}$ in Eq. (9.109) for $\tau_{\rm GR}$:

$$\frac{1}{\tau_{\rm GR}} = -\frac{\dot{E}_{c,r}}{2E_{c,r}}. \tag{9.118}$$

For an unstable mode, because ω_i and ω_r have opposite signs, both $E_{c,r}$ and $\dot{E}_{c,r}$ are positive.

The effect of viscosity. The shear and bulk viscosity of neutron-star matter and mutual friction in a superfluid interior can each suppress the growth of the CFS instability. Bulk viscosity is large at high temperatures, shear viscosity and mutual friction at lower temperatures, leaving a window of opportunity in which a star with large enough angular velocity can be unstable. This temperature window has been determined in Newtonian or post-Newtonian gravity and for uniformly rotating stars. According to these results, based on assumptions discussed below, the f-mode CFS instability can occur only in nascent neutron stars that rotate close to the mass-shedding limit [420, 170, 327, 326, 418, 737]; because general relativity and differential rotation enhance the instability, the actual temperature window may be significantly larger than found in the above references.

The effect of viscosity on the dynamical evolution of nonaxisymmetric perturbations can be estimated from Eqs. (9.116) and Eq. (7.125), giving the contributions to $\dot{E}_{c,r}$ from gravitational radiation and from shear and bulk viscosity, respectively:

$$\dot{E}_{c,r} = \dot{E}_{c,r\ \rm GR} + \dot{E}_{c,r\ \rm s} + \dot{E}_{c,r\ \rm b}, \tag{9.119}$$

where $\dot{E}_{c,r\ \mathrm{s}}$ and $\dot{E}_{c,r\ \mathrm{b}}$ are given by Eq. (7.125). The corresponding damping times τ_s and τ_b for shear and bulk viscosity can be estimated from the Newtonian forms of $\dot{E}_{c,r}$,

$$\frac{1}{\tau_s} := -\frac{\dot{E}_{c,r\ s}}{2E_{c,r}} = -\frac{1}{2E_{c,r}} \int 2\eta\delta\sigma_{ab}\delta\sigma^{ab}dV, \quad \frac{1}{\tau_b} = -\frac{1}{2E_{c,r}} \int \zeta\delta\theta^2 dV, \tag{9.120}$$

with $\delta\sigma_{ab} = \nabla_a\delta v_b + \nabla_b\delta v_a - \frac{2}{3}\delta_{ab}\nabla_c\delta v^c$ and $\delta\theta = \nabla_a\delta v^a$. Equations (9.109) and (9.119) imply the additive relation

$$\frac{1}{\tau} = \frac{1}{\tau_{\mathrm{GR}}} + \frac{1}{\tau_{\mathrm{s}}} + \frac{1}{\tau_{\mathrm{b}}}. \tag{9.121}$$

When these damping times are long compared to the oscillation time, their contributions can be computed separately, with $E_{c,r}$ approximated by its zero-viscosity value. A mode that is unstable in the absence of viscosity has $\tau_{GR} < 0$, and it remains unstable if $\tau < 0$ – that is, if the growth time $|\tau_{\mathrm{GR}}|$ is shorter than the viscous damping time, $1/(\tau_{\mathrm{s}}^{-1} + \tau_{\mathrm{b}}^{-1})$. The curves in Fig. 9.3 show the value of the critical angular velocity $\Omega_c(T)$ for which $1/\tau = 0$ at a given temperature T.

Bulk viscosity. Bulk viscosity arises from nuclear reactions driven by the changing density of an oscillating fluid element, with neutrons decaying as the fluid element expands and protons capturing electrons as it contracts. Because of the finite weak interaction time, the interaction incompletely restores beta equilibrium among neutrons, protons, and electrons in one oscillation time. This creates a phase lag between density and pressure perturbations and hence a nonzero value of the work done by a fluid element in one oscillation [587]: With m_0 the baryon mass of a fluid element, the energy lost per cycle is

$$\Delta E = -\oint p dV = -\left|\oint p\frac{m_0}{\rho^2}d\rho\right|. \tag{9.122}$$

This is the dissipation due to bulk viscosity. The effect of bulk viscosity on the instability depends crucially on the *proton fraction* x_p. If x_p is lower than a critical value ($\sim\frac{1}{9}$), the Fermi momentum of neutrons is too high compared to that of protons and electrons for the reactions to conserve energy and momentum without a collision with a spectator nucleon: Only *modified URCA* processes are allowed,

$$e + p + N \to n + N + \nu_e, \tag{9.123a}$$

$$n + N \to p + N + e + \bar{\nu}_e, \tag{9.123b}$$

where N stands for a nucleon. For the modified URCA reaction, the bulk viscosity coefficient is ([587], as corrected in [172])

$$\zeta = 6\times 10^{25}\rho_{15}^2 T_9^6 \left(\frac{\omega_r}{1\mathrm{Hz}}\right)^{-2} \quad [\mathrm{g\ cm^{-1}\ s^{-1}}], \tag{9.124}$$

where $\rho_{15} = \rho/(10^{15}\,\mathrm{g\ cm^{-3}})$ and $T_9 = T/(10^9\,\mathrm{K})$ and where a high-frequency approximation, roughly valid up to 10^{10} K, is used [747]. Bulk viscosity dominates shear viscosity above a few times 10^9 K and completely suppresses the instability in uniformly rotating stars at a few times 10^{10} K [327, 326, 737].

For some equations of state and at sufficiently high densities, the proton fraction becomes larger than $\sim \frac{1}{9}$, allowing *direct URCA* reactions to take place [392]:

$$e + p \to n + \nu_e, \tag{9.125a}$$

$$n \to p + e + \bar{\nu}_e, \tag{9.125b}$$

and the corresponding reactions with electrons replaced by muons. The viscosity coefficient, $\zeta_{\rm D}$, for direct URCA reactions is several orders of magnitude larger than for modified URCA [290, 747]

$$\zeta_{\rm D} = 6.5 \times 10^{32} q \left(\frac{x_p}{0.1}\right)^{1/3} \rho_{15}^{1/3} T_9^4 \left(\frac{\omega_r}{1\text{Hz}}\right)^{-2} \quad [\text{g cm}^{-1}\ \text{s}^{-1}], \tag{9.126}$$

where q is an EOS- and density-dependent parameter of order unity. In this case, the instability is affected by bulk viscosity at temperatures as low as a few times 10^8 K and is completely suppressed for uniformly rotating stars above roughly 10^9 K [747]. The impact of direct URCA reactions on the instability, however, is weaker than the difference in timescales would suggest for two reasons. First, below the superfluid transition temperature T_s, direct URCA reactions are significantly suppressed, and neutrino emission in direct URCA reactions would rapidly cool a nascent star, with temperature dropping below 10^9 K within a few minutes. Second, in the most favorable scenario for the f-mode instability, the case of a rapidly and differentially rotating proto–neutron star described by a stiff EOS and with $T/|W|$ near 0.2, the maximum density will be low, due to centrifugal forces, which could prohibit the direct URCA process from operating.

At temperatures above roughly 10^{10} K, another complication appears. In both Eqs. (9.124) and (9.126), a high-frequency approximation was assumed, which is no longer valid above that temperature. A revised calculation, also taking into account neutrino absorption (at such high temperatures matter is expected to be opaque to neutrinos [384, 82]), reveals that the modified URCA bulk viscosity no longer rises, but is reduced by an order of magnitude between 10^{10} K, and 10^{11} K, allowing the instability to operate in very hot proto–neutron stars [383]. Furthermore, the bulk viscosity due to direct URCA reaction is also decreasing above 10^{10}K and at above a few times 10^{10} K it may dramatically drop well below the bulk viscosity due to the modified URCA reaction. In the direct URCA case, however, this is not relevant for the f-mode instability, because of the initial rapid cooling of the star.

Shear viscosity. Shear viscosity in the interior of a neutron star has the character of the viscosity of a gas: Momentum transfer is dominated not by the chemical bonds that determine the viscosity of ordinary liquids but by the momentum carried by the particles. Viscosity *increases* with an increase in the mean free path of particles, because a larger mean free path allows particles to transfer momentum between fluid elements having larger differences in their fluid velocities. The viscosity of each particle species is then limited by the dominant interaction that limits its mean free path.

If the core of a neutron star is below the density at which hyperons, free quarks, or pion or kaon condensates are present, then it consists of strongly degenerate neutrons, protons, electrons, and muons. Two main types of collisions in npeμ matter determine the shear viscosity coefficient η in a neutron-star core. Momentum transfer due to neutrons dominates the shear viscosity at temperatures above the superfluid transition temperature, with its value fixed by neutron-neutron and neutron-proton scattering. On the other hand, below the neutron-superfluid transition temperature, momentum transfer by neutron-baryon scattering is suppressed, and collisions of electrons and muons between themselves and with protons, mediated by the electromagnetic force, dominate the viscosity. Computations of the shear viscosity coefficient, as a function of density and temperature, depend on the assumed high-density equation of state and can be found in [219, 19, 63, 630].

Shear viscosity increases as the temperature drops. For neutron stars hotter than the superfluid transition temperature T_s, a first estimate of the neutron-neutron shear viscosity coefficient is [219]

$$\eta_n = 2 \times 10^{18} \rho_{15}^{9/4} T_9^{-2} \quad [\text{g cm}^{-1}\ \text{s}^{-1}], \tag{9.127}$$

Below the superfluid transition temperature, electron-electron scattering determines the shear viscosity in the superfluid core [170]. An initial expression for the shear viscosity used by most references below,

$$\eta_e = 6 \times 10^{18} \rho_{15}^{2} T_9^{-2} \quad [\text{g cm}^{-1}\ \text{s}^{-1}], \tag{9.128}$$

has been revised by Shternin and Yakovlev [630] to account for departures from a Fermi liquid, with an analytic expression

$$\eta_e = 2.5 \times 10^{18} \left(\frac{x_p}{0.1}\rho_{15}\right)^2 T_9^{-5/3} \quad [\text{g cm}^{-1}\ \text{s}^{-1}]. \tag{9.129}$$

Using the above estimates, computations in the Newtonian limit and in the post-Newtonian approximation [327, 326, 737, 418, 12] show that the CFS instability is only somewhat affected by shear viscosity at a few times 10^9 K, the most relevant temperatures for a young neutron star; and it is completely suppressed for uniformly rotating stars only at temperatures of less than about 10^6 K. Even without including the effect of mutual friction, this makes it difficult for the f-mode CFS instability to be excited in recycled neutron stars in LMXBs (although the revised viscosity somewhat eases the constraint [12]). However, shear viscosity does not significantly affect the development of the instability in rapidly rotating, nascent neutron stars.

Mutual friction. As emphasized at the beginning of this section, below the superfluid transition temperature T_s, *mutual friction*, caused by the scattering of electrons off the cores of neutron vortices, is likely to suppress the f-mode instability.

Uncertainties. We began this section with a scenario describing an f-mode instability driven by gravitational waves that spin down a nascent neutron star. The sources of uncertainty in this scenario deserve emphasis. Calculations of the superfluid transition temperature T_s "range from vanishingly small to almost 10^{11} K" [518], with the upper end more than high enough to eliminate the instability. Bulk viscosity would be greatly enhanced if the core's constituents included hyperons or free quarks, as well as by direct URCA reactions. The saturation amplitude remains to be determined, with initial nonlinear calculations in an early stage [346]; work by Alford, Mahmoodifar, and Schwenzer [13] shows a strong amplitude dependence of bulk viscosity that may also contribute to saturation. Finally, the effect of crust formation as the star cools has not been studied for the f-mode instability.

9.6.2 Instability of axial modes

In contrast to the low-m f-modes, for which the onset of instability is at large values of Ω, axial modes have zero frequency in nonrotating stars and are unstable in perfect-fluid models for all values of m and for all angular velocities Ω. The discovery by Andersson [16] that the r-modes are generically unstable to the CFS gravitational-radiation–driven instability follows immediately from the opposite signs of ω_r and ω_i in Eqs. (8.29) and (8.30). The result was proved in a relativistic context, independent of the assumption of discrete normal modes, by Friedman and Morsink [239], who showed for arbitrary m and Ω that there are axial perturbations with negative canonical energy. Because the r-modes involve only a perturbed velocity field, to lowest order in Ω, gravitational radiation through *current multipoles* dominates over that produced by mass multipoles [428, 27].

The essential differences in the character of r-modes and f-modes lead to correspondingly sharp differences in their growth times and in the damping due to mutual friction and, to a lesser extent, to bulk viscosity.

The growth time of the unstable $l = m$ r-modes can be much shorter than that of the f-modes at high rotation rates, because the frequency, $\omega_i = -\dfrac{(l-1)(l+2)}{(l+1)}m\Omega$, of the r-mode is of order Ω, where the frequency of the unstable f-mode is of order Ω–Ω_c and is therefore smaller. Again, the frequency dependence $\tau_{\rm GR} \propto \omega_i^{-2l}$ means that a moderate difference in frequency leads to a large difference in growth time. The contrast is most striking for the $m = 2$ mode: The ω_i^{-2l} dependence of the growth time on frequency means that it is the dominantly unstable r-mode, whereas the proximity of Ω_c to Ω_K for the $m = 2$ f-mode means that it is at best competitive with the higher m modes.

Bulk viscosity and mutual friction each vanish when there is no change in the density, and r-modes involve no change in density to leading order in $O(\Omega)$. For mutual friction, the dependence on density stems from the fact that the difference in velocity between the neutron and proton superfluids requires a departure from β-equilibrium. It is the smaller value of mutual friction (see, e.g., [307]) that may allow the r-mode to be unstable in old neutron stars spun up by accretion. In place

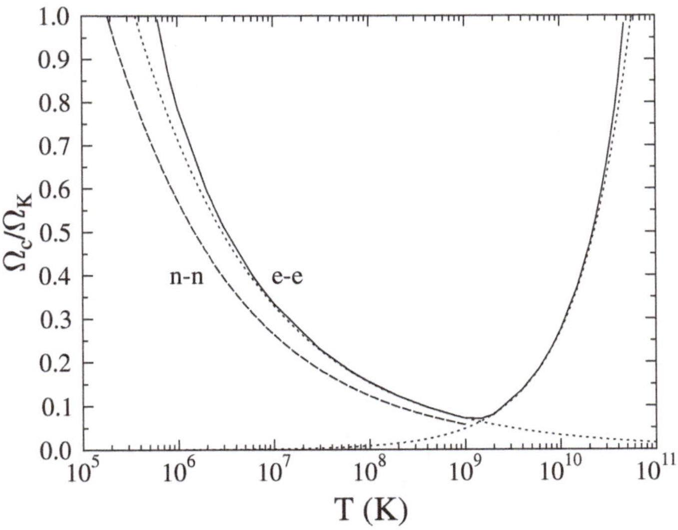

Figure 9.4. Critical angular velocity for the onset of the r-mode instability as a function of temperature (for a $1.5\,M_\odot$ neutron star model). The solid line corresponds to the $O(\Omega^2)$ result using electron-electron shear viscosity and modified URCA bulk viscosity. The dashed line corresponds to the case of neutron-neutron shear viscosity. Dotted lines are $O(\Omega)$ approximations. (Figure from [369], reproduced with permission © ESO.)

of the critical-frequency diagram of Fig. 9.3, one has a similar diagram (Figs. 9.4 and 9.5) without the shaded region associated with mutual friction. The shear viscosity is computed from e-e and n-n scattering in the superfluid core.

The figures appear to imply instability in the 10^7–10^8 K range characteristic of accreting neutron stars for stars with angular velocity as small as $0.2\Omega_K$, but there are major uncertainties. In particular, the viscosity curves shown ignore shear viscosity in the boundary layer (Ekman layer) between the crust and the interior fluid, and they assume a npeμ core. If the crust is nearly rigid, the shear viscosity in the boundary layer could be dominant, and, as noted in the f-mode discussion, hyperons or free quarks in the core could allow nonleptonic interactions that greatly enhance the bulk viscosity.

The uncertainty associated with these dissipative alternatives allows alternate scenarios for the behavior of the instability in accreting neutron stars: A star could be spun up by accretion until the amplitude of the unstable mode is large enough that the angular momentum radiated in gravitational waves balances that gained in accretion; or a runaway instability could arise from a decreasing viscosity as the mode's dissipated energy heats the star. These possibilities are discussed at the end of this section, following a more detailed review of the radiation-reaction growth times and the viscous damping times of unstable r-modes.

Radiation-reaction growth times. We again use Eq. (9.109) to estimate the growth time $\tau_{\rm GR}$, with E_c and $\dot{E}_c$ given by Eqs. (9.112) and (9.116). This time, however, we

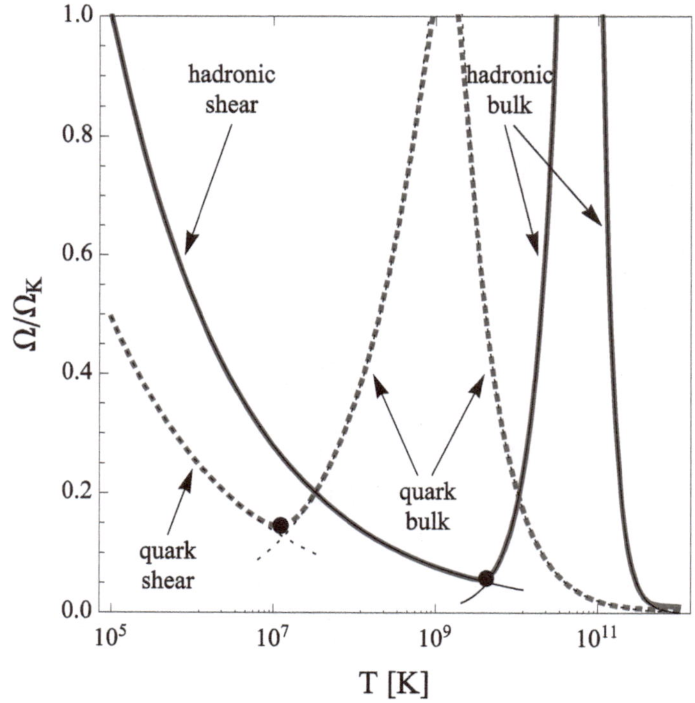

Figure 9.5. A different version of Fig. 9.4, contrasting the critical angular velocity for the onset of the r-mode instability when only nucleons are present (solid curve) to the angular velocity in a strange quark star (dotted curve). For a hybrid star with the entire core a bag of up, down, and strange quarks, the curves are essentially those of the quark star, whereas a hybrid star in which quarks comprise less than half the core has an instability region similar to that of a star with only nucleons. (Reprinted figure with permission from Alford, Mahmoodifar, and Schwenzer [12]. Copyright 2011 by the American Physical Society.)

will see that it is the current multipole J_{ll}, not the mass multipole, that contributes to $\dot{E}_c$ at lowest nonvanishing order in Ω.

The uniform-density Newtonian model of Section 8.6.2 provides a simple analytical way to obtain the approximate growth time $\tau_{\rm GR}$ and damping time $\tau_{\rm s}$ due to shear viscosity of an unstable r-mode. At order Ω, the $l = m$ mode is purely axial, with perturbed velocity field given by Eq. (8.55),

$$\delta v^a = u_0 \frac{r^{l+1}}{R^l} \epsilon^{ba} \nabla_b Y_{ll}. \tag{9.130}$$

The canonical energy in the rotating frame then has the value

$$\begin{aligned} E_{c,r} &= \frac{1}{2} \int \rho \delta v_a^* \delta v^a dV \\ &= \frac{3}{8\pi} \frac{l(l+1)}{2l+3} M u_0^2. \end{aligned} \tag{9.131}$$

The power $\dot{E}_c$ radiated is quadratic in the current multipole δJ_{ll},

$$\dot{E}_c = -\omega_r \omega_i^{2l+1} N_l \left|\delta J_{ll}\right|^2, \tag{9.132}$$

and from Eq. (7.247), we have

$$\begin{aligned} \delta J_{ll} &= \frac{2}{l+1} \int r^l \, \rho \delta v_a \epsilon^{ba} r \nabla_b Y_{ll}^* dV \\ &= \frac{3}{2\pi} \frac{l}{2l+3} M R^l u_0. \end{aligned} \tag{9.133}$$

Finally, from Eqs. (9.131)–(9.133), the growth time of the unstable r-mode is given to lowest nonvanishing order in Ω by

$$\begin{aligned} \tau_{\rm GR} &= -2E_{c,r}/\dot{E}_{c,r} \\ &= -\frac{c^{2l+3}}{24G} \frac{[(2l+3)!!]^2}{(2l+3)(l-1)^{2l}} \left(\frac{l+1}{l+2}\right)^{2l+2} \frac{\Omega^{-2l-2}}{MR^{2l}}. \end{aligned} \tag{9.134}$$

Because mass multipoles have polar parity, the lowest mass multipole associated with an $l = m$ r-mode is $D_{l+1,l}$, and it is $O(\Omega)$, one order in Ω higher than J_{ll}:

$$\delta D_{l+1,l} \sim -\bar{\Omega} M R^{l+1}. \tag{9.135}$$

The contribution from $\delta D_{l+1,l}$ to $\dot{E}_c$ is proportional to $\omega_r \omega^{2(l+1)+2} |\delta D_{l+1,l}|^2$, four orders higher in Ω than the contribution from δJ_{ll}, and it therefore contributes to $\tau_{\rm GR}$ a term four orders higher in Ω, a term of order $O(\Omega^{-2l+2})$.

For the $l = m = 2$ r-modes, the growth time, given by Eq. (9.134), has the value

$$\tau_{\rm GR} = 1.4 \times 10^6 \left(\frac{10^3 \text{ s}^{-1}}{\Omega}\right)^6 \left(\frac{1.4 M_\odot}{M}\right) \left(\frac{10 \text{ km}}{R}\right)^4 \text{ s} \tag{9.136}$$

and becomes very short for rapidly rotating stars, because of the $1/\Omega^6$ dependence.

Viscous damping times. From Eqs. (9.120), (9.131), and (8.55), we obtain the damping time due to shear viscosity,

$$\tau_s = \frac{3}{4\pi(l-1)(2l+3)} \frac{M}{\eta R}. \tag{9.137}$$

Using Eq. (9.128) for the electron-electron shear viscosity coefficient, we can write the damping time for the $l = m = 2$ mode as

$$\tau_s = 3.6 \times 10^7 \left(\frac{R}{10\text{km}}\right)^5 \left(\frac{T}{10^9 \text{K}}\right)^2 \left(\frac{1.4 M_\odot}{M}\right) \text{ s}. \tag{9.138}$$

Because we have obtained a solution for $\delta \boldsymbol{v}$ only up to $O(\Omega)$, we cannot apply (9.120) to obtain the damping timescale due to bulk viscosity – to this order, $\boldsymbol{\nabla} \cdot \delta \boldsymbol{v} = 0$ for any EOS, and we would need to find numerical solutions at higher order (for a compressible EOS). Instead, we estimate the bulk viscosity timescale using the result of [170], which was derived for f-modes. To account for the fact that density variations for $l = m$ r-modes enter through an $l + 1$ term, we replace

l with $l+1$ in the formula derived in [170], obtaining the following approximate expression:

$$\tau_b \simeq \frac{3(2l+5)}{2\pi(l+1)^3}\frac{\Gamma^4 M}{\zeta R}. \tag{9.139}$$

For the purpose of estimating the effects of bulk viscosity only, Γ is taken to be equal to 5, that is, to correspond to a stiff (nearly uniform density), $N = 0.25$ polytrope. For the $l = m$ r-mode and using Eq. (9.124) for the bulk viscosity coefficient, we find[7]

$$\tau_b \simeq 4.6 \times 10^9 \times \left(\frac{R}{10\,\mathrm{km}}\right)^5 \left(\frac{\Omega}{10^3\,\mathrm{s}^{-1}}\right)^2 \left(\frac{1.4M_\odot}{M}\right) \left(\frac{10^9\mathrm{K}}{T}\right)^6 \mathrm{s}. \tag{9.140}$$

Figure 9.4 shows the critical angular velocity (in units of $\Omega_K \simeq 0.67\sqrt{\pi G\rho}$) as a function of temperature for a representative model with mass $M = 1.5M_\odot$ and radius $R = 12.47$ km. The solid curve is used for τ_{GR}. For comparison, both cases of electron-electron viscosity and neutron-neutron viscosity are shown. According to this scenario, a rapidly rotating neutron star born at a temperature of 10^{11} K would enter the instability region (which is $10^6\,\mathrm{K} < T < 10^{10}\,\mathrm{K}$) and lose angular momentum, slowing down to a rotational period of about 15 ms at $T \simeq 1.5 \times 10^9$ K.

For slow rotation rates, one can use the $O(\Omega)$ equation (9.134) to construct approximate expressions for the two parts of the curve in Fig. 9.4. The part of the critical curve where the shear viscosity dominates can be approximated by

$$\Omega_c^{(\mathrm{s})} \simeq 581 \left(\frac{10\mathrm{km}}{R}\right)^{3/2} \left(\frac{10^9\mathrm{K}}{T}\right)^{1/3} \mathrm{s}^{-1}, \tag{9.141}$$

whereas the bulk-viscosity–dominated part is described by

$$\Omega_c^{(\mathrm{b})} \simeq 362 \left(\frac{R}{10\mathrm{km}}\right)^{9/8} \left(\frac{T}{10^9\mathrm{K}}\right)^{3/4} \mathrm{s}^{-1}, \tag{9.142}$$

and both are shown as dotted curves in Fig. 9.4, in near agreement with the $O(\Omega^2)$ result, even for relatively rapid rotation.

Other mechanisms affecting viscosity. If the equation of state yields a significant fraction of hyperons [341, 427, 285, 493, 306] or free quarks [451, 452, 22, 339, 575] in the inner core, bulk viscosity is greatly enhanced, and it could entirely suppress any instability of young stars or limit the instability to the short time needed for the star to cool to about 10^{10} K. The main part of the enhancement is due to a set of nonleptonic weak interactions of the form

$$n + n \leftrightarrow p + \Sigma^-, \quad n + p \leftrightarrow p + \Lambda, \quad n + n \leftrightarrow n + \Lambda;$$

again the fact that oscillations do not maintain exact chemical equilibrium implies a phase lag between density and pressure perturbations and work done in each cycle.

[7] Notice that for $N = 1$ polytropes, τ_b can be an order of magnitude shorter than the estimate given here; see [25].

Hyperons and unconfined quarks require a soft equation of state, because nucleons change to hyperons only if the core has high enough density to push the Fermi level of neutrons above the hyperon threshold. The observation of a $2M_\odot$ neutron star has therefore diminished the likelihood of a hyperon core, particularly in stars with mass below $1.5M_\odot$, characteristic of pulsars at birth. The higher-mass accreting stars are better candidates for cores with hyperons or quarks.

After the formation of a crust, shear viscosity in the boundary layer at the crust-core interface can be substantially larger than the shear viscosity in the core [70, 23, 429, 569]. The initial computation used a rigid crust that did not participate in the r-mode oscillation, but the magnitude of the effect depends on the extent to which the r-mode penetrates the crust – by how much the crust can slip. A *slippage parameter* $\mathcal{S}$ measures the fractional difference in velocity of the normal fluid between the crust and the core and the fractional pinning of vortices in the crust [408, 351, 714, 262, 263]: In the case of negligible magnetic field contribution $\mathcal{S} = (v - v_0)/v_0$, where v is the fluid velocity immediately below the crust, v_0 the corresponding velocity if no crust were present.

r-Mode instability scenarios in accreting neutron stars. As Bildsten [68] and Andersson, Kokkotas, and Stergioulas [28] pointed out, while mutual friction appears to rule out any f-mode instability of accreting stars, accretion may well lead to unstable r-modes. Possible alternative pathways for the instability depend on the strength of dissipative mechanisms and the magnitude of the saturation amplitude set by nonlinear couplings (see Bondarescu, Teukolsky, and Wasserman [92] for a more detailed account and Chapter 10 for a review of work on the saturation amplitude.)

Alternative scenarios can be described as paths in the Ω-T plane. Each begins with a star spun up by accretion until its angular velocity exceeds the critical angular velocity Ω_c for the $l = m = 2$ mode. During the initial gradual spin-up, the temperature of the accreting star has a fixed value associated with the accretion rate, corresponding to the vertical path on the left in Fig. 9.6. Once the r-mode becomes unstable, it begins to grow exponentially, and the energy dissipated by viscosity heats the star, decreasing the shear (and boundary layer) viscosity.

The first class of scenarios, represented by the path in Fig. 9.6, are cycles that end with *thermal runaways* [647, 405]: With the shear decreasing, the unstable mode reaches saturation amplitude and spins down. The now stable star cools, is again gradually spun up by accretion, and the cycle repeats.

For a large saturation amplitude, the spin-down time is much shorter than the spin-up time, shorter by a factor of about 10^6, and the chance of observing a star while it is emitting gravitational waves is therefore correspondingly small (fast runaway). For a small saturation amplitude, of order 10^{-5}, however, the spin-down can last for a time nearly as long as the spin-up time, leaving open the possibility of detecting gravitational waves from rapidly rotating neutron stars in some LMXBs (slow runaway) [314].

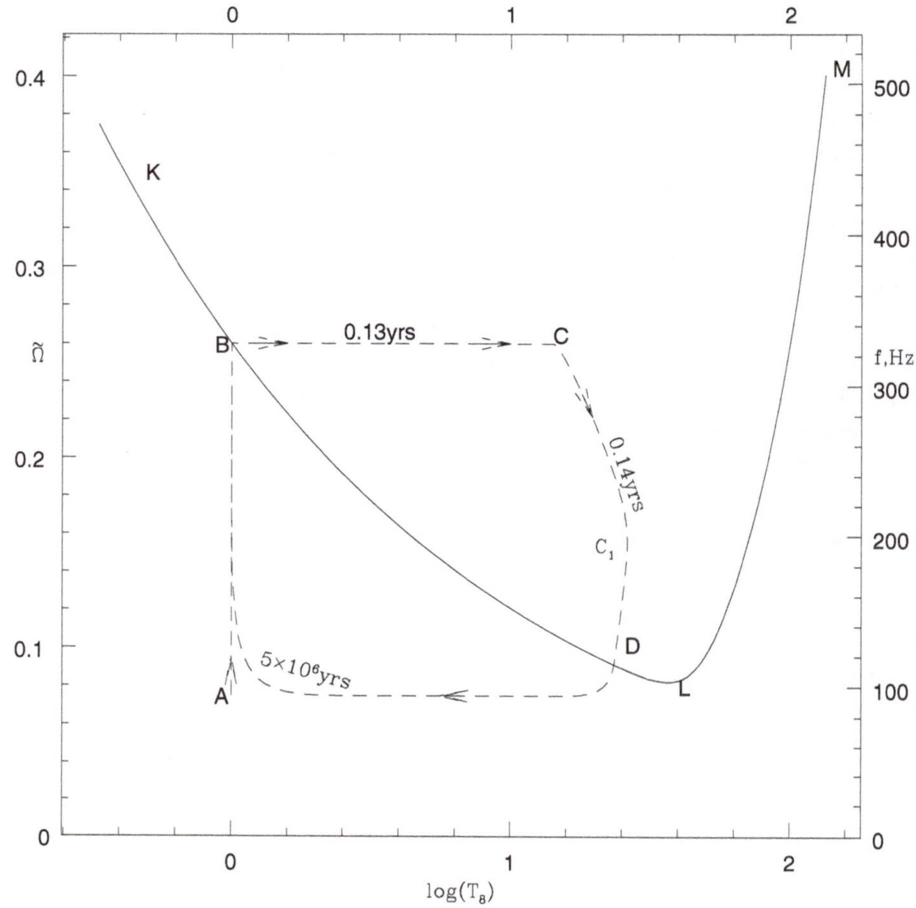

Figure 9.6. Cyclic runaway evolution of an unstable r-mode in an accreting neutron star. As in Fig. 9.4, line KLM is the critical angular velocity curve above which the mode is unstable. Leg AB represents the accretional spin-up of the neutron star to the critical angular velocity; BC, the heating stage in which the r-mode becomes unstable, grows, and heats the star; CD, a rapid spin-down in which the angular velocity decreases due to the emission of gravitational radiation; and DB, the neutron star cooling back to the equilibrium temperature with simultaneous spin-up by accretion, closing the cycle. (Figure from Levin [405]. Reproduced by permission of the AAS.)

A second class of scenarios are *steady-state* evolutions in which a star spins up until angular momentum gained in accretion balances angular momentum lost to gravitational waves. This idea was first suggested in the context of f-modes (before the understanding of mutual friction) by Papaloizou and Pringle [524] and examined in detail by Wagoner [713]. Its revival for r-modes relies on the enhanced bulk viscosity associated with hyperons or unconfined quarks [714, 22, 493]: With the bulk viscosity peak widened to lower temperature, the unstable star reaches the bulk viscosity curve before it has had time to spin down. Once there, an equilibrium state is established near the critical stability curve, with $-\tau_{\rm GR} = \tau_{\rm b}$, and the star evolves along the curve until the Papaloizou-Pringle-Wagoner angular momentum balance is reached.

Bondarescu, Teukolsky, and Wasserman [92] find the dependence of these scenarios on the hyperon superfluid transition temperature, the fraction of the star for which the proton fraction is above the critical level for direct URCA reactions, and the slippage factor $\mathcal{S}$ (see Chapter 10, Fig. 10.6). With the possible exception of rapid thermal runaways, the r-mode amplitude remains very small ($\sim 10^{-5}$), but the correspondingly long duration of the instability means that LMXBs are candidates for gravitational-wave detection by advanced LIGO-class interferometers [92, 720, 513].

The accumulation of gravitational waves from the r-mode instability in neutron stars throughout the visible universe contributes to the gravitational wave stochastic background. Owen et al. [514], Ferrari, Matarrese, and Schneider [216], and more recently Zhu, Fan, and Zhu [751] consider the detectability of this part of the background, with the last of these references taking into account the smaller expected amplitude of the instability.

Wind-up of magnetic fields. At second-order in perturbation theory, an r-mode includes an axisymmetric part with a nonzero δv_ϕ: That is, it leads to differential rotation that grows as the perturbation grows. By winding up a poloidal field, differential rotation can generate a strong toroidal magnetic field, and several authors [647, 567, 565] have considered a possible associated dissipation of the mode. For a small nonlinear saturation amplitude, of order 10^{-3} or less, work by Rezzolla et al. [566] finds a timescale to increase the magnetic field that is longer than the viscous damping time. If, however, the saturation amplitude should be unexpectedly large, of order 10^{-2}, and the initial magnetic field is larger than 10^{11} G, the critical angular velocity Ω_c for instability can be significantly increased. Using a similar formalism, subsequent work by Cuofano and Drago [167] modifies r-mode time evolutions by regarding the rate at which the magnetic field energy grows as an additional damping term. For an initial field of 10^9 G, they find the thermal runaway cycle is slightly modified, with a reduced maximum amplitude of order 10^{-5}; a smaller initial field appears unlikely to reduce the amplitude below the nonlinear saturation value.

9.7 Viscosity-driven secular instability

A different type of nonaxisymmetric instability in rotating stars is the instability driven by viscosity, which breaks the circulation of the fluid [570, 340]. The instability is suppressed by gravitational radiation, so it cannot act in the temperature window in which the CFS instability is active. The instability sets in when the frequency of an $l = -m$ (corotating) mode goes through zero in the rotating frame. In contrast to the CFS instability, the viscosity-driven instability is not generic in rotating stars. The $l = 2$ mode becomes unstable at high rotation rates for very stiff equations of state, and higher l-modes become unstable at still larger rotation rates.

To understand this family of instabilities, it is helpful to begin with an intuitive discussion of the way in which the bar-mode instability driven by viscosity differs from that driven by gravitational radiation. Gravitational radiation lowers the energy in an inertial frame while conserving circulation. Viscosity, on the other hand, lowers the energy in a comoving frame while conserving total angular momentum. That means when viscosity is present (and gravitational radiation is negligible), an equilibrium configuration is secularly unstable if there is a nearby configuration with the same angular momentum and with lower energy in the rotating frame.

When do nonaxisymmetric configurations have lower energy than axisymmetric equilibria? A nonrotating star minimizes its total energy by adopting a shape that maximizes its gravitational binding energy: Static stars are spherical. Slow rotation leads to an oblate figure, but the minimum energy configuration is still axisymmetric, again because a symmetric configuration maximizes the binding energy. As the rotation increases, however, the gravitational binding energy of an increasingly disklike axisymmetric equilibrium steadily decreases, until it eventually falls below the binding energy of two nearly spherical balls of fluid with the same angular momentum: The competition between a kinetic energy, minimized at fixed angular momentum by having the fluid as far as possible from the axis of rotation, and gravitational binding energy, maximized by concentrating the fluid into a ball, leads to a minimum total energy with two orbiting balls. This argument implies that, along a sequence of models with increasing rotation, there is some angular velocity above which there are nonaxisymmetric configurations with lower energy and the same angular momentum. The argument does not in itself show that there are *nearby* configurations with lower energy; but that is in fact how the bar-mode instability sets in. Beyond some point along a sequence of rotating models, bars have lower energy than oblate spheroids, because the star loses more binding energy by becoming flatter than by reducing its symmetry.

Along the Maclaurin sequence of uniformly rotating, uniform-density ellipsoids, the point beyond which there are lower-energy configurations with the same angular momentum is the bifurcation point at which the Jacobi sequence of triaxial ellipsoids branches off. Beyond that point, viscosity makes the sequence unstable by allowing the star to redistribute velocity while conserving total momentum.

In Newtonian polytropes, the instability occurs only for stiff polytropes of index $N < 0.808$ [340, 635]. In addition, the viscosity-driven bar mode appears at the same critical $T/|W|$ ratio as the bar mode driven by gravitational radiation [331]. Relativistic effects make the viscosity-driven instability weaker [83, 604, 186] (while the gravitational-radiation-driven instability becomes stronger, so that the two instabilities no longer occur at the same value of $T/|W|$). The most promising case for the onset of the viscosity-driven instability (in terms of the critical rotation rate) would be rapidly rotating strange stars [270], but the instability can appear only if its growth rate is larger than the damping rate due to the emission of gravitational radiation.

9.8 Dynamical and shear instabilities

Dynamical nonaxisymmetric instabilities. If a protoneutron star has a sufficiently high rotation rate (so that, e.g., $T/|W| > 0.27$ in the case of Maclaurin spheroids), it will be subject to a dynamical instability driven by hydrodynamics and gravity. Through the $l = 2$ mode, the instability will deform the star into a bar shape. This highly nonaxisymmetric configuration will emit strong gravitational waves with frequencies in the kHz regime. The development of the instability and the resulting waveform have been computed numerically in the context of Newtonian gravity by Houser, Centrella, and Smith [320] and in full general relativity by Shibata, Baumgarte, and Shapiro [611] (see Section 10.4.1).

The bar-mode instability could be excited in a hot proto–neutron star a few milliseconds after core bounce; alternatively, it could be excited a few tenths of seconds later, when the proto–neutron star cools as a result of neutrino emission and contracts further, with $T/|W|$ increasing during contraction (roughly as $\sim 1/R$) until the instability threshold at ~ 0.27 is reached. The amplitude of the emitted quadrupole gravitational waves is of order $h \sim MR^2\Omega^2/D$, where D is the distance to the source; expressed in terms of the ellipticity ϵ of the bar, we have

$$h \approx 9 \times 10^{-23} \left(\frac{\epsilon}{0.2}\right) \left(\frac{f}{3\text{kHz}}\right)^2 \left(\frac{15\text{Mpc}}{D}\right) M_{1.4} R_{10}^2, \tag{9.143}$$

where $M_{1.4}$ is the mass in units of $1.4M_\odot$ and R_{10} is radius in units of 10 km. For uniform rotation, the frequency f of quadrupole radiation is twice the rotational frequency Ω.

General relativity enhances the onset of the instability, lowering the threshold to $T/|W| \sim 0.24$ [611, 581]. A detailed investigation of the required initial conditions of the progenitor core, which can lead to the onset of the dynamical bar-mode instability in the formed proto–neutron star, was presented in [618]. The nonlinear evolution of the instability depends critically on the mass and rotation rate of the star, and the various outcomes are discussed in detail in Chapter 10.

Shear instabilities. Apart from the nonaxisymmetric dynamical instabilities that exist only at very high $T/|W|$, additional nonaxisymmetric instabilities related to the shear of differential rotation, *shear instabilities*, that appear at low $T/|W|$ (hence also the name *low-$T/|W|$ instabilities*) have been shown to exist under certain conditions. In the context of rotating stars, Centrella et al. [120] first found a one-armed ($m = 1$) instability in models with a polytropic index $N = 3$, strong differential rotation, and $T/|W| \sim 0.14$. More relevant for neutron stars, Shibata, Karino, and Eriguchi [614, 615] found an $m = 2$ instability for $T/|W|$ as low as 0.01, for models with polytropic index $N = 1$. Subsequently, such instabilities were observed to arise in simulations of rotating core collapse, in which a rapidly rotating proto–neutron star formed [509, 505], making them much more astrophysically relevant than the dynamical instabilities. The interest in these instabilities is strong, because they can radiate stronger gravitational waves than the postbounce burst signal itself.

Shear instabilities were shown to exist in differentially rotating disks [527], cylinders [43, 447], and self-gravitating stars [448]. The instabilities found in simulations of rotating compact stars are interpreted as shear instabilities, based on various properties. Watts, Andersson, and Jones [718] find that the instabilities depend essentially on two parameters, the ratio $T/|W|$ and the degree of differential rotation. In this 2-dimensional parameter space, there is a distinct region in which the instability occurs. For low degrees of differential rotation, the star is always stable to shear instabilities. Above a certain threshold, the instability appears at a certain value of $T/|W|$ only. As the degree of differential rotation becomes larger, the instability region widens to both smaller and larger $T/|W|$. For very strong differential rotation, the upper boundary of the instability region merges with the instability region of the dynamical nonaxisymmetric instability, whereas the lower boundary extends to very small rotation rates.

In [718], the above properties are explained as a result of the existence of *corotation points* – that is, of a point inside the star at which the pattern speed of an oscillation mode (in this case of the f-mode) matches the local angular velocity. Along a sequence of rotating models with the same degree of differential rotation, there exists a range of frequencies in which modes posses corotation points, forming a *corotation band.* This suggests that when a mode enters or exits the corotation band and the degree of differential rotation is above the threshold value for the shear instability, it becomes unstable or stable, respectively. Numerical simulations of selected models in [614] showed indeed that the growth timescale of the instability is shortest in the center of the band (somewhat longer than the dynamical timescale), getting longer toward the edges. Furthermore, as differential rotation becomes stronger, the corotation band widens, eventually encompassing both co- and counterrotating modes, as well as the dynamically unstable mode at high rotation rates. The main features of the explanation given in [718] were confirmed in detailed numerical simulations in general relativity by Corvino et al. [163], not only for the $m = 2$ mode but also for $m = 1$ and $m = 3$ modes.

An analysis of the shear instability in differentially rotating stars by Ou and Tohline [510] suggests that the growth rate of the instability is related to the location of the corotation point within a *resonant cavity* that forms due to the existence of a minimum in the *vortensity*[8] profile. Such a resonant cavity has been shown to be essential for the growth of the Rossby wave instability in accretion disks, trapping radially propagating waves [445, 526, 446, 411, 410] (in contrast, the growth of the Papaloizou-Pringle instability in accretion disks relies on a resonant cavity formed by the inner and outer edges of an accretion disk). For a given model, several modes can be simultaneously unstable, but with different growth rates. The more deeply

[8] The ratio of vorticity to density.

the corotation point for a given unstable mode is located inside the resonant cavity, the faster the instability will grow. In [163], it was shown that rapidly rotating models that are unstable only to the dynamical nonaxisymmetric instability also have a minimum in their vortensity profile, so that the latter property is not by itself a sufficient condition for the existence of shear instabilities. The nonlinear evolution of shear instabilities will be discussed in Chapter 10.

10
Nonlinear dynamics of rotating relativistic stars

In a multitude of astrophysical contexts in which a rotating relativistic star evolves about or away from an equilibrium configuration, linear perturbation theory does not adequately describe the star's evolution. Obvious examples are dynamical rotational instabilities; instability to collapse, leading to the formation of a rotating black hole; formation of compact stars after core bounce; and the large-amplitude oscillations of supramassive neutron stars that survive briefly after the merger of a double neutron-star system. In all these cases, a numerical treatment of the full Einstein equations is required, using the techniques developed in the field of *numerical relativity.* For CFS-unstable modes, nonlinear Newtonian perturbation theory has been used to find the saturation amplitude governing the perturbations; however, the nonlinear behavior of modes of rapidly rotating relativistic stars has so far been explored using fully nonlinear evolution codes. Even the linear modes themselves have been efficiently found by fully nonlinear evolutions of initial data describing small perturbations from equilibrium: Numerically solving the eigenvalue problem for outgoing normal modes of rotating stars has proved to be more difficult.

Both the modeling of rotating spacetimes with matter and the modeling of black-hole spacetimes with collapsing matter sources in multidimensions have been challenging tasks in numerical relativity. Nevertheless, through the adoption of advanced numerical techniques, several obstacles have been overcome. The difficulties encountered in such simulations can be traced back to inherent difficulties and complexities of the Einstein-Euler system, as well as the large computational resources needed for multidimensional simulations. For example, the precise numerical computation of the gravitational radiation emitted in the process of black hole formation is particularly challenging, as the energy released in gravitational waves is several orders of magnitude smaller than the total rest-mass energy of the system, requiring very high accuracy to be achieved over a large number of dynamical timescales.

10.1 Numerical methods for nonlinear simulations

10.1.1 Evolution of the equations of ideal hydrodynamics

In numerical work, the most widely adopted formulation of the hydrodynamical equations is the *conservative formulation* [461, 44], in which baryon conservation, the three components of the Euler equation, and energy conservation, (2.29), (2.30), and (2.31), are written in a first-order, flux-conserving and hyperbolic form. With the Euler equation's covariant divergence, $\nabla_\beta(T_\alpha{}^\beta\sqrt{|g|})\gamma^\alpha{}_a$, split into a coordinate divergence and a source term, its three components and the two conservation equations can be written as a system of five equations

$$\partial_t \mathbf{U} + \partial_j \mathbf{f}^j(\mathbf{U}) = \mathbf{s}(\mathbf{U}) \,, \tag{10.1}$$

where $\mathbf{U}$ is a set of five densities, and the fluxes $\mathbf{f}^j(\mathbf{U})$, and source terms $\mathbf{s}(\mathbf{U})$ are to be computed in terms of $\mathbf{U}$ and the metric [321, 222]. The *primitive* hydrodynamical variables, namely the rest-mass density ρ, fluid 3-velocity v^i, and specific internal energy e, are abandoned in favor of the densities $\mathbf{U}$.

We first give the explicit components of $\mathbf{U}$, $\mathbf{f}^j$, and $\mathbf{s}$ and then provide a short derivation of the equations. We use the 3+1 notation of Section 2.1, but not the covariant form of the conservation equations of that section. $\mathbf{U}$ comprises the following densities measured by an observer with 4-velocity n^α normal to a $t =$ constant hypersurface: the baryon density $-\rho u^\alpha n_\alpha\sqrt{\gamma}$; the spatial momentum density of Eq. (2.19),

$$j_a\sqrt{\gamma} = -T^\alpha{}_a n_\alpha\sqrt{\gamma} = \rho h W^2 v_i\sqrt{\gamma},$$

where W is the Lorentz factor defined in Eq. (2.26); and an energy density $\tau\sqrt{\gamma}$ obtained by subtracting the rest-mass density from the total energy density:

$$\tau := (T^{\alpha\beta}n_\alpha n_\beta + \rho u^\alpha n_\alpha) = \rho h W^2 - p - W\rho. \tag{10.2}$$

The five components of $\mathbf{U}$ are then

$$\mathbf{U} = \begin{pmatrix} W\rho \\ j_i \\ \tau \end{pmatrix} \sqrt{\gamma}. \tag{10.3}$$

The flux vectors $\mathbf{f}^j$ are given by

$$\mathbf{f}^j := \begin{bmatrix} (\alpha v^j - \beta^j)W\rho \\ (\alpha v^j - \beta^j)j_i + \alpha p\delta^j_i \\ (\alpha v^j - \beta^j)\tau + \alpha v^j p \end{bmatrix} \sqrt{\gamma}, \tag{10.4}$$

where the index i labels the flux of the ith component of the momentum density. The source terms $\mathbf{s}$ depend on the stress-energy tensor, as well as on the metric and its first derivatives, but not on derivatives of the primitive variables; the five components are

$$\mathbf{s} := \begin{bmatrix} 0 \\ \alpha T_\mu{}^\nu \Gamma^\mu{}_{\nu i} \\ \alpha(T^{\mu t}\partial_\mu\alpha - \alpha T^{\mu\nu}\Gamma^t{}_{\mu\nu}) \end{bmatrix} \sqrt{\gamma}, \tag{10.5}$$

where $\Gamma^\sigma{}_{\mu i}$ and $\Gamma^t{}_{\mu\nu}$ are 4-dimensional Christoffel symbols.

Derivation. Because the divergence of a vector density $\mathcal{A}^a = A^a\sqrt{\gamma}$ involves a coordinate divergence $D_a\mathcal{A}^a = \partial_i\mathcal{A}^i$ (see Eq. [A.47]), baryon conservation, Eq. (2.29), already has a flux-conserving form:

$$\partial_t(\rho u^t \alpha\sqrt{\gamma}) + \partial_i(\rho u^i \alpha\sqrt{\gamma}) = 0. \tag{10.6}$$

To express the components of the relativistic Euler equation in an analogous form involving the coordinate divergence of a flux term, it is quickest to begin with the original equation $\nabla_\beta(T_\alpha{}^\beta\sqrt{|g|})\gamma^\alpha{}_a = 0$, rather than its covariant 3+1 expression (2.30):

$$0 = \nabla_\nu(T_i{}^\nu\sqrt{|g|}) = \partial_\nu(T_i{}^\nu\sqrt{|g|}) - \sqrt{|g|}\Gamma^\mu{}_{\nu i}T_\mu{}^\nu. \tag{10.7}$$

Using Eqs. (2.27) and (2.19) to express $T_i{}^j$ in terms of j_i, we have

$$T_i{}^j\sqrt{|g|} = [j_i(\alpha v^j - \beta^j) + \alpha p]\sqrt{\gamma},$$

and the Euler equation takes the desired form

$$\partial_t(j_i\sqrt{\gamma}) + \partial_j\{[(\alpha v^j - \beta^j)j_i + \alpha p\delta_i^j]\sqrt{\gamma}\} = \alpha\sqrt{\gamma}\Gamma^\mu{}_{\nu i}T_\mu{}^\nu. \tag{10.8}$$

Finally, the energy conservation law is the linear combination, $-[n_\alpha\nabla_\beta T^{\alpha\beta} + \nabla_\alpha(\rho u^\alpha)]\sqrt{|g|} = 0$, of the energy conservation and baryon conservation equations, Eqs. (10.6) and (1.27). Writing

$$\begin{aligned} -\alpha n_\nu\nabla_\mu(T^{\mu\nu}\sqrt{|g|}) &= \alpha\partial_\mu(T^{t\mu}\sqrt{|g|}) \\ &= \partial_t(\alpha^2 T^{tt}\sqrt{\gamma}) + \partial_j(\alpha^2 T^{tj}\sqrt{\gamma}) - \alpha\partial_\mu\alpha T^{t\mu}\sqrt{\gamma} \\ &\quad + \alpha^2\Gamma^t{}_{\mu\nu}T^{\mu\nu}\sqrt{\gamma} \\ &= \partial_t[(h\rho W^2 - p)\sqrt{\gamma}] + \partial_j\{(\alpha v^j - \beta^j)(h\rho W^2 - p) \\ &\quad + \alpha v^j p]\sqrt{\gamma}\} - \alpha(T^{t\mu}\partial_\mu\alpha - \alpha T^{\mu\nu}\Gamma^t{}_{\mu\nu})\sqrt{\gamma}, \end{aligned} \tag{10.9}$$

and using Eq. (10.6), we obtain the last component of Eq. (10.1).

The system of hydrodynamical equations (10.1) is supplemented by an EOS, which can be a one-parameter EOS, such as a barotropic polytrope (1.119), (1.120), a two-parameter ideal gas "Γ-law" EOS[1]

$$p(\rho, e) = (\Gamma - 1)\rho e, \tag{10.10}$$

or a more general one-parameter or two-parameter EOS derived from microphysical considerations. In the case of the polytropic EOS (1.119, 1.120), the evolution equation for τ is redundant, because τ is given algebraically in terms of the rest-mass density ρ. In the case of the ideal gas EOS (10.10), on the other hand, the entropy increases at shocks, and the evolution equation for τ needs to be solved. Note that

[1] Much of the recent numerical relativity literature uses the term *ideal fluid EOS* or *gamma-law EOS* to refer to the relativistic form of the ideal gas equation of state $p = nkT = (\Gamma - 1)u$, with the specific internal energy e defined by Eq. (1.12), $u = \rho e$. The terminology is not consistent in the literature: Some of the cosmology literature uses *ideal fluid* to simply mean a perfect fluid, whereas the nonrelativistic hydrodynamics literature uses *ideal fluid* to mean an incompressible perfect fluid.

the polytropic EOS (1.119), (1.120) does not allow any transfer of kinetic energy to thermal energy; that is, it is not suitable for describing shock heating.

Additional details on numerical methods for general-relativistic hydrodynamics can be found in [222]. An important feature of the first-order hyperbolic form of the equations is that it has allowed the extension to a general relativistic context of the powerful numerical methods developed in classical hydrodynamics, in particular high-resolution shock-capturing (HRSC) schemes based, for example, on linearized Riemann solvers. Such schemes are essential for obtaining a stable evolution and a correct representation of shocks. Stability is important, because (under certain conditions) a stable scheme converges to a weak solution of the hydrodynamical equations, but a nonconservative scheme would converge to the wrong weak solution in the presence of a shock [318]. For an introduction to HRSC methods, see [404, 387, 699].

More specifically, for the numerical simulation of rotating stars, HRSC methods that employ a third-order piecewise parabolic method (PPM) reconstruction [149] were shown to be superior to other methods in maintaining a sharp angular velocity profile at the surface [225].

10.1.2 Spacetime evolution

For the evolution of the spacetime, one can use several different approaches, which either employ some approximation or attempt to solve the full system of Einstein equations.

IWM-CFC approximation

If one is not interested in accurate gravitational-wave extraction in a numerical simulation, then the IWM-CFC spatial conformal flatness approximation (Section 6.2.1) is suitable for obtaining a spacetime evolution with reasonable accuracy. The scheme has been used in dynamical studies of spacetimes describing pulsating, collapsing, or merging relativistic stars – for example [729, 728, 730, 189, 190, 501, 577, 578, 193, 500]. In certain cases, however, the above formulation suffers from nonuniqueness of solutions. This problem is circumvented by a slight reformulation, introducing an auxiliary variable [577, 162].

BSSN formulation

Evolution equations. The covariant nature of the Einstein equation leads to difficulties in constructing an appropriate coordinate representation that would allow for stable and accurate simulations, especially when black holes form (characterized by unavoidable physical singularities). A successful numerical approach to this problem relies on reformulating the original 3+1 (ADM) approach (Section 2.1) in order to achieve long-term stability. Building on the experience developed with lower-dimensional formulations, Nakamura, Oohara, and Kojima [491] presented a conformal traceless reformulation of the ADM system that subsequent studies (see, e.g., [620, 54, 608, 610, 9, 8, 224, 735]) showed to be robust for different classes of spacetimes. The most widespread variant was given in [620, 54] and is commonly referred to as the BSSN (or BSSNOK) formulation.

The conformal, trace-free reformulations of the 3+1 equations (2.18) and (2.23) make use of the conformal decomposition of the 3-metric introduced in Section 6.2.3, and of the trace-free part of the extrinsic curvature. Here, we follow closely the presentation of [54]. As in Eq. (6.72), the conformal three-metric $\tilde{\gamma}_{ab}$ is defined by

$$\tilde{\gamma}_{ab} = e^{-4\psi}\gamma_{ab}, \tag{10.11}$$

with the conformal factor chosen to satisfy $\tilde{\gamma} = {}^3\eta$; in Cartesian coordinates, $\tilde{\gamma} = 1$, and we have

$$e^{4\psi} = \gamma^{1/3}. \tag{10.12}$$

The trace-free part of the extrinsic curvature K_{ab}, defined by

$$A_{ab} = K_{ab} - \frac{1}{3}\gamma_{ab}K, \tag{10.13}$$

is assigned the conformal weight

$$\tilde{A}_{ab} = e^{-4\psi}A_{ab}. \tag{10.14}$$

Indices of tensors labeled by tildes are lowered and raised by $\tilde{\gamma}_{ab}$ and its inverse metric $\tilde{\gamma}^{ab}$.

The evolution equations for the conformal 3-metric $\tilde{\gamma}_{ab}$ and its related conformal factor ϕ are trivially written as

$$\frac{d}{dt}\tilde{\gamma}_{ab} = -2\alpha\tilde{A}_{ab}, \tag{10.15}$$

$$\frac{d}{dt}\psi = -\frac{1}{6}\alpha K, \tag{10.16}$$

whereas the trace of Eq. (2.23) gives the evolution equation for K,

$$\frac{d}{dt}K = -\gamma^{ab}D_aD_b\alpha + \alpha\left[\tilde{A}_{ab}\tilde{A}^{ab} + \frac{1}{3}K^2 + 4\pi\left(\rho_E + S_c{}^c\right)\right], \tag{10.17}$$

where $d/dt = \partial_t - \mathcal{L}_{\boldsymbol{\beta}}$, and the Hamiltonian constraint was used to eliminate the Ricci scalar. Several possibilities exist for constructing an evolution equation for the trace-free extrinsic curvature $\tilde{A}_{ab}$, such as

$$\frac{d}{dt}\tilde{A}_{ab} = e^{-4\psi}\left[-D_aD_b\alpha + \alpha\left(R_{ab} - 8\pi S_{ab}\right)\right]^{TF} \tag{10.18}$$

$$+\alpha\left(K\tilde{A}_{ab} - 2\tilde{A}_{ac}\tilde{A}^c_b\right), \tag{10.19}$$

which is again obtained from Eq. (2.23) (*TF* denotes the trace-free part of an expression). There are many ways to write several of the terms in the above equation. For example, one could eliminate the Ricci scalar R through the use of the Hamiltonian constraint.

We have seen that the Ricci tensor has the decomposition (6.83), $R_{ab} = \tilde{R}_{ab} + R^{\psi}_{ab}$, where the "conformal" part $\tilde{R}_{ab}$ is the Ricci tensor of the conformal metric $\tilde{\gamma}_{ab}$, whereas the "conformal-factor" part R^{ψ}_{ab}, given by Eq. (6.87), involves only derivatives of ψ. To proceed, one defines the "conformal connection functions"

$$\tilde{\Gamma}^i := \tilde{\gamma}^{jk}\tilde{\Gamma}^i_{jk} = -\tilde{\gamma}^{ij}{}_{,j}, \tag{10.20}$$

where the $\tilde{\Gamma}^i{}_{jk}$ are the Christoffel symbols of the conformal metric, and the last equality holds if $\tilde{\gamma} = 1$. The Ricci tensor then becomes

$$\begin{aligned}\tilde{R}_{ij} = &-\frac{1}{2}\tilde{\gamma}^{lm}\tilde{\gamma}_{ij,lm} + \tilde{\gamma}_{k(i}\partial_{j)}\tilde{\Gamma}^k + \tilde{\Gamma}^k\tilde{\Gamma}_{(ij)k} \\ &+\tilde{\gamma}^{lm}\left(2\tilde{\Gamma}^k_{l(i}\tilde{\Gamma}_{j)km} + \tilde{\Gamma}^k_{im}\tilde{\Gamma}_{klj}\right).\end{aligned} \tag{10.21}$$

If the $\tilde{\Gamma}^i$ are promoted to independent variables, then the expression for the Ricci tensor retains its elliptic character, bringing the $\tilde{\gamma}_{ab}, \tilde{A}_{ab}$ system a step closer to being hyperbolic. The price is that one must now evolve three additional quantities. A suitable form of the additional evolution equations is

$$\begin{aligned}\frac{\partial}{\partial t}\tilde{\Gamma}^i = &-2\tilde{A}^{ij}\alpha_{,j} + 2\alpha\Big(\tilde{\Gamma}^i_{jk}\tilde{A}^{kj} \\ &-\frac{2}{3}\tilde{\gamma}^{ij}K_{,j} - 8\pi\tilde{\gamma}^{ij}S_j + 6\tilde{A}^{ij}\psi_{,j}\Big) \\ &-\frac{1}{3}\tilde{\gamma}^{li}\beta^l{}_{,lj} + \tilde{\gamma}^{lj}\beta^i{}_{,lj},\end{aligned} \tag{10.22}$$

where a divergence of $\tilde{A}^{ab}$ was eliminated with the help of the momentum constraint.

With this reformulation, in addition to the evolution equations (10.15) and (10.19) for the conformal 3-metric $\tilde{\gamma}_{ab}$ and the conformal-traceless extrinsic curvature $\tilde{A}_{ab}$, there are evolution equations (10.16), (10.17), and (10.22), for the conformal factor ψ, the trace of the extrinsic curvature K, and the $\tilde{\Gamma}^i$s. This evolution system for $\{\psi, K, \tilde{\gamma}_{ab}, \tilde{A}_{ab}, \tilde{\Gamma}^i\}$ is first order in time and second order in space, but a first-order hyperbolic version (by reduction of its spatial derivatives) turns out to be strongly hyperbolic. In [9, 8] the properties of this conformal traceless formulation of the Einstein equations were compared to those of the original ADM system. In particular, in [9] a number of strongly gravitating systems, such as strong gravitational waves, black holes, boson stars, and relativistic stars, were considered. The comparison showed that the BSSN scheme is superior in terms of stability in long-term evolutions. Further improvements are achieved through *adjusted* BSSN schemes (see, e.g., [353]).

Gauge choices. A robust choice for a class of slicing conditions are hyperbolic K-driver conditions, of the form

$$(\partial_t - \beta^i\partial_i)\alpha = -f(\alpha)\,\alpha^2(K - K_0), \tag{10.23}$$

with $f(\alpha) > 0$ and $K_0 := K(t=0)$ (see [78, 11, 41]) . This is a generalization of many well-known slicing conditions. By setting $f = 1$, one recovers the "harmonic" slicing condition, whereas setting $f = q/\alpha$, with q an integer, yields the generalized "1+log" condition [78]. A computationally efficient choice is $f = 2/\alpha$.

For the spatial gauge, in [41] the "Gamma-driver" shift conditions were used [11, 10], which essentially drive the $\tilde{\Gamma}^i$ to constant values. In this respect, the "Gamma-driver" shift conditions are similar to the "Gamma-freezing" condition $\partial_t \tilde{\Gamma}^k = 0$, which, in turn, is closely related to the well-known minimal distortion shift condition [636]. The differences between these two conditions involve Christoffel symbols and are related to the fact that the minimal distortion condition is covariant, whereas the Gamma-freezing condition is not. In [41], the hyperbolic Gamma-driver condition,

$$\partial_t^2 \beta^i = F\,\partial_t \tilde{\Gamma}^i - \eta\,\partial_t \beta^i, \tag{10.24}$$

where F and η are, in general, positive functions of space and time is used. For the hyperbolic Gamma-driver conditions it is crucial to add a dissipation term with coefficient η to avoid strong oscillations in the shift. Experience has shown that by tuning the value of this dissipation coefficient, it is possible to almost freeze the evolution of the system at late times. In [41], constant values $F = \frac{3}{4}$ and $\eta = 3$ were chosen.

Generalized harmonic coordinates (GHC)

An entirely different approach to the BSSN formulation and its 3+1 framework is an evolution system based on *generalized harmonic coordinates* (GHC) with constraint damping, in which each coordinate x^μ satisfies a scalar wave equation of the form

$$H^\mu = \Box x^\mu = \frac{1}{\sqrt{|g|}}\partial_\nu(\sqrt{|g|}g^{\mu\nu}) =: -\Gamma^\mu, \tag{10.25}$$

where the H^μ are specified functions on spacetime, or are specified functions of the coordinates and the metric, or are obtained from initial data by hyperbolic equations of their own (see [244, 254, 546] and references therein).[2]

Define constraint functions C_μ by

$$C_\mu := g_{\mu\nu}(H^\nu + \Gamma^\nu), \tag{10.26}$$

and let C^α and H^α be the vectors whose components with respect to a fixed set of coordinates x^μ are C^μ and H^μ. The coordinates are generalized harmonic only after one has found a metric that satisfies the gauge constraints $C_\mu = 0$.

In numerical evolutions, one enforces these by adding to the Einstein equation a constraint-damping term. To understand how this works, it is helpful to write the Ricci tensor as a sum of two expressions: a hyperbolic expression that remains

[2] Given a solution to the Einstein equation, one can regard *any* coordinates as generalized harmonic coordinates for some H^μ: That is, one can simply use Eq. (10.25) to define H^μ. In practice H^μ (or an equation determining it) is independently specified, and that specification fixes the coordinates.

when the gauge constraints $C_\mu = 0$ are satisfied, and the additional term $\nabla_{(\mu}C_{\nu)}$, involving second derivatives of the metric, that is nonvanishing when $C_\mu \neq 0$:

$$R_{\mu\nu} \equiv -\frac{1}{2}g^{\sigma\tau}\partial_\sigma\partial_\tau g_{\mu\nu} - \partial_\sigma g_{\tau(\mu}\partial_{\nu)}g^{\sigma\tau} - \nabla_{(\mu}H_{\nu)} - \Gamma^\sigma_{\nu\tau}\Gamma^\tau_{\mu\sigma} \\ + \nabla_{(\mu}C_{\nu)}. \tag{10.27}$$

Then, when the gauge constraints are satisfied, the components of the Einstein equation, $R_{\mu\nu} - 8\pi\left(T_{\mu\nu} - \frac{1}{2}g_{\mu\nu}T\right) = 0$, have the form

$$\frac{1}{2}g^{\sigma\tau}\partial_\sigma\partial_\tau g_{\mu\nu} + \partial_\sigma g_{\tau(\mu}\partial_{\nu)}g^{\sigma\tau} + \nabla_{(\mu}H_{\nu)} + \Gamma^\sigma_{\nu\tau}\Gamma^\tau_{\mu\sigma} = -8\pi\left(T_{\mu\nu} - \frac{1}{2}g_{\mu\nu}T\right). \tag{10.28}$$

On the other hand, if Eq. (10.28) is satisfied but the gauge constraints are not exactly enforced, then the identity (10.27) implies $R_{\alpha\beta} - 8\pi(T_{\alpha\beta} - \frac{1}{2}g_{\alpha\beta}T) = \nabla_{(\alpha}C_{\beta)}$, or

$$E_{\alpha\beta} := G_{\alpha\beta} - 8\pi T_{\alpha\beta} = \nabla_{(\alpha}C_{\beta)} - \frac{1}{2}g_{\alpha\beta}\nabla_\gamma C^\gamma. \tag{10.29}$$

This equation implies a time evolution of the gauge constraints, and it is an evolution in which constraint violations grow in time.

To avoid this outcome, one adds to Eq. (10.29) artificial terms – the terms in square brackets below – that damp the growth of the constraints; that is, one numerically solves the equations

$$\frac{1}{2}g^{\sigma\tau}\partial_\sigma\partial_\tau g_{\mu\nu} + \partial_\sigma g_{\tau(\mu}\partial_{\nu)}g^{\sigma\tau} + \nabla_{(\mu}H_{\nu)} + \Gamma_\tau\Gamma^\tau_{\mu\nu} + \Gamma^\sigma_{\nu\tau}\Gamma^\tau_{\mu\sigma} \\ = -8\pi\left(T_{\mu\nu} - \frac{1}{2}g_{\mu\nu}T\right) + \kappa\left[n_{(\mu}C_{\nu)} - \frac{1}{2}g_{\mu\nu}n^\tau C_\tau\right], \tag{10.30}$$

where κ is an adjustable parameter controlling the strength of constraint damping. When Eq. (10.30) is satisfied, $E_{\alpha\beta}$ is given by

$$E_{\alpha\beta} = \nabla_{(\alpha}C_{\beta)} - \frac{1}{2}g_{\alpha\beta}\nabla_\gamma C^\gamma - \kappa n_{(\alpha}C_{\beta)}. \tag{10.31}$$

The Bianchi identity, together with the equation of motion $\nabla_\beta T^{\alpha\beta} = 0$, gives $\nabla_\beta E^{\alpha\beta} = 0$; the right side of Eq. (10.31) then has vanishing divergence, implying that C^α satisfies the hyperbolic equation

$$\nabla^\beta\nabla_\beta C^\alpha + R^\alpha{}_\beta C^\beta = 2\kappa\nabla_\beta\left[n^{(\beta}C^{\alpha)}\right], \tag{10.32}$$

where the Ricci identity $[\nabla_\beta, \nabla_\alpha]C^\beta = R_{\alpha\beta}C^\beta$ was used. By enforcing the constraint $C^\alpha = 0$, Eq. (10.32), together with the evolution equation (10.30), implies the Einstein equation $E_{\alpha\beta} = 0$ and thus, in particular, the vector constraint $E_{\alpha\beta}n^\beta = 0$ (the Hamiltonian and momentum constraints of the 3+1 decomposition).

In Pretorius's vacuum code [545], H_t is required to satisfy a hyperbolic equation of its own. When H_α is projected along and normal to n^α, one obtains expressions for the gauge constraints in terms of the lapse α and shift β^a – that is, in terms of the gauge degrees of freedom of the 3+1 formalism. In terms of the 3+1 variables introduced in Section 2.1, the decomposition of $C_\alpha = 0$ is given by

$$H_\mu n^\mu = -K - n^\mu \partial_\mu (\ln \alpha), \tag{10.33a}$$

$$H_\mu \gamma^{\mu i} = -\bar{\Gamma}^i_{jk}\gamma^{jk} + \partial_j (\ln \alpha)\gamma^{ij} + \frac{1}{\alpha} n^\mu \partial_\mu \beta^i, \tag{10.33b}$$

where the $\bar{\Gamma}^i_{jk}$ are Christoffel symbols associated with the spatial metric γ_{ab}. Hence, the projection $H_\alpha n^\alpha$ is related to the time derivative of the lapse function, whereas the projection $H_\alpha \gamma^{\alpha a}$ is related to the time derivative of the shift vector.

The evolution system is then closed by choosing evolution equations for the source functions H_μ:

$$\Box H_t = -\xi_1 \frac{\alpha - 1}{\alpha^\eta} + \xi_2 H_{t,\alpha} n^\alpha, \tag{10.34}$$

$$H_i = 0, \tag{10.35}$$

where $\Box$ is again the scalar-wave operator. Notice that (10.34) is a damped wave equation, in which the parameter ξ_2 controls the damping, while the two parameters ξ_1 and η in the forcing term are designed to prevent the lapse from deviating too far from its flat-space value (which is helpful for the evolution of a black hole interior).

The second-order equation (10.32) can further be reduced to a first-order harmonic system in order to apply a variety of numerical methods; see, for example, [430, 674].

Other formulations

A new formulation of the Einstein equations based on a conformal and traceless decomposition of the covariant form of the Z4 system, which combines the advantages of a conformal decomposition (such as the one used in the BSSN formulation) with the advantages of a constraint-damped formulation (such as the generalized harmonic one), was presented by Alic et al. [14].

Bonazolla et al. [85] presented a constrained scheme in a generalization of the Dirac gauge to spherical coordinates, in which the ten Einstein equations are reduced to a system of five quasilinear elliptic equations (including the Hamiltonian and momentum constraints) coupled to two quasilinear scalar wave equations.

10.2 Stable equilibrium

Full 3+1 codes used to model stellar oscillations of rotating stars are initially tested for accuracy in modeling the stable equilibria themselves. Initial data for equilibrium models of stationary, axisymmetric stars can be prepared in the quasi-isotropic form (1.59) and (1.66) of the metric and can then be transformed to the (typically Cartesian) coordinates that underly the grid of the 3+1 code. The translation from

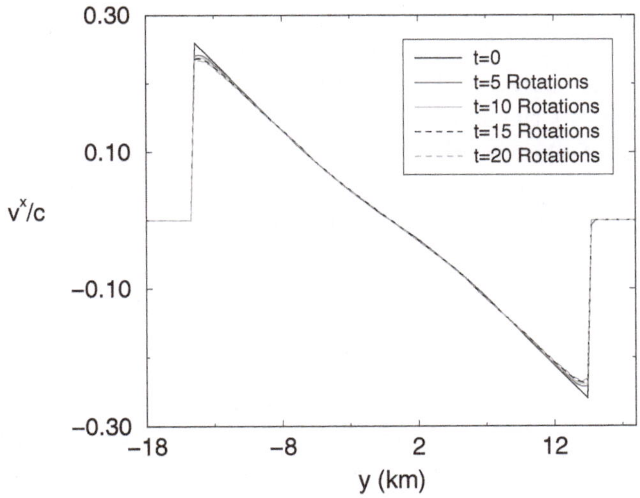

Figure 10.1. Time evolution of the rotational velocity profile for a stationary, rapidly rotating relativistic star (in a fixed spacetime evolution), using an HRSC scheme with third-order PPM reconstruction and a Cartesian grid with 116^3 grid points. The initial rotational profile is preserved to high accuracy, even after 20 rotational periods. (Figure from [663].)

the quasi-isotropic metric potentials of Eq. (1.59) to initial data associated with the 3+1 form (2.10) has the following properties:

- The shift vector has the single nonvanishing component, $\beta^\phi = -\omega$, where ω is the metric function in Eq. (1.59) describing the dragging of inertial frames by rotation. This shift vector satisfies the *minimal distortion shift* condition [636].
- The lapse function is $\alpha = e^\nu$, where ν is the metric function appearing in the g_{tt} component of (1.59) and satisfies the *maximal slicing* condition, $K = 0$ [636].
- The Eulerian observers defined in the 3+1 split, whose worldlines are normal to the $t =$ const. hypersurfaces, coincide with the zero–angular-momentum-observers (ZAMOs) defined in Sec. 1.2 for stationary, axisymmetric spacetimes.
- In evolutions of initial data close to an equilibrium configuration, the frequencies of discrete quasinormal modes can be seen as peaks in the spectrum of evolved variables; the spectrum is obtained by a Fourier transform with respect to coordinate time t at a fixed coordinate location [225].

The long-term stable evolution of rotating relativistic stars in 2- and 3-dimensional simulations has become possible through the use of HRSC methods (see [222] for a review). The method of choice uses a third-order PPM reconstruction [149], which was shown to be superior to other methods in accurately preserving a sharp rotational profile in long-term simulations [225, 663]. Figure 10.1 shows the rotational profile of a rapidly rotating, relativistic star in equilibrium at up to 20 rotational periods, when simulated with the third-order PPM reconstruction (in a fixed spacetime evolution) [663]. Throughout the star, the profile is maintained to high accuracy, except near the surface, where the sharp discontinuity of the rotational velocity is somewhat smoothed out. Notice that in such simulations, one must use a thin atmosphere (with density many orders of magnitude smaller than the central density of the star) in order to be able to apply the HRSC methods efficiently in the whole numerical grid (which also includes a region where there should be vacuum). Two- and 3-dimensional, fully coupled hydrodynamical and spacetime simulations of rotating relativistic stars in equilibrium were obtained in [608, 224, 197, 610].

10.3 Nonlinear pulsations

Pulsations of rotating relativistic stars are traditionally studied, when possible, as a time-independent, linear eigenvalue problem, but numerical relativity also allows the study of such pulsations via numerical simulations. An advantage of evolving the full, nonlinear system of equations in time is that one is not limited to linear pulsations; also, as we noted earlier, numerical evolutions have been an efficient way to find the linear modes.

Nonlinear couplings. The most fundamental nonlinear effect one can observe in a simulation is the appearance of *combination frequencies* of the linear pulsation modes, which is a general property of nonlinear systems [385].

To lowest order, these frequencies arise as linear sums and differences of different linear modes, including self-couplings. If the system has eigenfrequencies ω_i, the nonlinearity of the equations excites modes at frequencies $\omega_i \pm \omega_j$, with amplitudes proportional to the product of the amplitudes of the combining frequencies. This general behavior of a nonlinear system is not difficult to understand: It is due to the fact that, in a perturbative expansion, the second-order equations have a part quadratic in the first-order perturbation. Consider, for example, a family of solutions $\{v^a(\lambda), h_N(\lambda), \Phi(\lambda)\}$ to the Newtonian Euler equation

$$\partial_t v_a + v^b \nabla_b v_a + \nabla_a (h_N + \Phi) = 0,$$

with the linear perturbation δv^a, for example, denoted by $\overset{1}{v}{}^a$, and higher-order terms written as

$$v^a(\lambda) = \overset{0}{v}{}^a + \lambda \overset{1}{v}{}^a + \lambda^2 \overset{2}{v}{}^a + \cdots = v^a(0) + \lambda \frac{d}{d\lambda} v^a \bigg|_{\lambda=0} + \frac{1}{2} \lambda^2 \frac{d^2}{d\lambda^2} v^a \bigg|_{\lambda=0} + \cdots .$$

Then the linear perturbation, $\{\overset{1}{v}{}^a, \overset{1}{h}_N, \overset{1}{\Phi}\}$, satisfies

$$\partial_t \overset{1}{v}_a + \overset{1}{v}{}^b \nabla_b \overset{0}{v}_a + \overset{0}{v}{}^b \nabla_b \overset{1}{v}_a + \nabla_a (\overset{1}{h}_N + \overset{1}{\Phi}) = 0,$$

whereas the second-order perturbation satisfies an inhomogeneous equation whose source is the term $\overset{1}{v}{}^b \nabla_b \overset{1}{v}_a$, quadratic in the first-order velocity field:

$$\partial_t \overset{2}{v}_a + \overset{2}{v}{}^b \nabla_b \overset{0}{v}_a + \overset{0}{v}{}^b \nabla_b \overset{2}{v}_a + \nabla_a (\overset{2}{h}_N + \overset{2}{\Phi}) = -\overset{1}{v}{}^b \nabla_b \overset{1}{v}_a. \tag{10.36}$$

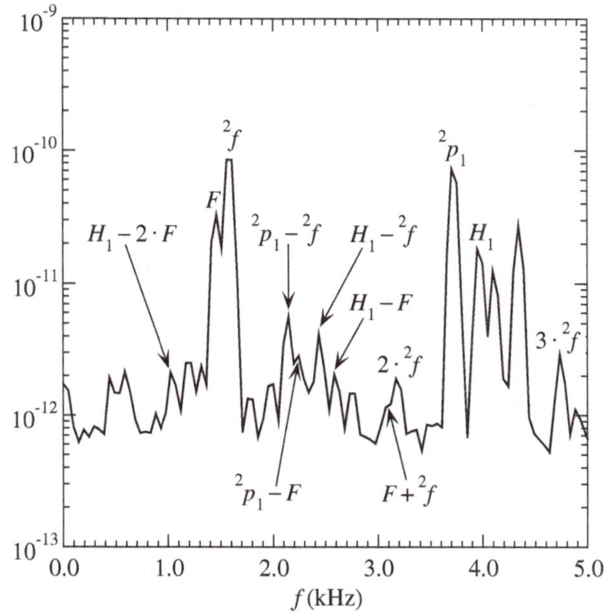

Figure 10.2. Fourier transform of the time evolution of the rest-mass density at a certain point in a nonrotating, relativistic star, which was perturbed by adding both $l = 0$ and $l = 2$ perturbations to the equilibrium initial data. Several linear modes (F, H_1, ${}^2f$, ${}^2p_1$) as well as several nonlinear combination frequencies ($2 \cdot {}^2f$, $3 \cdot {}^2f$, $H_1 - F$, ${}^2p_1 - F$, etc.) can be identified. (Figure from [193]. Reproduced by permission of John Wiley and Sons.)

With $\overset{1}{v}{}^a$ a sum of modes with harmonic time dependence $\cos(\omega_i + \phi_i)t$, the quadratic term is a sum of terms with time dependence $\cos(\omega_i \pm \omega_j + \phi_{ij})t$, for some constant phases ϕ_i and ϕ_{ij}. Because the left side of Eq. (10.36) is a time-independent linear operator acting on the second-order perturbation, it has a solution that is again a sum of terms with time dependence $\cos(\omega_i \pm \omega_j)t$. Nonlinear combination frequencies can be most easily identified in simulations of nonrotating stars, where the additional couplings due to rotation are absent. Figure 10.2 shows the Fourier transform of the time evolution of the rest-mass density at a certain point in a nonrotating, relativistic star, perturbed by adding both $l = 0$ and $l = 2$ perturbations to the equilibrium initial data [193]. Knowing the frequencies of the linear radial F- and H_1-modes and of the nonradial ${}^2f$- and ${}^2p_1$-modes, several combination frequencies can be identified, such as the self-couplings $2 \cdot {}^2f$ and $3 \cdot {}^2f$ and the linear sums $H_1 - F$ and $H_1 - 2 \cdot F$. Other examples are combinations such as ${}^2p_1 - F$, $H_1 - {}^2f$, and so on. A large number of additional peaks can be seen in Fig. 10.2, and essentially all of those should correspond to combination frequencies of the excited linear modes.

It is interesting to note that several of the combination frequencies seen here have frequencies much smaller than the fundamental radial mode F (which, in the linear approximation, has the lowest frequency). These can fall into the frequency range of the inertial modes for rotating models. Thus, in rotating models, further nonlinear

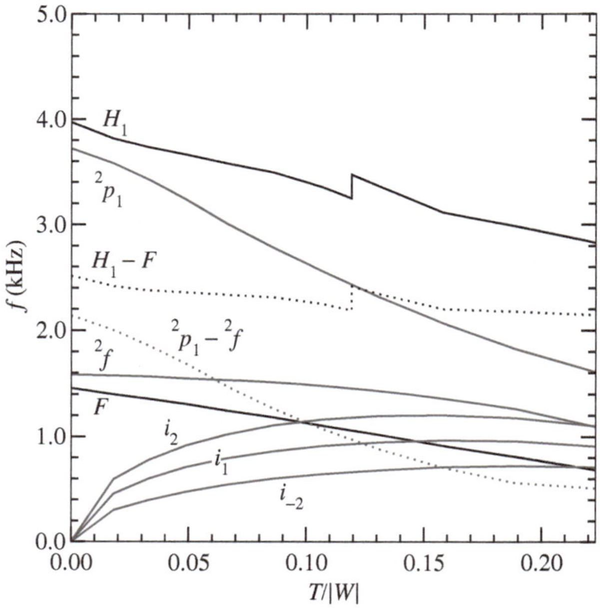

Figure 10.3. Frequencies of various linear modes and the two nonlinear combination frequencies $H_1 - F$ and ${}^2p_1 - {}^2f$ for sequence A. At specific rotation rates the frequencies of linear and nonlinear modes coincide, which can lead to nonlinear 3-mode resonance. (Figure from [193]. Reproduced by permission of John Wiley and Sons.)

interactions between radial modes and inertial modes can be expected. Identifying nonlinear combination frequencies in simulations of rotating models is more cumbersome, but one can at least distinguish them from the linear modes by comparing simulations performed with different initial amplitudes of the perturbations.

Nonlinear 3-mode resonances. The presence of nonlinear combination frequencies opens up the possibility of 3-mode resonances when the star is rotating. The reason such 3-mode resonances can take place is the fact that the effect of rotation on the different modes varies. As was discussed in Chapter 8, higher-order modes are typically affected more strongly by rotation than are lower-order modes, which results in avoided crossings between mode sequences. Rotation also influences the frequencies of the various combination frequencies to a different degree. Thus, at certain rotation rates, a nonlinear combination frequency can have the same frequency as a linear mode. Examples of such cases are demonstrated in Fig. 10.3, which shows the frequencies of the linear modes ($F, H_1, {}^2f, {}^2p_1$ and several axisymmetric inertial modes, i_2, i_1, i_{-2}) and of the two nonlinear combination frequencies ${}^2p_1 - {}^2f$ and $H_1 - F$, as a function of the rotational parameter $T/|W|$ [193]. The ${}^2p_1 - {}^2f$ combination frequency is crossing the frequency of the fundamental quasiradial F-mode at $T/|W| \sim 0.1$, whereas the $H_1 - F$ combination frequency is crossing the frequency of the ${}^2p_1$-mode at $T/|W| \sim 0.12$. We also note that at any rotation rate one can expect several nonlinear combination frequencies to coincide in frequency with some of the infinitely many inertial modes that exist in the inertial mode range. At such crossings, resonance effects could potentially take place, and even parametric instabilities could set in. Significant amounts of energy from one mode could be transferred to other modes.

In nonrotating or slowly rotating models, a parametric instability can take place only under special conditions that would allow the two modes to be in resonance. In contrast, in rapidly rotating models, rotational shifting of the frequency of different modes broadens the range of parameters for which interesting resonances could take place. The most interesting case of nonlinear couplings would be if pulsational energy from the quasiradial modes, which radiate only weakly in gravitational waves, could be transferred nonlinearly to more strongly radiating nonradial modes. Because during core bounce a significant amount of kinetic energy is stored in the radial modes of pulsation, the transfer of even a small percentage of this kinetic energy to a nonradial mode could result in the emission of strong gravitational waves. This scenario was first studied in [143] and [202], where possible parametric instabilities were investigated using a Newtonian, slowly rotating collapse model (for related work, see [530, 534]).

Nonlinear energy transfer between modes has also been demonstrated to occur during the nonlinear development of the r-mode instability (see next section) and during the minicollapse in rotating stars encountering a phase transition in the EOS during spin-down [1, 188].

As is detailed in the next section, 3-mode couplings and parametric resonances also play a significant role in the saturation of nonlinear dynamical and secular instabilities. In addition, it was shown in [662] that a particular nonlinear combination frequency (a coupling between the quasiradial and quadrupole oscillations) could be identified in the gravitational wave spectrum produced in the postmerger phase of binary neutron star mergers.

10.4 Nonlinear development of rotational instabilities

We review in the following subsections work on the nonlinear evolution of secular and dynamical instabilities introduced in Chapter 8. The exponential growth of small-amplitude unstable modes can initially be described perturbatively in terms of a corresponding growth in nonlinear couplings, like the quadratic term on the right of Eq. (10.36), that drive other modes – *daughter* modes. The subsequent behavior depends on the amplitude of the daughter modes when the rate at which energy flows to them approaches the growth rate of the original linear mode. Mode saturation is an energy balance with a mode amplitude small compared to unity (measured, for example, by $\delta p/p$ or $\delta v/v$). If the energy loss to other modes is weak enough to allow a growth to unit amplitude, the coherence of the dominant mode may be destroyed, with other modes and/or a turbulent flow persisting.

Modes most relevant for saturation are typically those whose frequencies are nearly in resonance: For a 3-mode interaction, this occurs when a quadratic driving term has nearly the same frequency as that of the mode it drives. With a suitable choice of signs, this is a requirement that $\omega_1 + \omega_2 + \omega_3$ not be large compared to $1/\tau_v$, with τ_v the viscous damping time of the daughter modes. At a critical amplitude of the parent mode, parametric instability occurs, with the daughter modes themselves driven unstable by quadratic terms that couple each to the parent mode.

10.4.1 Dynamical instabilities in rapidly rotating stars

The classical, nonaxisymmetric dynamical instabilities can occur in very rapidly rotating stars, with $T/|W| \gtrsim 0.23$–0.25. Depending on the compactness of the star, either the $m = 2$ (bar) mode or the $m = 3$ mode is the fastest growing. The instability initially develops exponentially, but soon the amplitude enters the nonlinear regime, so that fully nonlinear simulations are required to determine its outcome. The development of the instability has been studied in several Newtonian studies (e.g., [320, 696, 200, 319, 637, 555, 102, 495, 435]), post-Newtonian studies [581], and fully relativistic studies [611, 619, 40, 455].

In the fully relativistic study of [40], a number of differentially rotating models were considered, having a polytropic EOS with $\Gamma = 2$ and and masses between $1M_\odot$ and $2.5M_\odot$, while (1.125) was used as the rotation law. For these models, the threshold for the onset of the $m = 2$ instability varied only slightly between $T/|W| \sim 0.25$ and $T/|W| \sim 0.26$, whereas the $m = 3$ mode dominated only for models with masses $M \gtrsim 2.0M_\odot$ and $T/|W|$ between about 0.23 and 0.25.

Simulations of rotating core collapse indicate that even the most rapidly rotating progenitor cores have difficulty acquiring enough rotational energy for the ensuing proto–neutron star to be dynamically unstable (see [505] for a discussion and references). It is, however, useful to understand the saturation mechanism of this instability (through nonlinear couplings), because the same (or a similar) saturation mechanism may also be relevant for other instabilities that are more likely to occur (Section 10.4.2).

The $m = 2$ instability. As was demonstrated in [40, 455], the mechanism by which the $m = 2$ instability saturates and is subsequently suppressed depends critically on both the mass and rotation rate of the star.

- *Low-mass stars with $T/|W|$ only somewhat larger than critical.* For low-mass stars ($M \sim 1.5M_\odot$), with $T/|W|$ only somewhat larger than about 0.23 to 0.25, both the $m = 2$ and $m = 3$ modes initially grow exponentially, but the $m = 3$ mode has a slow growth rate. As the two modes grow, an $m = 1$ mode emerges from nonlinear coupling and grows exponentially. After several dynamical timescales, the $m = 2$ mode saturates at a maximum amplitude, after which its amplitude starts decreasing, as other modes continue to grow. At maximum amplitude, the star may form spiral arms. After several more dynamical timescales, the $m = 1$ mode overtakes first the $m = 3$ mode and then the

$m = 2$ mode. After that critical point, the amplitude of the $m = 2$ mode falls significantly, leaving a now-dominant $m = 1$ that saturates at a maximum amplitude. For stars with $T/|W|$ only slightly larger than critical, the timescale on which the $m = 1$ mode dominates is of order several tens of milliseconds. For stars with somewhat larger $T/|W|$, this timescale is reduced to about 10 ms.

- *Low-mass stars with $T/|W|$ appreciably larger than critical.* When $T/|W|$ is as large as 0.28 in stars of mass $\sim 1.5M_\odot$, the instability again grows exponentially and saturates within a few dynamical timescales. Now, however, other modes grow too slowly to overtake the $m = 2$ mode. Instead, its amplitude is reduced significantly by a rapid conversion of rotational energy into internal energy, over several tens of dynamical periods. This conversion can take place when the spiral arms expand supersonically and merge together, creating a surrounding disk where kinetic energy is dissipated by shock heating [320].
- *Massive stars.* When $M \sim 2.0M_\odot$, the amplitude of the $m = 2$ mode is rapidly reduced after saturation, even for stars very near the critical threshold for onset of the instability. In [455] this is attributed to a combination of the strength of the gravitational field and the conversion of rotational energy into internal energy.

In Newtonian simulations of the dynamical bar-mode instability presented in [363], the growth of other nonaxisymmetric modes is attributed to 3-mode resonances. In addition, vertical oscillations, with frequency twice as large as the frequency of the $m = 2$ mode, were observed and attributed to a parametric resonance [580].

The $m = 3$ instability. In [455], stars with masses $M \gtrsim 2.0M_\odot$ and with $T/|W|$ between about 0.23 and 0.25 possessed a dominant $m = 3$ mode that becomes exponentially unstable until saturation. Because of nonlinear couplings, other modes also grow. In these simulations, it appeared that the $m = 3$ instability would also be subsequently overtaken by the $m = 1$ mode, which would result in a significant reduction of its amplitude. The $m = 3$ instability has also been observed in Newtonian core-collapse simulations that produced a remnant with $T/|W| \gtrsim 0.27$ (see, e.g., [555], although in that case a 5% $m = 3$ perturbation was added to the initial data).

10.4.2 Shear instabilities in differentially rotating stars

In the nonlinear evolution of the shear (or low-$T/|W|$) instabilities discussed in Chapter 9, one observes that when the dominant unstable mode enters the nonlinear regime, it induces a growth of other nonaxisymmetric patterns. For example, the $m = 1$ instability can cause an accompanying exponential growth of $m = 2$ oscillations, which saturate at smaller amplitudes than the $m = 1$ mode itself. If more than one mode is unstable, then the nonlinear evolution has common features with what was already mentioned in Section 10.4.1 regarding the classical dynamical instabilities, that is, the dominant mode grows first and saturates, but, at the same time, the other unstable modes grow exponentially, with a smaller growth rate [510, 163]. At some point, the amplitude of the slower-growing modes can overtake the amplitude of the initially dominant one. For example, a star can initially feature a dominant $m = 2$ deformation, but at late times it can be dominated by an $m = 1$ (one-armed) mode. The gravitational-wave signal emitted by such a star will be thus nonmonochromatic and will include the imprint of the different dominant modes at different times.

Another interesting aspect of the nonlinear development of shear instabilities has been studied using Newtonian simulations in [122]. An investigation of the evolution of vorticity in these simulations suggests that the dominant unstable mode saturates at a nonlinear amplitude because of pressure waves carrying angular momentum outside the corotation radius. This seems to lead to a turbulent flow in the region outside the corotation radius, which is attributed to the onset of a Kelvin-Helmholtz shear instability.

Three-dimensional simulations by Ott et al. [509] demonstrated the growth of the shear instability in PNSs formed after rotating core collapse, having $T/|W| \sim 0.09$–0.13, whereas in [590] instabilities were found in a PNS with $T/|W| \sim 0.04$–0.05. In the simulations, the instability appears after tens of milliseconds following core bounce, and within about 10 ms it saturates at a nonlinear amplitude. The energy emitted in gravitational waves is roughly $10^{-7} M_\odot c^2$, with frequency at around 900 Hz. This is more than the gravitational-wave energy emitted due to oscillations excited at core bounce (see Section 10.7). Shear instabilities are thus promising for gravitational-wave detection, and it will be important to determine the parameter range in which they occur (dependence on rotation rate and differential rotation law) as well as their behavior in competition with other shear instabilities, such as the magnetorotational instability.

10.4.3 Secular instabilities

f-Mode instability. The f-mode gravitational-radiation (CFS) instability sets in only in rapidly rotating compact stars (see Chapter 8). The instability grows on a secular timescale that can be much longer than the rotational period of the star, but which becomes shorter as one considers faster initial rotation rates. The nonlinear evolution of the instability was initially studied under various simplifying assumptions. In the ellipsoidal approximation to rotating Newtonian stars (and adding a gravitational radiation reaction force in the evolution equations), Lai and Shapiro [384] obtained a complete description of the main phases of the nonlinear development of the instability. Starting from a differentially rotating star with a very large $T/|W| > 0.14$, they found that the f-mode saturates at a large, nonlinear amplitude. The star then slows down because of angular momentum loss due to

gravitational-wave emission and is driven toward becoming a Dedekind ellipsoid (see Chapter 8) that no longer radiates gravitational waves.

During this phase, the f-mode instability could become an excellent source of gravitational waves, depending on its nonlinear saturation amplitude. In the simple ellipsoidal approximation of [384], the nonaxisymmetric pattern radiates gravitational waves, sweeping through the sensitivity window of second-generation terrestrial interferometers, from 1 kHz down to about 100 Hz, with an amplitude that could become detectable out to a distance of more than 100 Mpc. In reality, this estimate will be reduced significantly, as the f-mode is expected to saturate at lower amplitudes, because of nonlinear couplings to other modes.

The nonlinear development of the f-mode instability has also been modeled in 3-dimensional, hydrodynamical simulations (in a Newtonian framework) by Ou, Tohline, and Lindblom [511] and by Shibata and Karino [613], essentially confirming the qualitative results obtained in [384]. In [511], the initial state was chosen to be a uniformly rotating, $N = 0.5$ polytropic model with $T/|W| \sim 0.18$. The development of the instability was accelerated by artificially increasing the gravitational-radiation–reaction force by a factor of several thousand. The star was driven to a Dedekind-like state, but after about ten dynamical timescales, its shape was destroyed by growing short-wavelength motions. The latter could be related to the onset of elliptic strain instabilities in stars with nonaxisymmetric flow patterns, and particularly in stars that resemble Riemann S-type ellipsoids [397, 400] (see [510] and discussion therein).

In [613] a differentially rotating, N = 1 polytropic model with large $T/|W| \sim 0.2$–0.26 was chosen as the initial equilibrium state. The development of the f-mode instability was again artificially accelerated to make the simulation feasible. The instability saturated at a nonlinear amplitude, and the star became ellipsoidal, with a moderately large ellipticity. No short-wavelength instability was observed at the end of the simulation. In addition, the model did not appear to settle into a Dedekind-like state, suggesting that the star could continue to radiate gravitational waves until reaching a stationary, axisymmetric shape. Clearly, a much more detailed study of the nonlinear development of the f-mode instability is necessary in order to determine the parameter ranges in which the different outcomes indicated in [511] and [613] are possible. In addition, it will be important to study the development of the instability on a realistic timescale, in order to allow nonlinear interactions with other, stable modes to take place. Such nonlinear interactions could set a much lower saturation amplitude than that observed in the simulations described above (see [346] for a first perturbative study of the f-mode saturation amplitude, done in the relativistic Cowling approximation). The simultaneous change of the background flow, due to magnetic braking of differential rotation, also needs to be taken into account.

r-Mode instability. The astrophysical significance of the gravitational-radiation–driven (CFS) r-mode instability depends crucially on its nonlinear development. First attempts to estimate the maximum saturation amplitude relied on hydrodynamical simulations, in which the r-mode was either set at an initial amplitude of order unity [663] or was first driven to large amplitude by a radiation-reaction term artificially amplified by a factor of several thousand [432, 433]. In the relativistic simulation of [663], the large-amplitude initial r-mode in a rapidly rotating polytropic neutron star model was evolved over several tens of hydrodynamical timescales, with only a small loss in amplitude.

This study demonstrated that r-modes are discrete in isentropic relativistic stars and that their nonlinear saturation does not occur on a hydrodynamical timescale. Saturation timescales longer than the dynamical (but shorter than the gravitational-radiation–reaction timescale) could not be investigated in [663], because of the severe computational requirements. In the Newtonian simulation of [432], the artificial acceleration of the gravitational-radiation reaction decreased the growth time of the r-mode in a rapidly rotating polytropic model from about 40 s to 10 ms, leading to saturation at a large amplitude of $O(1)$, due to surface shocks. Because the artificial acceleration was used throughout the simulation, nonlinear mode couplings occurring on timescales in between the dynamical timescale and the gravitational-radiation reaction timescale could also not be taken into account. The weak coupling between the r-mode and other inertial/hybrid modes on short, hydrodynamical timescales was explained in a second-order perturbative study by Schenk et al. [591], who found that such couplings vanish to zeroth order in the rotation rate.

Aiming at studying the nonlinear development of r-modes on timescales longer than the hydrodynamical one, Gressman et al. [279] followed an approach similar to that of [432, 433], using an accelerated gravitational-radiation reaction to drive the fundamental $l = m = 2$ r-mode oscillation to large amplitudes in a rapidly rotating, Newtonian polytropic model; however, the gravitational-radiation reaction was switched off before any surface shocks appeared. When the star was then evolved on a timescale longer than in [663], there was a sudden breakdown of the r-mode oscillation. The sudden breakdown indicated a transfer of energy to other modes, through nonlinear mode couplings, suggesting that the actual saturation amplitude of the r-mode was smaller than the amplitude at which gravitational-radiation reaction was switched off. The breakdown was also accompanied by a redistribution of the angular momentum of the (initially uniformly rotating star), leading to a differentially rotating background. The distribution of the r-mode amplitude density $\tilde{\alpha}(x)$ in the equatorial plane (see [279] for definition) at specific times before and after the breakdown is shown in Fig. 10.4.

The left panel of Fig. 10.5 shows the evolution of the r-mode amplitude during the accelerated phase (dashed line) and after gravitational-radiation reaction was switched off (solid lines). The smaller the amplitude at which gravitational-radiation reaction is switched off, the longer it takes for the breakdown to appear. Because the actual growth time of the unstable r-mode is much longer than the tens of milliseconds shown in Fig. 10.5, the actual saturation amplitude should be much smaller than $O(1)$. In the right panel of Fig. 10.5, the Fourier transform of the evolution of the velocity component v^z (one of the main velocity components

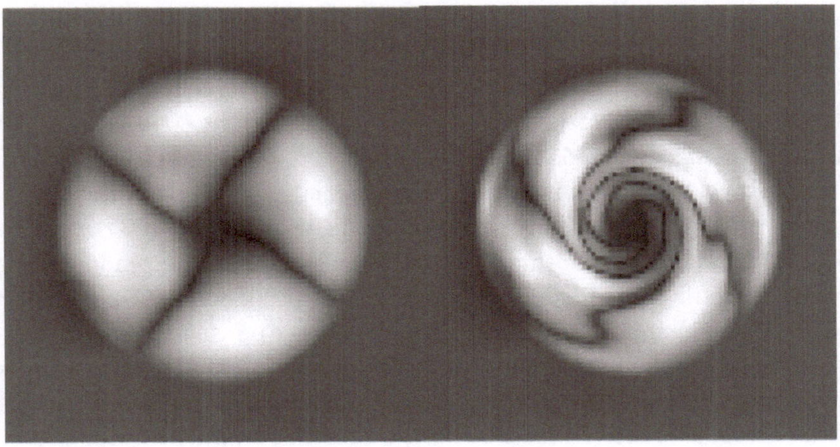

Figure 10.4. Breakdown of a large-amplitude r-mode oscillation, after it is artificially driven to large, nonlinear amplitudes. The amplitude density $\tilde{\alpha}(x)$ is shown in the equatorial plane before (left) and after (right) the breakdown. (Figure from [279].)

involved in the r-mode oscillation) at a point along the x-axis is shown at two different times, at early times (dashed line) and at late times, but before the breakdown (solid line). It is evident that although initially the main contribution to the Fourier transform consists only of the r-mode frequency (main peak), at late times several additional peaks are clearly noticeable. In particular, energy appears to be transferred to several nonlinear combination frequencies (self-couplings at integer

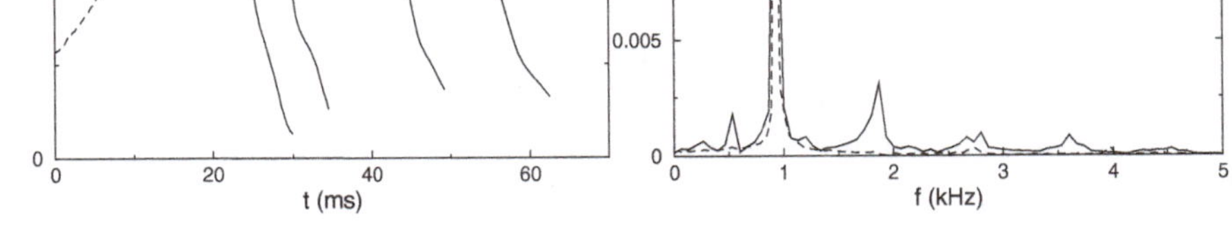

Figure 10.5. (Left) Evolution of the r-mode amplitude: an unstable mode is first allowed to grow artificially fast, until it reaches various maximum amplitudes (dashed line). The simulation then proceeds on a natural timescale (solid lines). The decrease in amplitude is due to the r-mode breaking down because it has exceeded its real saturation amplitude. (Right) FFT amplitude of the evolution of v^z at a point in the equatorial plane at an early time (dashed line), when it is dominated by the r-mode, and at a later time (solid line), when additional modes have appeared because of the r-mode breakdown. (Figure from [279].)

multiples of the frequency) as well as to several modes of smaller frequency, which can safely be assumed to be different inertial modes. That the breakdown of a large-amplitude r-mode can indeed be explained by 3-mode couplings (between the r-mode and two specific inertial modes) was confirmed in a second-order perturbative study by Brink, Teukolsky, and Wasserman [98] (see below) and by Lin and Suen in [416], where further analysis of the simulations in [279] provided strong evidence for a particular resonant 3-mode coupling of the $l = m = 2$ r-mode to two $m = -1$ inertial modes, whose sum of the frequencies equaled the r-mode frequency.

The existence of a small nonlinear saturation amplitude for the r-mode (of order 10^{-1} to 10^{-4}) was first demonstrated in a second-order perturbative approach that takes into account 3-mode couplings, by Arras et al. [37] (using a formalism developed initially by Schenk et al. [591] and Morsink [487]). In [37] it was also suggested that if the damping of inertial modes is small, then the saturation amplitude would be set by a steady-state cascade of energy to a large number of modes. Subsequent work [98, 97, 99] found that the lowest parametric instability threshold is set by the interaction of the r-mode with the $(n = 13, m = 3)$ and $(n = 14, m = 1)$ inertial modes, whereas the second-lowest parametric instability threshold involves the $(n = 15, m = 7)$ and $(n = 14, m = 5)$ inertial modes. When the r-mode is assumed to be interacting nonlinearly with a very large number of inertial modes (in nearly resonant triplets), the r-mode amplitude fluctuates with time, but remains below the amplitude set by the second-lowest parametric instability threshold, which is of order 10^{-4} (the amplitude set by the lowest parametric instability threshold is also of the same order).

This discussion of a saturation amplitude set by nonlinear couplings needs the caveat that, for a star spun up by accretion to the steady state described in Section 9.6.2, bulk viscosity from hyperon interactions could preempt nonlinear interactions in setting an upper limit on the amplitude: Wagoner [714] showed that, for central temperature $T_c \lesssim 3 \times 10^9$ and a core with hyperons, an accreting star can reach a steady state with r-mode amplitude of order 10^{-5} to 10^{-6}, somewhat below the lowest parametric instability threshold found in [99]. Bondarescu, Teukolsky, and Wasserman [92] incorporated this alternative, together with neutrino cooling, shear viscosity, and dissipation within the core-crust boundary layer, in a detailed study of the nonlinear development of the r-mode instability in accreting neutron stars. The coupled equations for the 3-mode system (including the 3-mode coupling that sets the lowest parametric instability threshold) were evolved numerically, in conjunction with equations governing the spin and temperature evolution. The simulations showed, for high enough bulk viscosity, the mode amplitudes settling into a quasistationary state within a period of a few years after the spin frequency of the star has increased above the r-mode stability curve. Quasistationarity is due to thermal equilibrium between viscous heating and neutrino cooling from bulk viscosity of a hyperon core. Because these quasistationary states are algebraic solutions of the 3-mode amplitude equations, Bondarescu et al. were able to reduce the eight evolution equations to two equations governing the rotational frequency and the temperature of the star. Several different evolutionary trajectories were found,

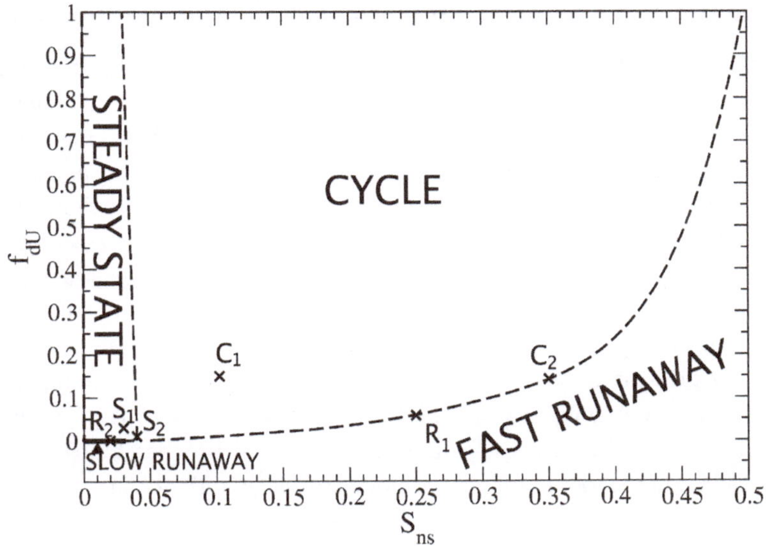

Figure 10.6. The figure shows how the behavior of an unstable r-mode depends on shear and bulk viscosity, when there is a hyperon core. Bulk viscosity, measured by the fraction $f_{\rm dU}$ of the star in which direct URCA reactions can occur, increases along the vertical axis. Shear viscosity *decreases* along the horizontal axis as the slippage $S_{\rm ns}$ at the boundary layer between core and crust increases. (Reprinted figure with permission from Bondarescu, Teukolsky, and Wasserman [92]. Copyright 2011 by the American Physical Society.)

depending on the magnitude of shear (boundary-layer) viscosity and on bulk viscosity (neutrino emission from hyperon direct URCA reactions) in the manner shown in Fig. 10.6. The trajectories are either steady state or cyclical; cyclical trajectories that spin down to stability while maintaining thermal equilibrium at each angular velocity are labeled "Cycle" in the figure. Evolutions for which thermal equilibrium is not reached are the runaway trajectories of the figure. The maximum amplitude was of order 10^{-5} in all cases.

Nonlinear effects in the development of the r-mode instability have also been considered in the case of newborn, isolated compact stars [93]. As in [92], a particular triplet of modes was used, the r-mode and two near-resonant inertial modes. Saturation was set by the parametric instability threshold for this triplet. As in the case of accreting compact stars, a multitude of different evolutionary trajectories was found, depending on the internal physics of the compact star, mainly on the competition between neutrino cooling and viscous heating. When the cooling is balanced by dissipative heating, as the star spins down, it oscillates about a quasisteady thermal equilibrium. For low hyperon bulk viscosity, or when the rate of cooling is low, the energy dissipated in the two inertial modes is not sufficient to stop the growth of the r-mode amplitude, and several parametric instability thresholds are passed. An accurate modeling of this behavior would require the inclusion of multiple mode triplets. On the other hand, in the strong viscosity limit it was found that the parametric instability threshold becomes independent of the frequency detuning.

10.5 Damping of nonlinear oscillations due to formation of shocks

Large-amplitude, nonlinear pulsations in a star may lead to shock formation in its outer layers. When these shocks propagate away from the star, a low-density envelope is created, in which subsequent shocks can dissipate their kinetic energy. In this way, pulsational kinetic energy is lost into shock heating, leading to a gradual damping of pulsations. Because this process requires a star to oscillate with large amplitude, it is difficult to construct exact initial data that correspond to this situation. One way to excite nonlinear oscillations in a relativistic star would be to start with a marginally stable white dwarf model, which would collapse and form a compact star at high densities. Such a simulation requires a special treatment of the numerical grid, as it has to cover several length scales during contraction. A simpler way to induce nonlinear oscillations in a compact star is provided by the nonlinearities of the Einstein equation and is described in what follows.

For any EOS, there exist equilibrium models at very high densities that are radially unstable. Any small perturbation of these models will cause them either to collapse to a black hole, or to expand and *migrate* to the stable branch of equilibrium models, at smaller central densities. Which of the two paths the star will follow depends entirely on the sign of the perturbation. In [224], the migration of a nonrotating, unstable star to the stable branch was studied in a 3-dimensional, fully relativistic simulation. The initial model was constructed with an $N = 1$ polytropic EOS, with polytropic constant $K = 100$ (in units with $c = G = M_\odot = 1$).

The model had a central rest-mass density $\rho_c = 8.0 \times 10^{-3}$ $(4.9 \times 10^{15}\,\mathrm{g/cm^3})$, larger than that of the maximum-mass stable model, a gravitational mass of $M = 1.447 M_\odot$, and a rest mass of $M_0 = 1.535 M_\odot$. The initially unstable star migrated to the stable branch of equilibrium models at smaller central densities (see the inset of Fig. 10.7), due to small perturbations induced by the truncation errors of the numerical scheme.[3] An analogous behavior was observed in numerical simulations of relativistic boson stars [600, 308].

Figure 10.7 shows the evolution of the central rest-mass density ρ_c normalized to its initial value and up to a final time of 4.26 ms. On a very short dynamical timescale of about 0.5 ms, the star has expanded to much smaller densities, acquiring a large radial momentum. The star then enters large-amplitude, nonlinear oscillations around a new, stable equilibrium. Because the unstable and stable models have rather different degrees of compactness, the migration to the stable branch is accompanied by the conversion of a significant amount of gravitational binding energy into (initially) pulsational kinetic energy. When the simulation is performed using the two-parameter ideal-gas EOS $p = p(\rho, e)$ of Eq. (10.10), the fluid variables p and ρ are effectively evolved separately via Eq. (10.3), as mentioned in Section 10.1.1. An evolution with shocks then does not conserve entropy, and the oscillations are gradually damped, their energy lost to shock heating. In the evolution depicted by a solid line in Fig. 10.7, the star oscillates about a central

[3] In the particular numerical code, a collapse to a black hole could be induced by adding perturbations of the opposite sign.

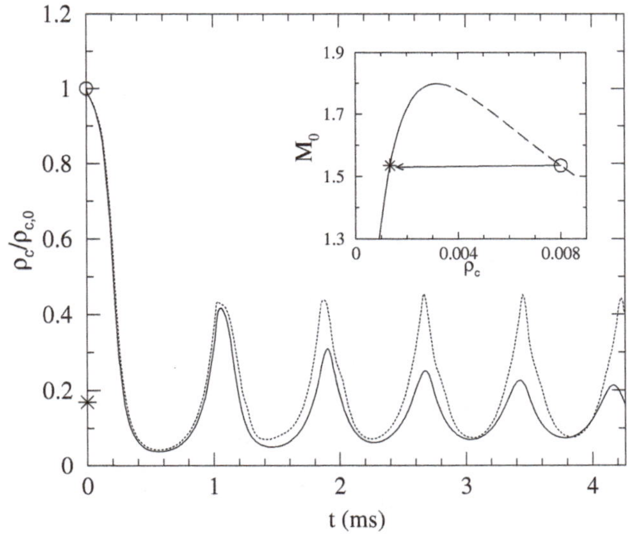

Figure 10.7. Evolution of the central rest-mass density ρ_c (normalized to its initial value, $\rho_{c,0}$) during the migration of an unstable, nonrotating relativistic star to a stable model with the same rest mass, but slightly smaller central density (indicated by an asterisk on the left vertical axis). The star first expands and then oscillates nonlinearly around the new stable equilibrium. The solid line corresponds to a simulation with a two-parameter ideal-gas EOS that allows shock heating, whereas the dotted line corresponds to a simulation with a barotropic EOS. The inset shows the path of the star in a rest-mass versus central rest-mass density plot. (Figure from [224].)

density close to that of a stable model with the same rest mass (the latter is indicated with an asterisk on the vertical axis of Fig. 10.7). In contrast, when the *barotropic* EOS (Eqs. [1.119], [1.120]) is enforced during the simulation (dotted line), the star behaves like a compressed spring that is allowed to expand, oscillating with a nearly constant amplitude. In this case, gravitational binding energy is periodically converted into kinetic energy and vice versa, whereas any dissipation due to shocks is suppressed. As the oscillations are highly nonlinear, the restoring force is different at higher densities than at lower densities, and the oscillations are therefore far from being sinusoidal.

The decrease in the oscillation amplitude, when the two-parameter EOS is used, is due to the dissipation of kinetic energy via shock heating. At the end of the first expansion (i.e., at the first minimum in Fig. 10.7), the star has expanded almost to the edge of the numerical grid. At this point, its outer layers have formed a low-density envelope around the high-density inner regions, and the star then starts to contract. Figure 10.8 displays the supersonic infall of the envelope at $t = 0.84$ ms (solid lines), at which time the inner regions are contracting homologously. After this "point of last good homology," the high-density inner regions reach their maximum infall velocity and then start slowing down. The infalling low-density envelope forms a shock when accreting onto the stalled inner regions (dotted lines at $t = 0.98$ ms in Fig. 10.8). After the inner regions bounce, they expand, and

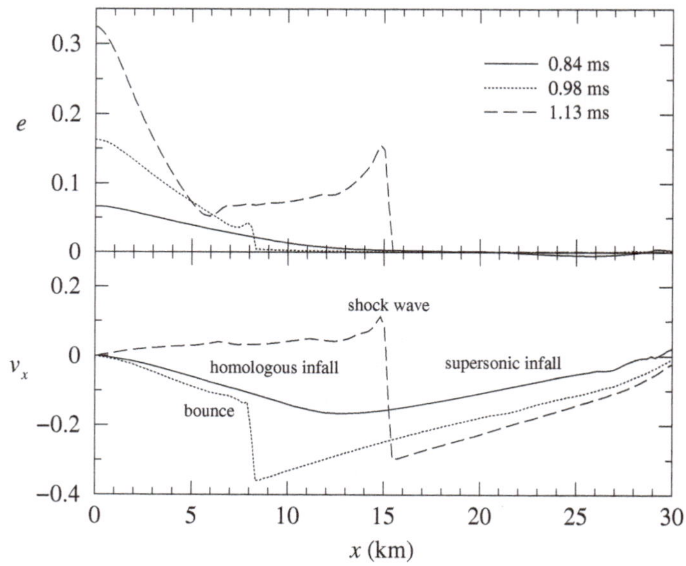

Figure 10.8. Shock formation in the outer layers of an oscillating star, during the first bounce near equilibrium density. The top and bottom panels show the specific internal energy e and radial velocity v_x, respectively, at three different times: the homologous infall phase, the inner core bounce, and the outward shock propagation. Shock heating leads to damping of the nonlinear oscillations. (Figure from [224].)

pressure waves feed the shock wave with kinetic energy, causing it to move rapidly outward (dashed lines at $t = 1.13$ ms in Fig. 10.8). In this way, the shock wave is dissipating the initial binding energy of the star so that the amplitude of the central density oscillations decreases with time. The above process is very similar to the core bounce in neutron star formation (see, for instance, the description in [481]), except for the fact that here the outer envelope is created during the first rapid expansion of the initially unstable star.

10.6 Mass-shedding–induced damping

Another striking nonlinear effect is the mass-shedding–induced damping of oscillations in stellar models that rotate at or near the mass-shedding limit [661, 193]. This damping mechanism can become important for pulsations in uniformly or weakly differentially rotating stars, for example when large-amplitude nonlinear pulsations are excited due to the sudden development of a phase transition [1, 188]. In [661], the mass-shedding–induced damping was demonstrated for the first time in uniformly rotating models using fixed-spacetime evolutions, that is, in the Cowling approximation. The same mechanism was shown to operate even when the spacetime is evolved in the IWM-CFC approximation [193], although the damping rate is much smaller in the latter case.

The damping mechanism works as follows: As the star approaches the mass-shedding limit, $\Omega = \Omega_K$, the effective gravity near its equator diminishes, exactly

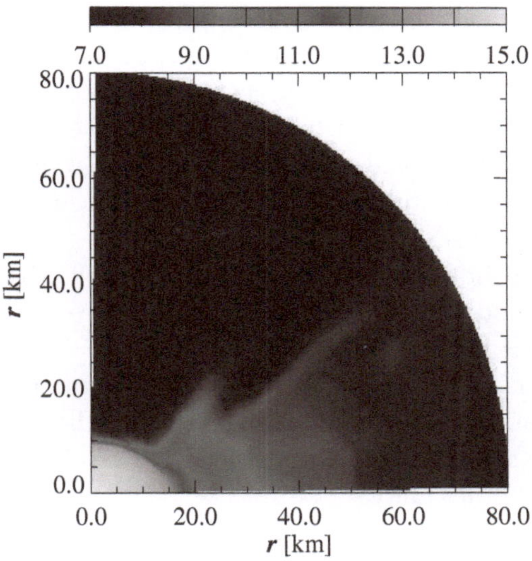

Figure 10.9. Distribution of the logarithm of the rest-mass density, $\log \rho$ (g cm^{-3}), in the meridional plane for an oscillating star that is rotating near its mass-shedding limit, at $t = 4.4$ ms. The vertical axis coincides with the rotation axis. Shocks are created after each pulsational period and travel through the atmosphere, creating a matter envelope in which pulsational energy is dissipated. (Figure from [193]. Reprinted by permission of John Wiley and Sons.)

vanishing when the fluid at the equator is in Keplerian orbit. A small radial pulsation then suffices to cause mass shedding after each oscillation period. As a result, a low-density envelope is created around the star. This envelope is initially concentrated in the regions close to the stellar equator, but with each oscillation period, more and more mass is shed in the form of shock waves, and the envelope expands outward and away from the equatorial plane. Because in rotating stars every pulsation mode has a radial velocity component, the damping affects all modes.

Figure 10.9 shows the distribution of the rest-mass density ρ in the meridional plane (at time $t = 4.4$ ms) when an oscillating star that rotates near its mass-shedding limit is evolved with the two-parameter EOS (10.10). Several oscillations have already occurred, creating a low-density, high-entropy envelope, which extends out to a few radii from the star. Near the equatorial plane, the envelope has a rest-mass density of $\sim 10^{-4}$ to 10^{-8} times the central rest mass density while it is mainly limited to angles less than $\sim 45^\circ$ away from the equatorial plane. Through this envelope, further consecutive shocks (created after each expansion phase of the pulsations) propagate outward. Notice that in the example shown in Fig. 10.9, a very small amplitude of radial pulsations was chosen. For larger pulsation amplitudes, the strength of the shocks and the extend of the high-entropy envelope will increase nonlinearly. When matter is shed from the star, kinetic energy is carried away to the expense of the pulsational energy. Consequently, the pulsations of the star are gradually damped, depending on the strength of mass shedding.

In an equilibrium model rotating at the mass-shedding limit, the effective gravity at the equatorial surface vanishes because of a delicate balance among the pressure gradient, the gravitational force, and the centrifugal force. If the spacetime is held fixed during the time evolution (which is equivalent to the Cowling approximation) [661], then the gravitational force cannot react to local variations of density caused by the pulsations. In this (unrealistic) case, matter can be more easily ejected from the stellar equatorial surface at each pulsation.

The mass-shedding–induced damping of pulsations could have significant implications for unstable modes that grow on secular timescales due to the gravitational-radiation–driven CFS instability. In particular, the $l = m = 2$ f-mode becomes unstable only near the mass-shedding limit in uniformly rotating stars [665, 486]. A detailed investigation is required to determine whether the rate of mass-shedding–induced damping is shorter than the secular growth rate of the instability at amplitudes smaller than of order unity.

In [1], the mass-shedding–induced damping was observed to be important for the evolution of rapidly rotating stars that undergo phase transitions.

10.7 Postbounce oscillations

One of the possible ways in which gravitational waves could be emitted during a core-collapse supernova event are postbounce oscillations that are initiated because of the sudden bounce of the core at high densities, when the equation of state becomes stiff. Although only radial ($l = 0$) modes are excited in a spherically symmetric collapse, the collapse of a rotating star excites quasiradial modes, each including $l > 0$ multipoles. A strong bounce would seem to be an ideal mechanism for exciting a number of modes with initially large amplitude. The large initial amplitude could, in principle, be sustained on a longer than dynamical timescale, because it would be damped primarily through shear viscosity (or through nonlinear mode-mode couplings). Detailed simulations, however, have shown that strong nonlinear effects at the time of bounce can severely reduce the initial oscillation amplitude on a dynamical timescale.

The detailed modeling of stellar core collapse and postbounce oscillation is a multiscale, multiphysics problem, involving length scales from the extended presupernova stellar core (thousands of kilometers) down to small-scale turbulence in the postbounce flow (on the order of meters). A complete model, resolving all length scales in 3 dimensions and fully including general relativity, magnetohydrodynamics, multidimensional Boltzmann neutrino transport, and a microphysical treatment of the EOS and of the nuclear and neutrino interactions is still beyond reach. In addition, as long as the high-density EOS, the pre-supernova stellar structure, and the angular momentum configuration are not well constrained, extensive parameter studies are required. Over several decades, a large number of numerical simulations have been carried out, with increasing degree of complexity and sophistication, attempting to accurately model a core-collapse event (see Ott [505] and references therein for a historical account).

The numerical simulation of rotating core collapse demonstrated that some features of spherically symmetric collapse also hold approximately for rotating cores: From the beginning of collapse, the collapsing core separates into a subsonically, homologously ($v \propto r$) contracting *inner core* and a supersonically infalling *outer core*. The initial size of the proto–neutron star (the matter that is dynamically relevant for the postbounce oscillations) is set by the mass of the inner core at core bounce. In some of these studies, at least three different types of collapse and bounce dynamics were identified. Type I models undergo core bounce governed by the stiffening of the nuclear EOS at nuclear density and display a strong damping of postbounce oscillations. Type II models, on the other hand, are affected significantly by rotation and undergo core bounce dominated by centrifugal forces at densities below nuclear. In this case, the core continues to oscillate with a large amplitude on a timescale larger than the dynamical one (type-II models are related to *fizzlers* [605, 695, 323], collapsed cores that could become temporarily or permanently stabilized by centrifugal forces before core bounce or black-hole formation). If one considers a very soft subnuclear EOS or very efficient electron capture, then the mass of the homologously collapsing inner core turns out to be small, and such type III models are also characterized by strongly damped postbounce oscillations. Regarding the efficiency of gravitational-wave emission due to postbounce oscillations, the advantage of type II models having a small damping rate (which would increase efficiency) is lost because in such cases the bounce happens at smaller density (which decreases efficiency).

Two- and 3-dimensional simulations of rotating core collapse that included general relativity, two different microphysical finite-temperature EOSs, and an approximate scheme to account for electron capture and neutrino losses during collapse [508, 507, 191, 192] showed that for a wide range of progenitor models and precollapse rotational configurations, only the type-I postbounce dynamics appears (strong nonlinear damping of postbounce pulsations). This finding reduces significantly the expected gravitational-wave signal from postbounce oscillations, compared to the case where these would damp on a gravitational-radiation–emission timescale. In other 3-dimensional, Newtonian simulations [107, 108, 506] that extended to several thousand hydrodynamical timescales after core bounce, evidence was found for the nonlinear growth of core g-modes through accretion downstreams. If the amplitude of g-mode oscillations can indeed become as large as suggested in the above simulations (and is not limited by effects, such as nonlinear couplings to higher-order modes, that could not be resolved in the above simulations), then this mechanism would emit stronger gravitational waves than the initial postbounce oscillations.

10.8 Collapse of rotating relativistic stars to Kerr black holes

Rotating relativistic stars can be subject to an axisymmetric (quasiradial) instability to collapse (see Chapter 8). A star could be driven into the unstable region by accretion in a binary system (or postbounce accretion after core collapse); by magnetic breaking of an isolated, supramassive millisecond pulsar; or by the cooling of a differentially rotating, hot proto–neutron star. Fully relativistic, 3-dimensional simulations of the collapse of supramassive polytropic models [612, 41] showed that the onset of the dynamical instability appears at only somewhat larger densities than the secular instability detected by the Friedman, Ipser, and Sorkin criterion (Section 9.2). Once the star becomes dynamically unstable, the collapse to a Kerr black hole becomes inevitable.

The numerical investigations of black-hole formation (beyond spherical symmetry) started in the early 1980s with the pioneering work of Nakamura [490], who worked in the (2+1)+1 formulation of the Einstein equations in cylindrical coordinates. Nakamura used a "hypergeometric" slicing condition that prevents the grid points from reaching the singularity when a black hole forms and introduced regularity conditions to avoid divergences at coordinate singularities. The simulations followed the collapse of a $10M_\odot$ "core" of a massive star (with different amounts of rotational energy) up to the formation of a rotating black hole, without extracting information on the emitted gravitational radiation. In a subsequent series of papers, Bardeen, Stark, and Piran [48, 652, 653, 651] reported results of axisymmetric simulations that gave first estimates of the gravitational radiation associated with the formation of rotating black holes. These simulations were based on a restriction of the 3+1 Einstein-Euler formalism to two spatial dimensions, using the radial gauge and a mixture of singularity-avoiding polar and maximal slicings. The initial model was a spherically symmetric relativistic polytrope in which gravitational collapse was induced by lowering the pressure by a prescribed percentage. Simultaneously, an approximate rigid-body rotation was added to the initial data. With such a setup, the energy ΔE carried away through gravitational waves from the collapse to a Kerr black hole was found to be $\Delta E/Mc^2 < 7 \times 10^{-4}$.

If the initial star's angular momentum is smaller than the maximum angular momentum $J = M^2$ of a Kerr black hole, its collapse is expected to lead to prompt black-hole formation. Shibata [609] used an axisymmetric code to investigate the effects of rotation on the criterion for prompt adiabatic collapse of rigidly and differentially rotating polytropes to a black hole, finding that the criterion for black-hole formation depends strongly on the amount of angular momentum, but only weakly on its (initial) distribution. The effects of shock heating when using a nonisentropic EOS are important in preventing prompt collapse to black holes in the case of large rotation rates. The mass of a possible disk left over from the collapse appears to be negligible (less than 10^{-3} of the initial stellar mass) [612, 610, 196].

In Baiotti et al. [41], the collapse of a supramassive, dynamically unstable rotating polytrope was followed in detail (in 3 dimensions and full general relativity) until well after the formation of a Kerr black hole. The event horizon was detected by evolving a null surface back in time, and several different methods for measuring its mass and angular momentum were compared. These simulations traced the complete transition from one exact (unstable) equilibrium solution of Einstein's equations to another (stable) exact equilibrium and were made possible by the implementation of excision techniques within the black hole horizon.

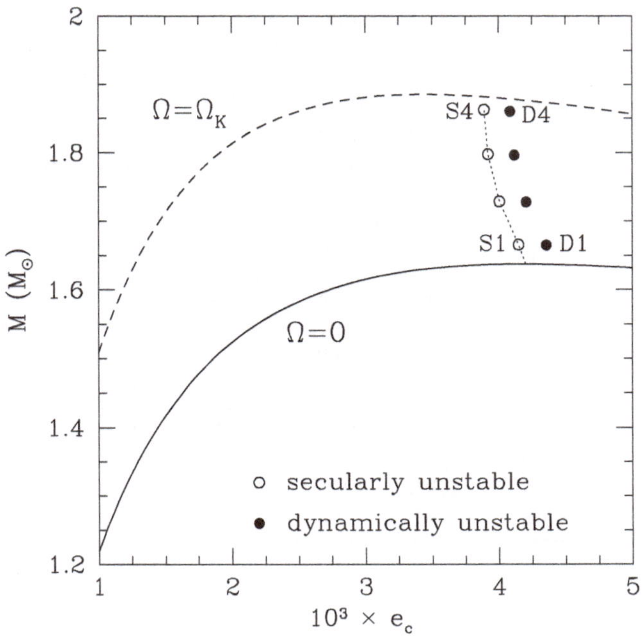

Figure 10.10. Gravitational mass vs. central energy density for equilibrium models constructed with the polytropic EOS, for $\Gamma = 2$ and polytropic constant $K = 100$. The solid, dashed, and dotted lines correspond to the sequence of nonrotating models, the sequence of models rotating at the mass-shedding limit, and the sequence of models that are at the onset of the secular instability to axisymmetric perturbations. Also shown are a number of secularly unstable (open circles) and dynamically unstable models (filled circles). (Figure from [41].)

Initial data for the fully relativistic, 3-dimensional dynamical simulations are usually constructed with an independent, 2-dimensional numerical code, such as the **rns** code (Section 4.1.4), that computes accurate stationary equilibrium models of axisymmetric and rapidly rotating relativistic stars. The data are then transformed from the polar-coordinate grid of the 2-dimensional code to the Cartesian grid of the 3+1 code. In [41], the focus was on initial models constructed with the polytropic EOS (1.119, 1.120), choosing $\Gamma = 2$ and a polytropic constant $K = 100$ (in units of $c = G = M_\odot = 1$). More specifically, four models along the secular-instability line (approximated as the turning-point line), were chosen, with polar-to-equatorial axis ratios of roughly 0.95, 0.85, 0.75, and 0.65; these models are labeled S1–S4, respectively, in Fig. 10.10. Four additional models were defined by increasing the central energy density of the secularly unstable models by 5%, keeping the same axis ratio. These models, labeled D1–D4 in Fig. 10.10, were found (as expected) to be dynamically unstable in the simulations that were performed.

Figure 10.10 shows the gravitational mass as a function of the central energy density for equilibrium models constructed with the chosen polytropic EOS. The solid, dashed, and dotted lines correspond respectively to the sequence of nonrotating models, the sequence of models rotating at the mass-shedding limit, and

Table 10.1. *Equilibrium properties of a number of initial stellar models. The different columns refer respectively to the central rest-mass density ρ_c, the ratio of the polar to equatorial coordinate radii r_p/r_e, the gravitational mass M, the circumferential equatorial radius R, the angular velocity Ω, the angular momentum in terms of J/M^2, the ratio of rotational kinetic energy to gravitational binding energy $T/|W|$, and the "height" of the corotating ISCO, h_+. For all models, $\Gamma = 2$ and $K = 100$. Units are set by $G = c = M_\odot = 1$.*

Model	$\rho_c^\dagger$	r_p/r_e	M	R	$\Omega^\ddagger$	J/M^2	$T/\|W\|^\ddagger$	h_+
S1	3.154	0.95	1.666	7.82	1.69	0.207	1.16	1.18
S2	3.066	0.85	1.729	8.30	2.83	0.363	3.53	0.51
S3	3.013	0.75	1.798	8.90	3.49	0.470	5.82	0.04
S4	2.995	0.65	1.863	9.76	3.88	0.545	7.72	–
D1	3.280	0.95	1.665	7.74	1.73	0.206	1.16	1.26
D2	3.189	0.85	1.728	8.21	2.88	0.362	3.52	0.58
D3	3.134	0.75	1.797	8.80	3.55	0.468	5.79	0.10
D4	3.116	0.65	1.861	9.65	3.95	0.543	7.67	–

$^\dagger \times 10^{-3}$
$^\ddagger \times 10^{-2}$

the sequence of models that are at the onset of the secular instability to axisymmetric perturbations. Furthermore, the secularly and dynamically unstable initial models used in the collapse simulations are shown as open and filled circles, respectively. Table 10.1 summarizes the main equilibrium properties of the initial models. The height of the corotating ISCO is defined as $h_+ = R_+ - R$, where R_+ is the circumferential radius at the corotating ISCO. Note that in those models for which a value of h_+ is not reported, all circular geodesic orbits outside the stellar surface are stable.

All simulations presented in [41] were computed using a uniformly spaced computational grid for which symmetry conditions were imposed across the equatorial plane. Different spatial resolutions were used to check convergence, up to a resolution of $288^2 \times 144$ cells. The outer boundary was set at about twice the initial stellar equatorial radius for D1 and at $\sim$1.4 times the initial stellar equatorial radius for D4. The hydrodynamics equations were solved employing the Marquina flux formula [194] and a third-order PPM reconstruction [149]. The components of the Einstein field equation, on the other hand, were evolved using the iterated Crank-Nicolson evolution scheme proposed by Choptuik (see, e.g., [683]), the "1+log" slicing condition, and the "Gamma-driver" shift conditions (Section 10.1.2). Finally, both polytropic and ideal-gas EOSs were used, although no significant difference was found in the dynamics when comparing the two cases. This is consistent with the relatively rapid collapse of uniformly rotating initial models, in which strong shocks do not develop.

Dynamics of the matter

Given an initial dynamically unstable stellar model, simple roundoff errors would produce an evolution leading either to gravitational collapse to a black hole or to migration to the stable branch of the equilibrium configurations [224]. (Although both evolutions are, in principle, possible, one expects collapse to occur in nature only for stars on the high-density side of the instability line.) In general, however, leaving the onset of the dynamical instability to the cumulative effect of numerical truncation produces instability growth times that are dependent on the grid resolution used. For this reason, the collapse in the simulations of [41] was induced by a slight reduction of the pressure in the initial configuration. As in the earlier simulations by Shibata [609], this was done uniformly throughout the star by using a polytropic constant K for the evolution that was slightly smaller than the one used to compute the initial data. A small change of order 2% in the polytropic constant was sufficient to induce collapse. To ensure second-order convergence from the start of the simulation, the Hamiltonian and momentum constraints were solved again after imposing the pressure reduction (details on solving the constraints using the York-Lichnerowicz conformal transverse-traceless decomposition can be found in [156]). Strict second-order convergence was, however, lost when excision was introduced after the black hole was formed, although the code remained convergent at a somewhat lower rate.

The dynamics resulting from the collapse of models S1–S4 and D1–D4 was qualitatively similar. Specifically, model D4, which, being rapidly rotating, was already rather flattened initially (i.e., $r_p/r_e = 0.65$) and had the largest J/M^2 among the dynamically unstable models (cf. Fig. 10.10 and Table 10.1), represents the most interesting case. In Figs. 10.11 and 10.12, some representative snapshots of isocontour levels of the rest-mass density (spaced logarithmically) during the collapse of this model are shown. Vectors indicate the velocity field, and a reference length is shown in the bottom panels. The time of the different snapshots is given in milliseconds. Note that a region around the singularity that has formed in the lower panel is excised from the computational domain and is indicated as an area filled with squares. Also shown with a thick dashed line is the coordinate location of the apparent horizon.

As the collapse proceeds, the large angular velocity of the initial model produces significant deviations from a spherical infall. Indeed, the parts of the star around the rotation axis that are experiencing smaller centrifugal forces collapse more promptly and, as a result, the configuration increases its oblateness. At about $t = 0.64$ ms, the collapse of model D4 produced an apparent horizon. Soon after this, the central regions of the computational domain were excised, allowing the simulation to continue. The dynamics of the matter at this stage is shown in the lower panel of Fig. 10.11, which refers to $t = 0.67$ ms and where both the location of the apparent horizon (thick dashed line) and of the excised region (area filled with squares) are shown. By this time, the star has flattened considerably, and all of the matter near the rotation axis has fallen inside the apparent horizon, but a disk of low-density matter remains near the equatorial plane, orbiting at very high velocities $\gtrsim 0.2c$. For this most rapidly rotating initial model, a centrifugal hangup appeared to halt collapse in the equatorial regions, but only for a short period lasting about 0.2 ms. The appearance of an effective barrier preventing a purely radial infall of

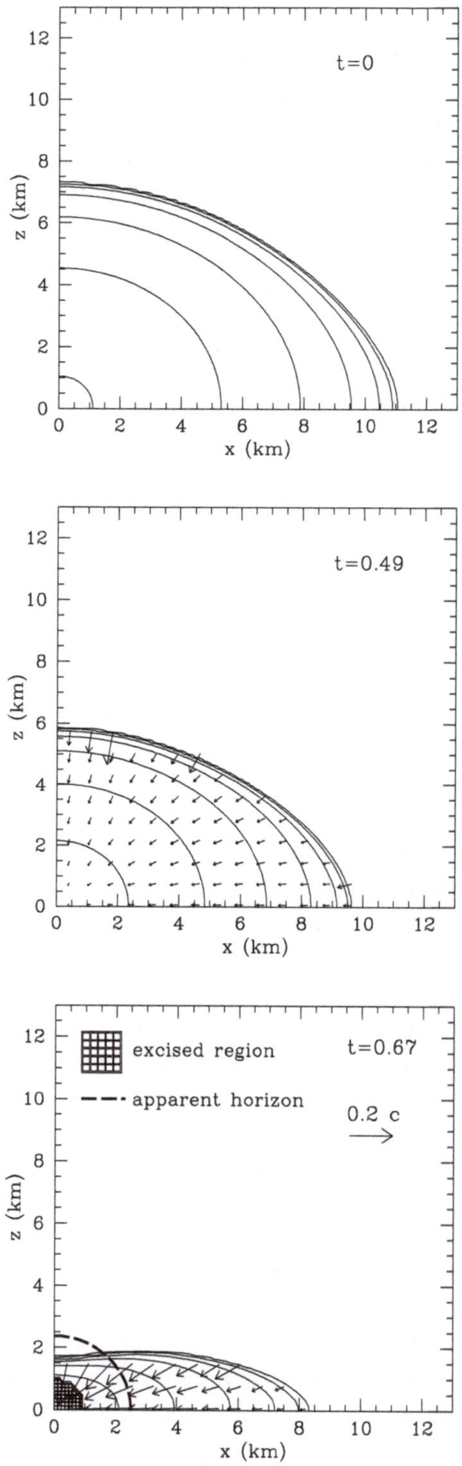

Figure 10.11. Collapse sequence for the rapidly rotating model D4 (see text for details). (Figure from [41].)

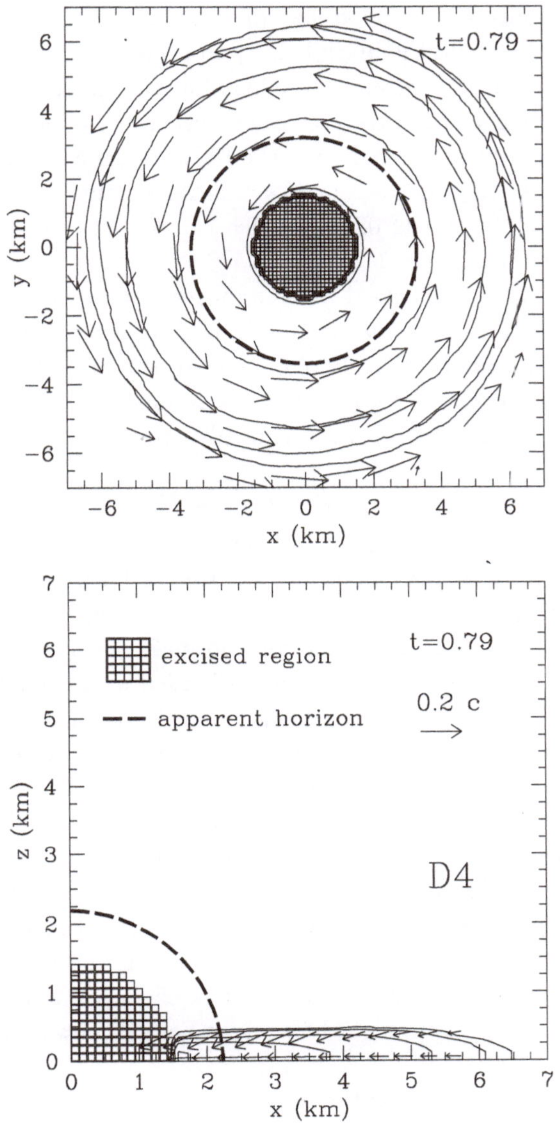

Figure 10.12. Magnified view of the final stages of the collapse of model D4 (see text for details). (Figure from [41].)

matter far from the rotational axis is the consequence of the larger initial angular momentum of the collapsing matter and of the pressure wave originating from the faster collapse along the rotational axis, compared to slowly rotating initial models.

Figure 10.12 shows that at the later time $t = 0.79$ ms, a disk forms outside the apparent horizon. At this stage the disk is considerably flattened but also has large radial inward velocities. It is, therefore, not dynamically stable and accretes rapidly onto the black hole. Note that as the area of the apparent horizon increased, so did the excised region, which was allowed to grow accordingly. By a time

$t = 0.85$ ms, essentially all of the stellar matter was within the trapped surface of the apparent horizon, and the black hole approached the Kerr solution. A simple kinematic explanation can be given for the instability of the disk formed during this oblate collapse and comes from relating the position of the outer edge of the disk when this first forms, with the location of the ISCO of the newly formed Kerr black hole. Measuring accurately the mass and spin of the rotating black hole reveals that the ISCO was located at $x = 11.08$ km, which is larger than the outer edge of the disk (cf. lower panel of Fig. 10.11).

The simulations in [41] confirmed that no massive and stable disks can form for initial models of neutron stars that are uniformly rotating. This, in turn, implies that the collapse of a rapidly rotating old and cold neutron star cannot lead to the formation of the central engine required in some gamma-ray burst models, namely a rotating black hole surrounded by a centrifugally supported, self-gravitating torus.

Dynamics of the spacetime

In order to investigate the formation of a black hole in a numerical simulation, one can use horizon finders, which compute the *apparent* horizon and the *event* horizon. The apparent horizon, which is defined as the outermost trapped surface (a closed surface on which all outgoing photons normal to it have zero expansion), is calculated at every time step, and its location is used to set up the excised region inside the horizon. In contrast, the event horizon, which is an expanding null surface composed of photons that will eventually find themselves trapped, is computed a posteriori, once the simulation is finished, by reconstructing the full spacetime from the 4-dimensional data each simulation produces. In stationary black-hole systems, where no mass-energy falls into the black hole, the apparent and event horizons coincide, but generally (in dynamical spacetimes) the apparent horizon lies inside the event horizon. In [41] a level-set method [187] was used to locate the event horizon, which rapidly grew to its asymptotic value after formation. With a temporal gap of $\sim 10M$ after the formation of the event horizon, the apparent horizon was found, and then it rapidly approached the event horizon, always remaining within it.

A simple method for computing the mass of the black hole is to note that, for a stationary Kerr black hole, the mass can be found directly in terms of the event-horizon geometry as

$$M = \frac{C_{\rm eq}}{4\pi} \, , \tag{10.37}$$

where $C_{\rm eq} \equiv \int_0^{2\pi} \sqrt{g_{\phi\phi}} d\phi$ is the proper equatorial circumference. As the black hole settles down to the Kerr solution, $C_{\rm eq}$ should tend to the correct value ($M_{\rm ADM}$, modulo the energy emitted in gravitational waves). The computation of the black hole mass, M, coming from the use of (10.37), is shown in Fig. 10.13, for models D1 and D4. The level-set approach of [187] needs initial guesses for the null surface, which converge exponentially to the correct event-horizon surface for earlier times. The duration of the evolution after the black hole has settled down is long enough

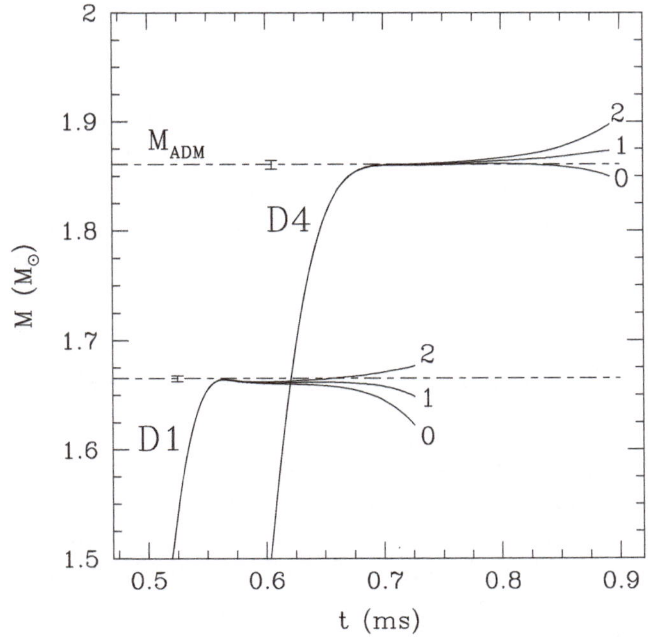

Figure 10.13. Evolution of the event-horizon mass $M = C_{eq}/4\pi$ for models D1 and D4. Different lines refer to the different initial guesses for the null surface and are numbered 0, 1, and 2. Note that they all converge exponentially to the expected value ($\sim M_{\rm ADM}$) for decreasing times. (Figure from [41].)

that late-time error in the guessed horizon yields negligible error in determining an earlier but still settled horizon. In Fig. 10.13, different lines refer to the different initial guesses and are numbered "0," "1," and "2," respectively. All choices lead to the same mass determination. Additional methods for determining the mass and angular momentum of black holes formed during collapse were presented in detail in [41]. A comparison among the different methods showed that an approach utilizing the *dynamical-horizons* framework (see [38] and references therein) is simple to implement and yields estimates that are accurate and more robust than those of other methods.

The gravitational waves extracted from such simulations [40] confirmed that the emitted energy scales with the fourth power in the dimensionless angular momentum of the initial model, but the largest emitted energy (for the most rapidly rotating model) was found to be several orders of magnitude smaller than what was found in the approximate simulations of Stark and Piran [652]. The difference is due to the approximations made in [652] when constructing initial data, while [41, 40] used highly accurate initial data constructed with the `rns` code [664].

Collapse of differentially rotating models

Giacomazzo, Rezzollo, and Stergioulas [260] extended the above results to the case of differentially rotating initial data. For models with $J/M^2 < 1$, that is, sub-Kerr models, both the dynamics of the collapse and the consequent emission of gravitational waves resemble those seen for uniformly rotating stars, although with an overall decrease in the efficiency of gravitational-wave emission. For $J/M^2 > 1$, no supra-Kerr models were found to be dynamically unstable to collapse (although their existence was not strictly excluded), which suggests that generic conditions for a supra-Kerr progenitor do not lead to a naked singularity (when stable supra-Kerr models were artificially depleted of their pressure, the resulting collapse still did not produce a naked singularity but led to a another differentially rotating equilibrium).

10.9 Collapse of rotating supermassive stars

Supermassive stars are possible progenitors of supermassive black holes; they are modeled as polytropes with $N \sim 3$. In spherical symmetry, every general relativistic polytrope with index $N = 3$ is unstable to radial oscillations [126] – in turn, there exists a critical $N_c < 3$ for which the star is marginally stable. Rotation can increase this critical value again [229]. Collapse is initiated in unstable models when a quasiradial perturbation grows on a dynamical timescale, which means that the model has already passed a turning point along a sequence of constant angular momentum, implying the onset of a secular instability to collapse [238]. Numerical simulations have confirmed the existence of such a turning point along the cooling sequence of models of supermassive stars [579], whereas in [577] the collapse of rotating supermassive stars, modeled as relativistic $N = 3$ polytropes, was studied under the assumption that these were differentially rotating. If a supermassive star is indeed differentially rotating as it evolves along the cooling sequence, it might encounter a nonaxisymmetric instability [75, 496, 497].

Zink et al. [753] found that differentially rotating $N = 3$ polytropes exist that can have a quasitoroidal shape (in addition to the usual quasispherical models). Moreover, it was found that such models may be unstable to nonaxisymmetric instabilities, leading to fragmentation of the star. Because the initial models were marginally stable against collapse, the resulting fragments did, indeed, subsequently collapse. In one case, where a one-armed instability was dominant, the time evolution of the resulting fragment was followed (using adaptive mesh refinement) until the apparent horizon of a black hole was detected, demonstrating for the first time such a new path to black hole formation. The dynamical instability triggering black hole formation in the above setup may be related to low-$T/|W|$ instabilities, which seem to be associated with the existence of corotation points (points where the frequency of a mode equals the local angular velocity of the fluid) (see [718, 717, 719] and Chapter 8).

Because the parameter space of possible initial models is large, and given that 3-dimensional simulations of this kind are still quite expensive in terms of computational resources, one can restrict attention to several isolated sequences, where just one initial model parameter is varied, to gain evidence on its systematic effects, and to a plane in parameter space defined by a constant central rest-mass density and a fixed parameter $\Gamma = 1 + 1/N = 4/3$ in the polytropic EOS. As long as one is concerned with the question of stability of modes along a sequence, the consideration of models of the same central density is not overly restrictive as far as the development of the instability is concerned, whereas the nature of the final remnant might be rather sensitive to this assumption. For supermassive stars, an event horizon could form before thermonuclear reactions become important, depending on the metallicity and mass of the progenitor [247]. Because of this, it is conceivable that the type of evolution in [753] can be used as an approximate model of supermassive black hole formation. The detection of gravitational waves may uncover unexpected processes involving black hole formation, and in that case it is useful to have a general understanding of possible dynamical scenarios.

10.9.1 Numerical setup

In [753], equilibrium models of supermassive stars were constructed as general relativistic, differentially rotating, quasitoroidal polytropes. To construct equilibrium polytropes of this kind, an extended version of the `rns` code was used [664]. Equilibrium models exist in a 4-dimensional parameter space $(\Gamma, \rho_c, A, r_p/r_e)$. The `rns` code solves the equilibrium equations iteratively, starting from a nonrotating model. When two different branches of models are present (i.e., both a quasispherical branch and a quasitoroidal branch), it was usually necessary to select a number of intermediate attractors as trial fields, in order to converge to the desired model. Some models were thus constructed by first obtaining a specific quasitoroidal model, and then moving in parameter space along the quasitoroidal branch to the target model. Whereas in [753] a *reference model* was studied, in [754] sequences in ρ_c, Γ, and r_p/r_e containing this model were studied, focusing on the important case $\Gamma = 4/3$, because it approximately represents a radiation-pressure–dominated star.

To examine the indirect collapse by fragmentation of a polytrope with quasitoroidal shape, Zink et al. [753] chose a model with the same rotation law and central density as in Saijo's [577] sequence of models, but with a ratio of polar to equatorial coordinate radius $r_\mathrm{p}/r_\mathrm{e} = 0.24$ (hereafter called the *reference model*; see Table 10.2 for its detailed properties). The ratio of rotational kinetic energy to gravitational binding energy for this model was $T/|W| = 0.227$, whereas its compactness was $M/R = 0.085$, which underlines the necessity of using general relativity for modeling the late stages of supermassive star evolution.

All simulations in [753, 754] were performed in full general relativity. The only assumption on symmetry was a reflection invariance with respect to the equatorial plane of the star. The gauge freedom was fixed by the generalized "1+log" slicing condition for the lapse function, with $f(\alpha) = 2/\alpha$, and by the hyperbolic-like condition suggested in [610] for the shift vector. In addition, adaptive mesh refinement was used, following the location of the density maximum and adding additional grid patches as the density became larger. During evolution, the ideal fluid EOS was used.

Table 10.2. *Parameters and integral quantities of the reference quasitoroidal polytrope used in [753, 754] (in units of $c = G = K = 1$). The first four quantities are parameters, whereas Ω_e is the angular velocity at the equator and Ω_K is the associated Keplerian velocity of the same model.*

Polytropic index Γ	Γ	4/3
Central rest-mass density	ρ_c	3.38×10^{-6}
Degree of differential rotation	A/r_e	1/3
Coordinate axis ratio	r_p/r_e	0.24
Density ratio	ρ_{max}/ρ_c	16.71
ADM mass	M	7.003
Rest mass	M_0	7.052
Compactness	M/R	0.085
Dimensionless angular momentum	J/M^2	1.064
Kinetic to binding energy	$T/\|W\|$	0.227
(See caption)	Ω_e/Ω_K	0.467

After mapping the initial equilibrium model to a hierarchy of Cartesian grids, a small cylindrical density perturbation of the form

$$\delta\rho = \frac{\rho}{r_e} \sum_{m=1}^{4} \lambda_m B \varpi^m \sin(m\phi) \tag{10.38}$$

was added. Here, $m \in \{1, 2, 3, 4\}$, λ_m is either 0 or 1, ϖ is the cylindrical radius and $B \sim 10^{-3}$ is the amplitude of the perturbation. When a perturbation with $\lambda_i = \delta_{ij}$, $j \in \{1, 2, 3, 4\}$, leads to an instability with the same number of nodal lines in the equatorial plane, this instability is called an $m = j$ *mode.* In order to determine the amplitude of a specific mode in the equatorial plane, a projection of the rest-mass density onto Fourier modes at the coordinate radius of the initial density maximum in the equatorial plane was performed.[4] In this way, one can compute weighted averages

$$C_m = \frac{1}{2\pi} \int_0^{2\pi} \rho(\varpi, \varphi, z = 0)\, e^{im\varphi}\, d\varphi \tag{10.39}$$

and normalized real mode amplitudes

$$A_m = \frac{|C_m|}{C_0} \tag{10.40}$$

(see Tohline, Durisen, and McCollough [696]).

10.9.2 Dynamical evolution of the reference model

Simulations of the reference model in Table 10.2 were performed with several different numerical setups. The evolution of the moduli of the Fourier components

[4] These quantities are not gauge invariant, but they provide a useful way of characterizing the representation of the instability within a given choice of coordinates.

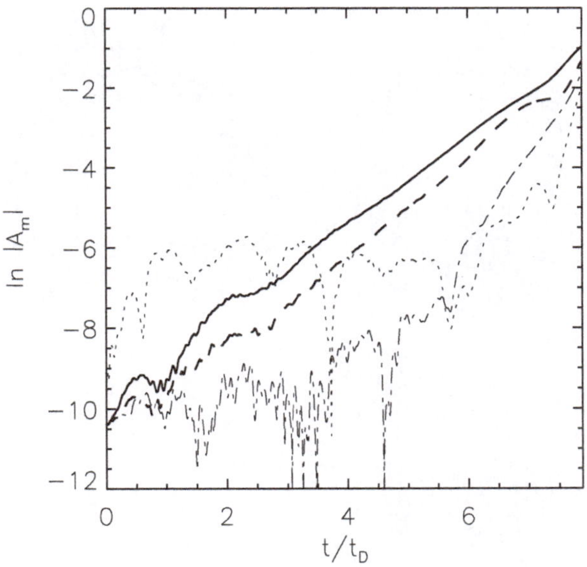

Figure 10.14. Time evolution of Fourier mode amplitudes (see text for definition) for the reference model. Shown are the $m = 1$ (thick solid line), $m = 2$ (thick dashed line), $m = 3$ (thin dash-dotted line), and $m = 4$ (thin dotted line) mode amplitudes. (Figure from [753].)

(when perturbations with $\lambda_1 = \lambda_2 = \lambda_3 = \lambda_4 = 1$ are imposed) is shown in Fig. 10.14. It is evident that, initially, the $m = 4$ component (*thin dotted line*) induced by the Cartesian grid is dominant. However, the star is unstable to the $m = 1$ (thick solid line) and $m = 2$ (thick dashed line) modes, which consequently grow into the nonlinear regime, their e-folding times being rather close. In units of the dynamical timescale (defined as $t_{\rm D} = r_{\rm e}\sqrt{r_{\rm e}/M}$), the e-folding times were $\approx 0.93 t_{\rm D}$ for $m = 1$ and $\approx 0.84 t_{\rm D}$ for $m = 2$, respectively. Mode frequencies were $\approx 3.05/t_{\rm D}$ for $m = 1$ and $\approx 3.31/t_{\rm D}$ for $m = 2$, respectively.

To determine whether a black hole is formed by a fragment, it is necessary to cover the fragment with significantly higher resolution than affordable by fixed mesh refinement. Hence, in [753] a simplified adaptive mesh refinement scheme was implemented to follow the system to black hole formation: In this scheme, a tracking function (provided by the location of a density maximum) was used to construct a locally fixed hierarchy of grids moving with the fragment. Additional refinement levels were switched on during contraction, until an apparent horizon was found. Because the e-folding times for $m = 1$ and $m = 2$ turned out to be similar, the number of fragments produced by the dominant nonaxisymmetric instability and their subsequent nonlinear evolution depended sensitively on the initial perturbation.

The time evolution of the density in the equatorial plane, when only an $m = 1$ perturbation was imposed initially, is shown in Fig. 10.15. Although the initial model is axisymmetric, it has already developed a strong $m = 1$ deviation from axisymmetry at $t = 6.43 t_{\rm D}$, which consequently evolves into a collapsing off-center fragment. At $t = 7.45 t_{\rm D}$, an apparent horizon was found. The horizon was centered

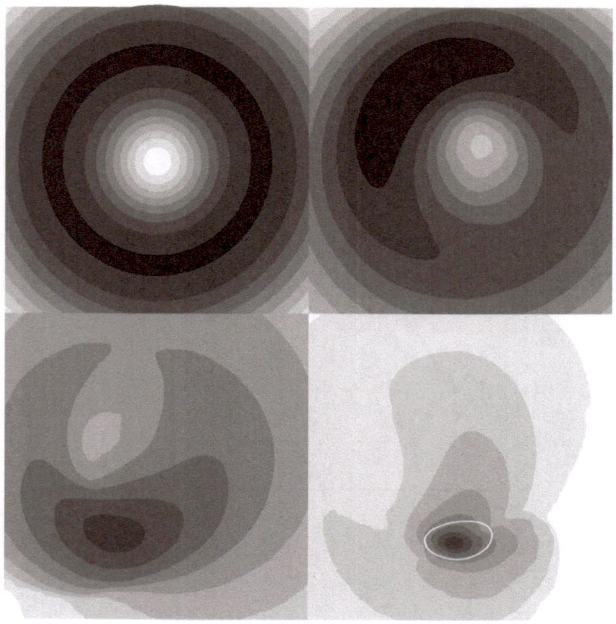

Figure 10.15. Time evolution of the rest-mass density in the equatorial plane, when only an $m = 1$ perturbation is imposed initially [753]. Shown are isocontours of the logarithm of the rest-mass density. The four snapshots were taken at $t/t_{\rm D} = 0$, 6.43, 7.14, and 7.45, respectively, and show the formation and collapse of the fragment produced by the $m = 1$ instability. The last slice contains an apparent horizon demarked by the white, closed line. Note that the shades of gray used for illustration are adapted to the current maximal density at each time, and that darker shades denote higher densities. (Figure from [753].)

on the collapsing fragment at a coordinate radius of $r_{\rm AH} \approx 0.16 r_{\rm e}$ and had an irreducible mass of $M_{\rm AH} \approx 0.24 M_{\rm star}$. Its coordinate representation was significantly deformed: its shape was close to ellipsoidal, with an axis ratio of ~2:1.1:1. The apparent horizon was covered by three refinement levels and 50 to 100 grid points along each axis.

In contrast, when only an $m = 2$ perturbation was imposed initially, the evolution of the density in the equatorial plane was as in the snapshots shown in Fig. 10.16. Two orbiting and collapsing fragments formed. However, even with the adaptive mesh refinement method, constraint violations prevented the simulation from continuing until formation of apparent horizons. Cell-based adaptive mesh refinement and a better choice of gauge might be required in this case.

10.9.3 Sequence of $\Gamma = 4/3$ polytropes with fixed central rest-mass density

The influence of certain parameters on the stability properties of the relativistic quasitoroidal polytropes can be studied by following sequences of models

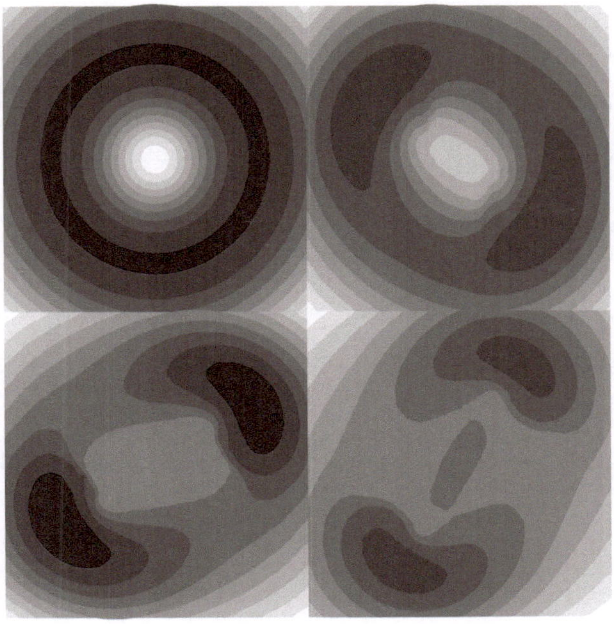

Figure 10.16. As in Fig. 10.15, but for an $m = 2$ initial perturbation. In this case, two fragments formed. Constraint violations forced a termination of the simulation before apparent horizons could be located. (Figure from [753].)

(which contain the reference model), along which specific parameters are varied. In [754], models in a 2-dimensional parameter subspace, with fixed $\Gamma = 4/3$ and $\rho_c = 10^{-7}$, which differ in the rotation law parameter, A, and in the axis ratio, r_p/r_e, were studied. The choice of the central density did not seem to affect the almost exponential development of a nonaxisymmetric unstable mode in the linear regime in any considerable way, even for very compact quasitoroidal polytropes. Models with $\rho_c = 10^{-7}$ are already quite compact, with $M/R \approx 0.01 \ldots 0.1$, and $M/r_p \approx 0.014 \ldots 0.5$.

Figure 10.17 gives an overview of the stability properties of the selected models. The Roman numerals “I” to “III” refer to the highest m with an unstable mode – that is, in addition to the reference polytrope, which belongs to the class “II,” there exist models that are unstable to an $m = 3$ perturbation, and models that appear to be stable against $m = 2$. Models denoted with an “A” were found to be unstable to an axisymmetric mode and to collapse before any nonaxisymmetric instability developed. Finally, models marked with “(I)” could be either stable or long-term unstable with a growth time $\tau \gg t_{dyn}$. Each model was evolved for up to $10\, t_{dyn}$ to determine its stability. This limit is arbitrary but imposed by the significant resource requirements of such simulations. If no mode amplitude exceeds the level of the $m = 4$ noise during this time, the model is marked with a “(I)” (this does not imply that the model is actually stable; it could be unstable to, e.g., an $m = 1$ mode with slow growth rate). The additional lines in Fig. 10.17 are approximate

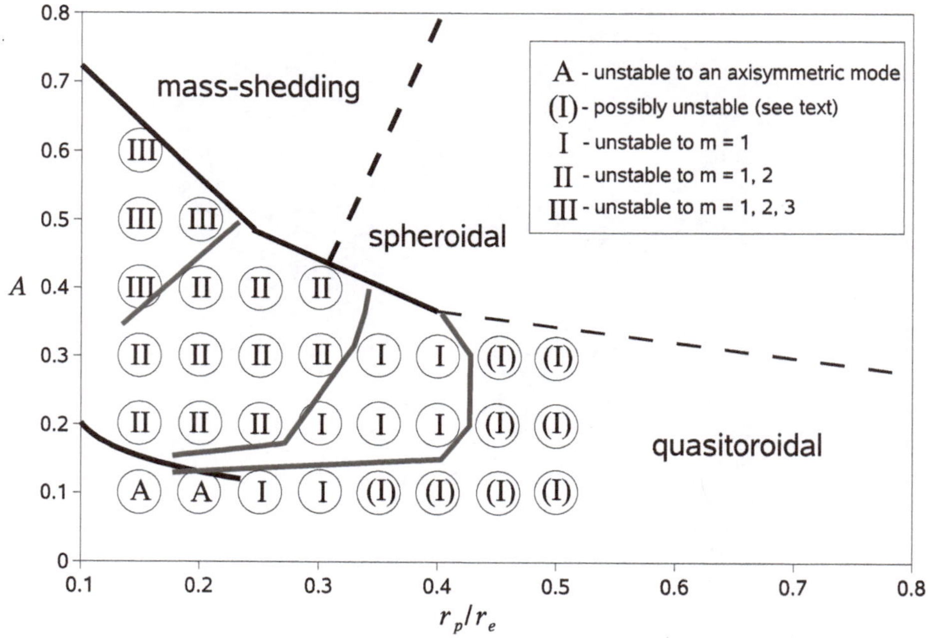

Figure 10.17. Stability of quasitoroidal models with $\rho_c = 10^{-7}$. A Latin number denotes the highest azimuthal order of the unstable modes, i.e. "I" for $m = 1$ unstable, "II" for $m = 1, 2$ unstable, and "III" for $m = 1, 2, 3$ unstable. Models denoted by "(I)" are either long-term unstable with growth times $\tau \gg t_{dyn}$, or stable, and models denoted by "A" exhibit an axisymmetric instability. The line in the lower left corner is the approximate location of the sequence $J/M^2 = 1$, and the three lines inside the quasitoroidal region are the approximate locations of sequences with $T/|W| = 0.14$ (right), $T/|W| = 0.18$ (middle), and $T/|W| = 0.26$ (left). (Figure from [754].)

isolines of the functions $T/|W|$ for the values 0.14, 0.18, and 0.26 and of the function J/M^2 for the value 1. As long as the models do not rotate too differentially, $T/|W|$ still seems to be a reasonable indicator of the nonaxisymmetric stability of the polytropes, even though they are quasitoroidal and relativistic.

Not every model that developed a nonaxisymmetric instability showed properties of a prompt collapse. Figure 10.18 differentiates between those models where prompt collapse was detected (as evident from the evolution of the minimum of the lapse function), labeled "C," and those models, labeled "B," that bounced without prompt collapse (within the finite simulation time). Based on this classification, it is reasonable to expect prompt black hole formation for models labeled "C," whereas the final outcome for models labeled "B" is unclear, as a delayed collapse cannot be excluded.

In order to determine whether the nonaxisymmetric, dynamical instability is a corotation-type instability, one can compare the local angular velocity $\Omega(\varpi)$ to the mode pattern speed $1/m\, d\phi/dt$ (approximately valid for the whole star). In Fig. 10.19, the angular velocity distribution and the approximate pattern speeds of

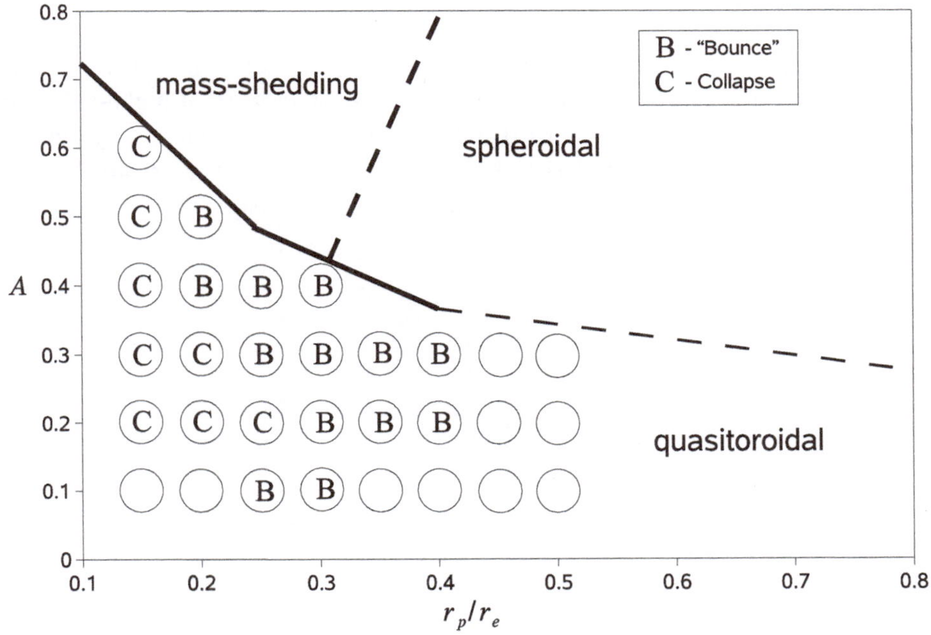

Figure 10.18. Prompt collapse (C) versus bounce (B) in remnants of the models in Fig. 10.17, which are unstable with respect to nonaxisymmetric modes. Models showing a bounce at finite densities within the finite simulation time could still show a delayed collapse to a black hole on longer timescales. (Figure from [754].)

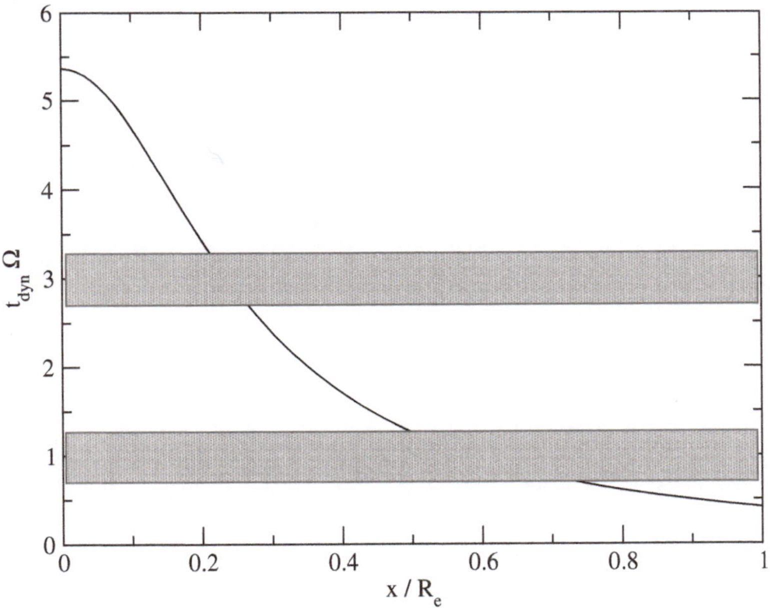

Figure 10.19. Angular velocity of the reference polytrope in the equatorial plane, and approximate pattern speeds (with error bar) of the $m = 1$ mode (upper rectangle), and the $m = 2$ mode (lower rectangle). Both modes have corotation points inside the star. (Figure from [754].)

the $m = 1$ and $m = 2$ modes are shown for the reference model in [753, 754]. Both modes have corotation points: the $m = 1$ mode near the radius of highest density at $0.25\, R_e$, and the $m = 2$ mode near $0.55\, R_e$, indicating that the nonaxisymmetric instability in quasitoroidal models of supermassive stars must be related to the low-$T/|W|$ instabilities found in differentially rotating compact stars.

Appendix A

Lie derivatives, forms, densities, and integration

A.1 Lie derivatives

Lie derivatives arise naturally in the context of fluid flow and are a tool that can simplify calculations and aid one's understanding of relativistic fluids.

Begin, for simplicity, in a Newtonian context, with a stationary fluid flow with 3-velocity $\mathbf{v}(\mathbf{r})$. A function f is said to be *dragged along* by the fluid flow, or *Lie derived* by the vector field $\mathbf{v}$ that generates the flow, if the value of f is constant on a fluid element, that is, constant along a fluid trajectory $\mathbf{r}(t)$:

$$\frac{d}{dt} f[\mathbf{r}(t)] = \mathbf{v} \cdot \nabla f = 0. \tag{A.1}$$

The *Lie derivative* of a function f, defined by

$$\mathcal{L}_{\mathbf{v}} f = \mathbf{v} \cdot \nabla f, \tag{A.2}$$

is the directional derivative of f along $\mathbf{v}$, the rate of change of f measured by a comoving observer.

Consider next a vector that joins two nearby fluid elements, two points $\mathbf{r}(t)$ and $\bar{\mathbf{r}}(t)$ that move with the fluid: Call the connecting vector $\lambda\mathbf{w}$, so that for small λ the fluid elements are nearby: $\lambda\mathbf{w} = \bar{\mathbf{r}}(t) - \mathbf{r}(t)$. Then $\lambda\mathbf{w}$ is said to be *dragged along* by the fluid flow, as shown in Fig. A.1. In the figure, the endpoints of $\mathbf{r}(t_i)$ and $\bar{\mathbf{r}}(t_i)$ are labeled $\mathrm{r_i}$ and $\bar{\mathrm{r}}_\mathrm{i}$.

A vector field $\mathbf{w}$ is *Lie derived* by $\mathbf{v}$ if, for small λ, $\lambda\mathbf{w}$ is dragged along by the fluid flow. To make this precise, we are requiring that the equation

$$\mathbf{r}(t) + \lambda\mathbf{w}(\mathbf{r}(t)) = \bar{\mathbf{r}}(t) \tag{A.3}$$

be satisfied to $O(\lambda)$. Taking the derivative of both sides of the equation with respect to t at $t = 0$, we have

$$\begin{aligned} \mathbf{v}(\mathbf{r}) + \lambda\mathbf{v} \cdot \nabla\mathbf{w}(\mathbf{r}) &= \mathbf{v}(\bar{\mathbf{r}}) = \mathbf{v}[\mathbf{r} + \lambda\mathbf{w}(\mathbf{r})] \\ &= \mathbf{v}(\mathbf{r}) + \lambda\mathbf{w} \cdot \nabla\mathbf{v}(\mathbf{r}) + O(\lambda^2), \end{aligned} \tag{A.4}$$

which holds if and only if

$$[\mathbf{v}, \mathbf{w}] \equiv \mathbf{v} \cdot \nabla\mathbf{w} - \mathbf{w} \cdot \nabla\mathbf{v} = 0. \tag{A.5}$$

The commutator $[\mathbf{v}, \mathbf{w}]$ is the *Lie derivative* of $\mathbf{w}$ with respect to $\mathbf{v}$, written

$$\mathcal{L}_{\mathbf{v}}\mathbf{w} = [\mathbf{v}, \mathbf{w}]. \tag{A.6}$$

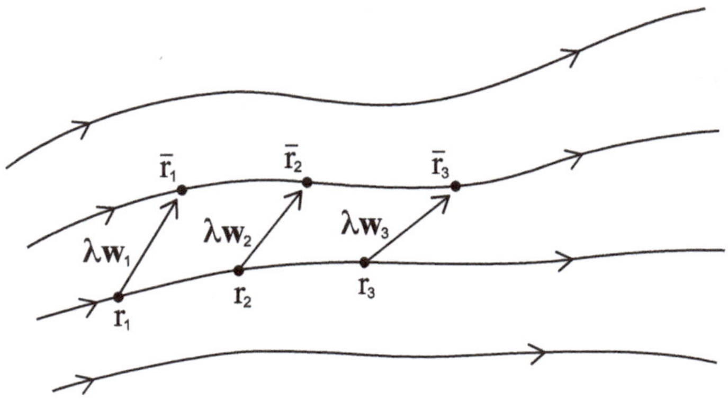

Figure A.1. Two nearby fluid elements move along the flow lines, their successive positions labeled r_i and $\bar{r}_i$. A vector field $\lambda\mathbf{w}$ is said to be dragged along by the flow when, as shown here, it connects successive positions of two nearby fluid elements.

Then $\mathbf{w}$ is Lie derived by $\mathbf{v}$ when $\mathcal{L}_\mathbf{v}\mathbf{w} = \mathbf{0}$. The Lie derivative $\mathcal{L}_\mathbf{v}\mathbf{w}$ compares the change in the vector field $\mathbf{w}$ in the direction of $\mathbf{v}$ to the change that would occur if $\mathbf{w}$ were dragged along by the flow generated by $\mathbf{v}$. In a curved spacetime, the Lie derivative of a function f is again its directional derivative,

$$\mathcal{L}_\mathbf{u} f = u^\alpha \nabla_\alpha f. \tag{A.7}$$

If u^α is the 4-velocity of a fluid, generating the fluid trajectories in spacetime, $\mathcal{L}_\mathbf{u} f$ is commonly termed the convective derivative of f. The Newtonian limit of u^α is the 4-vector $\boldsymbol{\partial}_t + \mathbf{v}$, and $\mathcal{L}_\mathbf{u} f$ has as its limit the Newtonian convective derivative $(\partial_t + \mathbf{v}\cdot\nabla)f$, again the rate of change of f measured by a comoving observer. (Now the flow is arbitrary, not the stationary flow of our earlier Newtonian discussion.)

A connecting vector is naturally a contravariant vector, the tangent to a curve joining nearby points in a flow; in a curved spacetime, the Lie derivative of a contravariant vector field is again defined by Eq. (A.6),

$$\mathcal{L}_\mathbf{u} w^\alpha = u^\beta \nabla_\beta w^\alpha - w^\beta \nabla_\beta u^\alpha. \tag{A.8}$$

We have used a fluid flow generated by a 4-velocity u^α to motivate a definition of Lie derivative; the definition, of course, is the same in any dimension and for any vector fields:

$$\mathcal{L}_\mathbf{v} w^a = v^b \nabla_b w^a - w^b \nabla_b v^a. \tag{A.9}$$

Although the covariant derivative operator ∇ appears in the above expression, it is in fact independent of the choice of derivative operator. This is immediate from the symmetry $\Gamma^i_{jk} = \Gamma^i_{(jk)}$, which implies that the components have in any chart the form

$$\mathcal{L}_\mathbf{v} w^i = v^j \partial_j w^i - w^j \partial_j v^i. \tag{A.10}$$

We now extend the definition of Lie derivative to arbitrary tensors.[1] To do this, we endow a comoving observer with a frame of vectors that, like the observer, are dragged along by the motion of the fluid, vectors $\mathbf{e}_0, \cdots, \mathbf{e}_3$ with $\mathcal{L}_{\mathbf{u}}\mathbf{e}_\mu = 0$. Then the Lie derivative of a tensor is just given by the comoving derivative $\mathbf{u} \cdot \boldsymbol{\partial} T^{\mu\cdots\nu}{}_{\sigma\cdots\tau}$ of its components along the dragged-along frame. We begin with the Lie derivative of a covector (now in any dimension). Let e^a be a vector that is Lie derived by v^a, and let σ_a be a covector field. Then $\mathcal{L}_{\mathbf{u}}\sigma_a$ is defined by requiring that the Lie derivative of σ_a have component along e^a given by

$$(\mathcal{L}_{\mathbf{v}}\sigma_a)e^a = \mathcal{L}_{\mathbf{v}}(\sigma_a e^a). \tag{A.11}$$

Because the component $\sigma_a e^a$ is a scalar, its Lie derivative is the directional derivative $v \cdot \nabla(\sigma \cdot e) = v \cdot \partial(\sigma \cdot e)$. From Eq. (A.11), we have

$$\begin{aligned}(\mathcal{L}_{\mathbf{v}}\sigma_a)e^a &= v^b\nabla_b(\sigma_a e^a) = (v^b\nabla_b\sigma_a)e^a + \sigma_a v^b\nabla_b e^a \\ &= (v^b\nabla_b\sigma_a)e^a + \sigma_b e^a\nabla_a v^b,\end{aligned}$$

where our assumption $\mathcal{L}_{\mathbf{v}}e^a = 0$, together with a change in dummy indices, was used to obtain the final term on the right side. Because the value of e^a at any one point along the trajectory is arbitrary, we have

$$\mathcal{L}_{\mathbf{v}}\sigma_a = v^b\nabla_b\sigma_a + \sigma_b\nabla_a v^b. \tag{A.12}$$

The reader may want to verify that this definition also follows from the Leibnitz rule, the requirement that, for any vector w^a,

$$\mathcal{L}_{\mathbf{v}}(\sigma_a w^a) = (\mathcal{L}_{\mathbf{v}}\sigma_a)w^a + \sigma_a\mathcal{L}_{\mathbf{v}}w^a.$$

Again it is easy to check that the definition (A.12) is independent of the choice of derivative operator, that the components in any chart are given by

$$\mathcal{L}_{\mathbf{v}}\sigma_i = v^j\partial_j\sigma_i + \sigma_j\partial_i v^j. \tag{A.13}$$

Finally, the Lie derivative of an arbitrary tensor $T^{a_1\cdots a_m}{}_{b_1\cdots b_n}$ again follows from the Leibnitz rule or, equivalently, from the requirement that its components along a dragged-along frame are the comoving derivatives $v \cdot \partial T^{i\cdots j}{}_{k\cdots l}$: For any vectors $e_1^a, \ldots, e_n^a$ and covectors $\omega_a^1, \ldots, \omega_a^m$ Lie derived by v^a,

$$(\mathcal{L}_{\mathbf{v}}T^{a\cdots b}{}_{c\cdots d})e_1^c\cdots e_n^d\omega_a^1\cdots\omega_b^m = \mathcal{L}_{\mathbf{v}}(T^{a\cdots b}{}_{c\cdots d}e_1^c\cdots e_n^d\omega_a^1\cdots\omega_b^m). \tag{A.14}$$

A short calculation, analogous to that following Eq. (A.11), yields

$$\begin{aligned}\mathcal{L}_{\mathbf{v}}T^{a_1\cdots a_m}{}_{b_1\cdots b_n} &= v^c\nabla_c T^{a_1\cdots a_m}{}_{b_1\cdots b_n} \\ &\quad - T^{c\cdots a_m}{}_{b_1\cdots b_n}\nabla_c v^{a_1} - \cdots - T^{a_1\cdots c}{}_{b_1\cdots b_n}\nabla_c v^{a_m} \\ &\quad + T^{a_1\cdots a_m}{}_{c\cdots b_n}\nabla_{b_1}v^c + \cdots + T^{a_1\cdots a_m}{}_{b_1\cdots c}\nabla_{b_n}v^c,\end{aligned} \tag{A.15}$$

[1] In keeping with the spirit of a comoving observer, we are using a definition of $\mathcal{L}_{\mathbf{v}}$ that involves a passive transformation – a Lie dragged frame of the observer – instead of the active dragging of the tensor by the diffeo generated by the vector field. The equivalent active picture is presented at the end of this section.

independent of the derivative operator, and with components in a chart given by

$$\begin{aligned}\mathcal{L}_{\mathbf{v}} T^{i_1\cdots i_m}{}_{j_1\cdots j_n} &= v^k \partial_k T^{i_1\cdots i_m}{}_{j_1\cdots j_n} \\ &\quad - T^{k\cdots i_m}{}_{j_1\cdots j_n} \partial_k v^{i_1} - \cdots - T^{i_1\cdots k}{}_{j_1\cdots j_n} \partial_k u^{i_m} \\ &\quad + T^{i_1\cdots i_m}{}_{k\cdots j_n} \partial_{j_1} v^k + \cdots + T^{i_1\cdots i_m}{}_{j_1\cdots k} \partial_{j_n} v^k . \end{aligned} \qquad (A.16)$$

Action of diffeos[2] *and relation to Lie derivatives.* We began by using the flow generated by a velocity field $\mathbf{v}$ to motivate the definition (A.6) of Lie derivative of a vector field $\mathbf{w}$. It is useful to see formally the way in which any vector field generates a flow and to use that flow to give a geometrical definition of Lie derivative. Note first that the trajectory (worldline) of a fluid element is an *integral curve* of the vector field u^α, where:

Definition. An integral curve $c(\lambda)$ of a vector field ξ^a is a curve whose tangent vector at each point $P = c(\lambda_0)$ is $\xi^a(P)$.

In a chart $\{x^i\}$, the tangent ξ^i to an curve $c(\lambda)$ has components $\frac{d}{d\lambda}c^i(\lambda)$; the statement that $c(\lambda)$ is an integral curve has the form $\frac{d}{d\lambda}c^i(\lambda) = \xi^i[c(\lambda)]$.

Proposition. *Any smooth vector field ξ^a in an n-dimensional manifold M has a family of* integral curves, *one through each point of M.*[3]

Example 1. As noted, the velocity field u^α of a fluid has as its integral curves the fluid trajectories parameterized by proper time. The 3-dimensional vector field $\mathbf{v}$ of a stationary Newtonian flow has as its integral curves the flow lines, parameterized by Newtonian time.

Example 2. The vector field $\boldsymbol{\partial}_\phi = x\boldsymbol{\partial}_y - y\boldsymbol{\partial}_x$ has as integral curves the lines of constant t, r, θ,

$$\lambda \to (t, r, \theta, \phi + \lambda)$$

Note that when a vector field vanishes at P (e.g., $\boldsymbol{\partial}_\phi$ vanishes on the symmetry axis $x = y = 0$) the integral curve simply stays at P: $c(\lambda) = P$.

We can view the 4-dimensional flow of a fluid as a family of smooth maps of the fluid to itself in the following way: In a given proper time τ each point P in the fluid moves along the fluid trajectory through P from $c(0) = P$ to the point $c(\tau)$. As in the case of the 4-velocity, we can use the integral curves of any vector field to define a family ψ_λ of diffeos of a manifold to itself (for a star, the fluid has a boundary, and the map ψ_τ is from the support of the fluid to itself). We do this as follows.

[2] A smooth, 1-1 map of a manifold onto itself or onto another manifold is called a *diffeo* (or diffeomorphism).

[3] This result is equivalent to the existence theorem for solutions to ordinary differential equations, proved, for example, in Coddington and Levinson [146].

For each point P let $c(\lambda)$ be the integral curve of ξ^a for which $P = c(0)$. For a fixed value λ, define the map ψ_λ by

$$\psi_\lambda(P) = c(\lambda). \tag{A.17}$$

That is, ψ_λ maps each point P to the point a parameter distance λ from P along the integral curve through P. The vector field ξ^a is said to *generate* the family ψ_λ of diffeos. In a chart $\{x^i\}$, we have

$$\xi^i(x) = \frac{d}{d\lambda}\psi^i_\lambda(x)\bigg|_{\lambda=0}. \tag{A.18}$$

Example 3. The vector field $\boldsymbol{\partial}_\phi$ generates the family of diffeos

$$(t, r, \theta, \phi) \to (t, r, \theta, \phi + \lambda),$$

rotations by λ in the x-y plane about the axis where $\boldsymbol{\partial}_\phi$ vanishes.

We can now repeat for manifolds the relation with which we began this section, between the flow – the diffeos – generated by a vector field and the Lie derivative. We again need the action of a diffeo ψ on a tensor. First a quick, coordinate-dependent derivation, then a more revealing geometric definition of the action:

The action of ψ is the active transformation corresponding to the passive transformation of a tensor's components under a change of coordinates. That is, if one fixes the spacetime and changes the coordinate system from $\{x^i\}$ to $\{\bar{x}^i = \chi^i(x^1, \cdots, x^n)\}$, the components of a tensor $T^a{}_b$ at the same point change from $T^i{}_j$ to

$$\frac{\partial\chi^i}{\partial x^k}\frac{\partial\chi^{-1\,l}}{\partial x^j}T^k{}_l. \tag{A.19}$$

Here the fixed point P has coordinates $\{x^i(P)\}$ in one chart, coordinates $\{\bar{x}^i(P)\}$ in the other. For an active transformation, in which one fixes the coordinate system and maps each point P to $\psi(P)$, the change in the components of a tensor $T^a{}_b$ follow from (A.19) if we define $\bar{x}^i$ by $x^i(\psi(P)) = \bar{x}^i(P)$. Then, denoting by $\psi T^a{}_b$ the transformed tensor, we have

$$\psi T^i{}_j\big|_{\psi(P)} = \frac{\partial\psi^i}{\partial x^k}\frac{\partial\psi^{-1\,l}}{\partial x^j}T^k{}_l(P). \tag{A.20}$$

The tensor T at P is said to be dragged along by the diffeo ψ to the tensor ψT at $\psi(P)$, and the action of ψ on tensors is often written $\psi_* T$.

The better, geometric definition of ψT is closely tied to our intuitive understanding of a fluid flow. As in the flat-space discussion at the beginning of this section, we start with functions and vectors. A function f of a fluid is conserved if the fluid drags the function along with it. Define the dragged-along function ψf by requiring that $\psi f[\psi(P)] = f(P)$, or

$$\psi f(P) := f[\psi^{-1}(P)]. \tag{A.21}$$

A function f on a fluid is Lie derived by the fluid flow if

$$\frac{d}{d\tau}\psi_\tau f \equiv -\mathcal{L}_{\mathbf{u}} f = 0, \tag{A.22}$$

where ψ_τ is again the family of diffeos generated by the fluid velocity u^α. Then, for any smooth vector field ξ^a and function f on a manifold M, one can define the Lie derivative $\mathcal{L}_{\boldsymbol{\xi}} f$ by

$$\mathcal{L}_{\boldsymbol{\xi}} f = -\frac{d}{d\lambda}\psi_\lambda f, \tag{A.23}$$

where ψ_λ is the family of diffeos generated by ξ^a.

To extend the action of diffeos to vector fields, we consider a curve $c(s)$ joining two fluid elements, and let w^α be its tangent vector. For small s, sw^α can be regarded as a connecting vector joining fluid elements a distance s apart. For a fixed proper time τ, write $\psi \equiv \psi_\tau$. The fluid flow maps the curve $c(s)$ to a curve $\psi[c(s)]$, dragged along a proper time τ with the fluid flow. The tangent vector w^α to c at a point P is in this way dragged along to a vector ψw^α tangent to $\psi \circ c$ at $\psi(P)$, a proper distance τ from P. Its components are

$$\psi w^\mu|_{\psi(P)} = \frac{d}{ds}\psi^\mu[c(s)]\bigg|_{s=0} = \partial_\nu\psi^\mu\frac{dc^\nu}{ds}\bigg|_{s=0} = \partial_\nu\psi^\mu w^\nu(P). \tag{A.24}$$

Equivalently, $\psi w^\mu(P) = \partial_\nu\psi^\mu w^\nu[\psi^{-1}(P)]$.

More generally, any diffeo ψ of a manifold M to itself drags a vector field w^a to a vector field ψw^a, with

$$\psi w^i(P) = \partial_j\psi^i w^j[\psi^{-1}(P)]. \tag{A.25}$$

A vector field w^a is Lie derived by a vector field ξ^a if the family of diffeos ψ_λ generated by ξ^a leave w^a unchanged, if $\psi_\lambda w^a = w^a$; and the Lie derivative is again given by

$$\mathcal{L}_{\mathbf{v}} w^a = -\frac{d}{d\lambda}\psi_\lambda w^a. \tag{A.26}$$

To check that this agrees with our earlier definition, we use Eq. (A.18) and $\psi_0 =$ identity, writing

$$\begin{aligned}\frac{d}{d\lambda}\left(\psi_\lambda w^i|_P\right)\big|_{\lambda=0} &= \frac{d}{d\lambda}\left[\frac{\partial\psi^i_\lambda}{\partial x^j}w^j\left(\psi_\lambda^{-1}(P)\right)\right]_{\lambda=0}\\ &= \frac{\partial}{\partial x^j}\left(\frac{d\psi^i_\lambda}{d\lambda}\right)_{\lambda=0} w^j(P) + \frac{\partial\psi^i_\lambda}{\partial x^j}\bigg|_{\lambda=0}\frac{d}{d\lambda}w^j\left(\psi_\lambda^{-1}(P)\right).\end{aligned}$$

Using

$$\frac{d}{d\lambda}w^i\left(\psi_\lambda^{-1}(P)\right) = \frac{\partial w^i}{\partial x^j}\frac{d\psi^j_{-\lambda}(P)}{d\lambda} = -\xi^j\partial_j w^i \quad\text{and}\quad \frac{\partial\psi^i_\lambda}{\partial x^j}\bigg|_{\lambda=0} = \delta^i_j,$$

we have

$$\frac{d}{d\lambda}\psi_\lambda w^i|_{\lambda=0} = \partial_j\xi^i w^j - \xi^j\partial_j w^i$$
$$= -\mathcal{L}_{\boldsymbol{\xi}} w^i.$$

The action of a diffeo on a covector σ_a can be found by writing it, at a point P, as the gradient of a function. The covector field $\nabla_a f$ is in this way dragged to a covector field $\psi\nabla_a f = \nabla_a(\psi f) = \nabla_a[f\circ\psi^{-1}]$, with components given by

$$\nabla_i(\psi f)|_P = \partial_i(f\circ\psi^{-1})|_P = \partial_j f|_{\psi^{-1}(P)}\partial_i\psi^{-1\,j}.$$

Then ψ drags any covector field σ_a to the covector field $\psi\sigma_a$ with components

$$\psi\sigma_i(P) = \partial_i\psi^{-1\,j}\sigma_j[\psi^{-1}(P)]. \tag{A.27}$$

Because a tensor is a sum of tensor products of vectors and covectors, a tensor field $T^{a\cdots b}{}_{c\cdots d}$ is dragged to a tensor field $\psi T^{a\cdots b}{}_{c\cdots d}$, with components

$$\psi T^{i\cdots j}{}_{k\cdots l}(P) = \partial_m\psi^i\cdots\partial_n\psi^j\partial_k\psi^{-1\,p}\cdots\partial_l\psi^{-1\,q}T^{m\cdots n}{}_{p\cdots q}[\psi^{-1}(P)]. \tag{A.28}$$

Finally, the Lie derivative of any tensor is given by

$$\mathcal{L}_{\boldsymbol{\xi}} T^{a\cdots b}{}_{c\cdots d} = -\frac{d}{d\lambda}\psi_\lambda T^{a\cdots b}{}_{c\cdots d}, \tag{A.29}$$

and a calculation essentially identical to that for a vector field verifies that the definition yields Eq. (A.23).

A.2 Integration, forms, and densities

A.2.1 Introduction to integration on manifolds

In flat space, the area of a parallelogram spanned by the vectors $\mathbf{A},\mathbf{B}$ is $|\mathbf{A}\times\mathbf{B}| = |\epsilon_{ab}A^aB^b|$; and the volume spanned the vectors $\mathbf{A},\mathbf{B},\mathbf{C}$ is $|\mathbf{A}\times\mathbf{B}\cdot\mathbf{C}| = |\epsilon_{abc}A^aB^bC^c|$. Similarly, in Minkowski space, requiring that the volume spanned by four orthonormal vectors t^α, x^α, y^α, z^α be 1 implies that a parallelepiped Ω spanned by any four vectors A^α, B^α, C^α, D^α is

$$|\Omega| = |\epsilon_{\alpha\beta\gamma\delta}A^\alpha B^\beta C^\gamma D^\delta|.$$

The vectors are positively oriented if $\epsilon_{\alpha\beta\gamma\delta}A^\alpha B^\beta C^\gamma D^\delta > 0$.

The volume of an arbitrary region Ω is obtained by adding volumes of infinitesimal parallelepipeds spanned by vectors along the coordinate axes, e_0^α, e_1^α, e_2^α, e_3^α with lengths Δx^0, Δx^1, Δx^2, Δx^3:

$$\Delta^4 V = \epsilon_{0123}\Delta x^0\Delta x^1\Delta x^2\Delta x^3$$
$$= \frac{1}{4!}\epsilon_{\mu\nu\sigma\tau}\Delta x^\mu\Delta x^\nu\Delta x^\sigma\Delta x^\tau(-1)^\Pi,$$

where $(-1)^\Pi = 1$ when $\mu,\nu\,\sigma,\tau$ is an even permutation Π of $0,1,2,3$ and $(-1)^\Pi = -1$ for an odd permutation. Because $\epsilon_{\alpha\beta\gamma\delta}A^\alpha B^\beta C^\gamma D^\delta$ is a scalar, the volume of a region of flat space is given in any chart by

$$|\Omega| = \int_\Omega \epsilon_{0123}\, dx^0 dx^1 dx^2 dx^3 \equiv \int_\Omega d^4V.$$

The Jacobian, $\left|\frac{\partial x}{\partial x'}\right|$, that relates the volume element in two different coordinate systems arises from the coordinate transformation of the totally antisymmetric tensor $\epsilon_{\alpha\beta\gamma\delta}$:

$$\epsilon_{0'1'2'3'} = \frac{\partial x^\mu}{\partial x'^0}\frac{\partial x^\nu}{\partial x'^1}\frac{\partial x^\sigma}{\partial x'^2}\frac{\partial x^\tau}{\partial x'^3}\epsilon_{\mu\nu\sigma\tau} = \left|\frac{\partial x}{\partial x'}\right| \epsilon_{0123}.$$

A curved space is locally flat in the sense that, in a locally inertial coordinate system, the metric components are flat up to quadratic order in the coordinates. By demanding that the volumes of small regions, to first order in the length of a side, are those measured by a locally inertial observer using her local Minkowski metric, one uniquely picks out the volume element

$$d^4V = \epsilon_{0123}\, dx^0 dx^1 dx^2 dx^3. \tag{A.30}$$

Because the totally antisymmetric tensor ϵ_{0123} has the value $\sqrt{|g|}$, the volume element can be written in the equivalent form[4]

$$d^4V = \sqrt{|g|}\, dx^0 dx^1 dx^2 dx^3, \tag{A.31}$$

and in n dimensions,

$$d^nV = \epsilon_{1\cdots n} dx^1 \cdots dx^n = \sqrt{|g|}\, dx^1 \cdots dx^n. \tag{A.32}$$

In index notation, one writes

$$d^4V = \epsilon_{a\cdots b} dS^{a\cdots b}$$

and thinks of $dS^{a\cdots b}$ having components

$$\text{“}\ dS^{\mu\nu\sigma\tau} = \pm\frac{1}{4!} dx^\mu dx^\nu dx^\sigma dx^\tau.\text{”}$$

The integral,

$$\int_\Omega f d^nV = \int_\Omega f\, \epsilon_{1\cdots n}\, dx^1 \cdots dx^n = \int_\Omega f \sqrt{|g|}\, dx^1 \cdots dx^n, \tag{A.33}$$

over a region Ω is is well defined (that is, its value is independent of the choice of coordinates), because, under a change of coordinates, the integrand on the right side is multiplied by the Jacobian $\left|\frac{\partial x}{\partial x'}\right|$.

[4] The equality $\epsilon_{0123} = \sqrt{|g|}$ can be obtained as follows. In an orthonormal frame (e.g., in locally inertial coordinates), ϵ_{0123} and $\sqrt{|g|}$ each have the value 1. Under a coordinate transformation from x^μ to x'^μ, $\sqrt{|g|}$, like ϵ_{0123}, is multiplied by the absolute value of the Jacobian $|\partial x/\partial x'|$, implying that for any positively oriented coordinate system, $\sqrt{|g|} = \epsilon_{0123}$.

A.2.2 Forms and densities

The tensor $\epsilon_{a\cdots b}$ and the quantity $\sqrt{|g|}$ that appear in the alternative ways of writing an integral are, respectively, an example of a form and a scalar density. As we will see, there is a duality between forms and densities that underlies a equivalence between Stokes's theorem and Gauss's theorem. We define forms and densities, present the duality that relates them, and go on to the corresponding duality relating the integral theorems.

Forms

Definition. A *p-form* $\sigma_{a\cdots b}$ is an antisymmetric, covariant tensor with p indices. In particular, a scalar f is a 0-form, a covariant vector A_a is a 1-form, and an antisymmetric 2-index tensor F_{ab} is a 2-form.

Definition. The exterior derivative $d\sigma$ of a p-form σ is the $p+1$ form

$$(d\sigma)_{ab\cdots c} = (p+1)\nabla_{[a}\sigma_{b\ldots c]}. \tag{A.34}$$

The factor $p+1$ is the number of independent ways of distributing the $p+1$ indices between ∇ and σ. The antisymmetry implies that $d\sigma$ is independent of the derivative operator; in any chart it has components

$$(d\sigma)_{ij\cdots k} = (p+1)\partial_{[i}\sigma_{j\ldots k]}. \tag{A.35}$$

Antisymmetry and the commutativity of partial derivatives imply for any form σ

$$d^2\sigma = 0. \tag{A.36}$$

Lie derivatives and exterior derivatives commute,

$$\mathcal{L}_{\mathbf{v}} d\sigma = d\mathcal{L}_{\mathbf{v}}\sigma, \tag{A.37}$$

and the two derivatives satisfy the Cartan identity,

$$\mathcal{L}_{\mathbf{v}}\sigma = v\cdot(d\sigma) + d(v\cdot\sigma), \text{where } (v\cdot\sigma)_{a\cdots b} := v^c\sigma_{ca\cdots b}. \tag{A.38}$$

Each relation can be proved by induction on p (the number of indices of σ). For an n-form in n dimensions, the second relation can be written as

$$\mathcal{L}_{\mathbf{v}}\sigma_{a\cdots b} = \nabla_c(\sigma_{a\cdots b}v^c), \tag{A.39}$$

with special case

$$\mathcal{L}_{\mathbf{v}}\epsilon_{a\cdots b} = \nabla_c(\epsilon_{a\cdots b}v^c) = \epsilon_{a\cdots b}\nabla_c v^c. \tag{A.40}$$

In n dimensions, any nonzero n-form is functionally proportional to any other, because each has only one independent component. In particular, because any n-form σ is given by $\sigma_{a\cdots b} = f\epsilon_{a\cdots b}$ for some scalar f, the integral of an n-form is well defined, given in any coordinate system by

$$\int_\Omega \sigma_{a\cdots b}dS^{a\cdots b} = \int \sigma_{1\cdots n}dx^1\cdots dx^n.$$

(Again the integral is well defined because a change of coordinates multiplies the value of a p-form by the Jacobian of the transformation.)

Densities

A *scalar density* $\mathfrak{f}$ (of weight 1), by definition, transforms under a change of coordinates in the same way as one component of an n-form in n dimensions: $\mathfrak{f} \rightarrow \left|\frac{\partial x}{\partial x'}\right| \mathfrak{f}$. Just as one can write any n-form as a multiple of $\epsilon_{a \cdots b}$ (once one is given a metric), one can write any scalar density as a scalar multiple of $\sqrt{|g|}$:

$$\mathfrak{f} = f\sqrt{|g|}, \tag{A.41}$$

for some scalar f (namely, $\mathfrak{f}/\sqrt{|g|}$). One can analogously introduce vector and tensor densities by transformation laws that differ from those of vectors and tensors by the Jacobian of the transformation: The change of the components of a vector density under a coordinate transformation is given by $\mathfrak{j}^i \rightarrow \left|\frac{\partial x}{\partial x'}\right| \frac{\partial x'^i}{\partial x^k} \mathfrak{j}^k$. Again, one can write any vector density in the form $\mathfrak{j}^a = j^a \sqrt{|g|}$, with j^a a vector field.

The Lie derivative of a scalar or tensor density can be deduced from this fact using Eq. (2.50), $\mathcal{L}_{\mathbf{v}} g = 2g\nabla_a v^a$ (a metric-free derivation uses the geometric definition of Lie derivative and the fact that the action of a diffeo on a density differs from its action on a tensor by the inverse Jacobian of the diffeo): We have

$$\mathcal{L}_{\mathbf{v}} \sqrt{|g|} = \sqrt{|g|}\nabla_a v^a = \nabla_a(\sqrt{|g|}v^a). \tag{A.42}$$

Finally, Eqs. (A.2) and (A.42) imply that the Lie derivative of a general scalar density $\mathfrak{f}$ is

$$\mathcal{L}_{\mathbf{v}} \mathfrak{f} = \nabla_a(\mathfrak{f} v^a). \tag{A.43}$$

Duality. A duality between p-forms $\sigma_{a_1 \cdots a_p}$ and antisymmetric tensor densities $\mathcal{A}^{a_1 \cdots a_q}$, with $q = n - p$ indices (n the dimension of the space) is given by

$$\mathcal{A}^{a_1 \cdots a_q} = \frac{1}{p!}\sqrt{|g|}\epsilon^{a_1 \cdots a_q b_1 \cdots b_p} \sigma_{b_1 \cdots b_p}. \tag{A.44}$$

The inverse relation is $\sigma_{a_1 \cdots a_p} = \pm\frac{1}{q!}|g|^{-1/2}\epsilon_{a_1 \cdots a_p b_1 \cdots b_q} \mathcal{A}^{b_1 \cdots b_q}$, where the sign is positive for a positive definite metric, negative for a metric with Lorentz signature.

The divergence $\nabla \cdot \mathcal{A}$, of an antisymmetric density,

$$(\nabla \cdot \mathcal{A})^{a \cdots b} := \nabla_c \mathcal{A}^{ca \cdots b}, \tag{A.45}$$

is dual to the exterior derivative $d\sigma$ of the form σ dual to $\mathcal{A}$, and it is independent of the choice of ∇, having components

$$\nabla_k \mathcal{A}^{ki \cdots j} = \partial_k \mathcal{A}^{ki \cdots j}. \tag{A.46}$$

In particular, the equation for the divergence of a vector density $\mathcal{A}^a$,

$$\nabla_a \mathcal{A}^a = \partial_i \mathcal{A}^i, \tag{A.47}$$

is equivalent (once one has a metric) to the familiar relation $\nabla_a A^a = \frac{1}{\sqrt{|g|}}\partial_i(\sqrt{|g|}A^i)$.

The relations (A.36), (A.37), and (A.38) are dual to the relations

$$\nabla \cdot (\nabla \cdot \mathcal{A}) = 0, \tag{A.48}$$

$$\mathcal{L}_{\mathbf{v}} \nabla \cdot \mathcal{A} = \nabla \cdot \mathcal{L}_{\mathbf{v}} \mathcal{A}, \tag{A.49}$$

$$\mathcal{L}_{\mathbf{v}} \mathcal{A} = \nabla \cdot (v \wedge \mathcal{A}) + v \wedge \nabla \cdot \mathcal{A}, \tag{A.50}$$

where $(v \wedge \mathcal{A})^{ab\cdots c} := (q+1)v^{[a}\mathcal{A}^{b\cdots c]}$.

On an n-dimensional manifold, integrals are naturally defined for n-forms or for scalar densities, which can be said to be *dual* to their corresponding n-forms. That is, the integrals

$$\int \omega_{a\cdots b} dS^{a\cdots b} = \int \omega_{1\cdots n} dx^1 \cdots dx^n \quad \text{and} \quad \int \mathfrak{f} dx^1 \cdots dx^n \tag{A.51}$$

are well defined because under a change of coordinates, each integrand is multiplied by the Jacobian of the transformation. The mathematical literature adopts an index-free notation in which the integral of an n-form ω over an n-dimensional region Ω is written $\int_\Omega \omega$.

A.3 Gauss's theorem and Stokes's theorem

To understand the relation between differential and integral conservation laws for relativistic stars and black holes, one needs the generalization of Stokes's theorem and Gauss's theorem to manifolds. It is helpful to see quickly how these generalizations go before we move to a more formal presentation. The simplest version of Stokes's theorem is its 2-dimensional form, namely Green's theorem:

$$\int_S (\partial_x A_y - \partial_y A_x) dx\, dy = \int_c (A_x dx + A_y dy),$$

where c is a curve bounding the 2-surface S. The theorem involves the integral over a 2-surface of the antisymmetric tensor $\nabla_a A_b - \nabla_b A_a$. In 3 dimensions, the tensor is dual to the curl of $\mathbf{A}$: $(\nabla \times \mathbf{A})^a = \epsilon^{abc}\nabla_b A_c$; Stokes's generalization of Green's theorem can then be written either in the form

$$\int_S (\boldsymbol{\nabla} \times \mathbf{A}) \cdot d\mathbf{S} = \int_c \mathbf{A} \cdot \mathbf{dl}$$

or in terms of the antisymmetric tensor $\nabla_a A_b - \nabla_b A_a$

$$\int_S (\nabla_a A_b - \nabla_b A_a) dS^{ab} = \int_c A_a dl^a, \tag{A.52}$$

where, for an antisymmetric tensor F_{ab}, $F_{ab}dS^{ab}$ means $F_{12}dx^1dx^2 + F_{23}dx^2dx^3 + F_{31}dx^3dx^1$. Written in this form, the theorem is already correct in a curved spacetime. The reason is that the antisymmetric derivative $\nabla_a A_b - \nabla_b A_a$ has in curved space the same form it has in flat space: As we have seen, its components in any coordinate system are just $\partial_i A_j - \partial_j A_i$.

As a result, the flat-space proof of Green's theorem and Stokes's theorem, based on the fundamental theorem of calculus ($\int_a^b f'(x)dx = f(b) - f(a)$), holds in curved space as well: Let S be a coordinate square in a surface of constant x^3. Then

$$\int_S (\partial_1 A_2 - \partial_2 A_1)dx^1dx^2 = \int_c (A_1 dx^1 + A_2 dx^2),$$

with the boundary of the square traversed counterclockwise as seen from above the square.

Gauss's theorem,

$$\int_V \nabla_a A^a d^3x = \int_S A^a dS_a, \tag{A.53}$$

with S a surface bounding the volume V, again has a simple generalization to curved space. Although the divergence $\nabla_a A^a$ does not have components $\partial_i A^i$, the divergence of $\mathcal{A}^a := A^a\sqrt{g}$ does: $\nabla_a \mathcal{A}^a = \partial_i \mathcal{A}^i$. Again the flat-space proof of Gauss's theorem follows from an integration over a coordinate cube using the fundamental theorem of calculus for the integral over each coordinate; and again the integration over a coordinate cube has the identical form in curved space. If V is a coordinate cube,

$$\int_V \partial_i \mathcal{A}^i dx^1\, dx^2\, dx^3 = \int_S (\mathcal{A}^1 dx^2 dx^3 + \mathcal{A}^2 dx^1 dx^3 + \mathcal{A}^3 dx^1 dx^2);$$

any volume is approximated by an arbitrarily fine division into coordinate cubes.

A.3.1 Integrals on a submanifold

A metric $g_{\alpha\beta}$ on spacetime induces a 3-metric ${}^3g_{ab}$ on a 3-dimensional hypersurface V. This is easily seen as follows: A metric is an inner product, assigning to any two vectors u^α and v^α on M the scalar $g_{\alpha\beta}u^\alpha v^\beta$. A vector v^a on V can be identified with a vector v^α on M that has no component along the normal n_α to V: $v^\alpha n_\alpha = 0$. Then

$${}^3g_{ab}v^a w^b := g_{\alpha\beta}v^\alpha w^\beta. \tag{A.54}$$

More concretely, one can define ${}^3g_{ab}$ by requiring that in a coordinate system $\{x^0, x^1, x^2, x^3\}$ for which x^0 is constant on V,

$${}^3g_{ij} = g_{ij}. \tag{A.55}$$

The definitions agree, because in coordinates of this kind, v^α and w^α are vectors on V if and only if $v^0 = w^0 = 0$. (Because n_α is along $\nabla_\alpha x^0$ and $\nabla_\mu x^0 = \delta^0_\mu$, we have $v^\mu \nabla_\mu x^0 = v^0$.)

Given the 3-metric g_{ab}, the 3-form ϵ_{abc} on V is uniquely defined up to sign. When one chooses coordinates $\{x^0, x^1, x^2, x^3\}$ on a spacetime, one ordinarily chooses the tensor $\epsilon_{\alpha\beta\gamma\delta}$ for which the coordinate system has positive orientation, for which $\epsilon_{0123} = \sqrt{|g|}$, not $-\sqrt{|g|}$, and one calls the direction of increasing x^0 the future. When $\epsilon_{0123} > 0$ and when x^0 is constant on V and increasing to the future, we will say the corresponding positive orientation V is the one for which $\epsilon_{123} = \sqrt{^3g}$.

In discussing hypersurfaces in spacetime, it is convenient to avoid the distinction between tensors on V and tensors on M. In the same way that a vector v^a on V is uniquely identified with a vector v^α on M for which $v^\alpha n_\alpha = 0$, we can identify the 3-metric $^3g_{ab}$ and 3-form ϵ_{abc} with tensors $^3g_{\alpha\beta}$ and $\epsilon_{\alpha\beta\gamma}$ on M that are orthogonal to n_α and whose spatial indices coincide with those of $^3g_{ab}$ and ϵ_{abc}.

$$^3g_{\alpha\beta} = g_{\alpha\beta} + n_\alpha n_\beta, \qquad \epsilon_{\alpha\beta\gamma} = n^\delta \epsilon_{\delta\alpha\beta\gamma}, \tag{A.56}$$

where n^α is the future pointing unit normal to V, namely

$$n_\mu = \frac{-\delta^0_\mu}{(-g^{00})^{1/2}}. \tag{A.57}$$

We then have $n^0 = \sqrt{-g^{00}} > 0$, and as we show next, the spatial components of $\epsilon_{\alpha\beta\gamma}$ are $\epsilon_{ijk} = \sqrt{^3g}$.

It follows immediately from Eq. (A.56) that $^3g_{\alpha\beta}$ satisfies the defining conditions $^3g_{\alpha\beta}n^\beta = 0$ and $^3g_{ab}u^a v^b = {}^3g_{\alpha\beta}u^\alpha v^\beta = g_{\alpha\beta}u^\alpha v^\beta$, for u^α and v^α vectors on V (equivalently $^3g_{\alpha\beta}$ has spatial components g_{ij}).

The spatial components of $\epsilon_{\alpha\beta\gamma}$ are

$$\epsilon_{\mu ijk} n^\mu = \left(-g^{00}\right)^{1/2} \epsilon_{0ijk},$$

or $\epsilon_{123} = \sqrt{gg^{00}}$. Now $g^{00} = \dfrac{|\text{minor of } g_{00}|}{g}$, and the minor of g_{00} is the matrix of the spatial components $g_{ij} = {}^3g_{ij}$ of the metric. Thus

$$\epsilon_{123} = \sqrt{^3g}. \tag{A.58}$$

Warning. Although contravariant tensors $T^{\alpha\cdots\beta}$ are tensors on V if and only if their zeroth components all vanish, the zeroth components of covariant tensors $T_{\alpha\cdots\beta}$ on V will not generally vanish, because in general $g_{0i} \neq 0$. The information about the 3-dimensional objects is carried by the spatial components, and the spatial components of the corresponding 3-dimensional and 4-dimensional tensors coincide.

One writes

$$\int dV = \int_V \epsilon_{\alpha\beta\gamma} dS^{\alpha\beta\gamma}, \tag{A.59}$$

for the volume of V, where one thinks of $dS^{\alpha\beta\gamma}$ as antisymmetric, with nonzero components $\pm\frac{1}{3!}dx^i dx^j dx^k$, with the sign depending on whether i, j, k is an even or odd permutation of $1, 2, 3$. In coordinates $\{x^0, x^1, x^2, x^3\}$ with V an $x^0 =$ constant surface,

$$\int dV = \int \sqrt{^3g}\, d^3x = \int \sqrt{gg^{00}}\, dx^1 dx^2 dx^3.$$

For any 3-form,

$$\int \omega_{\alpha\beta\gamma} dS^{\alpha\beta\gamma} = \int \omega_{123} dx^1 dx^3 dx^3.$$

If S is a 2-dimensional surface with unit normals n^α and m^α, and we pick the order of the two normals correctly, its area is given in terms of $\epsilon_{\alpha\beta} = \epsilon_{\alpha\beta\gamma\delta} n^\alpha m^\beta$ by

$$\int_S dS = \int_S \epsilon_{\alpha\beta} dS^{\alpha\beta}. \tag{A.60}$$

In a chart x^0, x^1, x^2, x^3, with S a surface of constant x^0 and x^1, we have

$$\epsilon_{23} = \sqrt{^2g} = \left[g_{22}g_{33} - (g_{23})^2\right]^{1/2},$$

and $\int dS = \int \sqrt{^2g} dx^2 dx^3$.

Example. The magnetic flux through a surface S of radius R in t, r, θ, ϕ coordinates is

$$\int_S F_{ij} dS^{ij} = \int_S F_{\theta\phi} d\theta d\phi.$$

Finally, the integral of a scalar along a curve c is $f = \int_c f dl$, with dl proper length along c; the corresponding line integral of a 1-form along c is

$$\int_c A_\alpha dl^\alpha = \int_c A_\alpha t^\alpha dl,$$

where t^α is the unit tangent to c.

Summary of notation

$$\int f d^4V = \int f\epsilon_{\alpha\beta\gamma\delta} dS^{\alpha\beta\gamma\delta} = \int f\sqrt{|g|}\ d^4x$$

$$\int \omega = \int \omega_{\alpha\beta\gamma\delta} dS^{\alpha\beta\gamma\delta} = \int \omega_{0123} d^4x$$

$$\int f dV = \int f\epsilon_{\alpha\beta\gamma} dS^{\alpha\beta\gamma} = \int f\sqrt{{}^3g}\ d^3x = \int_{x^0\ \text{constant}} f\sqrt{gg^{00}}\ d^3x$$

$$\int \omega = \int \omega_{\alpha\beta\gamma} dS^{\alpha\beta\gamma} = \int_{x^0\ \text{constant}} \omega_{123} dx^1 dx^2 dx^3$$

$$\int f dS = \int f\epsilon_{\alpha\beta} dS^{\alpha\beta} = \int f\sqrt{{}^2g}\ d^2x = \int_{x^0, x^1\ \text{constant}} f\left[g_{22}g_{33} - (g_{23})^2\right]^{1/2} d^2x$$

$$\int \omega = \int \omega_{\alpha\beta} dS^{\alpha\beta} = \int_{x^0, x^1\ \text{constant}} \omega_{23} dx^2 dx^3$$

$$\int_c f dl = \int_{x^0, x^1, x^2\ \text{constant}} f\sqrt{g_{33}}\ dx^3$$

$$\int_c \omega = \int_c \omega_\alpha dl^\alpha = \int_{x^0, x^1, x^2\ \text{constant}} \omega_3 dx^3$$

Gauss's theorem (divergence theorem)

In $\mathbb{R}^n$, in Cartesian coordinates, the integral of a divergence over an n-cube can be expressed as a surface integral after an integration by parts:

$$\begin{aligned}\int_V \partial_i A^i d^n x &= \int_V \partial_1 A^1 dx^1 dx^2 \cdots dx^n + \cdots + \int_V \partial_n A^n dx^n dx^1 \cdots dx^{n-1} \\ &= \int_{\partial_{1+} V} A^1 dx^2 \cdots dx^n - \int_{\partial_{1-} V} A^1 dx^2 \cdots dx^n + \cdots \\ &\quad + \int_{\partial_{n+} V} A^n dx^1 \cdots dx^{n-1} - \int_{\partial_{n-} V} A^n dx^1 \cdots dx^{n-1} \\ &= \int_{\partial V} A^i dS_i \quad (\partial V \text{ means the boundary of } V)\end{aligned} \tag{A.61}$$

where $dS_i = \pm\epsilon_{ij\cdots k} dx^j \cdots dx^k \dfrac{1}{(n-1)!}$, with

$dS_1 = +dx^2 \cdots dx^n$ for x^1 increasing outward,

$dS_1 = -dx^2 \cdots dx^n$, for x^1 increasing inward.

More generally the integral over any volume V in $\mathbb{R}^n$ of a divergence is related to a surface integral by

$$\int_V \partial_i A^i\ d^n V = \int_{\partial V} A^i dS_i = \int_{\partial V} A^i n_i dS$$

where n_i is the unit outward normal to S (along the gradient of an outwardly increasing scalar) and dS the area element of S. In curved space the analogous result follows immediately from the form of the divergence of a vector density:

$$\int_\Omega \nabla_a A^a d^n V = \int_\Omega \partial_i \left(\sqrt{|g|} A^i\right) d^n x. \tag{A.62}$$

To generalize Gauss's theorem to an n-dimensional manifold M with metric g_{ab}, let Ω be an n-dimensional submanifold with smooth boundary $\partial\Omega$. Define a surface element dS_a on $\partial\Omega$ by requiring that, in any chart $x^1, \ldots, x^n$ for which x^1 is constant on $\partial\Omega$ and increasing outward,

$$dS_i = \nabla_i x^1 \sqrt{|g|}\, dx^2 \cdots dx^n = \delta_i^1 \sqrt{|g|}\, dx^2 \cdots dx^n. \tag{A.63}$$

This definition involves no choice of orientation (in fact, Ω need not be orientable, as long as the outward normal to $\partial\Omega$ is well defined.)

Gauss's Theorem in Curved Space. For any smooth vector field A^a on M,

$$\int_\Omega \nabla_a A^a d^n V = \int_{\partial\Omega} A^a dS_a. \tag{A.64}$$

Sketch of proof. When Ω is a coordinate cube, the steps in Eq. (A.61) go through as written, with A^i replaced by $\mathcal{A}^i = A^i \sqrt{|g|}$:

$$\begin{aligned}\int_\Omega \partial_i \mathcal{A}^i d^n x &= \int_{\partial_{1+} V} \mathcal{A}^1 dx^2 \cdots dx^n - \int_{\partial_{1-} V} \mathcal{A}^1 dx^2 \cdots dx^n + \cdots \\ &\quad + \int_{\partial_{n+} V} \mathcal{A}^n dx^1 \cdots dx^{n-1} - \int_{\partial_{n-} V} \mathcal{A}^n dx^1 \cdots dx^{n-1} \\ &= \int_{\partial\Omega} A^i dS_i.\end{aligned}$$

Decomposing the volume of integration into a set of coordinate cubes and taking the limit as the size of each cube shrinks to zero yields

$$\int_\Omega \nabla_a A^a d^n V = \int_{\partial\Omega} A^a dS_a; \tag{A.65}$$

as usual, surface terms from cubes that share a surface cancel, because the outward normal to one cube is the inward normal to the adjacent cube.

This form is correct for a region in a space with a metric, independent of the signature of the metric. When $\partial\Omega$ has a unit outward normal n_a (along the gradient of a scalar that increases outward), one can write dS_a in the form $dS_a = n_a dS$. In this case,

$$\int \nabla_a A^a d^n V = \int A^a n_a dS. \tag{A.66}$$

Example. The integral form of baryon mass conservation $\nabla_\alpha(\rho u^\alpha) = 0$ is

$$0 = \int_\Omega \nabla_\alpha(\rho u^\alpha) d^4V = \int_{\partial\Omega} \rho u^\alpha dS_\alpha$$
$$= \int_{V_2} \rho u^\alpha dS_\alpha - \left| \int_{V_1} \rho u^\alpha dS_\alpha \right|.$$

Here the fluid is taken to have finite spatial extent, and the spacetime region Ω is bounded by the initial and final spacelike hypersurfaces V_1 and V_2. In a coordinate system for which V_1 and V_2 are surfaces of constant t, with t increasing to the future, we have $dS_\mu = \nabla_\mu t\sqrt{|g|}d^3x = \delta^t_\mu\sqrt{|g|}d^3x$ on V_2, $dS_\mu = -\delta^t_\mu\sqrt{|g|}d^3x$ on V_1, and

$$\int_\Omega \nabla_\alpha(\rho u^\alpha) d^4V = \int_{V_2} \rho u^t \sqrt{|g|} d^3x - \int_{V_1} \rho u^t \sqrt{|g|} d^3x. \tag{A.67}$$

If, on a slicing of spacetime, one chooses on each hypersurface V a surface element dS_α along $+\nabla_\alpha t$, the conservation law is then

$$M_0 = \int_V \rho u^\alpha dS_\alpha = \text{constant}. \tag{A.68}$$

Note that the fact that one can write the conserved quantity associated with a current j^α in the form

$$\int_V j^\alpha dS_\alpha = \int_V j^t \sqrt{|g|}\ d^3x,$$

means that there is no need to introduce n_α and $\sqrt{{}^3g}$ to evaluate the integral. This fact is *essential* if one is evaluating an integral $\int j^\alpha dS_\alpha$ over a null surface, where there is no unit normal. The flux of energy or of baryons across the horizon of a Schwarzschild black hole, for example, can be computed in Eddington-Finkelstein or Kruskal coordinates: In ingoing Eddington-Finkelstein coordinates v, r, θ, ϕ, the horizon is a surface of constant r, and we have

$$\int j^\alpha dS_\alpha = \int j^r \sqrt{|g|}\ dv d\theta d\phi.$$

Generalized divergence theorem

The key to Gauss's theorem is the fact that the divergence of a vector density has the form $\partial_i\left(\sqrt{|g|}\ A^i\right)$ or $\partial_i \mathcal{A}^i$. This is true of any q-index antisymmetric tensor density, $\mathcal{A}^{a\cdots b} = A^{a\cdots b}\sqrt{|g|}$, and an analogous theorem holds. Because the text uses only the case of a two-index antisymmetric tensor, the electromagnetic field $F^{\alpha\beta}$, and because the way one extends the proof will be clear, we will give the generalization in detail for this case. The theorem now relates an integral over an $n-1$-dimensional submanifold S of M to an integral over its $(n-2)$-dimensional boundary ∂S:

$$\int_S \nabla_b A^{ab} dS_a = \int_{\partial S} A^{ab} dS_{ab}, \tag{A.69}$$

where the sign of dS_a and the meaning of dS_{ab} are defined as follows.

Let $x^1, \ldots, x^n$ be a positively oriented chart on a subset of M for which x^1 is constant on S, and with x^2 constant on ∂S and increasing outward. Choose the sign of dS_a by $dS_a = \epsilon_{ab\cdots c} dS^{b\cdots c}$ or, equivalently, by requiring $dS_a = \nabla_a x^1 \sqrt{|g|} dx^2 \cdots dx^n$. The volume element dS_{ab} is similarly chosen to satisfy, in our oriented chart,

$$dS_{ab} = \nabla_{[a} x^1 \nabla_{b]} x^2 \sqrt{|g|} dx^3 \cdots dx^n. \tag{A.70}$$

Then in the coordinates' domain,

$$\int_S \nabla_b A^{ab} dS_a = \int_S \nabla_b (A^{ab} \nabla_a x^1 \sqrt{|g|}) dx^2 \cdots dx^n. \tag{A.71}$$

But the last integrand is just the divergence of the vector density $\tilde{A}^b = A^{ab} \nabla_a x^1 \sqrt{|g|}$, and we have already proved Gauss's law for this case:

$$\int_S \nabla_b \tilde{A}^b \sqrt{|g|} dx^2 \cdots dx^n = \int_{\partial S} \tilde{A}^b d\tilde{S}_b, \tag{A.72}$$

where, in each chart, $d\tilde{S}_b = \nabla_b x^2 \sqrt{|g|} dx^3 \cdots dx^n$. Finally, $\tilde{A}^b dS_b = A^{ab} dS_{ab}$, whence Eq. (A.72) is identical to the generalized divergence theorem, Eq. (A.69).

Example (Electric charge). Let V be a ball containing a charge Q. The 4-dimensional form of Gauss's law relates the charge $Q = \int_V j^\alpha dS_\alpha$ in V to the electric flux $\int F^{\alpha\beta} dS_{\alpha\beta}$ through the 2-dimensional surface of V:

$$Q = \int_V j^\alpha dS_\alpha = \frac{1}{4\pi} \int_V \nabla_\beta F^{\alpha\beta} dS_\alpha = \frac{1}{4\pi} \int_{\partial V} F^{\alpha\beta} dS_{\alpha\beta}. \tag{A.73}$$

Pick positively oriented coordinates t, r, θ, ϕ for which V is a t =constant surface and ∂V an r = constant surface. Then

$$Q = \frac{1}{4\pi} \int F^{tr} \sqrt{|g|} d\theta d\phi. \tag{A.74}$$

Flat space:

$$F^{tr} = E^r = \frac{Q}{r^2} \Longrightarrow \quad \int_V j^\alpha dS_\alpha = \frac{1}{4\pi} \int \left(\frac{Q}{r^2}\right) (r^2 \sin\theta) d\theta d\phi = \frac{Q}{4\pi} \int d\Omega = Q.$$

For a q-index antisymmetric tensor $A^{a\cdots bc}$, the generalized divergence theorem takes the form $\int_S \nabla_c A^{a\cdots bc} dS_{a\cdots b} = \int_{\partial S} A^{a\cdots bc} dS_{a\cdots bc}$.

A.3.2 Stokes's theorem

The divergence $\nabla_a A^a$ of a vector on an n-dimensional manifold is dual to the exterior derivative $(d\omega)_{ab\cdots c} = n \nabla_{[a} \omega_{b\cdots c]}$ of the $(n-1)$-form ω dual to A^a. That is, with

$$\omega_{b\cdots c} := A^a \epsilon_{ab\cdots c}, \tag{A.75}$$

we have

$$\epsilon_{ab\cdots c} \nabla_d A^d = (d\omega)_{ab\cdots c}. \tag{A.76}$$

Because both sides of the equation are n-forms, one need check only one component: $n\nabla_{[1}\omega_{2\cdots n]} = \nabla_1(A^d\epsilon_{d2\cdots n}) - \nabla_2(A^d\epsilon_{d13\cdots n}) - \cdots - \nabla_n(A^d\epsilon_{d2\cdots n-11}) = \epsilon_{12\cdots n}\nabla_d A^d$. The corresponding dual of the generalized divergence theorem above is called Stokes's Theorem.

Stokes's Theorem. *Let ω be an $(n-1)$-form on an n-dimensional manifold S with boundary ∂S. Then*

$$\int_S d\omega = \int_{\partial S} \omega. \tag{A.77a}$$

In index notation,

$$\int_S (d\omega)_{ab\cdots c} dS^{ab\cdots c} = \int_{\partial S} \omega_{b\cdots c} dS^{b\cdots c}. \tag{A.77b}$$

The theorem implicitly assumes an orientation for ∂S obtained from that of S by requiring that, if $x^1, \cdots, x^n$ is a positively oriented chart on S with ∂S a surface of constant x^1 and x^1 increasing outward, then $x^2, \cdots, x^n$ is a positively oriented chart for ∂S.

Proof. This dual of Gauss's theorem follows quickly from Eq. (A.62), in the form

$$\int_S \nabla_d A^d \epsilon_{ab\cdots c} dS^{ab\cdots c} = \int_{\partial S} A^a dS_a. \tag{A.78}$$

Define A^a by Eq. (A.75), and note that, with the orientation chosen above, $dS_a = \epsilon_{ab\cdots c} dS^{b\cdots c}$. Then

$$\int_S \nabla_d A^d \epsilon_{ab\cdots c} dS^{ab\cdots c} = \int_{\partial S} A^a \epsilon_{ab\cdots c} dS^{b\cdots c}, \tag{A.79}$$

and, from Eqs. (A.75) and (A.76), the result follows:

$$\int_S (d\omega)_{ab\cdots c} dS^{ab\cdots c} = \int_{\partial S} \omega_{b\cdots c} dS^{b\cdots c}. \tag{A.80}$$

□

Example: Stokes's theorem in three dimensions. Let A_a be a 3-vector. $(dA)_{ab} = \nabla_a A_b - \nabla_b A_a$;

$$\int_S (\nabla_a A_b - \nabla_b A_a) dS^{ab} = \int_c A_a dl^a,$$

and c is the curve bounding S. As noted at the beginning of this section, this is equivalent to the usual form of Stokes's theorem in vector calculus:

$$\int_S \vec{\nabla} \times \vec{A} \cdot d\vec{S} = \int_S (\epsilon^{abc}\nabla_b A_c)\, \epsilon_{ade} dS^{de} = \int_S (\nabla_a A_b - \nabla_b A_a) dS^{ab} = \int_c \vec{A} \cdot d\vec{l}.$$

A.3.3 Diffeomorphism invariance

The usual invariance of an integral under a coordinate transformation has as its active equivalent the invariance of an integral under a diffeo. Let $\omega_{a...b}$ be an n-form on an n-dimensional volume V. With $\psi(V)$ the image of V and $\psi\omega_{a...b}$ the dragged n-form, the invariance relation is

$$\int_{\psi(V)} \psi\omega = \int_V \omega. \tag{A.81}$$

As in the case of coordinate transformations, invariance under diffeos follows from the fact that the components of $\omega_{a...b}$ change by a Jacobian. Intuitively, the invariance follows from the physical equivalence of diffeo-related tensors on diffeo-related domains.

A corollary for a family of diffeos, ψ_λ, is used in Eq. (1.52) to obtain the relation between conservation of vorticity and conservation of circulation:
Let ξ^a be the generator of the family ψ_λ. Then

$$\frac{d}{d\lambda}\int_{\psi_\lambda(V)} \omega = \int_V \mathcal{L}_{\boldsymbol{\xi}}\omega. \tag{A.82}$$

The proof is immediate from Eq. (A.29), starting from Eq. (A.81) in the form

$$\int_{\psi_\lambda(V)} \omega = \int_V \psi_{-\lambda}\omega : \quad \frac{d}{d\lambda}\int_{\psi_\lambda(V)} \omega = \int_V \frac{d}{d\lambda}\psi_{-\lambda}\omega = \int_V \mathcal{L}_{\boldsymbol{\xi}}\omega.$$

In particular, for ψ_τ the family of diffeos describing the flow of a fluid with 4-velocity u^α, we have the equality used in Eq. (1.52),

$$\frac{d}{d\tau}\int_{c_\tau} hu_\alpha dl^\alpha = \int_c \mathcal{L}_{\mathbf{u}}(hu_\alpha)dl^\alpha, \tag{A.83}$$

where c is a closed curve in the fluid and $c_\tau = \psi_\tau(c)$.

Appendix B

The Newtonian limit of the two-potential formalism

B.1 Introduction

Because of its importance for astrophysics in a variety of contexts, the problem of the oscillations of rotating Newtonian stellar models has received considerable attention. Even so, up until very recently all attempts to obtain solutions to the equations governing the oscillations of rotating models had been unsuccessful. The lone exception is Clement's [145] analysis of certain axisymmetric normal modes.

In large part, the reason why the Newtonian oscillation problem has posed such difficulty is that the eigenequations for normal-mode pulsations are traditionally written down in terms of the Lagrangian displacement vector ξ^a. This leads to a complicated eighth-order system of equations for four dependent variables. This system has proved intractable for generally nonaxisymmetric modes. Although it has been used to develop variational principles (e.g., Lynden-Bell and Ostriker [450], Friedman and Schutz [241, 242]) for estimating normal-mode eigenfrequencies, it has yielded little information concerning the eigenfunctions themselves. And in a variety of contexts, detailed knowledge of the eigenfunctions is needed. For example, knowledge of the normal-mode eigenfunctions, in addition to the eigenfrequencies, is required for accurate assessment of the combined effects of gravitational radiation reaction and of viscous dissipation on the evolution and stability of millisecond pulsars, and for understanding the interaction between a pulsating star and a surrounding accretion disk.

We shall describe here a method developed in [331, 453, 325, 326, 327] for solving the normal-mode equations of rapidly rotating Newtonian stellar models. This method involves a reformulation of the stellar-pulsation equations in terms of two potential functions: $\delta\Phi$, the Eulerian perturbation of the gravitational potential; and δU, the difference between the Eulerian perturbation of the enthalpy and $\delta\Phi$. When the eigenequations are rewritten in terms of δU and $\delta\Phi$, a relatively simple fourth-order system of equations is obtained. This system has been solved successfully for both the eigenfrequencies and eigenfunctions of normal modes.

B.2 The two-potential formalism

The basic equations governing the evolution of a Newtonian fluid configuration are the continuity equation, Euler's equation, and Poisson's equation:

$$\frac{\partial \rho}{\partial t} + \nabla_a(\rho v^a) = 0, \tag{B.1}$$

$$\rho\left(\frac{\partial v^a}{\partial t} + v^b\nabla_b v^a\right) = -\nabla^a p + \rho\nabla^a\Phi, \tag{B.2}$$

$$\nabla^a\nabla_a\Phi = -4\pi G\rho. \tag{B.3}$$

Here the variables ρ, v^a, p, and Φ are the mass density, velocity vector, pressure, and gravitational potential, respectively; G is Newton's constant; and ∇_a is the standard Euclidean covariant derivative. Throughout this subsection, tensor indices are raised and lowered with the 3-dimensional Euclidean metric g_{ab} and its inverse g^{ab}. (In Cartesian coordinates, g_{ab} is the identity matrix.)

We are interested in the pulsations of an equilibrium stellar model that is axisymmetric and rotating, perhaps differentially, about its z-axis. Hence the unperturbed velocity field is of the form $v^a = \Omega\phi^a$, where Ω is the equilibrium angular velocity and ϕ^a is a rotational Killing vector field that satisfies

$$\nabla_a\phi_b + \nabla_b\phi_a = 0. \tag{B.4}$$

Our demand that meridional circulation be absent from the equilibrium state implies that Eq. (B.2) becomes

$$\frac{1}{2}\Omega^2\nabla_a\varpi^2 = \frac{\nabla_a p}{\rho} - \nabla_a\Phi, \tag{B.5}$$

where ϖ is the standard cylindrical radial coordinate.

Given an equilibrium configuration, we focus on small perturbations of its structure and motions away from equilibrium. We obtain the evolution equations governing these perturbations by linearizing Eqs. (B.1)–(B.3) in the perturbations. This yields the perturbed equations

$$\frac{\partial \delta\rho}{\partial t} + v^a\nabla_a\delta\rho + \nabla_a(\rho\delta v^a) = 0, \tag{B.6}$$

$$\frac{\partial \delta v^a}{\partial t} + v^b\nabla_b\delta v^a + \delta v^b\nabla_b v^a = -\frac{\nabla^a\delta p}{\rho} + \delta\rho\frac{\nabla^a p}{\rho^2} + \nabla^a\delta\Phi, \tag{B.7}$$

$$\nabla_a\nabla^a\delta\Phi = -4\pi G\delta\rho. \tag{B.8}$$

As usual, the symbol δ denotes the Eulerian perturbation of a quantity. Quantities not preceded by a δ are understood to be equilibrium values. We complete the system of perturbed equations by assuming that the Lagrangian change in pressure is proportional to the Lagrangian change in the density:

$$\Delta p = \delta p + \xi^a \nabla_a p = \frac{\Gamma_1 p}{\rho} \Delta \rho = \frac{\Gamma_1 p}{\rho} (\delta \rho + \xi^a \nabla_a \rho), \tag{B.9}$$

where Γ_1 is the adiabatic index, and ξ^a is the Lagrangian displacement. The vector ξ^a is related to the velocity perturbation by the relation

$$\delta v^a = \frac{\partial \xi^a}{\partial t} + v^b \nabla_b \xi^a - \xi^b \nabla_b v^a. \tag{B.10}$$

The traditional analysis of the perturbation equations involves eliminating δv^a in favor of ξ^a. The alternative method that has been developed recently proceeds in a different way, by eliminating ξ^a in favor of δv^a, and by then eliminating δv^a itself in a way that we now describe.

We focus on the normal-mode solutions to the above perturbation equations, that is, those solutions with time dependence $e^{i\omega t}$ and azimuthal-angle dependence $e^{im\phi}$, where ω is the mode frequency and m is an integer. For these modes, Eq. (B.10) yields an algebraic expression for ξ^a in terms of δv^b (or vice versa):

$$\xi_a = -i \left(\frac{g_{ab}}{\sigma} - \frac{i\phi_a \nabla_b \Omega}{\sigma^2} \right) \delta v^b, \tag{B.11}$$

where $\sigma \equiv \omega + m\Omega$. Eq. (B.9) now takes the form

$$\delta \rho = \frac{\rho}{\Gamma_1 p} \delta p + \frac{i\rho^2}{\sigma} A_a \delta v^a, \tag{B.12}$$

where

$$A_a := \frac{\nabla_a \rho}{\rho^2} - \frac{\nabla_a p}{\Gamma_1 \rho p}. \tag{B.13}$$

Note that $A_a = 0$ for barotropic configurations, which have adiabatic index $\Gamma_1 = \Gamma := d \log p / d \log \rho$. For general configurations, such as those in which the pressure is not a unique function of the density, A_a does not vanish. With $\delta\rho$ eliminated via Eq. (B.12) , Eq. (B.9) has the representation

$$i\sigma Q^{-1}_{ab} \delta v^b := (i\sigma g_{ab} + 2\nabla_b v_a - \phi_a \nabla_b \Omega - \frac{i}{\sigma} \nabla_a p A_b) \delta v^b \tag{B.14}$$

$$= -\nabla_a \delta U - \rho(\delta U + \delta \Phi) A_a, \tag{B.15}$$

where

$$\delta U := \frac{\delta p}{\rho} - \delta \Phi. \tag{B.16}$$

Notice that Eq. (B.15) is algebraic in δv^a and can be solved for δv^a as long as Q^{-1}_{ab} is invertible. In this case the solution to Eq. (B.15) for δv^a is

$$\delta v^a = iQ^{ab} \nabla_b \delta U + i\rho(\delta U + \delta \Phi) Q^{ab} A_b, \tag{B.17}$$

where

$$Q^{ab} = \frac{\lambda}{\sigma^3}\Big[(\sigma^2 - A^c\nabla_c p)g^{ab} - 2\omega^a\Omega^b + i\sigma\phi^a\nabla^b\Omega \tag{B.18}$$

$$- 2i\sigma\Omega\nabla^a\phi^b + \nabla^a p A^b + \frac{i}{\sigma}\phi^a\phi^{bc}\nabla_c p A^d\omega_d \tag{B.19}$$

$$- \frac{2i}{\sigma}\phi^b\phi^{ac}A_c\Omega^d\nabla_d p\Big]. \tag{B.20}$$

Here

$$\det Q^{-1\,a}{}_b \equiv \frac{\sigma^3}{\lambda} = \frac{1}{\sigma}(\sigma^4 - \sigma^2 A^a\nabla_a p - 2\sigma^2\Omega^a\omega_a + 2A^a\omega_a\Omega^b\nabla_b p); \tag{B.21}$$

$\omega^a = \epsilon^{abc}\nabla_b v_c$ is the fluid vorticity; ϵ_{abc} is the antisymmetric tensor (components $= \pm 1$ in Cartesian coordinates); $\Omega^a = \Omega z^a$ is the fluid angular-velocity vector, where z^a is a unit vector parallel to the rotation axis; and $\phi^{ab} = \epsilon^{abc}\phi_c/\phi^d\phi_d$.

We use Eq. (B.17) to eliminate δv^a in favor of δU and $\delta\Phi$. We also eliminate $\delta\rho$ by combining Eq. (B.6), (B.16), and (B.12), which yields

$$\delta\rho = \Psi_1(\delta U + \delta\Phi) - \frac{\rho^2}{\sigma}A_a Q^{ab}\nabla_b\delta U, \tag{B.22}$$

where

$$\Psi_1 = \frac{\rho^2}{\Gamma_1 p} - \frac{\rho^3}{\sigma}A_a Q^{ab}A_b. \tag{B.23}$$

We have not yet used Eq. (B.6) and (B.8) . It is these that now provide the fundamental set of coupled eigenequations for the two potentials δU and $\delta\Phi$. Using Eqs. (B.16), (B.17), and (B.22) to eliminate δp, $\delta\rho$, and δv^a, we are able to re-express Eqs. (B.6) and (B.8) as the fourth-order system

$$\nabla_a(\rho Q^{ab}\nabla_b\delta U) + \Psi_3\delta U = -\rho^2 Q^{ab}A_b\nabla_a\delta\Phi - \Psi_2\delta\Phi, \tag{B.24}$$

$$\nabla^a\nabla_a\delta\Phi + 4\pi G\,\Psi_1\delta\Phi = \frac{4\pi G\rho^2}{\sigma}A_a Q^{ab}\nabla_b\delta U - 4\pi G\,\Psi_1\delta U, \tag{B.25}$$

where

$$\Psi_2 = \sigma\Psi_1 + \nabla_a(\rho^2 Q^{ab}A_b), \tag{B.26}$$

$$\Psi_3 = \Psi_2 - \frac{m\lambda}{\sigma^3}\rho^2\left(\sigma A^a\nabla_a\Omega + \frac{4\sigma\Omega}{\varpi}A^a\nabla_a\varpi + \frac{1}{\sigma}A_a\phi^{ab}\nabla_b p A^c\omega_c\right). \tag{B.27}$$

It is easy to verify that Eqs. (B.24) and (B.25) are real equations for $\delta U(z,\varpi)$ and $\delta\Phi(z,\varpi)$, where $\delta U = \delta U(z,\varpi)e^{i\omega t+im\varphi}$ and $\delta\Phi = \delta\Phi(z,\varpi)e^{i\omega t+im\varphi}$.

We complete the specification of the eigenvalue problem by imposing appropriate boundary conditions at the stellar surface and at $r = \infty$, where r is the spherical radial coordinate. At the stellar surface, the appropriate boundary condition is that the Lagrangian change in the pressure vanish, namely,

$$\Delta p = \delta p + \xi^a \nabla_a p = 0 \tag{B.28}$$

at the stellar surface. It follows from Eqs. (B.11), (B.16), and (B.17) that this condition can be rewritten as

$$\delta U + \delta\Phi + \frac{\nabla_a p}{\sigma} Q^{ab} \left[\frac{1}{\rho} \nabla_b \delta U + \rho A_b (\delta U + \delta\Phi) \right] = 0. \tag{B.29}$$

The remaining boundary condition is that $\delta\Phi \to 0$ sufficiently rapidly as $r \to \infty$. For a perturbation with φ-coordinate dependence $e^{im\varphi}$, examination of the expansion of $\delta\Phi$ as a power series in negative powers of r reveals that this condition can be expressed as

$$\frac{\partial}{\partial r}\delta\Phi + \sum_{l \geq |m|} \frac{(l+1)(2l+1)(l-m)!}{2r(l+m)!} P_l^m(\mu) \int_{-1}^{1} \delta\Phi(r, \mu') P_l^m(\mu') d\mu' = 0, \tag{B.30}$$

where $\mu = \cos\theta$, θ is the polar angular coordinate, and P_l^m is the associated Legendre function.

B.2.1 The numerical method of solution

We next describe briefly a numerical method that has been developed and used to solve the eigenequations directly for ω, δU, and $\delta\Phi$. For more details, the reader is referred to the extensive discussion presented elsewhere [325].

In the method of Ipser and Lindblom, the equations describing the equilibrium of a stellar model and the eigenequations (B.24), (B.25), (B.29), and (B.30) describing a normal mode of pulsation are first written out in spherical coordinates $r, \mu = \cos\theta, \varphi$, and are then represented as difference equations on a finite grid. The chosen grid consists of uniformly-spaced points along $(2L-1)$ radial spokes that emanate from the origin at the zeros μ_i of the Legendre polynomial P_{2L-1}. This choice of the angular location of spokes permits one to accurately represent integrals of a function over μ, and its various derivatives with respect to μ, as weighted averages of the values of the function at the various μ_i. Radial derivatives of a function are represented in terms of standard three-point difference formulae. For convenience, the boundary condition (B.29) is imposed not only at the last point inside the star along each spoke but also at all points outside the star out to the last grid point along the spoke. This amounts to a smooth extension of the definition of δU into the exterior region, and it simplifies the numerical analysis. The boundary condition (B.30) is imposed at the last grid point along each spoke.

When this procedure is completed, the eigensystem takes the form of a set of coupled linear equations for the values of δU and $\delta\Phi$ at the grid points:

$$\sum_b B_a{}^b \delta U_b = \sum_b C_a{}^b \delta\Phi_b, \tag{B.31}$$

$$\sum_b D_a{}^b \delta\Phi_b = \sum_b E_a{}^b \delta U_b, \tag{B.32}$$

where the indices label the grid points. Equations (B.31) and (B.32) can be combined to yield a single equation for either δU_a or $\delta\Phi_a$, but it turns out to be more efficient to leave them as separate equations and to solve them iteratively as follows. Given an estimate $\delta U^{(i)}$, $\delta\Phi^{(i)}$, $\omega^{(i)}$ for the eigenfunctions and eigenvalue (taken, perhaps, from the solution obtained for a slightly different stellar model), Eq. (B.32) is solved for the new estimate $\delta\Phi^{(i+1)}$, which is then used in Eq. (B.31) to find $\delta U^{(i+1)}$. Next, a suitable average $\langle\delta U\rangle$ of the values δU_a is used to compute the quantity $z^{(i)} \equiv \langle\delta U^{(i-1)}\rangle/\langle\delta U^{(i)}\rangle$, which monitors the average change of δU from one iteration to the next. The quantity $z^{(i)}$ is used to update the eigenvalue in a way that attempts to predict the value that will make $z^{(i+1)}$ as close to unity as possible:

$$\omega^{(i+1)} = \omega^{(i)} + s(1 - z^{(i)})\frac{\omega^{(i)} - \omega^{(i-1)}}{z^{(i)} - z^{(i-1)}}.$$

Here s is a convergence factor ≤ 1 chosen to maximize stability of the iteration process. Iteration continues until convergence is achieved.

Bibliography

[1] E. B. Abdikamalov, H. Dimmelmeier, L. Rezzolla, and J. C. Miller. Relativistic simulations of the phase-transition-induced collapse of neutron stars. *Mon. Not. R. Astron. Soc.*, 392:52–76, 2009.

[2] M. A. Abramowicz. Theory of level surfaces inside relativistic, rotating stars. II. *Acta Astronom.*, 24:45–52, 1974.

[3] M. A. Abramowicz. Rayleigh and Solberg criteria reversal near black holes: the optical geometry explanation. arXiv:astro-ph/0411718, November 2004.

[4] M. A. Abramowicz and A. R. Prasanna. Centrifugal force reversal near a Schwarzschild black-hole. *Mon. Not. R. Astron. Soc.*, 245:720–728, 1990.

[5] S. Akiyama, J. A. Wheeler, D. L. Meier, and I. Lichtenstadt. The magnetorotational instability in core-collapse supernova explosions. *Astrophys. J.*, 584:954–970, 2003.

[6] S. Akiyama and J. C. Wheeler. The nonmonotonic dependence of supernova and compact remnant formation on progenitor rotation. *Astrophys. J.*, 629:414–421, 2005.

[7] M. Alcubierre. *Introduction to 3+1 Numerical Relativity.* Number 140 in International Series of Monographs in Physics. Oxford University Press, New York, 2008.

[8] M. Alcubierre, B. Allen, B. Brügmann, E. Seidel, and W.-M. Suen. Towards an understanding of the stability properties of the 3+1 evolution equations in general relativity. *Phys. Rev. D*, 62:124011, 2000.

[9] M. Alcubierre, B. Bruegmann, T. Dramlitsch, J. A. Font, P. Papadopoulos, E. Seidel, N. Stergioulas, and R. Takahashi. Towards a stable numerical evolution of strongly gravitating systems in general relativity: The conformal treatments. *Phys. Rev. D*, 62:044034, 2000.

[10] M. Alcubierre, B. Brügmann, P. Diener, M. Koppitz, D. Pollney, E. Seidel, and R. Takahashi. Gauge conditions for long-term numerical black hole evolutions without excision. *Phys. Rev. D*, 67:084023, 2003.

[11] M. Alcubierre, B. Brügmann, D. Pollney, E. Seidel, and R. Takahashi. Black hole excision for dynamic black holes. *Phys. Rev. D*, 64:061501, 2001.

[12] M. Alford, S. Mahmoodifar, and K. Schwenzer. Viscous damping of r-modes: Small amplitude instability. *Phys. Rev. D*, 85:024007, 2012.

[13] M. G. Alford, S. Mahmoodifar, and K. Schwenzer. Large amplitude behavior of the bulk viscosity of dense matter. *J. Phys. G: Nucl. Part. Phys.*, 37(12):125202, December 2010.

[14] D. Alic, C. Bona-Casas, C. Bona, L. Rezzolla, and C. Palenzuela. A conformal and covariant formulation of the Z4 system with constraint-violation damping. arXiv:1106.2254v1, 2011.

[15] V. A. Ambartsumyan and G. S. Saakyan. The degenerate superdense gas of elementary particles. *Soviet Astron.*, 4:187–191, 1960.

[16] N. Andersson. A new class of unstable modes of rotating relativistic stars. *Astrophys. J.*, 502:708–713, 1998.

[17] N. Andersson and G. L. Comer. On the dynamics of superfluid neutron star cores. *Mon. Not. R. Astron. Soc.*, 328:1129–1143, 2001.

[18] N. Andersson and G. L. Comer. Probing neutron star superfluidity with gravitational-wave data. *Phys. Rev. Lett.*, 87:241101, 2001.

[19] N. Andersson, G. L. Comer, and K. Glampedakis. How viscous is a superfluid neutron star core? *Nucl. Phys. A*, 763:212–229, 2005.

[20] N. Andersson, G. L. Comer, and D. Langlois. Oscillations of general relativistic superfluid neutron stars. *Phys. Rev. D*, 66:104002, 2002.

[21] N. Andersson, K. Glampedakis, and B. Haskell. Oscillations of dissipative superfluid neutron stars. *Phys. Rev. D*, 79:103009, 2009.

[22] N. Andersson, D. I. Jones, and K. D. Kokkotas. Strange stars as persistent sources of gravitational waves. *Mon. Not. R. Astron. Soc.*, 337:1224–1232, 2002.

[23] N. Andersson, D. I. Jones, K. D. Kokkotas, and N. Stergioulas. R-mode runaway and rapidly rotating neutron stars. *Astrophys. J.*, 534:L75–L78, 2000.

[24] N. Andersson and K. D. Kokkotas. Towards gravitational wave asteroseismology. *Mon. Not. R. Astron. Soc.*, 299:1059–1068, 1998.

[25] N. Andersson and K. D. Kokkotas. The r-mode instability in rotating neutron stars. *Int. J. Mod. Phys. D*, 10:381–441, 2001.

[26] N. Andersson, K. D. Kokkotas, and B. F. Schutz. A new numerical approach to the oscillation modes of relativistic stars. *Mon. Not. R. Astron. Soc.*, 274:1039–1048, 1995.

[27] N. Andersson, K. D. Kokkotas, and B. F. Schutz. Gravitational radiation limit on the spin of young neutron stars. *Astrophys. J.*, 510:846–853, 1999.

[28] N. Andersson, K. D. Kokkotas, and N. Stergioulas. On the relevance of the r-mode instability for accreting neutron stars and white dwarfs. *Astrophys. J.*, 516:307–314, 1999.

[29] N. Andersson, T. Sidery, and G. L. Comer. Mutual friction in superfluid neutron stars. *Mon. Not. R. Astron. Soc.*, 368:162–170, 2006.

[30] A. M. Anile. *Relativistic Fluids and Magneto-Fluids.* Cambridge University Press, Cambridge, UK, 1989.

[31] M. Ansorg, A. Kleinwächter, and R. Meinel. Highly accurate calculation of rotating neutron stars. Detailed description of the numerical methods. *Astron. Astrophys.*, 405:711–721, 2003.

[32] M. Ansorg, D. Gondek-Rosińska, and L. Villain. On the solution space of differentially rotating neutron stars in general relativity. *Mon. Not. R. Astron. Soc.*, 396:2359–2366, 2009.

[33] M. Ansorg, A. Kleinwächter, and R. Meinel. Highly accurate calculation of rotating neutron stars. *Astron. Astrophys.*, 381:L49–L52, 2002.

[34] N. V. Ardeljan, G. S. Bisnovatyi-Kogan, and S. G. Moiseenko. Magnetorotational supernovae. *Mon. Not. R. Astron. Soc.*, 359:333–344, 2005.

[35] W. D. Arnett and R. L. Bowers. A microscopic interpretation of neutron star structure. *Astrophys. J. Suppl. Ser.*, 33:415–436, 1977.

[36] R. Arnowitt, S. Deser, and C. W. Misner. The dynamics of general relativity. In L. Witten, editor, *Gravitation: An Introduction to Current Research*, pages 227–265, Wiley: New York, 1973.

[37] P. Arras, É. É. Flanagan, S. M. Morsink, A. K. Schenk, S. A. Teukolsky, and I. Wasserman. Saturation of the r-mode instability. *Astrophys. J.*, 591:1129–1151, 2003.

[38] A. Ashtekar. Black hole dynamics in general relativity. *Pramana*, 69:77–92, 2007.

[39] A. Ashtekar and A. Magnon-Ashtekar. On conserved quantities in general relativity. *J. Math. Phys.*, 20:793–800, 1979.

[40] L. Baiotti, R. De Pietri, G. M. Manca, and L. Rezzolla. Accurate simulations of the dynamical bar-mode instability in full general relativity. *Phys. Rev. D*, 75:044023, 2007.

[41] L. Baiotti, I. Hawke, P. J. Montero, F. Löffler, L. Rezzolla, N. Stergioulas, J. A. Font, and E. Seidel. Three-dimensional relativistic simulations of rotating neutron-star collapse to a Kerr black hole. *Phys. Rev. D*, 71:024035, 2005.

[42] S. Balberg and A. Gal. An effective equation of state for dense matter with strangeness. *Nucl. Phys. A*, 625:435–472, 1997.

[43] E. Balbinski. A new shear instability in rotating, self-gravitating, perfect fluids. *Mon. Not. R. Astron. Soc.*, 216:897–908, 1985.

[44] F. Banyuls, J. A. Font, J. M. Ibáñez, J. M. Martí, and J. A. Miralles. Numerical 3+1 general relativistic hydrodynamics: A local characteristic approach. *Astrophys. J.*, 476:221–231, 1997.

[45] J. M. Bardeen. A variational principle for rotating stars in general relativity. *Astrophys. J.*, 162:71–95, 1970.

[46] J. M. Bardeen. Rapidly rotating stars, disks, and black holes. In C. DeWitt and B. S. DeWitt, editors, *Black Holes*, Les Houches 1972, pages 241–289, Gordon & Breach, New York, 1973.

[47] J. M. Bardeen, J. L. Friedman, B. F. Schutz, and R. Sorkin. A new criterion for secular instability of rapidly rotating stars. *Astrophys. J.*, 217:L49–L53, 1977.

[48] J. M. Bardeen and T. Piran. General relativistic axisymmetric rotating systems: Coordinates and equations. *Phys. Rep.*, 96:205–250, 1983.

[49] J. M. Bardeen, K. S. Thorne, and D. W. Meltzer. A catalogue of methods for studying the normal modes of radial pulsation of general-relativistic stellar models. *Astrophys. J.*, 145:505–513, 1966.

[50] J. M. Bardeen and R. V. Wagoner. Uniformly rotating disks in general relativity. *Astrophys. J.*, 158:L65–L69, 1969.

[51] J. M. Bardeen and R. V. Wagoner. Relativistic disks. I. Uniform rotation. *Astrophys. J.*, 167:359–423, 1971.

[52] A. P. Barnes, P. G. Le Floch, B. G. Schmidt, and J. M. Stewart. The Glimm scheme for perfect fluids on plane-symmetric Gowdy spacetimes. *Class. Quant. Grav.*, 21:5043–5074, 2004.

[53] D. Baumgart and J. L. Friedman. Zero frequency modes of the Maclaurin spheroids. *Proc. R. Soc. Lond. A*, 405:65–72, 1986.

[54] T. W. Baumgarte and S. L. Shapiro. On the numerical integration of Einstein's field equations. *Phys. Rev. D*, 59:024007, 1999.

[55] T. W. Baumgarte and S. L. Shapiro. *Numerical Relativity: Solving Einstein's Equations on the Computer.* Cambridge University Press, New York, 2010.

[56] T. W. Baumgarte, S. L. Shapiro, and M. Shibata. On the maximum mass of differentially rotating neutron stars. *Astrophys. J.*, 528:L29–L32, 2000.

[57] T. W. Baumgarte, S. L. Shapiro, and S. A. Teukolsky. Computing the delayed collapse of hot neutron stars to black holes. *Astrophys. J.*, 458:680–691, 1996.

[58] G. Baym and E. Chandler. The hydrodynamics of rotating superfluids. I. Zero-temperature, nondissipative theory. *J. Low Temp. Phys.*, 50:57–87, 1983.

[59] R. Beig. Arnowitt-Deser-Misner energy and g_{00}. *Phys. Lett. A*, 69:153–155, 1978.

[60] M. Bejger, T. Bulik, and P. Haensel. Constraints on the dense matter equation of state from the measurements of PSR J0737-3039A moment of inertia and PSR J0751+1807 mass. *Mon. Not. R. Astron. Soc.*, 364:635–639, 2005.

[61] J. D. Bekenstein and E. Oron. New conservation laws in general-relativistic magnetohydrodynamics. *Phys. Rev. D*, 18:1809–1819, 1978.

[62] J. D. Bekenstein and E. Oron. Interior magnetohydrodynamic structure of a rotating relativistic star. *Phys. Rev. D*, 19:2827–2837, 1979.

[63] O. Benhar and M. Valli. Shear viscosity of neutron matter from realistic nucleon-nucleon interactions. *Phys. Rev. Lett.*, 99:232501, 2007.

[64] E. Berti and N. Stergioulas. Approximate matching of analytic and numerical solutions for rapidly rotating neutron stars. *Mon. Not. R. Astron. Soc.*, 350:1416–1430, 2004.

[65] E. Berti, F. White, A. Maniopoulou, and M. Bruni. Rotating neutron stars: an invariant comparison of approximate and numerical space-time models. *Mon. Not. R. Astron. Soc.*, 358:923–938, 2005.

[66] H. R. Beyer and K. D. Kokkotas. On the r-mode spectrum of relativistic stars. *Mon. Not. R. Astron. Soc.*, 308:745–750, 1999.

[67] H. R. Beyer and B. G. Schmidt. Newtonian stellar oscillations. *Astron. Astrophys.*, 296:722–726, 1995.

[68] L. Bildsten. Gravitational radiation and rotation of accreting neutron stars. *Astrophys. J.*, 501:L89–L93, 1998.

[69] L. Bildsten and C. Cutler. Tidal interactions of inspiraling compact binaries. *Astrophys. J.*, 400:175–180, 1992.

[70] L. Bildsten and G. Ushomirsky. Viscous boundary-layer damping of r-modes in neutron stars. *Astrophys. J.*, 529:L33–L36, 2000.

[71] G. S. Bisnovatyi-Kogan and S. G. Moiseenko. Core-collapse supernovae: Magnetorotational explosions and jet formation. *Prog. Theor. Phys. Supp.*, 172:145–155, 2008.

[72] J. K. Blackburn and S. Detweiler. Close black-hole binary systems. *Phys. Rev. D*, 46:2318–2333, 1992.

[73] S. I. Blinnikov, I. D. Novikov, T. V. Perevodchikova, and A. G. Polnarev. Exploding neutron stars in close binaries. *Soviet Astron. Letters*, 10:177–179, 1984.

[74] M. Bocquet, S. Bonazzola, E. Gourgoulhon, and J. Novak. Rotating neutron star models with a magnetic field. *Astron. Astrophys.*, 301:757–775, 1995.

[75] P. Bodenheimer and J. P. Ostriker. Rapidly rotating stars. VIII. Zero-viscosity polytropic sequences. *Astrophys. J.*, 180:159–170, 1973.

[76] A. R. Bodmer. Collapsed nuclei. *Phys. Rev. D*, 4:1601–1606, 1971.

[77] I. Bombaci, M. Prakash, M. Prakash, P. J. Ellis, J. M. Lattimer, and G. E. Brown. Newborn hot neutron stars. *Nucl. Phys. A*, 583:623–628, 1995.

[78] C. Bona, J. Massó, E. Seidel, and J. Stela. New formalism for numerical relativity. *Phys. Rev. Lett.*, 75:600–603, 1995.

[79] C. Bona, C. Palenzuela-Luque, and C. Bona-Casas. *Elements of Numerical Relativity and Relativistic Hydrodynamics: From Einstein' s Equations to Astrophysical Simulations*, 2nd edition. Number 783 in Lecture Notes in Physics. Springer-Verlag, Berlin, 2009.

[80] A. Bonanno, L. Rezzolla, and V. Urpin. Mean-field dynamo action in protoneutron stars. *Astron. Astrophys.*, 410:L33–L36, 2003.

[81] S. Bonazzola. The virial theorem in general relativity. *Astrophys. J.*, 182:335–340, 1973.

[82] S. Bonazzola, J. Frieben, and E. Gourgoulhon. Spontaneous symmetry breaking of rapidly rotating stars in general relativity. *Astrophys. J.*, 460:379–389, 1996.

[83] S. Bonazzola, J. Frieben, and E. Gourgoulhon. Spontaneous symmetry breaking of rapidly rotating stars in general relativity: Influence of the 3D-shift vector. *Astron. Astrophys.*, 331:280–290, 1998.

[84] S. Bonazzola and E. Gourgoulhon. A virial identity applied to relativistic stellar models. *Class. Quant. Grav.*, 11:1775–1784, 1994.

[85] S. Bonazzola, E. Gourgoulhon, P. Grandclément, and J. Novak. Constrained scheme for the einstein equations based on the Dirac gauge and spherical coordinates. *Phys. Rev. D*, 70:104007, 2004.

[86] S. Bonazzola, E. Gourgoulhon, and J.-A. Marck. Relativistic formalism to compute quasiequilibrium configurations of nonsynchronized neutron star binaries. *Phys. Rev. D*, 56:7740–7749, 1997.

[87] S. Bonazzola, E. Gourgoulhon, and J.-A. Marck. Numerical approach for high precision 3-D relativistic star models. *Phys. Rev. D*, 58:104020, 1998.

[88] S. Bonazzola, E. Gourgoulhon, Grandclément P., and J.-A. Marck. Darwin-Riemann problems in general relativity. *Prog. Theor. Phys. Supp.*, 136:216–234, 1999.

[89] S. Bonazzola, E. Gourgoulhon, M. Salgado, and J.-A. Marck. Axisymmetric rotating relativistic bodies: A new numerical approach for "exact" solutions. *Astron. Astrophys.*, 278:421–443, 1993.

[90] S. Bonazzola and G. Maschio. Models of rotating neutron stars in general relativity. In R. D. Davies and F. Graham-Smith, editors, *The Crab Nebula, Proceedings from IAU Symposium no. 46 held at Jodrell Bank, England, August 5–7, 1970*, number 46 in International Astronomical Union Symposium, page 346. Reidel, Dordrecht, 1971.

[91] S. Bonazzola and S. Schneider. An exact study of rigidly and rapidly rotating stars in general relativity with application to the Crab pulsar. *Astrophys. J.*, 191:195–290, 1974.

[92] R. Bondarescu, S. A. Teukolsky, and I. Wasserman. Spin evolution of accreting neutron stars: Nonlinear development of the r-mode instability. *Phys. Rev. D*, 76:064019, 2007.

[93] R. Bondarescu, S. A. Teukolsky, and I. Wasserman. Spinning down newborn neutron stars: Nonlinear development of the r-mode instability. *Phys. Rev. D*, 79:104003, 2009.

[94] J. M. Bowen and J. W. York. Time-symmetric initial data for black holes and black-hole collisions. *Phys. Rev. D*, 21:2047–2056, 1980.

[95] R. H. Boyer and R. Lindquist. A variational principle for a rotating relativistic fluid. *Phys. Lett.*, 20:504–506, 1966.

[96] K. Brecher and G. Caporaso. Obese "neutron" stars. *Nature*, 259:377, 1976.

[97] J. Brink, S. A. Teukolsky, and I. Wasserman. Nonlinear coupling network to simulate the development of the r-mode instability in neutron stars. I. Construction. *Phys. Rev. D*, 70:124017, 2004.

[98] J. Brink, S. A. Teukolsky, and I. Wasserman. Nonlinear couplings of r-modes: Energy transfer and saturation amplitudes at realistic timescales. *Phys. Rev. D*, 70:121501, 2004.

[99] J. Brink, S. A. Teukolsky, and I. Wasserman. Nonlinear coupling network to simulate the development of the r mode instability in neutron stars. II. Dynamics. *Phys. Rev. D*, 71:064029, 2005.

[100] D. J. Brown. On variational principles for gravitating perfect fluids. *Ann. Phys.*, 248:1–33, 1996.

[101] G. E. Brown and H. A. Bethe. A scenario for a large number of low-mass black holes in the galaxy. *Astrophys. J.*, 423:659–664, 1994.

[102] J. D. Brown. Gravitational waves from the dynamical bar instability in a rapidly rotating star. *Phys. Rev. D*, 62:084024, 2000.

[103] L. Burderi and N. D'Amico. Probing the equation of state of ultradense matter with a submillisecond pulsar search experiment. *Astrophys. J.*, 490:343–352, 1997.

[104] A. Burrows and B. A. Fryxell. An instability in neutron stars at birth. *Science*, 258:430–434, 1992.

[105] A. Burrows, J. Hayes, and B. A. Fryxell. On the nature of core-collapse supernova explosions. *Astrophys. J.*, 450:830–850, 1995.

[106] A. Burrows and J. M. Lattimer. The birth of neutron stars. *Astrophys. J.*, 307:178–196, 1986.

[107] A. Burrows, E. Livne, L. Dessart, C. D. Ott, and J. Murphy. A new mechanism for core-collapse supernova explosions. *Astrophys. J.*, 640:878–890, 2006.

[108] A. Burrows, E. Livne, L. Dessart, C. D. Ott, and J. Murphy. Features of the acoustic mechanism of core-collapse supernova explosions. *Astrophys. J.*, 655:416–433, 2007.

[109] E. M. Butterworth. On the structure and stability of rapidly rotating fluid bodies in general relativity. II – The structure of uniformly rotating pseudopolytropes. *Astrophys. J.*, 204:561–572, 1976.

[110] E. M. Butterworth and J. R. Ipser. On the structure and stability of rapidly rotating fluid bodies in general relativity. I – The numerical method for computing structure and its application to uniformly rotating homogeneous bodies. *Astrophys. J.*, 204:200–223, 1976.

[111] M. G. Calkin. An action principle for magnetohydrodynamics. *Can. J. Phys.*, 41:2241–51, 1963.

[112] A. Campolattaro and K. S. Thorne. Nonradial pulsation of general-relativistic stellar models. V. Analytic analysis for l = 1. *Astrophys. J.*, 159:847–858, 1970.

[113] M. Cantor. Spaces of functions with asymptotic conditions on R^n. *Indiana Univ. Math. J.*, 24:897–902, 1975.

[114] B. W. Carroll, E. G. Zweibel, C. J. Hansen, P. N. McDermott, M. P. Savedoff, J. H. Thomas, and H. M. Van Horn. Oscillation spectra of neutron stars with strong magnetic fields. *Astrophys. J.*, 305:767–783, 1986.

[115] B. Carter. Killing horizons and orthogonally transitive groups in spacetime. *J. Math. Phys.*, 10:70–81, 1969.

[116] B. Carter. The commutation property of a stationary, axisymmetric system. *Commun. Math. Phys.*, 17:233–238, 1970.

[117] B. Carter. Black hole equilibrium states. In C. DeWitt and B. S. DeWitt, editors, *Black Holes*, Les Houches Summer School of Theoretical Physics, pages 57–214. Gordon and Breach, New York, 1973.

[118] B. Carter. Elastic perturbation theory in general relativity and variational principle for a rotating solid star. *Commun. Math. Phys.*, 30:261–286, 1973.

[119] B. Carter and H. Quintana. Stationary elastic rotational deformation of a relativistic neutron star model. *Astrophys. J.*, 202:511–522, 1975.

[120] J. M. Centrella, K. C. B. New, L. L. Lowe, and J. D. Brown. Dynamical rotational instability at low T/W. *Astrophys. J.*, 550:L193–L196, 2001.

[121] P. Cerdá-Durán, J. A. Font, and H. Dimmelmeier. General relativistic simulations of passive-magneto-rotational core collapse with microphysics. *Astron. Astrophys.*, 474:169–191, 2007.

[122] P. Cerdá-Durán, V. Quilis, and J. A. Font. AMR simulations of the low $T/|W|$ bar-mode instability of neutron stars. *Comput. Phys. Commun.*, 177:288–297, 2007.

[123] P. Cerdá-Durán, N. Stergioulas, and J. A. Font. Alfvén QPOs in magnetars in the anelastic approximation. *Mon. Not. R. Astron. Soc.*, 397:1607–1620, 2009.

[124] D. Chakrabarty. The spin distribution of millisecond X-ray pulsars. In R. Wijnands et al., editors, *A Decade of Accreting Millisecond X-Ray Pulsars*, volume 1068 of *AIP Conference Proceedings*, pages 67–74, 2008.

[125] F. Chambat. On the non-existence of some heterogeneous equivalents of ellipsoidal Dedekind figures. *Astron. Astrophys.*, 292:76–81, 1994.

[126] S. Chandrasekhar. The dynamical instability of gaseous masses approaching the Schwarzschild limit in general relativity. *Astrophys. J.*, 140:417–433, 1964.

[127] S. Chandrasekhar. The post-Newtonian equations of hydrodynamics in general relativity. *Astrophys. J.*, 142:1488–1512, 1965.

[128] S. Chandrasekhar. The post-Newtonian effects of general relativity on the equilibrium of uniformly rotating bodies. III, the deformed figures of the Jacobi ellipsoids. *Astrophys. J.*, 148:621–644, 1967.

[129] S. Chandrasekhar. *Ellipsoidal Figures of Equilibrium.* Yale University Press, New Haven, CT, 1969.

[130] S. Chandrasekhar. Solutions of two problems in the theory of gravitational radiation. *Phys. Rev. Lett.*, 24:611–615, 1970.

[131] S. Chandrasekhar. The post-Newtonian effects of general relativity on the equilibrium of uniformly rotating bodies. VI, the deformed figures of the Jacobi ellipsoids (continued). *Astrophys. J.*, 467:455–463, 1971.

[132] S. Chandrasekhar and G. Contopoulos. The virial theorem in general relativity in the post-Newtonian approximation. *Proc. Natl. Acad. Sci. USA*, 49:608–613, 1963.

[133] S. Chandrasekhar and D. D. Elbert. The deformed figures of the Dedekind ellipsoids in the post-Newtonian approximation to general relativity. *Astrophys. J.*, 458:699–713, 1978.

[134] S. Chandrasekhar and V. Ferrari. On the non-radial oscillations of slowly rotating stars induced by the Lense-Thirring effect. *Proc. R. Soc. Lond. A*, 433:423–440, 1991.

[135] S. Chandrasekhar and V. Ferrari. On the non-radial oscillations of spherical relativistic stars. *Proc. R. Soc. Lond. A*, 432:247–279, 1991.

[136] S. Chandrasekhar and J. L. Friedman. On the stability of axisymmetric systems to axisymmetric perturbations in general relativity, I. The equations governing nonstationary, stationary, and perturbed systems. *Astrophys. J.*, 175:379–405, 1972.

[137] S. Chandrasekhar and J. L. Friedman. On the stability of axisymmetric systems to axisymmetric perturbations in general relativity. II. A criterion for the onset of instability in uniformly rotating configurations and the frequency of the fundamental mode in case of slow rotation. *Astrophys. J.*, 176:745–768, 1972.

[138] Y. Choquet-Bruhat and J. W. York. The Cauchy problem. In A. Held, editor, *General Relativity and Gravitation, One Hundred Years after the Birth of Albert Einstein*, volume 1, page 99. Plenum Press, New York, 1980.

[139] P. T. Chruściel. On the relation between the Einstein and the Komar expressions for the energy of the gravitational field. *Ann. Inst. Henri Poincaré (A)*, 42:267–282, 1985.

[140] P. T. Chruściel, J. Jezierski, and J. Kijowski. *Hamiltonian Field Theory in the Radiating Regime.* Springer-Verlag, Berlin, 2001.

[141] A. I. Chugunov and M. E. Gusakov. Non-radial superfluid modes in oscillating neutron stars. *Mon. Not. R. Astron. Soc.*, 418:L54–L58, 2011.

[142] A. I. Chugunov and C. J. Horowitz. Breaking stress of neutron star crust. *Mon. Not. R. Astron. Soc.*, 407:L54–L58, 2010.

[143] J. P. A. Clark. Unpublished work toward a PhD thesis, Yale University, 1979.

[144] J. P. A. Clark and D. M. Eardley. Evolution of close neutron star binaries. *Astrophys. J.*, 215:311–322, July 1977.

[145] M. J. Clement. Normal modes of oscillation for rotating stars. I – The effect of rigid rotation on four low-order pulsations. *Astrophys. J.*, 249:746–760, 1981.

[146] E. A. Coddington and N. Levinson. *Theory of Ordinary Differential Equations.* McGraw-Hill, New York, 1955.

[147] A. Colaiuda, H. Beyer, and K. D. Kokkotas. On the quasi-periodic oscillations in magnetars. *Mon. Not. R. Astron. Soc.*, 396:1441–1448, 2009.

[148] A. Colaiuda, V. Ferrari, L. Gualtieri, and J. A. Pons. Relativistic models of magnetars: structure and deformations. *Mon. Not. R. Astron. Soc.*, 385:2080–2096, 2008.

[149] P. Colella and P. R. Woodward. The piecewise parabolic method (PPM) for gas-dynamical simulations. *J. Comput. Phys.*, 54:174–201, 1984.

[150] M. Colpi, S. L. Shapiro, and S. A. Teukolsky. Explosion of a rotating neutron star near the minimum mass. *Astrophys. J.*, 369:422–439, 1991.

[151] G. L. Comer and D. Langlois. Hamiltonian formulation for multiconstituent relativistic perfect fluids. *Class. Quant. Grav.*, 10:2317–2327, 1993.

[152] G. L. Comer, D. Langlois, and L. M. Lin. Quasi-normal modes of general relativistic superfluid neutron stars. *Phys. Rev. D*, 60:104025, 1999.

[153] N. Comins. On secular instabilities of rigidly rotating stars in general relativity – I. Theoretical formalism. *Mon. Not. R. Astron. Soc.*, 189:233–253, 1979.

[154] N. Comins. On secular instabilities of rigidly rotating stars in general relativity – II. Numerical results. *Mon. Not. R. Astron. Soc.*, 189:255–272, 1979.

[155] N. Comins and B. F. Schutz. On the ergoregion instability. *Proc. R. Soc. Lond. A*, 364:211–226, 1978.

[156] G. Cook. Initial data for numerical relativity. *Living Rev. Relativity*, 3:5, 2000.

[157] G. B. Cook, S. L. Shapiro, and S. A. Teukolsky. Spin-up of a rapidly rotating star by angular momentum loss – effects of general relativity. *Astrophys. J.*, 398:203–223, 1992.

[158] G. B. Cook, S. L. Shapiro, and S. A. Teukolsky. Rapidly rotating neutron stars in general relativity: Realistic equations of state. *Astrophys. J.*, 424:823–845, 1994.

[159] G. B. Cook, S. L. Shapiro, and S. A. Teukolsky. Rapidly rotating polytropes in general relativity. *Astrophys. J.*, 422:227–242, 1994.

[160] G. B. Cook, S. L. Shapiro, and S. A. Teukolsky. Testing a simplified version of Einstein's equations for numerical relativity. *Phys. Rev. D*, 53:5533–5540, 1996.

[161] J. N. Cook, S. L. Shapiro, and B. C. Stephens. Magnetic braking and viscous damping of differential rotation in cylindrical stars. *Astrophys. J.*, 599:1272, 2003.

[162] I. Cordero-Carrión, P. Cerdá-Durán, H. Dimmelmeier, J. L. Jaramillo, J. Novak, and E. Gourgoulhon. An improved constrained scheme for the Einstein equations: an approach to the uniqueness issue. *Phys. Rev. D*, 79:024017, 2009.

[163] G. Corvino, L. Rezzolla, S. Bernuzzi, R. De Pietri, and B. Giacomazzo. On the shear instability in relativistic neutron stars. *Class. Quant. Grav.*, 27:114104, 2010.

[164] R. Courant and D. Hilbert. *Methods of Mathematical Physics; vol. II: Partial Differential Equations.* Interscience, New York, 1962.

[165] T. G. Cowling. The non-radial oscillations of polytropic stars. *Mon. Not. R. Astron. Soc.*, 101:367–375, 1941.

[166] F. Crawford, V. M. Kaspi, and J. F. Bell. A search for sub-millisecond pulsations in unidentified FIRST and NVSS radio sources. In M. Kramer, N. Wex, and N. Wielebinski, editors, *Pulsar Astronomy – 2000 and Beyond*, volume 202 of *ASP Conf. Series*, page 31, ASP, San Francisco, 2000.

[167] C. Cuofano and A. Drago. Magnetic fields generated by r-modes in accreting millisecond pulsars. *Phys. Rev. D*, 82:084027, 2010.

[168] C. Cutler. Post-Newtonian effects on the modes of rotating stars. *Astrophys. J.*, 374:248–254, 1991.

[169] C. Cutler. Gravitational waves from neutron stars with large toroidal B fields. *Phys. Rev. D*, 66:084025, 2002.

[170] C. Cutler and L. Lindblom. The effect of viscosity on neutron star oscillations. *Astrophys. J.*, 314:234–241, 1987.

[171] C. Cutler and L. Lindblom. Post-Newtonian frequencies for the pulsations of rapidly rotating neutron stars. *Astrophys. J.*, 385:630–641, 1992.

[172] C. Cutler, L. Lindblom, and R. J. Splinter. Damping times for neutron star oscillations. *Astrophys. J.*, 363:603–611, 1990.

[173] N. D'Amico. The Bologna submillisecond pulsar survey. In M. Kramer, N. Wex, and N. Wielebinski, editors, *Pulsar Astronomy – 2000 and Beyond*, volume 202 of *ASP Conf. Series*, page 27, ASP, San Francisco, 2000.

[174] T. Damour and G. Schäfer. Higher-order relativistic periastron advances and binary pulsars. *Nuovo Cimento B*, 101:127–176, 1988.

[175] B. Datta. Recent developments in neutron star physics. *Fund. Cosmic Phys.*, 12:151–239, 1988.

[176] B. Datta, S. S. Hasan, P. K. Sahu, and A. R. Prasanna. Radial modes of rotating neutron stars in the Chandrasekhar-Friedman formalism. *Int. J. Mod. Phys. D.*, 7:49, 1998.

[177] D. J. Dean and M. Hjorth-Jensen. Pairing in nuclear systems: from neutron stars to finite nuclei. *Rev. Mod. Phys.*, 75:607–656, 2003.

[178] P. G. deGennes. *Superconductivity of Metals and Alloys.* Addison-Wesley, Reading, MA, 1989.

[179] P. B. Demorest, T. Pennucci, S. M. Ransom, M. S. E. Roberts, and J. W. T. Hessels. A two-solar-mass neutron star measured using Shapiro delay. *Nature*, 467:1081–1083, 2010.

[180] L. Dessart, A. Burrows, R. Livne, and C. D. Ott. Multidimensional radiation/hydrodynamic simulations of proto-neutron star convection. *Astrophys. J.*, 645:534–550, 2006.

[181] S. Detweiler. Periodic solutions of the Einstein equations for binary systems. *Phys. Rev. D*, 50:4929–4943, 1994.

[182] S. L. Detweiler and J. R. Ipser. A variational principle and a stability criterion for the non-radial modes of pulsation of stellar models in general relativity. *Astrophys. J.*, 185:685–708, 1973.

[183] S. L. Detweiler and L. Lindblom. On the evolution of the homogeneous ellipsoidal figures. *Astrophys. J.*, 213:193–199, 1977.

[184] S. L. Detweiler and L. Lindblom. On the nonradial pulsations of general relativistic stellar models. *Astrophys. J.*, 292:12–15, 1985.

[185] M. Dey, I. Bombaci, J. Dey, S. Ray, and B. C. Samanta. Strange stars with realistic quark vector interaction and phenomenological density-dependent scalar potential. *Phys. Lett. B*, 438:123, 1998.

[186] T. Di Girolamo and M. Vietri. Post-Newtonian treatment of bar mode instability in rigidly rotating equilibrium configurations for polytropic stars. *Astrophys. J.*, 581:519–549, 2002.

[187] P. Diener. A new general purpose event horizon finder for 3D numerical spacetimes. *Class. Quant. Grav.*, 20:4901–4917, 2003.

[188] H. Dimmelmeier, M. Bejger, P. Haensel, and L. J. Zdunik. Dynamic migration of rotating neutron stars due to a phase transition instability. *Mon. Not. R. Astron. Soc.*, 396:2269–2288, 2009.

[189] H. Dimmelmeier, J. A. Font, and E. Müller. Relativistic simulations of rotational core collapse I. Methods, initial models, and code tests. *Astron. Astrophys.*, 388:917–935, 2002.

[190] H. Dimmelmeier, J. A. Font, and E. Müller. Relativistic simulations of rotational core collapse II. collapse dynamics and gravitational radiation. *Astron. Astrophys.*, 393:523–542, 2002.

[191] H. Dimmelmeier, C. D. Ott, H.-T. Janka, A. Marek, and E. Müller. Generic gravitational-wave signals from the collapse of rotating stellar cores. *Phys. Rev. Lett.*, 98:251101, 2007.

[192] H. Dimmelmeier, C. D. Ott, A. Marek, and H.-T. Janka. Gravitational wave burst signal from core collapse of rotating stars. *Phys. Rev. D*, 78:064056, 2008.

[193] H. Dimmelmeier, N. Stergioulas, and J. A. Font. Non-linear axisymmetric pulsations of rotating relativistic stars in the conformal flatness approximation. *Mon. Not. R. Astron. Soc.*, 368:1609–1630, 2006.

[194] R. Donat and A. Marquina. Capturing shock reflections: An improved flux formula. *J. Comput. Phys.*, 125:42–58, 1996.

[195] M. D. Duez, Y. T. Liu, S. L. Shapiro, M. Shibata, and B. C. Stephens. Evolution of magnetized, differentially rotating neutron stars: Simulations in full general relativity. *Phys. Rev. D*, 73:104015, 2006.

[196] M. D. Duez, Y. T. Liu, S. L. Shapiro, and B. C. Stephens. General relativistic hydrodynamics with viscosity: Contraction, catastrophic collapse, and disk formation in hypermassive neutron stars. *Phys. Rev. D*, 69:104030, 2004.

[197] M. D. Duez, P. Marronetti, S. L. Shapiro, and T. W. Baumgarte. Hydrodynamic simulations in 3+1 general relativity. *Phys. Rev. D*, 67:024004, 2003.

[198] V. Duez, J. Braithwaite, and S. Mathis. On the stability of non-force-free magnetic equilibria in stars. *Astrophys. J. Lett.*, 724:L34–L38, 2010.

[199] R. C. Duncan and C. Thompson. Formation of very strongly magnetized neutron stars – implications for gamma-ray bursts. *Astrophys. J.*, 392:L9–L13, 1992.

[200] R. H. Durisen, R. A. Gingold, J. E. Tohline, and A. P. Boss. Dynamic fission instabilities in rapidly rotating N = 3/2 polytropes – a comparison of results from finite-difference and smoothed particle hydrodynamics codes. *Astrophys. J.*, 305:281–308, 1986.

[201] J. Dyson and B. F. Schutz. Perturbations and stability of rotating stars. I – Completeness of normal modes. *Proc. R. Soc. Lond. A*, 368:389–410, November 1979.

[202] D. M. Eardley. Theoretical models for sources of gravitational waves. In *Gravitational Radiation, Proceedings of the Advanced Study Institute, Les Houches, France*, volume 124 of *Lecture Notes in Physics*, pages 257–296. North-Holland, Amsterdam, 1983.

[203] C. Eckart. The thermodynamics of irreversible processes III. Relativistic theory of the simple fluid. *Phys. Rev.*, 58:919–924, 1940.

[204] R. T. Edwards, W. van Strate, and M. Bailes. A search for submillisecond pulsars. *Astrophys. J.*, 560:365–370, 2001.

[205] R. I. Epstein. Lepton-driven convection in supernovae. *Mon. Not. R. Astron. Soc.*, 188:305–325, 1979.

[206] R. I. Epstein. Acoustic properties of neutron stars. *Astrophys. J.*, 333:880–894, 1988.

[207] Y. Eriguchi, I. Hachisu, and K. Nomoto. Structure of rapidly rotating neutron stars. *Mon. Not. R. Astron. Soc.*, 266:179–185, 1994.

[208] Y. Eriguchi and E. Müller. Structure of rapidly rotating axisymmetric stars. I – A numerical method for stellar structure and meridional circulation. *Astron. Astrophys.*, 248:435–447, 1991.

[209] F. J. Ernst. New formulation of the axially symmetric gravitational field problem. *Phys. Rev.*, 167:1175–1177, 1968.

[210] F. J. Ernst. New formulation of the axially symmetric gravitational field problem. II. *Phys. Rev.*, 168:1415–1417, 1968.

[211] H. Ertel. Über Hydrodynamische Wirbelsätze. *Physikal. Zeitschr. Leipzig*, 15:526–529, 1942.

[212] Z. B. Etienne, Y. T. Liu, S. L. Shapiro, and T. W. Baumgarte. General relativistic simulations of black-hole-neutron-star mergers: Effects of black-hole spin. *Phys. Rev. D*, 79(4):044024, 2009.

[213] E. Farhi and R. L. Jaffe. Strange matter. *Phys. Rev. D*, 30:2379–2390, 1984.

[214] V. Ferrari, L. Gualtieri, and J. A. Pons. Unstable g-modes in proto–neutron stars. *Class. Quant. Grav.*, 24:5093–5102, 2007.

[215] V. Ferrari, L. Gualtieri, J. A. Pons, and A. Stavridis. Rotational effects on the oscillation frequencies of newly born proto–neutron stars. *Mon. Not. R. Astron. Soc.*, 350:763–768, 2004.

[216] V. Ferrari, S. Matarrese, and R. Schneider. Stochastic background of gravitational waves generated by a cosmological population of young, rapidly rotating neutron stars. *Mon. Not. R. Astron. Soc.*, 303:258–264, 1999.

[217] L. S. Finn. G-modes in zero-temperature neutron stars. *Mon. Not. R. Astron. Soc.*, 227:265–293, 1987.

[218] L. S. Finn. Relativistic stellar pulsations in the Cowling approximation. *Mon. Not. R. Astron. Soc.*, 232:259–275, 1988.

[219] E. Flowers and N. Itoh. Transport properties of dense matter. *Astrophys. J.*, 206:218–242, 1976.

[220] E. Flowers and N. Itoh. Transport properties of dense matter. II. *Astrophys. J.*, 230:847–858, 1979.

[221] G. Fodor, C. Hoenselaers, and Z. Perjés. Multipole moments of axisymmetric systems in relativity. *J. Math. Phys.*, 30:2252–2257, 1989.

[222] J. A. Font. Numerical hydrodynamics and magnetohydrodynamics in general relativity. *Living Rev. Relativity*, 11:7, 2008.

[223] J. A. Font, H. Dimmelmeier, A. Gupta, and N. Stergioulas. Axisymmetric modes of rotating relativistic stars in the Cowling approximation. *Mon. Not. R. Astron. Soc.*, 325:1463–1470, 2001.

[224] J. A. Font, T. Goodale, S. Iyer, M. Miller, L. Rezzolla, E. Seidel, N. Stergioulas, W.-M. Suen, and M. Tobias. Three-dimensional general relativistic hydrodynamics II: Long-term dynamics of single relativistic stars. *Phys. Rev. D*, 65:084024, 2002.

[225] J. A. Font, N. Stergioulas, and K. D. Kokkotas. Nonlinear hydrodynamical evolution of rotating relativistic stars: Numerical methods and code tests. *Mon. Not. R. Astron. Soc.*, 313:678–688, 2000.

[226] Y. Fourés-Bruhat (Choquet-Bruhat). Sur l' intégration des équations de la relativité générale. *J. Rational Mech. Anal.*, 5:951, 1956.

[227] W. A. Fowler. Erratum: Massive stars, relativistic polytropes, and gravitational radiation. *Rev. Mod. Phys.*, 36:1104–1104, 1964.

[228] W. A. Fowler. Massive stars, relativistic polytropes, and gravitational radiation. *Rev. Mod. Phys.*, 36:545–554, 1964.

[229] W. A. Fowler. The stability of supermassive stars. *Astrophys. J.*, 144:180–200, 1966.

[230] B. Friedman and V. R. Pandharipande. Hot and cold, nuclear and neutron matter. *Nucl. Phys. A*, 361:502–520, 1981.

[231] J. L. Friedman. Generic instability of rotating relativistic stars. *Commun. Math. Phys.*, 62:247–278, 1978.

[232] J. L. Friedman. How fast can pulsars spin? In N. Asby, D. F. Bartlett, and W. Wyssed, editors, *General Relativity and Gravitation, 1989. Proceedings of the 12th International Conference on General Relativity and Gravitation*, pages 21–40, Cambridge University Press, Cambridge, UK, 1990.

[233] J. L. Friedman. Upper limit on the rotation of relativistic stars. In A.S. Fruchter and et al., editors, *Millisecond Pulsars. A Decade of Surprise*, volume 72 of *ASP Conference Series*, pages 177–185, Astronomical Society of the Pacific, San Francisco, 1995.

[234] J. L. Friedman and J. R. Ipser. On the maximum mass of a uniformly rotating neutron star. *Astrophys. J.*, 314:594–597, 1987.

[235] J. L. Friedman and J. R. Ipser. Rapidly rotating relativistic stars. *Phil. Trans. R. Soc. Lond. A*, 340:391–422, 1992.

[236] J. L. Friedman, J. R. Ipser, and L. Parker. Rapidly rotating neutron star models. *Astrophys. J.*, 304:115–139, 1986. Erratum: *Astrophys. J.*, 351:705, 1990.

[237] J. L. Friedman, J. R. Ipser, and L. Parker. Implications of a half-millisecond pulsar. *Phys. Rev. Lett.*, 62:3015–3019, 1989.

[238] J. L. Friedman, J. R. Ipser, and R. D. Sorkin. Turning-point method for axisymmetric stability of rotating relativistic stars. *Astrophys. J.*, 325:722–724, 1988.

[239] J. L. Friedman and S. M. Morsink. Axial instability of rotating relativistic stars. *Astrophys. J.*, 502:714–720, 1998.

[240] J. L. Friedman and B. F. Schutz. On the stability of relativistic systems. *Astrophys. J.*, 200:204–220, 1975.

[241] J. L. Friedman and B. F. Schutz. Lagrangian perturbation theory of nonrelativistic fluids. *Astrophys. J.*, 221:937–957, 1978.

[242] J. L. Friedman and B. F. Schutz. Secular instability of rotating Newtonian stars. *Astrophys. J.*, 222:281–296, 1978.

[243] J. L. Friedman, K. Uryū, and M. Shibata. Thermodynamics of binary black holes and neutron stars. *Phys. Rev. D*, 65:065035, 2002.

[244] H. Friedrich. On the hyperbolicity of Einstein's and other gauge field equations. *Commun. Math. Phys.*, 100:525–543, 1985.

[245] C. L. Fryer and A. Heger. Core-collapse simulations of rotating stars. *Astrophys. J.*, 541:1033–1050, 2000.

[246] M. Fujimoto. Nonlinear motions of rotating gaseous ellipsoids. *Astrophys. J.*, 170:143–152, 1971.

[247] G. M. Fuller, S. E. Woosley, and T. A. Weaver. The evolution of radiation-dominated stars. I – Nonrotating supermassive stars. *Astrophys. J.*, 307:675–686, 1986.

[248] M. Gabler. *Coupled Core-Crust-Magnetosphere Oscillations of Magnetars.* PhD thesis, Technische Universität München – Max-Planck-Institut für Astrophysik, 2011.

[249] M. Gabler, P. Cerdá-Durán, J. A. Font, E. Müller, and N. Stergioulas. Magneto-elastic oscillations and the damping of crustal shear modes in magnetars. *Mon. Not. R. Astron. Soc.*, 410:L37–L41, 2011.

[250] E. Gaertig, K. Glampedakis, K. D. Kokkotas, and B. Zink. The f-mode instability in relativistic neutron stars. *Phys. Rev. Lett.*, 107:101102, 2011.

[251] E. Gaertig and K. D. Kokkotas. Oscillations of rapidly rotating relativistic stars. *Phys. Rev. D*, 78:064063, 2008.

[252] E. Gaertig and K. D. Kokkotas. Relativistic g-modes in rapidly rotating neutron stars. *Phys. Rev. D*, 80:064026, 2009.

[253] E. Gaertig and K. D. Kokkotas. Gravitational wave asteroseismology with fast rotating neutron stars. *Phys. Rev. D*, 83:064031, 2011.

[254] D. Garfinkle. Harmonic coordinate method for simulating generic singularities. *Phys. Rev. D*, 65:044029, 2002.

[255] U. H. Gerlach and U. K. Sengupta. Gauge-invariant perturbations on most general spherically symmetric space-times. *Phys. Rev. D*, 19:2268–2272, 1979.

[256] R. Geroch. Multipole moments. II. Curved space. *J. Math. Phys.*, 11:2580–2588, 1970.

[257] R. Geroch. Relativistic theories of dissipative fluids. *J. Math. Phys.*, 36:4226–4241, 1995.

[258] R. Geroch and L. Lindblom. Causal theories of dissipative relativistic fluids. *Ann. Phys.*, 207:394–416, 1991.

[259] P. Ghosh and F. K. Lamb. Accretion by rotating magnetic neutron stars. II – Radial and vertical structure of the transition zone in disk accretion. *Astrophys. J.*, 232:259–276, 1979.

[260] B. Giacomazzo, L. Rezzolla, and N. Stergioulas. Collapse of differentially rotating neutron stars and cosmic censorship. *Phys. Rev. D*, 84:024022, 2011.

[261] G. Gibbons and J. Stewart. Absence of asymptotically flat solutions to Einstein's equations which are periodic and empty near infinity. In W. B. Bonnor, J. N. Islam, and M. A. H. MacCallum, editors, *Classical General Relativity*, pages 77–94. Cambridge University Press, Cambridge, UK, 1984.

[262] K. Glampedakis and N. Andersson. Crust-core coupling in rotating neutron stars. *Phys. Rev. D*, 74:044040, 2006.

[263] K. Glampedakis and N. Andersson. Ekman layer damping of r modes revisited. *Mon. Not. R. Astron. Soc.*, 371:1311–1321, 2006.

[264] N. K. Glendenning. Limiting rotational period of neutron stars. *Phys. Rev. D*, 46:4161–4168, 1992.

[265] N. K. Glendenning. *Compact Stars, Nuclear Physics, Particle Physics, and General Relativity*. Springer-Verlag, New York, 1997.

[266] N. K. Glendenning and F. Weber. Nuclear solid crust on rotating strange quarks stars. *Astrophys. J.*, 400:647–658, 1992.

[267] D. Gondek, P. Haensel, and J. L. Zdunik. Radial pulsations and stability of protoneutron stars. *Astron. Astrophys.*, 325:217–227, 1997.

[268] D. Gondek-Rosińska, T. Bulik, J. L. Zdunik, E. Gourgoulhon, S. Ray, J. Dey, and M. Dey. Rotating compact strange stars. *Astron. Astrophys.*, 363:1005–1012, 2000.

[269] D. Gondek-Rosińska and Gourgoulhon E. Jacobi-like bar mode instability of relativistic rotating bodies. *Phys. Rev. D*, 66:044021, 2002.

[270] D. Gondek-Rosińska, E. Gourgoulhon, and P. Haensel. Are rotating strange quark stars good sources of gravitational waves? *Astron. Astrophys.*, 412:777–790, 2003.

[271] E. Gourgoulhon. *3+1 Formalism in General Relativity: Bases of Numerical Relativity*, volume 846 of *Lecture Notes in Physics*. Springer-Verlag, Berlin, 2012.

[272] E. Gourgoulhon and S. Bonazzola. Noncircular axisymmetric stationary spacetimes. *Phys. Rev. D*, 48:2635–2652, 1993.

[273] E. Gourgoulhon and S. Bonazzola. A formulation of the virial theorem in general relativity. *Class. Quant. Grav.*, 11:443–452, 1994.

[274] E. Gourgoulhon, P. Haensel, R. Livine, E. Paluch, S. Bonazzola, and J.-A. Marck. Fast rotation of strange stars. *Astron. Astrophys.*, 349:851–862, 1999.

[275] E. Gourgoulhon, Ch. Markakis, K. Uryū, and Y. Eriguchi. Magnetohydrodynamics in stationary and axisymmetric spacetimes: A fully covariant approach. *Phys. Rev. D*, 83:104007, 2011.

[276] J. O. Goussard, P. Haensel, and J. L. Zdunik. Rapid uniform rotation of protoneutron stars. *Astron. Astrophys.*, 321:822–834, 1997.

[277] J. O. Goussard, P. Haensel, and J. L. Zdunik. Rapid differential rotation of protoneutron stars and constraints on radio pulsar periods. *Astron. Astrophys.*, 330:1005–1016, 1998.

[278] P. Grandclément, S. Bonazzola, E. Gourgoulhon, and J.-A. Marck. A multidomain spectral method for scalar and vectorial Poisson equations with noncompact sources. *J. Comput. Phys.*, 170:231–260, 2001.

[279] P. Gressman, L.-M. Lin, W.-M. Suen, N. Stergioulas, and J. L. Friedman. Nonlinear r-modes in neutron stars: instability of an unstable mode. *Phys. Rev. D*, 66:041303, 2002.

[280] J. Groah, J. Smoller, and B. Temple. *Shock Wave Interactions in General Relativity*. Springer-Verlag, New York, 2007.

[281] C. Gundlach and J. M. Martín-García. Gauge-invariant and coordinate-independent perturbations of stellar collapse: The interior. *Phys. Rev. D*, 61(8):084024, 2000.

[282] N. Gürlebeck and D. Petroff. The axisymmetric case for the post-Newtonian Dedekind ellipsoids. *Astrophys. J.*, 722:1207–1215, 2010.

[283] Y. Gürsel. Multipole moments for stationary systems: The equivalence of the Geroch-Hansen formulation and the Thorne formulation. *Gen. Rel. Grav.*, 15:737–754, 1983.

[284] P. Haensel. Equation of state of dense matter and maximum mass of neutron stars. In J.-M. Hameury and C. Motch, editors, *Final Stages of Stellar Evolution*, volume 7 of *EAS Publication Series*, pages 249–282. EDP Sciences, Les Ulis, France, 2003.

[285] P. Haensel, K. P. Levenfish, and D. G. Yakovlev. Bulk viscosity in superfluid neutron star cores. III. Effects of σ^- hyperons. *Astron. Astrophys.*, 381:1080–1089, 2002.

[286] P. Haensel and B. Pichon. Experimental nuclear masses and the ground state of cold dense matter. *Astron. Astrophys.*, 283:313, 1994.

[287] P. Haensel, A. Y. Potekhin, and D. G. Yakovlev. *Neutron Stars 1: Equation of State and Structure*. Springer-Verlag, 2007.

[288] P. Haensel and M. Prószyński. Pion condensation in cold dense matter and neutron stars. *Astrophys. J.*, 258:306–320, 1982.

[289] P. Haensel, M. Salgado, and S. Bonazzola. Equation of state of dense matter and maximum rotation frequency of neutron stars. *Astron. Astrophys.*, 296:746–751, 1995.

[290] P. Haensel and R. Schaeffer. Bulk viscosity of hot-neutron-star matter from direct URCA processes. *Phys. Rev. D*, 45:4708–4712, 1992.

[291] P. Haensel and J. L. Zdunik. A submillisecond pulsar and the equation of state of dense matter. *Nature*, 340:617–619, 1989.

[292] P. Haensel, J. L. Zdunik, and F. Douchin. Equation of state of dense matter and the minimum mass of cold neutron stars. *Astron. Astrophys.*, 385:301–307, 2002.

[293] Y. Hagihara. *Theories of Equilibrium Figures of a Rotating Homogeneous Fluid Mass*. Number 186 in NASA SP. NASA, Washington, DC, 1969.

[294] R. O. Hansen. Multipole moments of stationary space-times. *J. Math. Phys.*, 15:46–52, 1974.

[295] A. Hanson, T. Regge, and C. Teitelboim. *Constrained Hamiltonian Systems*. Academia Nationale dei Lincei, 1976.

[296] B. K. Harrison, K. S. Thorne, M. Wakano, and J. A. Wheeler. *Gravitation Theory and Gravitational Collapse*. University of Chicago Press, Chicago, 1965.

[297] J. B. Hartle. Slowly rotating relativistic stars. I. Equations of structure. *Astrophys. J.*, 150:1005–1029, 1967.

[298] J. B. Hartle. Slowly-rotating relativistic stars. IV. Rotational energy and moment of inertia for stars in differential rotation. *Astrophys. J.*, 161:111–118, 1970.

[299] J. B. Hartle. Bounds on the mass and moment of inertia of nonrotating neutron stars. *Phys. Rep.*, 46:201–247, 1978.

[300] J. B. Hartle and J. L. Friedman. Slowly rotating relativistic stars. VIII. Frequencies of the quasi-radial modes of an $n = 3/2$ polytrope. *Astrophys. J.*, 196:653–660, 1975.

[301] J. B. Hartle and M. W. Munn. Slowly rotating relativistic stars. V. Static stability analysis of n = 3/2 polytropes. *Astrophys. J.*, 198:467–476, 1975.

[302] J. B. Hartle and A. G. Sabbadini. The equation of state and bounds on the mass of nonrotating neutron stars. *Astrophys. J.*, 213:831–835, 1977.

[303] J. B. Hartle and D. H. Sharp. Variational principle for the equilibrium of a relativistic, rotating star. *Astrophys. J.*, 147:317–333, 1967.

[304] J. B. Hartle and K. S. Thorne. Slowly rotating relativistic stars. II. Models for neutron stars and supermassive stars. *Astrophys. J.*, 153:807–834, 1968.

[305] M. Hashimoto, K. Oyamatsu, and Y. Eriguchi. Upper limit of the angular velocity of neutron stars. *Astrophys. J.*, 436:257–261, 1994.

[306] B. Haskell and N. Andersson. Superfluid hyperon bulk viscosity and the r-mode instability of rotating neutron stars. *Mon. Not. R. Astron. Soc.*, 408:1897–1915, 2010.

[307] B. Haskell, N. Andersson, and A. Passamonti. r-modes and mutual friction in rapidly rotating superfluid neutron stars. *Mon. Not. R. Astron. Soc.*, 397:1464–1485, 2009.

[308] S. H. Hawley and M. W. Choptuik. Boson stars driven to the brink of black hole formation. *Phys. Rev. D*, 62:104024, 2000.

[309] A. Heger, N. Langer, and S. E. Woosley. Presupernova evolution of rotating massive stars. I. Numerical method and evolution of the internal stellar structure. *Astrophys. J.*, 528:368–396, 2000.

[310] A. Heger, S. E. Woosley, N. Langer, and H. C. Spruit. Presupernova evolution of rotating massive stars and the rotation rate of pulsars. In A. Maeder and P. Eenens, editors, *Stellar Rotation*, number 215 in Proceedings of IAU Symposium, 2003.

[311] D. J. Hegyi. The upper mass limit for neutron stars including differential rotation. *Astrophys. J.*, 217:244–247, 1977.

[312] H. Heiselberg and V. R. Pandharipande. Recent progress in neutron star theory. *Annu. Rev. Nucl. Part. S.*, 50:481, 2000.

[313] J. W. T. Hessels, S. M. Ransom, I. H. Stairs, P. C. C. Freire, V. M. Kaspi, and F. Camilo. A radio pulsar spinning at 716 Hz. *Science*, 311:1901–1904, 2006.

[314] J. Heyl. Low-mass X-ray binaries may be important laser interferometer gravitational-wave observatory sources after all. *Astrophys. J.*, 574:L57–L60, 2002.

[315] W. A. Hiscock and L. Lindblom. Generic instabilities in first-order dissipative relativistic fluid theories. *Phys. Rev. D*, 31:725–733, 1985.

[316] D. Holm. Hamilton techniques for relativistic fluid dynamics and stability theory. In A. Anile and Y. Choquet-Bruhat, editors, *Relativistic Fluid Dynamics*, volume 1385 of *Lecture Notes in Mathematics*, pages 65–151. Springer-Verlag, Berlin, 1987.

[317] E. Høiland. *Avhandliger Norske Videnskaps-Akademi i Oslo, I, Math. Naturv. Klasse*, 11:1–24, 1941.

[318] T. Y. Hou and P. G. Lefloch. Why nonconservative schemes converge to wrong solutions: error analysis. *Math. Comput.*, 62:497–530, 1994.

[319] J. L. Houser and J. M. Centrella. Gravitational radiation from rotational instabilites in compact stellar cores with stiff equations of state. *Phys. Rev. D*, 54:7278–7297, 1996.

[320] J. L. Houser, J. M. Centrella, and S. C. Smith. Gravitational radiation from nonaxisymmetric instability in a rotating star. *Phys. Rev. Lett.*, 72:1314–1317, 1994.

[321] J. M. Ibáñez. Riemann solvers in relativistic hydrodynamics: Basics and astrophysical applications. *J. Korean Astron. Soc.*, 34:191–201, 2001.

[322] J. N. Imamura, J. L. Friedman, and R. H. Durisen. Secular stability limits for rotating polytropic stars. *Astrophys. J.*, 294:474–478, 1985.

[323] J. N. Imamura, B. K. Pickett, and R. H. Durisen. The nonlinear evolution of massive stellar core collapses that "fizzle." *Astrophys. J.*, 587:341–355, 2003.

[324] J. R. Ipser, M. B. Kislinger, and P. D. Morley. Unpublished 1975.

[325] J. R. Ipser and L. Lindblom. The oscillations of rapidly rotating Newtonian stellar models. *Astrophys. J.*, 355:226–240, 1990.

[326] J. R. Ipser and L. Lindblom. On the adiabatic pulsations of accretion disks and rotating stars. *Astrophys. J.*, 379:285–289, 1991.

[327] J. R. Ipser and L. Lindblom. The oscillations of rapidly rotating Newtonian stellar models. II – Dissipative effects. *Astrophys. J.*, 373:213–221, 1991.

[328] J. R. Ipser and L. Lindblom. On the pulsations of relativistic accretion disks and rotating stars – the Cowling approximation. *Astrophys. J.*, 389:392–399, 1992.

[329] J. R. Ipser and L. Lindblom. The oscillations and stability of rotating fluids: The two-potential formalism. *Ann. N. Y. Acad. Sci.*, 675:84–92, 1992.

[330] J. R. Ipser and R. A. Managan. On the existence and structure of inhomogeneous analogs of the Dedekind and Jacobi ellipsoids. *Astrophys. J.*, 250:362–372, 1981.

[331] J. R. Ipser and R. A. Managan. An Eulerian variational principle and a criterion for the occurrence of nonaxisymmetric neutral modes along rotating axisymmetric sequences. *Astrophys. J.*, 292:517–521, 1985.

[332] J. R. Ipser and R. H. Price. Nonradial pulsations of stellar models in general relativity. *Phys. Rev. D*, 43:1768–1773, 1991.

[333] J. R. Ipser and K. S. Thorne. Nonradial pulsation of general-relativistic stellar models. VI. Corrections. *Astrophys. J.*, 181:181–182, 1973.

[334] J. A. Isenberg. Waveless approximation theories of gravity. *Int. J. Mod. Phys. D.*, 17:265–273, 2008.

[335] G. L. Israel, T. Belloni, L. Stella, Y. Rephaeli, D. E. Gruber, P. Casella, S. Dall'Osso, N. Rea, M. Persic, and R. E. Rothschild. The discovery of rapid X-ray oscillations in the tail of the SGR 1806-20 hyperflare. *Astrophys. J.*, 628:L53–L56, 2005.

[336] W. Israel. Nonstationary irreversible thermodynamics: A causal relativistic theory. *Ann. Phys.*, 100:310–331, 1976.

[337] W. Israel and J. M. Stewart. Transient relativistic thermodynamics and kinetic theory. *Ann. Phys.*, 118:341, 1979.

[338] J. D. Jackson. *Classical Electrodynamics*, 3rd edition, Wiley, New York, 1998.

[339] P. Jaikumar, G. Rupak, and A. W. Steiner. Viscous damping of r-mode oscillations in compact stars with quark matter. *Phys. Rev. D*, 78:123007, 2008.

[340] R. A. James. The structure and stability of rotating gas masses. *Astrophys. J.*, 140:552–582, 1964.

[341] P. B. Jones. Comment on "Gravitational radiation instability in hot young neutron stars." *Phys. Rev. Lett.*, 86:1384–1384, 2001.

[342] P. Kaaret, Z. Prieskorn, J. J. M. in't Zand, S. Brandt, N. Lund, S. Mereghetti, D. Götz, E. Kuulkers, and J. A. Tomsick. Evidence of 1122 Hz X-ray burst oscillations from the neutron star X-ray transient XTE J1739-285. *Astrophys. J.*, 657:L97–L100, 2007.

[343] D. B. Kaplan and A. E. Nelson. Strange goings on in dense nucleonic matter. *Phys. Lett. B*, 175:57–63, 1986.

[344] S. Karino and Y. Eriguchi. Gravitational radiation reaction driven secular instability of f-mode oscillations in differentially rotating newtonian stars. *Astrophys. J.*, 578:413–419, 2002.

[345] W. Kastaun. Inertial modes of rigidly rotating neutron stars in Cowling approximation. *Phys. Rev. D*, 77:124019, 2008.

[346] W. Kastaun, B. Willburger, and K. D. Kokkotas. Saturation amplitude of the f-mode instability. *Phys. Rev. D*, 82:104036, 2010.

[347] W. Keil and H.-T. Janka. Hadronic phase transitions at supranuclear densities and the delayed collapse of newly formed neutron stars. *Astron. Astrophys.*, 296:145–163, 1995.

[348] W. Keil, H.-T. Janka, and E. Müller. Ledoux convection in protoneutron stars – a clue to supernova nucleosynthesis? *Astrophys. J.*, 473:L111, 1996.

[349] I. M. Khalatnikov. *Introduction to the Theory of Superfluidity*. Addison-Wesley, New York, 1958.

[350] J. Kijowski, A. Smólski, and A. Górnicka. Hamiltonian theory of self-gravitating perfect fluid and a method of effective deparametrization of Einstein's theory of gravitation. *Phys. Rev. D*, 41:1875–1884, 1990.

[351] J. B. Kinney and G. Mendell. r-modes in accreting neutron stars with magnetoviscous boundary layers. *Phys. Rev. D*, 67(2):024032, 2003.

[352] K. Kiuchi, M. Shibata, and S. Yoshida. Evolution of neutron stars with toroidal magnetic fields: Axisymmetric simulation in full general relativity. *Astrophys. J.*, 78:024029, 2008.

[353] K. Kiuchi and H. Shinkai. Numerical experiments of adjusted Baumgarte-Shapiro-Shibata-Nakamura systems for controlling constraint violations. *Phys. Rev. D*, 77:044010, 2008.

[354] K. Kiuchi and S. Yoshida. Relativistic stars with purely toroidal magnetic fields. *Phys. Rev. D*, 78:044045, 2008.

[355] C. S. Kochanek. Coalescing binary neutron stars. *Astrophys. J.*, 398:234–247, 1992.

[356] Y. Kojima. Equations governing the nonradial oscillations of a slowly rotating relativistic star. *Phys. Rev. D*, 46:4289–4303, 1992.

[357] Y. Kojima. Coupled pulsations between polar and axial modes in a slowly rotating relativistic star. *Prog. Theor. Phys.*, 90:977–990, 1993.

[358] Y. Kojima. Normal modes of relativistic stars in slow rotation limit. *Astrophys. J.*, 414:247–253, 1993.

[359] Y. Kojima. Chapter 4. The rotational effects of general relativity on the stellar pulsations. *Prog. Theor. Phys. Supp.*, 128:251–293, 1997.

[360] Y. Kojima. Quasi-toroidal oscillations in rotating relativistic stars. *Mon. Not. R. Astron. Soc.*, 293:49–52, 1998.

[361] Y. Kojima and M. Hosonuma. The r-mode oscillations in relativistic rotating stars. *Astrophys. J.*, 520:788–796, 1999.

[362] Y. Kojima and M. Hosonuma. Approximate equation relevant to axial oscillations on slowly rotating relativistic stars. *Phys. Rev. D*, 62:044006, 2000.

[363] Y. Kojima and M. Saijo. Amplification of azimuthal modes with odd wave numbers during dynamical bar-mode growth in rotating stars. *Phys. Rev. D*, 78:124001, 2008.

[364] K. Kokkotas and B. Schmidt. Quasi-normal modes of stars and black holes. *Living Rev. Relativity*, 2:2, 1999.

[365] K. D. Kokkotas and J. Ruoff. Instabilities of relativistic stars. In *2001: A relativistic spacetime odyssey*, 25th Johns Hopkins Workshop, 2002. Firenze, 2001.

[366] K. D. Kokkotas, J. Ruoff, and N. Andersson. w-Mode instability of ultracompact relativistic stars. *Phys. Rev. D*, 70:043003, 2004.

[367] K. D. Kokkotas and B. F. Shutz. Normal modes of a model radiating system. *Gen. Rel. Grav.*, 18:913–921, 1986.

[368] K. D. Kokkotas and B. F. Shutz. W-modes: A new family of normal modes for pulsating relativistic stars. *Mon. Not. R. Astron. Soc.*, 225:119–128, 1992.

[369] K. D. Kokkotas and N. Stergioulas. Analytic description of the r-mode instability in uniform density stars. *Astron. Astrophys.*, 341:110–116, 1999.

[370] A. Komar. Covariant conservation laws in general relativity. *Phys. Rev.*, 113:934–936, 1959.

[371] H. Komatsu, Y. Eriguchi, and I. Hachisu. Rapidly rotating general relativistic stars. I – Numerical method and its application to uniformly rotating polytropes. *Mon. Not. R. Astron. Soc.*, 237:355–379, 1989.

[372] H. Komatsu, Y. Eriguchi, and I. Hachisu. Rapidly rotating general relativistic stars. II – Differentially rotating polytropes. *Mon. Not. R. Astron. Soc.*, 239:153–171, 1989.

[373] S. Koranda, N. Stergioulas, and J. L. Friedman. Upper limit set by causality on the rotation and mass of uniformly rotating relativistic stars. *Astrophys. J.*, 488:799–806, 1997.

[374] K. Kotake, H. Sawai, S. Yamada, and K. Sato. Magnetorotational effects on anisotropic neutrino emission and convection in core-collapse supernovae. *Astrophys. J.*, 608:391–404, 2004.

[375] A. Kovetz. Schwarzschild's criterion for convective instability in general relativity. *Z. Astrophys.*, 66:446, 1967.

[376] C. Krüger, E. Gaertig, and K. D. Kokkotas. Oscillations and instabilities of fast and differentially rotating relativistic stars. *Phys. Rev. D*, 81:084019, 2010.

[377] W. Kundt and M. Trümper. Orthogonal decomposition of axi-symmetric stationary spacetimes. *Z. Physik*, 192:419–422, 1966.

[378] T. Kunihiro, T. Takatsuka, and R. Tamagaki. Neutral pion condensation in hot and dense nuclear matter. *Prog. Theor. Phys. Supp.*, 112:197, 1993.

[379] H. P. Künzle and J. M. Nester. Hamiltonian formulation of gravitating perfect fluids and the Newtonian limit. *J. Math. Phys.*, 25:1009–1018, 1984.

[380] W. G. Laarakkers and E. Poisson. Quadrupole moments of rotating neutron stars. *Astrophys. J.*, 512:282–287, 1999.

[381] B. D. Lackey, M. Nayyar, and B. J. Owen. Observational constraints on hyperons in neutron stars. *Phys. Rev. D*, 73:024021, 2006.

[382] D. Lai. Secular instability of g-modes in rotating neutron stars. *Mon. Not. R. Astron. Soc.*, 307:1001–1007, 1999.

[383] D. Lai. Secular bar-mode evolution and gravitational waves from neutron stars. In *Astrophysical Sources for Ground-Based Gravitational Wave Detectors*, volume 575 of *AIP Conference Proceedings*, pages 246–257, 2001.

[384] D. Lai and S. L. Shapiro. Gravitational radiation from rapidly rotating nascent neutron stars. *Astrophys. J.*, 442:259–272, 1995.

[385] L. Landau and E. M. Lifshitz. *Mechanics*, 2nd edition. Pergamon Press, New York, 1976.

[386] L. Landau and E. M. Lifshitz. *Fluid Mechanics*, 2nd edition. Pergamon Press, New York, 1987.

[387] C. B. Laney. *Computational Gasdynamics*. Cambridge University Press, Cambridge, UK, 1998.

[388] D. Langlois, D. M. Sedrakian, and B. Carter. Differential rotation of relativistic superfluid in neutron stars. *Mon. Not. R. Astron. Soc.*, 297:1189–1201, 1998.

[389] J. Lasota, P. Haensel, and M. A. Abramowicz. Fast rotation of neutron stars. *Astrophys. J.*, 456:300–304, 1996.

[390] J. M. Lattimer and M. Prakash. The physics of neutron stars. *Science*, 304:536–542, 2004.

[391] J. M. Lattimer and M. Prakash. Neutron star observations: Prognosis for equation of state constraints. *Phys. Rep.*, 442:109–165, 2007.

[392] J. M. Lattimer, M. Prakash, C. J. Pethick, and P. Haensel. Direct URCA process in neutron stars. *Phys. Rev. Lett.*, 66:2701–2704, 1991.

[393] J. M. Lattimer and B. F. Schutz. Constraining the equation of state with moment of inertia measurements. *Astrophys. J.*, 629:979–984, 2005.

[394] J. M. Lattimer and F. D. Swesty. A generalized equation of state for hot, dense matter. *Nucl. Phys. A*, 535:331–376, 1991.

[395] L. M. Lattimer, M. Prakash, D. Masak, and A. Yahil. Rapidly rotating pulsars and the equation of state. *Astrophys. J.*, 355:241–254, 1990.

[396] P. G. Le Floch and J. M. Stewart. Shock waves and gravitational waves in matter spacetimes with Gowdy symmetry. *Portugaliae Mathematica*, 62:349–370, 2005.

[397] N. Lebovitz and A. Lifschitz. New global instabilities of the Riemann ellipsoids. *Astrophys. J.*, 458:699–713, 1996.

[398] N. R. Lebovitz. On Schwarzschild's criterion for the stability of gaseous masses. *Astrophys. J.*, 142:229, 1965.

[399] N. R. Lebovitz and A. Lifschitz. Short-wavelength instabilities of Riemann ellipsoids. *Phil. Trans. R. Soc. Lond. A*, 354:927–950, 1996.

[400] N. R. Lebovitz and K. I. Saldanha. On the weakly nonlinear development of the elliptic instability. *Phys. Fluids*, 11:3374–3379, 1999.

[401] U. Lee. Nonradial oscillations of neutron stars with the superfluid core. *Astron. Astrophys.*, 303:515–525, 1995.

[402] U. Lee and S. Yoshida. r-Modes of neutron stars with superfluid cores. *Astrophys. J.*, 586:403–418, 2003.

[403] M. Leins, H.-P. Nollert, and M. H. Soffel. Nonradial oscillations of neutron stars: A new branch of strongly damped normal modes. *Phys. Rev. D*, 48:3467–3472, 1993.

[404] R. J. LeVeque. *Numerical Methods for Conservation Laws*, 2nd edition. Birkhäuser, Basel, 2005.

[405] Y. Levin. Runaway heating by r-modes of neutron stars in low-mass X-ray binaries. *Astrophys. J.*, 517:328–333, 1999.

[406] Y. Levin. QPOs during magnetar flares are not driven by mechanical normal modes of the crust. *Mon. Not. R. Astron. Soc.*, 368:L35–L38, 2006.

[407] Y. Levin. On the theory of magnetar QPOs. *Mon. Not. R. Astron. Soc.*, 377:159–167, 2007.

[408] Y. Levin and G. Ushomirsky. Crust-core coupling and r-mode damping in neutron stars: a toy model. *Mon. Not. R. Astron. Soc.*, 324:917–922, July 2001.

[409] D. Lewis. Bifurcation of liquid drops. *Nonlinearity*, 6:491–522, 1993.

[410] H. Li, S. A. Colgate, B. Wendroff, and R. Liska. Rossby wave instability of thin accretion disks. III. Nonlinear simulations. *Astrophys. J.*, 551:874–896, 2001.

[411] H. Li, J. M. Finn, R. V. E. Lovelace, and S. A. Colgate. Rossby wave instability of thin accretion disks. II. Detailed linear theory. *Astrophys. J.*, 533:1023–1034, 2000.

[412] A. Liapounoff. *Sur les Figures d'Equilibre Peu Differentes des Ellipsoides d'une Masse Liquide Homogene*. L'Académie Impériale des Sciences, St. Petersburg, Russia, 1906.

[413] A. Lichnerowicz. *Relativistic Hydrodynamics and Magnetohydrodynamics*. Benjamin, New York, 1967.

[414] L. Lichtenstein. *Gleichgewichtsfiguren Rotierender Fluessigkeiten*. Springer, Berlin, 1933.

[415] L.-M. Lin, N. Andersson, and G. L. Comer. Oscillations of general relativistic multi-fluid/multilayer compact stars. *Phys. Rev. D*, 78:083008, 2008.

[416] L.-M. Lin and W.-M. Suen. Non-linear r-modes in neutron stars: a hydrodynamical limitation on r-mode amplitudes. *Mon. Not. R. Astron. Soc.*, 370:1295–1302, 2006.

[417] L. Lindblom. Stationary stars are axisymmetric. *Astrophys. J.*, 208:873–880, 1976.

[418] L. Lindblom. Critical angular velocities of rotating neutron stars. *Astrophys. J.*, 438:265–268, 1995.

[419] L. Lindblom. The relaxation effect in dissipative relativistic fluid theories. *Ann. Phys.*, 247:1–18, 1996.

[420] L. Lindblom and S. Detweiler. The role of neutrino dissipation in gravitational collapse. *Astrophys. J.*, 232232:L101–L105, 1979.

[421] L. Lindblom and S. L. Detweiler. The quadrupole oscillations of neutron stars. *Astrophys. J. Suppl. Ser.*, 53:73–92, 1983.

[422] L. Lindblom, G. Mendell, and J. R. Ipser. Relativistic stellar pulsations with near-zone boundary conditions. *Phys. Rev. D*, 56:2118–2126, 1997.

[423] L. Lindblom and W. A. Hiscock. On the stability of rotating stellar models in general relativity theory. *Astrophys. J.*, 267:384–401, 1983.

[424] L. Lindblom and G. Mendell. The oscillations of superfluid neutron stars. *Astrophys. J.*, 421:689–704, 1994.

[425] L. Lindblom and G. Mendell. Does gravitational radiation limit the angular velocities of superfluid neutron stars? *Astrophys. J.*, 444:804–809, 1995.

[426] L. Lindblom and G. Mendell. R-modes in superfluid neutron stars. *Phys. Rev. D*, 61:104003, 2000.

[427] L. Lindblom and B. J. Owen. Effect of hyperon bulk viscosity on neutron-star r-modes. *Phys. Rev. D*, 65:063006, 2002.

[428] L. Lindblom, B. J. Owen, and S. M. Morsink. Gravitational radiation instability in hot young neutron stars. *Phys. Rev. Lett.*, 80:4843–4846, 1998.

[429] L. Lindblom, B. J. Owen, and G. Ushomirsky. Effect of a neutron-star crust on the r-mode instability. *Phys. Rev. D*, 62:084030, 2000.

[430] L. Lindblom, M. A. Scheel, L. E. Kidder, R. Owen, and O. Rinne. A new generalized harmonic evolution system. *Class. Quant. Grav.*, 23:S447–S462, 2006.

[431] L. Lindblom and R. J. Splinter. The accuracy of the relativistic cowling approximation. *Astrophys. J.*, 348:198–202, 1990.

[432] L. Lindblom, J. E. Tohline, and M. Vallisneri. Non-linear evolution of the r-modes in neutron stars. *Phys. Rev. Lett.*, 86:1152–1155, 2001.

[433] L. Lindblom, J. E. Tohline, and M. Vallisneri. Numerical evolutions of nonlinear r-modes in neutron stars. *Phys. Rev. D*, 65:084039, 2002.

[434] I.-S. Liu, I. Müller, and T. Ruggeri. Relativistic thermodynamics of gases. *Ann. Phys.*, 169:191–219, 1986.

[435] Y. T. Liu. Dynamical instability of new-born neutron stars as sources of gravitational radiation. *Phys. Rev. D*, 65:124003, 2002.

[436] Y. T. Liu, Z. Etienne, J. Faber, S. Shapiro, K. Taniguchi, and T. Baumgarte. Fully general relativistic simulations of black hole-neutron star mergers. In *APS Meeting Abstracts*, pages 10005–+, April 2008.

[437] Y. T. Liu and L. Lindblom. Models of rapidly rotating neutron stars: Remnants of accretion induced collapse. *Mon. Not. R. Astron. Soc.*, 324:1063–1073, 2001.

[438] Y. T. Liu and S. L. Shapiro. Magnetic braking in differentially rotating, relativistic stars. *Phys. Rev. D*, 69:044009, 2004.

[439] M. Livio, J. R. Buchler, and S. A. Colgate. Rayleigh-Taylor driven supernova explosions – a two-dimensional numerical study. *Astrophys. J.*, 238:L139–L143, 1980.

[440] M. Livio and J. E. Pringle. The rotation rates of white dwarfs and pulsars. *Astrophys. J.*, 505:339–343, 1998.

[441] K. H. Lockitch. *Stability and Rotational Mixing of Modes in Newtonian and Relativistic Stars*. PhD thesis, University of Wisconsin–Milwaukee. Relativistic stars, rotation, instability, 1999.

[442] K. H. Lockitch, N. Andersson, and J. L. Friedman. Rotational modes of relativistic stars: Analytic results. *Phys. Rev. D*, 63:024019, 2001.

[443] K. H. Lockitch, N. Andersson, and A. L. Watts. Regularizing the r-mode problem for nonbarotropic relativistic stars. *Class. Quant. Grav.*, 21:4661–4675, 2004.

[444] K. H. Lockitch, J. L. Friedman, and N. Andersson. Rotational modes of relativistic stars: Numerical results. *Phys. Rev. D*, 68:124010, 2003.

[445] R. V. E. Lovelace and R. G. Hohlfeld. Negative mass instability of flat galaxies. *Astrophys. J.*, 221:51–61, 1978.

[446] R. V. E. Lovelace, H. Li, S. A. Colgate, and A. F. Nelson. Rossby wave instability of Keplerian accretion disks. *Astrophys. J.*, 513:805–810, 1999.

[447] P. J. Luyten. The dynamical stability of differentially rotating, self-gravitating cylinders. *Mon. Not. R. Astron. Soc.*, 242:447–456, 1990.

[448] P. J. Luyten. Non-axisymmetric dynamical instability of differentially rotating self-gravitating masses. *Mon. Not. R. Astron. Soc.*, 245:614–622, 1990.

[449] N. D. Lyford, T. W. Baumgarte, and S. L. Shapiro. Effects of differential rotation on the maximum mass of neutron stars. *Astrophys. J.*, 583:410–415, 2003.

[450] D. Lynden-Bell and J. P. Ostriker. On the stability of differentially rotating bodies. *Mon. Not. R. Astron. Soc.*, 136:293–310, 1967.

[451] J. Madsen. How to identify a strange star. *Phys. Rev. Lett.*, 81:3311–3314, 1998.

[452] J. Madsen. Probing strange stars and color superconductivity by r-mode instabilities in millisecond pulsars. *Phys. Rev. Lett.*, 85:10–13, 2000.

[453] R. A. Managan. On the secular instability of axisymmetric rotating stars to gravitational radiation reaction. *Astrophys. J.*, 294:463–473, 1985.

[454] R. A. Managan. Nonaxisymmetric normal modes and secular instabilities of rotating stars. *Astrophys. J.*, 309:598–608, 1986.

[455] G. M. Manca, L. Baiotti, R. De Pietri, and L. Rezzolla. Dynamical non-axisymmetric instabilities in rotating relativistic stars. *Class. Quant. Grav.*, 24:S171–S186, 2007.

[456] V. S. Manko, J. Martín, and E. Ruiz. Six-parameter solution of the Einstein-Maxwell equations possessing equatorial symmetry. *J. Math. Phys.*, 36:3063–3073, 1995.

[457] V. S. Manko, J. Martin, E. Ruiz, N. R. Sibgatullin, and M. N. Zaripov. Metric of a rotating, charged, magnetized, deformed mass. *Phys. Rev. D*, 49:5144–5149, 1994.

[458] V. S. Manko, E. W. Mielke, and J. D. Sanabria-Gómez. Exact solution for the exterior field of a rotating neutron star. *Phys. Rev. D*, 61:081501, 2000.

[459] V. S. Manko, J. D. Sanabria-Gómez, and O. V. Manko. Nine-parameter electrovac metric involving rational functions. *Phys. Rev. D*, 62:044048, 2000.

[460] K. Martel and E. Poisson. Gravitational perturbations of the Schwarzschild spacetime: A practical covariant and gauge-invariant formalism. *Phys. Rev. D*, 71:104003, 2005.

[461] J. M. Marti, J. M. Ibánez, and J. A. Miralles. Numerical relativistic hydrodynamics: Local characteristic approach. *Phys. Rev. D*, 43:3794–3801, 1991.

[462] J. M. Martín-García and C. Gundlach. Gauge-invariant and coordinate-independent perturbations of stellar collapse. II. Matching to the exterior. *Phys. Rev. D*, 64(2):024012, 2001.

[463] Y. Masada, T. Sano, and K. Shibata. The effect of neutrino radiation on magnetorotational instability in proto-neutron stars. *Astrophys. J.*, 655:447–457, 2007.

[464] Y. Masada, T. Sano, and H. Takabe. Nonaxisymmetric magnetorotational instability in proto–neutron stars. *Astrophys. J.*, 641:447–457, 2006.

[465] A. K. M. Masood-Ul-Alam. Proof that static stellar models are spherical. *Gen. Rel. Grav.*, 39:55–85, 2007.

[466] P. N. McDermott. Density discontinuity g-modes. *Mon. Not. R. Astron. Soc.*, 245:508–513, 1990.

[467] P. N. McDermott, H. M. Van Horn, and C. J. Hansen. Nonradial oscillations of neutron stars. *Astrophys. J.*, 325:725–748, 1988.

[468] P. N. McDermott, H. M. Van Horn, and J. F. Scholl. Nonradial g-mode oscillations of warm neutron stars. *Astrophys. J.*, 268:837–848, 1983.

[469] R. Meinel, M. Ansorg, A. Kleinwächter, G. Neugebauer, and D. Petroff. *Relativistic Figures of Equilibrium*. Cambridge University Press, Cambridge, UK, 2008.

[470] A. Melatos and D. J. B. Payne. Gravitational radiation from an accreting millisecond pulsar with a magnetically confined mountain. *Astrophys. J.*, 623:1044–1050, 2005.

[471] G. Mendell. Magnetohydrodynamics in superconducting-superfluid neutron stars. *Mon. Not. R. Astron. Soc.*, 296:903–912, 1998.

[472] A. B. Migdal. Stability of the vacuum and limiting fields. *Zh. Eksp. Teor. Fiz.*, 61: 2209–2224, 1971.

[473] B. D. Miller. The effect of gravitational radiation-reaction on the evolution of the Riemann S-type ellipsoids. *Astrophys. J.*, 187:609–620, 1974.

[474] G. Miniutti, J. A. Pons, E. Berti, L. Gualtieri, and V. Ferrari. Non-radial oscillation modes as a probe of density discontinuities in neutron stars. *Mon. Not. R. Astron. Soc.*, 338:389–400, 2003.

[475] J. A. Miralles, J. A. Pons, and V. A. Urpin. Convective instability in proto–neutron stars. *Astrophys. J.*, 543:1001–1006, 2000.

[476] J. A. Miralles, J. A. Pons, and V. A. Urpin. Hydromagnetic instabilities in proto–neutron stars. *Astrophys. J.*, 574:356–363, 2002.

[477] J. A. Miralles, J. A. Pons, and V. A. Urpin. Anisotropic convection in rotating proto–neutron stars. *Astrophys. J.*, 420:245–249, 2004.

[478] C. M. Misner and H. S. Zapolsky. High-density behavior and dynamical stability of neutron star models. *Phys. Rev. Lett.*, 12:635–637, 1964.

[479] C. W. Misner. Gravitational field energy and g_{00}. *Phys. Rev.*, 130:1590–1594, 1963.

[480] C. W. Misner, K. S. Thorne, and J. A. Wheeler. *Gravitation*. W. H. Freeman, San Francisco, 1973.

[481] R. Moenchmeyer, G. Schaefer, E. Mueller, and R. E. Kates. Gravitational waves from the collapse of rotating stellar cores. *Astron. Astrophys.*, 246:417–440, 1991.

[482] V. Moncrief. Gravitational perturbations of spherically symmetric systems. I. The exterior problem. *Ann. Phys.*, 88:323–342, 1974.

[483] V. Moncrief. Hamiltonian formalism for relativistic perfect fluids. *Phys. Rev. D*, 16:1702–1705, 1977.

[484] I. A. Morrison, T. W. Baumgarte, and S. L. Shapiro. Effect of differential rotation on the maximum mass of neutron stars: Realistic nuclear equations of state. *Astrophys. J.*, 610:941–947, 2004.

[485] I. A. Morrison, T. W. Baumgarte, S. L. Shapiro, and V. R. Pandharipande. The moment of inertia of the binary pulsar J0737-3039A: Constraining the nuclear equation of state. *Astrophys. J.*, 617:L135–L138, 2004.

[486] S. Morsink, N. Stergioulas, and S. Blattnig. Quasi-normal modes of rotating relativistic stars – neutral modes for realistic equations of state. *Astrophys. J.*, 510:854–861, 1999.

[487] S. M. Morsink. Relativistic precession around rotating neutron stars: Effects due to frame-dragging and stellar oblateness. *Astrophys. J.*, 571:435–446, 2002.

[488] I. Müller. Zum Paradoxon der Wärmeleitungstheorie. *Z. Physik*, 198:329–344, 1967.

[489] A. Nagar and L. Rezzolla. TOPICAL REVIEW: Gauge-invariant non-spherical metric perturbations of Schwarzschild black-hole spacetimes. *Class. Quant. Grav.*, 22:167, 2005.

[490] T. Nakamura. General relativistic collapse of axially symmetric stars leading to the formation of rotating black holes. *Prog. Theor. Phys.*, 65:1876–1890, 1981.

[491] T. Nakamura, K. Oohara, and Y. Kojima. General relativistic collapse to black holes and gravitational waves from black holes. *Prog. Theor. Phys. Supp.*, 90:1, 1987.

[492] L. Naso, L. Rezzolla, A. Bonanno, and L. Paternó. Magnetic field amplification in proto-neutron stars. The role of the neutron-finger instability for dynamo excitation. *Astron. Astrophys.*, 479:167–176, 2008.

[493] M. Nayyar and B. J. Owen. R-modes of accreting hyperon stars as persistent sources of gravitational waves. *Phys. Rev. D*, 73:084001, 2006.

[494] G. Neugebauer and H. Herold. Gravitational fields of rapidly rotating neutron stars: Theoretical foundation. In J. Ehlers and G. Schäfer, editors, *Relativistic Gravity Research, Proceedings of the 81 WE-Heraeus-Seminar Held at the Physikzentrum Bad Honnef, Germany*, volume 410 of *Lecture Notes in Physics*, page 305. Springer, Berlin, 1992.

[495] K. C. B. New, J. M. Centrella, and J. E. Tohline. Gravitational waves from long-duration simulations of the dynamical bar instability. *Phys. Rev. D*, 62:064019, 2000.

[496] K. C. B. New and S. L. Shapiro. Evolution of differentially rotating supermassive stars to the onset of bar instability. *Astrophys. J.*, 548:439–446, 2001.

[497] K. C. B. New and S. L. Shapiro. The formation of supermassive black holes and the evolution of supermassive stars. *Class. Quant. Grav.*, 18:3965–3975, 2001.

[498] T. Nozawa, N. Stergioulas, E. Gourgoulhon, and Y. Eriguchi. Construction of highly accurate models of rotating neutron stars – comparison of three different numerical schemes. *Astron. Astrophys. Suppl. Ser.*, 132:431–454, 1998.

[499] M. Obergaulinger, P. Cerdá-Durán, E. Müller, and M. A. Aloy. Semi-global simulations of the magneto-rotational instability in core collapse supernovae. *Astron. Astrophys.*, 498:241–271, 2009.

[500] R. Oechslin and H.-T. Janka. Gravitational waves from relativistic neutron-star mergers with microphysical equations of state. *Phys. Rev. Lett.*, 12:121102, 2007.

[501] R. Oechslin, S. Rosswog, and F.-K. Thielemann. Conformally flat smoothed particle hydrodynamics: Application to neutron star mergers. *Phys. Rev. D*, 65:103005, 2002.

[502] K. Oohara, T. Nakamura, and M. Shibata. A way to 3D numerical relativity. *Prog. Theor. Phys. Supp.*, 128:183–249, 1997.

[503] J. R. Oppenheimer and G. M. Volkoff. On massive neutron cores. *Phys. Rev.*, 55:374–381, 1939.

[504] A. Oron. Relativistic magnetized star with poloidal and toroidal fields. *Phys. Rev. D*, 66:023006, 2002.

[505] C. D. Ott. TOPICAL REVIEW: The gravitational-wave signature of core-collapse supernovae. *Class. Quant. Grav.*, 26:063001, 2009.

[506] C. D. Ott, A. Burrows, L. Dessart, and E. Livne. A new mechanism for gravitational-wave emission in core-collapse supernovae. *Phys. Rev. Lett.*, 96:201102, 2006.

[507] C. D. Ott, H. Dimmelmeier, A. Marek, H.-T. Janka, I. Hawke, B. Zink, and E. Schnetter. 3D collapse of rotating stellar iron cores in general relativity including deleptonization and a nuclear equation of state. *Phys. Rev. Lett.*, 98:261101, 2007.

[508] C. D. Ott, H. Dimmelmeier, A. Marek, H.-T. Janka, B. Zink, I. Hawke, and E. Schnetter. Rotating collapse of stellar iron cores in general relativity. *Class. Quant. Grav.*, 24:S139–S154, 2007.

[509] C. D. Ott, S. Ou, J. E. Tohline, and A. Burrows. One-armed spiral instability in a low-$T/|W|$ postbounce supernova core. *Astrophys. J.*, 625:L119–L122, 2005.

[510] S. Ou and J. E. Tohline. Unexpected dynamical instabilities in differentially rotating neutron stars. *Astrophys. J.*, 651:1068–1078, 2006.

[511] S. Ou, J. E. Tohline, and L. Lindblom. Nonlinear development of the secular bar-mode instability in rotating neutron stars. *Astrophys. J.*, 617:490–499, 2004.

[512] B. J. Owen. Maximum elastic deformations of compact stars with exotic equations of state. *Phys. Rev. Lett.*, 95:211101, 2005.

[513] B. J. Owen. How to adapt broad-band gravitational-wave searches for r-modes. *Phys. Rev. D*, 82:104002, 2010.

[514] B.J. Owen, L. Lindblom, C. Cutler, B. F. Schutz, A. Vecchio, and N. Andersson. Gravitational radiation instability in hot young neutron stars. *Phys. Rev. D*, 58:084020, 1998.

[515] F. Özel, G. Baym, and T. Güver. Astrophysical measurement of the equation of state of neutron star matter. *Phys. Rev. D*, 82:101301, 2010.

[516] F. Özel and D. Psaltis. Reconstructing the neutron-star equation of state from astrophysical measurements. *Astrophys. J.*, 80:103003, 2009.

[517] F. Özel, D. Psaltis, S. Ransom, P. Demorest, and M. Alford. The massive pulsar PSR J1614-2230: Linking quantum chromodynamics, gamma-ray bursts, and gravitational wave astronomy. *Astrophys. J.*, 724:L199–L202, 2010.

[518] D. Page, Prakash M., Lattimer J. M., and A. W. Steiner. Rapid cooling of the neutron star in Cassiopeia A triggered by neutron superfluidity in dense matter. *Phys. Rev. Lett.*, 106:081101, 2011.

[519] D. N. Page. Neutron star explosions. *Phys. Lett. A*, 91:201, 1982.

[520] V. R. Pandharipande, C. J. Pethick, and V. Thorsson. Kaon energies in dense matter. *Phys. Rev. Lett.*, 75:4567–4570, 1995.

[521] V. R. Pandharipande, D. Pines, and R. A. Smith. Neutron star structure: Theory, observation, and speculation. *Astrophys. J.*, 208:550–566, 1976.

[522] V. R. Pandharipande and R. A. Smith. A model neutron solid with π^0 condensate. *Nucl. Phys. A*, 237:507–532, 1975.

[523] V. R. Pandharipande and R. A. Smith. Nuclear matter calculations with mean scalar fields. *Phys. Lett. B*, 59:15, 1975.

[524] J. Papaloizou and J. E. Pringle. Gravitational radiation and the stability of rotating stars. *Mon. Not. R. Astron. Soc.*, 184:501–508, 1978.

[525] J. Papaloizou and J. E. Pringle. Non-radial oscillations of rotating stars and their relevance to the short-period oscillations of cataclysmic variables. *Mon. Not. R. Astron. Soc.*, 182:423–442, 1978.

[526] J. C. B. Papaloizou and D. N. C. Lin. Nonaxisymmetric instabilities in thin self-gravitating rings and disks. *Astrophys. J.*, 344:645–668, 1989.

[527] J. C. B. Papaloizou and J. E. Pringle. The dynamical stability of differentially rotating discs. II. *Mon. Not. R. Astron. Soc.*, 213:799–820, 1985.

[528] A. Papapetrou. Eine rotationssymmetrische Lösung in der allgemeinen Relativitätstheorie. *Ann. Phys.*, 447:309–315, 1953.

[529] G. Pappas. Matching of analytical and numerical solutions for neutron stars of arbitrary rotation. In N. Stergioulas and C. Tsagas, editors, *Proceedings of the 13th Conference Recent Developments in Gravity*, volume 189 of *Journal of Physics: Conference Series*, page 012028, 2009.

[530] A. Passamonti, M. Bruni, L. Gualtieri, and C. F. Sopuerta. Coupling of radial and nonradial oscillations of relativistic stars: Gauge-invariant formalism. *Phys. Rev. D*, 71:024022, 2005.

[531] A. Passamonti, B. Haskell, and N. Andersson. Oscillations of rapidly rotating superfluid stars. *Mon. Not. R. Astron. Soc.*, 396:951–963, 2009.

[532] A. Passamonti, B. Haskell, N. Andersson, D. I. Jones, and I. Hawke. Oscillations of rapidly rotating stratified neutron stars. *Mon. Not. R. Astron. Soc.*, 394:730–741, 2009.

[533] A. Passamonti, A. Stavridis, and K. D. Kokkotas. Nonaxisymmetric oscillations of differentially rotating relativistic stars. *Phys. Rev. D*, 77:024029, 2008.

[534] A. Passamonti, N. Stergioulas, and A. Nagar. Gravitational waves from nonlinear couplings of radial and polar nonradial modes in relativistic stars. *Phys. Rev. D*, 75:084038, 2007.

[535] A. Patruno, B. Haskell, and C. D'Angelo. Gravitational waves and the maximum spin frequency of neutron stars. *Astrophys. J.*, 746:9, 2012.

[536] F. A. E. Pirani and G. Williams. Rigid motion in a gravitational field. *Séminaire Janet. Mechanique analytique et méchanique céleste*, 5:1–16, 1962.

[537] H. Poincaré. Sur l'équilibre d'une masse fluide animée d'un mouvement de rotation. *Acta Math.*, 7:259–380, 1885.

[538] J. A. Pons, L. Gualtieri, J. A. Miralles, and V. Ferrari. Relativistic r modes and shear viscosity: regularizing the continuous spectrum. *Mon. Not. R. Astron. Soc.*, 363:121–130, 2005.

[539] J.A. Pons, S. Reddy, P. J. Ellis, M. Prakash, and J. M. Lattimer. Kaon condensation in proto–neutron star matter. *Phys. Rev. C*, 62:035803, 2000.

[540] J.A. Pons, S. Reddy, M. Prakash, J. M. Lattimer, and J.A. Miralles. Evolution of proto–neutron stars. *Astrophys. J.*, 513:780–804, 1999.

[541] M. Prakash, I. Bombaci, M. Prakash, P. J. Ellis, J. M. Lattimer, and R. Knorren. Composition and structure of protoneutron stars. *Phys. Rep.*, 280:1–77, 1997.

[542] M. Prakash, J. M. Lattimer, J. A. Pons, A. W. Steiner, and S. Reddy. Evolution of a neutron star from its birth to old age. In D. Blaschke, N. K.. Glendenning, and A. Sedrakian, editors, *Physics of Neutron Star Interiors*, volume 578 of *Lecture Notes in Physics*, pages 364–423. Springer, 2001.

[543] W. H. Press and J. M. Bardeen. Nonconservation of the Newman-Penrose conserved quantities. *Phys. Rev. Lett.*, 27(19):1303–1306, Nov 1971.

[544] W. H. Press and S. A. Teukolsky. On the evolution of the secularly unstable, viscous Maclaurin spheroids. *Astrophys. J.*, 181:513–517, 1973.

[545] F. Pretorius. Evolution of binary black-hole spacetimes. *Phys. Rev. Lett.*, 95:121101, 2005.

[546] F. Pretorius. Numerical relativity using a generalized harmonic decomposition. *Class. Quant. Grav.*, 22:425–451, 2005.

[547] R. Price and K. S. Thorne. Non-radial pulsation of general-relativistic stellar models. II. Properties of the gravitational waves. *Astrophys. J.*, 155:163–182, 1969.

[548] R. H. Price and V. Husain. Model for the completeness of quasinormal modes of relativistic stellar oscillations. *Phys. Rev. Lett.*, 68:1973–1976, 1991.

[549] R. H. Price, C. Markakis, and J. L. Friedman. Iteration stability for simple Newtonian stellar systems. *J. Math. Phys.*, 50:073505, 2009.

[550] D. Priou. The perturbations of a fully general relativistic and rapidly rotating neutron star – I. Equations of motion for the solid crust. *Mon. Not. R. Astron. Soc.*, 254:435–452, 1992.

[551] R. Prix, G. L. Comer, and N. Andersson. Inertial modes of non-stratified superfluid neutron stars. *Mon. Not. R. Astron. Soc.*, 348:625–637, 2004.

[552] R. Prix and M. Rieutord. Adiabatic oscillations of non-rotating superfluid neutron stars. *Astron. Astrophys.*, 393:949–963, 2002.

[553] J. Provost, G. Berthomieu, and A. Rocca. Low frequency oscillations of a slowly rotating star – quasi-toroidal modes. *Astron. Astrophys.*, 94:126–133, 1981.

[554] H. Quaintrell, A. J. Norton, T. D. C. Ash, P. Roche, B. Willems, T. R. Bedding, I. K. Baldry, and R. P. Fender. The mass of the neutron star in Vela X-1 and tidally induced non-radial oscillations in GP Vel. *Astron. Astrophys.*, 401:313–323, 2003.

[555] M. Rampp, E. Müller, and M. Ruffert. Simulations of non-axisymmetric rotational core collapse. *Astron. Astrophys.*, 332:969–983, 1998.

[556] F. A. Rasio and S. L. Shapiro. Coalescing binary neutron stars. *Class. Quant. Grav.*, 16:R1–R29, 1999.

[557] J. W. S. Rayleigh. *Scientific Papers.* Cambridge University Press, Cambridge, UK, 1900.

[558] J. S. Read, B. D. Lackey, B. J. Owen, and J. L. Friedman. Constraints on a phenomenologically parametrized neutron-star equation of state. *Phys. Rev. D*, 79(12):124032, 2009.

[559] J. S. Read, C. Markakis, M. Shibata, K. Uryū, J. D. E. Creighton, and J. L. Friedman. Measuring the neutron star equation of state with gravitational wave observations. *Phys. Rev. D*, 79:124033, 2009.

[560] T. Regge and C. Teitelboim. Role of surface integrals in the Hamiltonian formulation of general relativity. *Ann. Phys.*, 88:286–318, 1974.

[561] T. Regge and J. A. Wheeler. Stability of a Schwarzschild singularity. *Phys. Rev.*, 108:1063–1069, 1957.

[562] M. Reinhardt and U. Geppert. The proto-neutron-star dynamo. Viability and impediments. *Astron. Astrophys.*, 435:201–206, 2005.

[563] A. Rendall. Theorems on existence and global dynamics for the Einstein equations. *Living Rev. Relativity*, 8:6, 2005.

[564] A. D. Rendall and B. G. Schmidt. Existence and properties of spherically symmetric static fluid bodies with a given equation of state. *Class. Quant. Grav.*, 8:985–1000, 1991.

[565] L. Rezzolla, F. K. Lamb, D. Marković, and S. L. Shapiro. Properties of r modes in rotating magnetic neutron stars. I. Kinematic secular effects and magnetic evolution equations. *Phys. Rev. D*, 64:104013, 2001.

[566] L. Rezzolla, F. K. Lamb, D. Marković, and S. L. Shapiro. Properties of r modes in rotating magnetic neutron stars. II. Evolution of the r modes and stellar magnetic field. *Phys. Rev. D*, 64:104014, 2001.

[567] L. Rezzolla, F. K. Lamb, and S. L. Shapiro. R-mode oscillations in rotating magnetic neutron stars. *Astrophys. J.*, 531:L139–L142, 2000.

[568] C. E. Rhoades and R. Ruffini. Maximum mass of a neutron star. *Phys. Rev. Lett.*, 32:324–327, 1974.

[569] M. Rieutord. Ekman layers and the damping of inertial r-modes in a spherical shell: Application to neutron stars. *Astrophys. J.*, 550:443–447, 2001.

[570] P. H. Roberts and K. Stewartson. On the stability of a Maclaurin spheroid of small viscosity. *Astrophys. J.*, 137:777–790, 1963.

[571] J. Ruoff and K. D. Kokkotas. On the r-mode spectrum of relativistic stars in the low-frequency approximation. *Mon. Not. R. Astron. Soc.*, 328:678–688, 2001.

[572] J. Ruoff and K. D. Kokkotas. On the r-mode spectrum of relativistic stars: the inclusion of the radiation reaction. *Mon. Not. R. Astron. Soc.*, 330:1027–1033, 2002.

[573] J. Ruoff, A. Stavridis, and K. D. Kokkotas. Evolution equations for the perturbations of slowly rotating relativistic stars. *Mon. Not. R. Astron. Soc.*, 332:676–688, 2002.

[574] J. Ruoff, A. Stavridis, and K. D. Kokkotas. Inertial modes of slowly rotating relativistic stars in the Cowling approximation. *Mon. Not. R. Astron. Soc.*, 339:1170–1182, 2003.

[575] G. Rupak and P. Jaikumar. Constraining phases of quark matter with studies of r-mode damping in compact stars. *Phys. Rev. C*, 82:055806, 2010.

[576] F. Ryan. Gravitational waves from the inspiral of a compact object into a massive, axisymmetric body with arbitrary multipole moments. *Phys. Rev. D*, 52:5707–5718, 1995.

[577] M. Saijo. The collapse of differentially rotating supermassive stars: Conformally flat simulations. *Astrophys. J.*, 615:866–879, 2004.

[578] M. Saijo. Dynamical bar instability in a relativistic rotational core collapse. *Phys. Rev. D*, 71:104038, 2005.

[579] M. Saijo, T. W. Baumgarte, S. L. Shapiro, and M. Shibata. Collapse of a rotating supermassive star to a supermassive black hole: Post-Newtonian simulations. *Astrophys. J.*, 569:349–361, 2002.

[580] M. Saijo and Y. Kojima. Faraday resonance in dynamical bar instability of differentially rotating stars. *Phys. Rev. D*, 77:063002, 2008.

[581] M. Saijo, M. Shibata, T. W. Baumgarte, and S. L. Shapiro. Dynamical bar instability in rotating stars: Effect of general relativity. *Astrophys. J.*, 548:919–931, 2001.

[582] H. Saio. R-mode oscillations in uniformly rotating stars. *Astrophys. J.*, 256:717–735, 1982.

[583] M. Salgado, S. Bonazzola, E. Gourgoulhon, and P. Haensel. High precision rotating neutron star models. I. Analysis of neutron star properties. *Astron. Astrophys.*, 291:155–170, 1994.

[584] M. Salgado, S. Bonazzola, E. Gourgoulhon, and P. Haensel. High precision rotating neutron star models. II. Large sample of neutron star properties. *Astron. Astrophys. Suppl. Ser.*, 108:455–459, 1994.

[585] R. Salmon. Hamiltonian fluid mechanics. *Annu. Rev. Fluid Mech.*, 20:225–256, 1988.

[586] L. Samuelsson and N. Andersson. Neutron star asteroseismology. Axial crust oscillations in the Cowling approximation. *Mon. Not. R. Astron. Soc.*, 374:256–268, 2007.

[587] R. F. Sawyer. Bulk viscosity of hot neutron-star matter and the maximum rotation rates of neutron stars. *Phys. Rev. D*, 39:3804–3806, 1989.

[588] R. F. Sawyer and D. J. Scalapino. Pion condensation in superdense nuclear matter. *Phys. Rev. D*, 7:953–964, 1973.

[589] G. Schäfer and A. Gopakumar. Minimal no-radiation approximation to Einstein's field equations. *Phys. Rev. D*, 69:021501, 2004.

[590] S. Scheidegger, T. Fischer, S. C. Whitehouse, and M. Liebendörfer. Gravitational waves from 3D MHD core collapse simulations. *Astron. Astrophys.*, 490:231–241, 2008.

[591] A. K. Schenk, P. Arras, É. É. Flanagan, S. A. Teukolsky, and I. Wasserman. Nonlinear mode coupling in rotating stars and the r-mode instability in neutron stars. *Phys. Rev. D*, 65:024001, 2002.

[592] K. Schöbel and M. Ansorg. Maximal mass of uniformly rotating homogeneous stars in Einsteinian gravity. *Astron. Astrophys.*, 405:405–408, 2003.

[593] B. L. Schumaker and K. S. Thorne. Torsional oscillations of neutron stars. *Mon. Not. R. Astron. Soc.*, 203:457–489, 1983.

[594] B. F. Schutz. Perfect fluids in general relativity: Velocity potentials and a variational principle. *Phys. Rev. D*, 2:2762–2773, 1970.

[595] B. F. Schutz. *A First Course in General Relativity.* Cambridge University Press, Cambridge, UK, 2009.

[596] B. F. Schutz and R. D. Sorkin. Variational aspects of relativistic field theories with applications to perfect fluids. *Ann. Phys.*, 107:1–43, 1977.

[597] B. F. Schutz, Jr. Taylor instabilities in relativistic stars. *Astrophys. J.*, 161:1173, 1970.

[598] B. F. Schutz, Jr. Linear pulsations and stability of differentially rotating stellar models. II. General-relativistic analysis. *Astrophys. J.*, 24:343, 1972.

[599] F. H. Seguin. The stability of nonuniform rotation in relativistic stars. *Astrophys. J.*, 197:745–765, 1975.

[600] E. Seidel and W.-M. Suen. Dynamical evolution of boson stars: Perturbing the ground state. *Phys. Rev. D*, 42:384–403, 1990.

[601] S. Setiawan, M. Ruffert, and H.-T. Janka. Three-dimensional simulations of non-stationary accretion by remnant black holes of compact object mergers. *Astron. Astrophys.*, 458:553–567, 2006.

[602] S. L. Shapiro. Differential rotation in neutron stars: Magnetic braking and viscous damping. *Astrophys. J.*, 544:397–408, 2000.

[603] S. L. Shapiro and S. A. Teukolsky. *Black Holes, White Dwarfs and Neutron Stars.* Wiley, New York, 1983.

[604] S. L. Shapiro and S. Zane. Bar mode instability in relativistic rotating stars: A post-Newtonian treatment. *Astrophys. J.*, 460:379–389, 1996.

[605] S. T. Shapiro and A. P. Lightman. Rapidly rotating, post-Newtonian neutron stars. *Astrophys. J.*, 207:263–278, 1976.

[606] H. Shen, H. Toki, K. Oyamatsu, and K. Sumiyoshi. Relativistic equation of state of nuclear matter for supernova explosion. *Prog. Theor. Phys.*, 100:1013–1031, 1998.

[607] M. Shibata. Relativistic formalism for computation of irrotational binary stars in quasiequilibrium states. *Phys. Rev. D*, 58:024012, 1998.

[608] M. Shibata. Fully general relativistic simulation of coalescing binary neutron stars: Preparatory tests. *Phys. Rev. D*, 60:104052, 1999.

[609] M. Shibata. Axisymmetric simulations of rotating stellar collapse in full general relativity – criteria for prompt collapse to black holes. *Prog. Theor. Phys.*, 104:325–358, 2000.

[610] M. Shibata. Collapse of rotating supramassive neutron stars to black holes: Fully general relativistic simulations. *Astrophys. J.*, 595:992–999, 2003.

[611] M. Shibata, T. W. Baumgarte, and S. L. Shapiro. The bar-mode instability in differentially rotating neutron stars: Simulations in full general relativity. *Astrophys. J.*, 542:453–463, 2000.

[612] M. Shibata, T. W. Baumgarte, and S. L. Shapiro. Stability and collapse of rapidly rotating, supramassive neutron stars: 3D simulations in general relativity. *Phys. Rev. D*, 61:044012, 2000.

[613] M. Shibata and S. Karino. Numerical evolution of secular bar-mode instability induced by the gravitational radiation reaction in rapidly rotating neutron stars. *Phys. Rev. D*, 70:084022, 2004.

[614] M. Shibata, S. Karino, and Y. Eriguchi. Dynamical instability of differentially rotating stars. *Mon. Not. R. Astron. Soc.*, 334:L27–L31, 2002.

[615] M. Shibata, S. Karino, and Y. Eriguchi. Dynamical bar-mode instability of differentially rotating stars: effects of equations of state and velocity profiles. *Mon. Not. R. Astron. Soc.*, 343:619–626, 2003.

[616] M. Shibata, Y. T. Liu, S. L. Shapiro, and B. C. Stephens. Magnetorotational collapse of massive stellar cores to neutron stars: Simulations in full general relativity. *Phys. Rev. D*, 74:104026, 2006.

[617] M. Shibata and M. Sasaki. Innermost stable circular orbits around relativistic rotating stars. *Phys. Rev. D*, 58:104011, 1998.

[618] M. Shibata and Y. Sekiguchi. Gravitational waves from axisymmetric rotating stellar core collapse to a neutron star in full general relativity. *Phys. Rev. D*, 69:084024, 2004.

[619] M. Shibata and Y. Sekiguchi. Three-dimensional simulations of stellar core collapse in full general relativity: Nonaxisymmetric dynamical instabilities. *Phys. Rev. D*, 71:024014, 2005.

[620] M. Shibata and Nakamura T. Evolution of three-dimensional gravitational waves: Harmonic slicing case. *Phys. Rev. D*, 52:5428–5444, 1995.

[621] M. Shibata and K. Taniguchi. Merger of binary neutron stars to a black hole: Disk mass, short gamma-ray bursts, and quasinormal mode ringing. *Phys. Rev. D*, 73(6):064027, March 2006.

[622] M. Shibata and K. Taniguchi. Merger of black hole and neutron star in general relativity: Tidal disruption, torus mass, and gravitational waves. *Phys. Rev. D*, 77(8):084015, 2008.

[623] M. Shibata, K. Taniguchi, and K. Uryū. Merger of binary neutron stars with realistic equations of state in full general relativity. *Phys. Rev. D*, 71(8):084021, April 2005.

[624] M. Shibata and K. Uryū. Merger of black hole-neutron star binaries: Non-spinning black hole case. *Phys. Rev. D*, 74(12):121503, 2006.

[625] M. Shibata and K. Uryū. Binary neutron star mergers in fully general relativistic simulations. *Phys. Rev. D*, 61:064001, 2000.

[626] M. Shibata, K. Uryū, and J. L. Friedman. Deriving formulations for numerical computation of binary neutron stars in quasicircular orbits. *Phys. Rev. D*, 70:044044, 2004.

[627] M. Shibata, K. Uryū, and K. Taniguchi. Merger of binary neutron stars with realistic equations of state in full general relativity. *Phys. Rev. D*, 71:084021, 2005.

[628] I. S. Shklovsky. Remarks on the nature of X-ray sources. *Soviet Astron.*, 9:224, 1965.

[629] I. S. Shklovsky. On the nature of the source of X-ray emission of SCO XR-1. *Astrophys. J.*, 148:L1, 1967.

[630] P. S. Shternin and D. G. Yakovlev. Shear viscosity in neutron star cores. *Phys. Rev. D*, 78:063006, 2008.

[631] N. R. Sibgatullin. *Oscillations and Waves in Strong Gravitational and Electromagnetic Fields* (Nauka, Moscow) 1984; English translation: Springer-Verlag, Berlin, 1991.

[632] N. R. Sibgatullin and R. A. Sunyaev. Disk accretion in the gravitational field of a rapidly rotating neutron star with a rotationally induced quadrupole mass distribution. *Astron. Lett.*, 24:774–787, 1998.

[633] N. R. Sibgatullin and R. A. Sunyaev. Energy release during disk accretion onto a rapidly rotating neutron star. *Astron. Lett.*, 26:699–724, 2000.

[634] W. Simon and R. Beig. The multipole structure of stationary space-times. *J. Math. Phys.*, 24:1163–1171, 1983.

[635] D. Skinner and L. Lindblom. On the viscosity-driven secular instability in rotating neutron star. *Astrophys. J.*, 461:920–926, 1996.

[636] L. Smarr and York, J. W. Jr. Kinematical conditions in the construction of spacetime. *Phys. Rev. D*, 17:2529–2551, 1978.

[637] S. C. Smith, J. L. Houser, and J. M. Centrella. Simulations of nonaxisymmetric instability in a rotating star: A comparison between Eulerian and smooth particle hydrodynamics. *Astrophys. J.*, 458:236–256, 1996.

[638] H. Solberg. Le mouvement d'inertie de l'atmosphere stable et son role dans la theorie des cyclones. In *Proces Verbaux de l' Association de Météorologie. International Union of*

Geodesy and Geophysics. 6th General Assembly (Edinburgh), volume 2, pages 66–82, Edinburgh, 1936. International Union of Geodesy and Geophysics.

[639] E. B. Sonin. Vortex oscillations and hydrodynamics of rotating superfluids. *Rev. Mod. Phys.*, 59:87–155, 1987.

[640] D. Soper. *Classical Field Theory*. John Wiley and Sons, New York, 1976.

[641] R. Sorkin. Conserved quantities as action variations. In J. W. Isenberg, editor, *High Energy Astrophysics*, volume 71 of *AMS Contemporary Mathematics*, pages 23–37, Providence, 1967. American Mathematical Society.

[642] R. D. Sorkin. A criterion for the onset of instability at a turning point. *Astrophys. J.*, 249:254–257, 1981.

[643] R. D. Sorkin. A stability criterion for many parameter equilibrium families. *Astrophys. J.*, 257:847–854, 1982.

[644] R. D. Sorkin. On stress-energy tensors. *Gen. Rel. Grav.*, 8:437–449, 1977.

[645] H. Sotani and K. D. Kokkotas. Alfvén polar oscillations of relativistic stars. *Mon. Not. R. Astron. Soc.*, 395:1163–1172, 2009.

[646] H. Sotani, K. D. Kokkotas, and N. Stergioulas. Alfvén quasi-periodic oscillations in magnetars. *Mon. Not. R. Astron. Soc.*, 385:L5–L9, 2008.

[647] H. C. Spruit. Gamma-ray bursts from X-ray binaries. *Astron. Astrophys.*, 341:L1–L4, 1999.

[648] H. C. Spruit. Origin of neutron star magnetic fields. In C. Bassa, Z. Wang, A. Cumming, and V. M. Kaspi, editors, *40 Years of Pulsars: Millisecond Pulsars, Magnetars and More*, volume 983 of *AIP Conference Proceedings*, pages 391–398, 2008.

[649] H. C. Spruit and E. S. Phinney. Birth kicks as the origin of pulsar rotation. *Nature*, 393:193, 1998.

[650] N. K. Spyrou and N. Stergioulas. Spin-down of relativistic stars with phase transitions and PSR J0537-6910. *Astron. Astrophys.*, 395:151–160, 2002.

[651] R. Stark and T. Piran. A general relativistic code for rotating axisymmetric configurations and gravitational radiation: Numerical methods and tests. *Comput. Phys. Rep.*, 5:221–264, 1987.

[652] R. F. Stark and T. Piran. Gravitational-wave emission from rotating gravitational collapse. *Phys. Rev. Lett.*, 55:891–894, 1985. Erratum: *Phys. Rev. Lett.* 55:891, 1985.

[653] R. F. Stark and T. Piran. A numerical computation of the gravitational radiation from rotating gravitational collapse. In R. Ruffini, editor, *Proceedings of the Fourth Marcel Grossmann Meeting on General Relativity, Part A*, pages 327–364. Elsevier Science, New York, 1986.

[654] A. Stavridis, A. Passamonti, and K. D. Kokkotas. Nonradial oscillations of slowly and differentially rotating compact stars. *Phys. Rev. D*, 75:064019, 2007.

[655] A. W. Steiner, J. M. Lattimer, and E. F. Brown. The equation of state from observed masses and radii of neutron stars. *Astrophys. J.*, 722:33–54, 2010.

[656] B. C. Stephens, M. D. Duez, Y. T. Liu, S. L. Shapiro, and M. Shibata. Collapse and black hole formation in magnetized, differentially rotating neutron stars. *Class. Quant. Grav.*, 24:S207–S219, 2007.

[657] B. C. Stephens, W. E. East, and F. Pretorius. Eccentric black-hole–neutron-star mergers. *Astrophys. J. Lett.*, 737:L5, 2011.

[658] N. Stergioulas. **rns**. Public domain code available at http://www.gravity.phys.uwm.edu/rns.

[659] N. Stergioulas. *The Structure and Stability of Rotating Relativistic Stars*. PhD thesis, University of Wisconsin–Milwaukee. 1996.

[660] N. Stergioulas. Rotating stars in relativity. *Living Rev. Relativ.*, 6:3, 2003.

[661] N. Stergioulas, T. A. Apostolatos, and J. A. Font. Non-linear pulsations in differentially rotating neutron stars: Mass-shedding–induced damping and splitting of the fundamental mode. *Mon. Not. R. Astron. Soc.*, 352:1089–1101, 2004.

[662] N. Stergioulas, A. Bauswein, K. Zagkouris, and H.-Th. Janka. Gravitational waves and non-axisymmetric oscillation modes in mergers of compact object binaries. *Mon. Not. R. Astron. Soc.*, 418:427–436, 2011.

[663] N. Stergioulas and J. A. Font. Nonlinear r-modes in rapidly rotating relativistic stars. *Phys. Rev. Lett.*, 86:1148–1151, 2001.

[664] N. Stergioulas and J. L. Friedman. Comparing models of rapidly rotating relativistic stars constructed by two numerical methods. *Astrophys. J.*, 444:306–311, 1995.

[665] N. Stergioulas and J. L. Friedman. Nonaxisymmetric neutral modes in rotating relativistic stars. *Astrophys. J.*, 492:301–322, 1998.

[666] N. Stergioulas, W. Kluźniak, and T. Bulik. Keplerian frequencies and innermost stable circular orbits of rapidly rotating strange stars. *Astron. Astrophys.*, 352:L116–L120, 1999.

[667] R. Stoeckly. Polytropic models with fast, nonuniform rotation. *Astrophys. J.*, 142:208–228, 1965.

[668] J. R. Stone, P. A. M. Guichon, and A. W. Thomas. Role of hyperons in neutron stars. arXiv:1012.2919v1, December 2010.

[669] K. Strobel, Ch. Schaab, and M. K. Weigel. Properties of non-rotating and rapidly rotating protoneutron stars. *Astron. Astrophys.*, 350:497–512, 1999.

[670] K. Strobel, F. Weber, and M. K. Weigel. Symmetric and asymmetric nuclear matter in the Thomas-Fermi model at finite temperatures. *Z. Naturforsch.*, 54a:83–90, 1999.

[671] T. E. Strohmayer. Density discontinuities and the g-mode oscillation spectra of neutron stars. *Astrophys. J.*, 417:273–278, 1993.

[672] K. Sumiyoshi, J. M. Ibáñez, and J. V. Romero. Thermal history and structure of rotating protoneutron stars with relativistic equation of state. *Astron. Astrophys. Suppl. Ser.*, 134:39–52, 1999.

[673] D. Swesty. Thermodynamically consistent interpolation for equation of state tables. *J. Comput. Phys.*, 127:118, 1996.

[674] B. Szilágyi, D. Pollney, L. Rezzolla, J. Thornburg, and J. Winicour. An explicit harmonic code for black-hole evolution using excision. *Class. Quant. Grav.*, 24:S275–S293, 2007.

[675] K. Takami, L. Rezzolla, and S. Yoshida. A quasi-radial stability criterion for rotating relativistic stars. *Mon. Not. R. Astron. Soc.*, 416:L1–L5, 2011.

[676] J.-L. Tassoul. *Theory of Rotating Stars.* Princeton University Press, Princeton, NJ, 1978.

[677] A. H. Taub. General relativistic variational principle for perfect fluids. *Phys. Rev.*, 94:1468–1470, 1954.

[678] A. H. Taub. Stability of general relativistic gaseous masses and variational principles. *Commun. Math. Phys.*, 15:235–254, 1969.

[679] R. J. Tayler. The adiabatic stability of stars containing magnetic fields – I. Toroidal fields. *Mon. Not. R. Astron. Soc.*, 161:365–380, 1973.

[680] C. Teichmüller, M. B. Fröb, and F. Maucher. Analytical approximation of the exterior gravitational field of rotating neutron stars. *Class. Quant. Grav.*, 28:155015, 2011.

[681] S. A. Teukolsky. Perturbations of a rotating black hole. I. Fundamental equations for gravitational, electromagnetic, and neutrino-field perturbations. *Astrophys. J.*, 185:635–648, 1973.

[682] S. A. Teukolsky. Irrotational binary neutron stars in quasi-equilibrium in general relativity. *Astrophys. J.*, 504:442–449, 1998.

[683] S. A. Teukolsky. Stability of the iterated Crank-Nicholson method in numerical relativity. *Phys. Rev. D*, 61:087501, 2000.

[684] C. Thompson and R. C. Duncan. Neutron star dynamos and the origins of pulsar magnetism. *Astrophys. J.*, 408:194–217, 1993.

[685] T. A. Thompson, E. Quataert, and A. Burrows. Viscosity and rotation in core-collapse supernovae. *Astrophys. J.*, 620:861–877, 2005.

[686] K. S. Thorne. Validity in general relativity of the Schwarzschild criterion for convection. *Astrophys. J.*, 144:201, 1966.

[687] K. S. Thorne. Relativistic stellar structure and dynamics. In C. DeWitt, Schatzman E., and P. Véron, editors, *High Energy Astrophysics*, volume III of *Les Houches Summer School of Theoretical Physics*, pages 259–441. Gordon and Breach, New York, 1967.

[688] K. S. Thorne. Nonradial pulsation of general-relativistic stellar models. III. Analytic and numerical results for neutron stars. *Astrophys. J.*, 158:1, 1969.

[689] K. S. Thorne. Nonradial pulsation of general-relativistic stellar models. IV. The weak-field limit. *Astrophys. J.*, 158:997–1019, 1969.

[690] K. S. Thorne. *General-Relativistic Astrophysics*, pages 149–216. University of Chicago Press, Chicago, 1978.

[691] K. S. Thorne. Multipole expansions of gravitational radiation. *Rev. Mod. Phys.*, 52:299–340, 1980.

[692] K. S. Thorne and A. Campolattaro. Non-radial pulsation of general-relativistic stellar models. I. Analytic analysis for $l \geq 2$. *Astrophys. J.*, 149:591–611, 1967.

[693] S. E. Thorsett and D. Chakrabarty. Neutron star mass measurements. I. Radio pulsars. *Astrophys. J.*, 512:288–299, 1999.

[694] F. X. Timmes, S. E. Woosley, and T. A. Weaver. The neutron star and black hole initial mass function. *Astrophys. J.*, 457:834–843, 1996.

[695] J. E. Tohline. The collapse of rotating stellar cores – equilibria between white dwarf and neutron star densities. *Astrophys. J.*, 285:721–728, 1984.

[696] J. E. Tohline, R. H. Durisen, and M. McCollough. The linear and nonlinear dynamic stability of rotating N = 3/2 polytropes. *Astrophys. J.*, 298:220–234, 1985.

[697] R. F. Tooper. General relativistic polytropic fluid spheres. *Astrophys. J.*, 140:434–459, 1964.

[698] R. F. Tooper. Adiabatic fluid spheres in general relativity. *Astrophys. J.*, 142:1541–1562, 1965.

[699] E. F. Toro. *Riemann Solvers and Numerical Methods for Fluid Dynamics: A Practical Introduction*, 2nd edition. Springer, New York, 1999.

[700] A. Trautman. Noether equations and conservation laws. *Commun. Math. Phys.*, 6:248–261, 1961.

[701] D. Tsang, J. S. Read, T. Hinderer, A. L. Piro, and R. Bondarescu. Resonant shattering of neutron star crusts. *Phys. Rev. Lett.*, 108:011102, 2012.

[702] W. Unno, Y. Osaki, H. Ando, H. Saio, and H. Shibahashi. *Nonradial Oscillations of Stars*. University of Tokyo Press, Tokyo, 1989.

[703] K. Uryū, F. Limousin, J. L. Friedman, E. Gourgoulhon, and M. Shibata. Nonconformally flat initial data for binary compact objects. *Phys. Rev. D*, 80:124004, 2009.

[704] K. Uryū and Y. Eriguchi. Existence of non-axisymmetric polytropes sustained by internal motions. *Mon. Not. R. Astron. Soc.*, 282:653–664, 1996.

[705] K. Uryū and Y. Eriguchi. A new numerical method for constructing quasi-equilibrium sequences of irrotational binary neutron stars in general relativity. *Phys. Rev. D*, 61:124023, 2000.

[706] K. Uryū, F. Limousin, J. L. Friedman, E. Gourgoulhon, and M. Shibata. Binary neutron stars: Equilibrium models beyond spatial conformal flatness. *Phys. Rev. Lett.*, 97:171101, 2006.

[707] G. Ushomirsky, C. Cutler, and L. Bildsten. Deformations of accreting neutron star crusts and gravitational wave emission. *Mon. Not. R. Astron. Soc.*, 319:902–932, 2000.

[708] M. Vavoulidis, K. D. Kokkotas, and A. Stavridis. Crustal oscillations of slowly rotating relativistic stars. *Mon. Not. R. Astron. Soc.*, 384:1711–1724, 2008.

[709] M. Vavoulidis, A. Stavridis, K. D. Kokkotas, and H. Beyer. Torsional oscillations of slowly rotating relativistic stars. *Mon. Not. R. Astron. Soc.*, 377:1553–1556, 2007.

[710] M. Vigelius and A. Melatos. Three-dimensional stability of magnetically confined mountains on accreting neutron stars. *Mon. Not. R. Astron. Soc.*, 386:1294–1308, 2008.

[711] C. Vilain. Virial theorem in general relativity. Consequences for stability of spherical systems. *Astrophys. J.*, 227:307–318, 1997.

[712] L. Villain, J. A. Pons, P. Cerdá-Durán, and E. Gourgoulhon. Evolutionary sequences of rotating protoneutron stars. *Astron. Astrophys.*, 418:283–294, 2004.

[713] R. V. Wagoner. Gravitational radiation from accreting neutron stars. *Astrophys. J.*, 278:345–348, 1984.

[714] R. V. Wagoner. Conditions for steady gravitational radiation from accreting neutron stars. *Astrophys. J.*, 578:L63–L66, 2002.

[715] R. M. Wald. *General Relativity*. University of Chicago Press, Chicago, 1984.

[716] A. L. Watts. Neutron starquakes and the dynamic crust. arXiv:1111.0514, 2011.

[717] A. L. Watts, N. Andersson, H. Beyer, and B. F. Schutz. The oscillation and stability of differentially rotating spherical shells: The normal-mode problem. *Mon. Not. R. Astron. Soc.*, 342:1156–1168, 2003.

[718] A. L. Watts, N. Andersson, and D. I. Jones. The nature of low $T/|W|$ dynamical instabilities in differentially rotating stars. *Astrophys. J.*, 618:L37–L40, 2005.

[719] A. L. Watts, N. Andersson, and R. L. Williams. The oscillation and stability of differentially rotating spherical shells: the initial-value problem. *Mon. Not. R. Astron. Soc.*, 350:927–938, 2004.

[720] A. L. Watts and B. Krishnan. Detecting gravitational waves from accreting neutron stars. *Adv. Space Res.*, 43:1049–1054, 2009.

[721] A. L. Watts and T. E. Strohmayer. Neutron star oscillations and QPOs during magnetar flares. *Adv. Space Res.*, 40:1446–1452, 2007.

[722] F. Weber and N. K. Glendenning. Exact versus approximate solution to Einstein's equations for rotating neutron stars. *Phys. Lett. B*, 265:1–5, 1991.

[723] F. Weber and N. K. Glendenning. Application of the improved Hartle method for the construction of general relativistic rotating neutron star models. *Astrophys. J.*, 390:541–549, 1992.

[724] F. Weber, N. K. Glendenning, and M. K. Weigel. Structure and stability of rotating relativistic neutron stars. *Astrophys. J.*, 373:579–591, 1991.

[725] S. Weinberg. *Gravitation and Cosmology: Principles and Applications of the General Theory of Relativity*. John Wiley and Sons, New York, 1972.

[726] N. E. White and W. Zhang. Millisecond X-ray pulsars in low-mass X-ray binaries. *Astrophys. J.*, 490:L87–L90, 1997.

[727] J. R. Wilson. Models of differentially rotating stars. *Astrophys. J.*, 176:195–204, 1972.

[728] J. R. Wilson. In R. A. Matzner, editor, *Texas Symposium on 3-Dimensional Numerical Relativity*, Austin, Texas, 1990. University of Texas.

[729] J. R. Wilson and G. J. Mathews. Relativistic hydrodynamics. In C. R. Evans, L. S. Finn, and D. W. Hobill, editors, *International Workshop on Frontiers in Numerical Relativity*, pages 306–314, Cambridge University Press, Cambridge, UK, 1989.

[730] J. R. Wilson and G. J. Mathews. Instabilities in close neutron star binaries. *Phys. Rev. Lett.*, 75:4161–4164, 1995.

[731] R. B. Wiringa, V. Fiks, and A. Fabrocini. Equation of state for dense nucleon matter. *Phys. Rev. C*, 38:1010–1037, 1988.

[732] E. Witten. Cosmic separation of phases. *Phys. Rev. D*, 30:272–285, 1984.

[733] G. A. E. Wright. Pinch instabilities in magnetic stars. *Mon. Not. R. Astron. Soc.*, 162:339–358, 1973.

[734] R. X. Xu and F. H. Busse. The birth of strange stars and their dynamo-originated magnetic fields. *Astron. Astrophys.*, 371:963–972, 2001.

[735] G. Yoneda and H. Shinkai. Advantages of a modified ADM formulation: Constraint propagation analysis of the Baumgarte-Shapiro-Shibata-Nakamura system. *Phys. Rev. D*, 66:124003, 2002.

[736] S. Yoshida. *R*-Modes of slowly rotating nonisentropic relativistic stars. *Astrophys. J.*, 558:263–269, 2001.

[737] S. Yoshida and Y. Eriguchi. Gravitational radiation driven secular instability of rotating polytropes. *Astrophys. J.*, 438:830–840, 1995.

[738] S. Yoshida and Y. Eriguchi. Neutral points of oscillation modes along equilibrium sequences of rapidly rotating polytropes in general relativity: Application of the Cowling approximation. *Astrophys. J.*, 490:779–784, 1997.

[739] S. Yoshida and Y. Eriguchi. A numerical study of normal modes of rotating neutron star models by the Cowling approximation. *Astrophys. J.*, 515:414–422, 1999.

[740] S. Yoshida and Y. Eriguchi. Quasi-radial modes of rotating stars in general relativity. *Mon. Not. R. Astron. Soc.*, 322:389–396, 2001.

[741] S. Yoshida and T. Futamase. R-mode instability of slowly rotating nonisentropic relativistic stars. *Phys. Rev. D*, 64(12):123001, 2001.

[742] S. Yoshida and Y. Kojima. Accuracy of the relativistic Cowling approximation in slowly rotating stars. *Mon. Not. R. Astron. Soc.*, 289:117–122, 1997.

[743] S. Yoshida and U. Lee. Relativistic r-modes in slowly rotating neutron stars: Numerical analysis in the Cowling approximation. *Astrophys. J.*, 567:1112–1120, 2002.

[744] S. Yoshida, L. Rezzolla, S. Karino, and Y. Eriguchi. Frequencies of f-modes in differentially rotating relativistic stars and secular stability limits. *Astrophys. J.*, 568:L41–L44, 2002.

[745] K. Zagkouris, H. Dimmelmeier, and N. Stergioulas. Axisymmetric modes of rotating relativistic stars with tabulated equations of state. In preparation.

[746] J.-P. Zahn. Instabilities and turbulence in rotating stars. In Zahn J.-P. and Zinn-Justin J., editors, *Astrophysical fluid dynamics*, number 47 in Session de l' Ecole d' Eté de Physique Théorique, Les Houches, pages 561–615. Elsevier Science, New York, 1993.

[747] J. L. Zdunik. Damping of GRR instability by direct URCA reactions. *Astron. Astrophys.*, 308:828–832, 1996.

[748] J. L. Zdunik, P. Haensel, and E. Gourgoulhon. The crust of rotating strange quark stars. *Astron. Astrophys.*, 372:535–543, 2001.

[749] J. L. Zdunik, P. Haensel, and E. Gourgoulhon. Recycling strange stars to millisecond periods. *Astron. Astrophys.*, 381:933–940, 2002.

[750] Ya. B. Zel'dovich and I. D. Novikov. *Relativistic Astrophysics*, volume 1. University of Chicago Press, Chicago, 1971.

[751] X.-J. Zhu, X.-L. Fan, and Z.-H. Zhu. Stochastic gravitational wave background from neutron star r-mode instability revisited. *Astrophys. J.*, 729:59, 2011.

[752] B. Zink, O. Korobkin, E. Schnetter, and N. Stergioulas. Frequency band of the f-mode Chandrasekhar-Friedman-Schutz instability. *Phys. Rev. D*, 81(8):084055, April 2010.

[753] B. Zink, N. Stergioulas, I. Hawke, C. D. Ott, E. Schnetter, and E. Müller. Formation of supermassive black holes through fragmentation of toroidal supermassive stars. *Phys. Rev. Lett.*, 96:121101, 2006.

[754] B. Zink, N. Stergioulas, I. Hawke, C. D. Ott, E. Schnetter, and E. Müller. Nonaxisymmetric instability and fragmentation of general relativistic quasitoroidal stars. *Phys. Rev. D*, 76:024019, 2007.

[755] T. Zwerger and E. Müller. Dynamics and gravitational wave signature of axisymmetric rotational core collapse. *Astron. Astrophys.*, 320:209–227, 1997.

Index

In the index, we have used boldface to indicate the page on which a definition of a term is mentioned.